MicroRNA Cancer Regulation

Advances in Experimental Medicine and Biology

Volume 774

For further volumes:
http://www.springer.com/series/5584

Ulf Schmitz • Olaf Wolkenhauer
Julio Vera
Editors

MicroRNA Cancer Regulation

Advanced Concepts, Bioinformatics and Systems Biology Tools

Springer

Editors
Ulf Schmitz
Department of Systems Biology
and Bioinformatics
University of Rostock
Rostock, Germany

Olaf Wolkenhauer
Department of Systems Biology
and Bioinformatics
University of Rostock
Rostock, Germany

Julio Vera
Department of Systems Biology
and Bioinformatics
University of Rostock
Rostock, Germany

ISSN 0065-2598
ISBN 978-94-007-5589-5 ISBN 978-94-007-5590-1 (eBook)
DOI 10.1007/978-94-007-5590-1
Springer Dordrecht Heidelberg New York London

Library of Congress Control Number: 2012956195

Printed on acid-free paper

Springer is part of Springer Science+Business Media (www.springer.com)

Preface

Background

Over the last decade, the interest in microRNAs (miRNAs) and their function as post-transcriptional regulators has grown significantly. It is now known that tumour growth in diverse cancer types and the development of metastasis can be mediated, if not caused, by dysfunctional regulation by miRNAs. There is subsequently an ever growing number of publications that link miRNAs to cancer (Fig. 1), focussing on the involvement in tumorigenesis and metastagenesis, including discussions about their role as biomarkers and potential role as therapeutic targets.

MiRNAs play diverse roles in the cell and are involved in many cell functions, like the cell cycle or proliferation, but they also regulate processes, like development and angiogenesis. Therefore, dysregulation of miRNA expression or alterations in the miRNA biogenesis can lead to pathogenesis, including multiple types of cancer. Important to note is that there exist more than 1,000 different miRNAs in human, each of them capable of regulating the expression of multiple genes. Thus, this abundant class of non-coding RNAs has a remarkable impact on gene regulation on a post-transcriptional level. With respect to cancer, some miRNAs are classified as tumour suppressors because they negatively regulate oncogenes. Others are *oncomirs* because they negatively regulate tumour suppressors. Recently, some miRNAs have also been associated with the development of metastasis and are typically referred to as *metastamir*.

The Systems Perspective on microRNAs in Cancer

Since the first miRNAs have been discovered, bioinformatics and systems biology approaches have paved the path for advancements in the field. Computational tools have been developed and are used for the identification of new miRNA genes, the prediction of potential mRNA targets, the functional categorization of target sets and for building mathematical models aiming for the interpretation of miRNA regulation in complex cellular networks, e.g. those associated with cancer.

It becomes therefore more and more evident that the study of single miRNA target interactions (one-on-one miRNA target regulation) is not appropriate. Instead, the joint repression of targets by multiple miRNAs and the multiplicity of targets affected by single miRNAs need to be investigated. One comes to

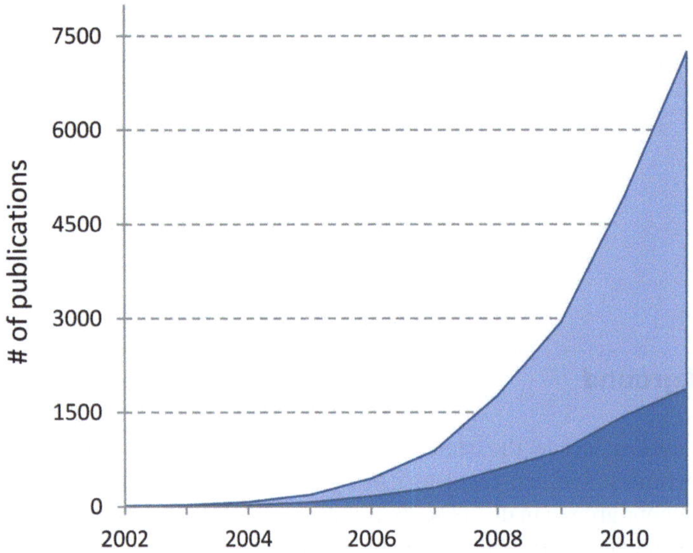

Fig. 1 Number of PubMed entries for 'microRNA+cancer' in total (*light blue*) and per year (*dark blue*)

the conclusion that understanding the role of miRNAs in cellular function and cancer is only possible when considering them as parts of complex regulatory networks.

Systems biology provides an appropriate approach to investigate the regulation of those complex miRNA-regulated networks composed of dozens to hundreds of proteins, genes and miRNAs. Moreover, a dynamical systems perspective becomes indispensable when dealing with networks enriched by non-linear motifs, including feedback and feedforward loops, which induce non-intuitive regulatory patterns like ultrasensitivity, bistability, or oscillations.

Scope of This Book and Target Audience

This edited volume reflects the current state of knowledge about the role of miRNAs in the formation and progression of solid tumours. The main focus lies on computational methods and applications, together with cutting edge experimental techniques that are used to approach all aspects of miRNA regulation in cancer. We are sure that the emergence of high-throughput quantitative techniques will make this integrative approach absolutely necessary in the near future.

This book will be a resource for researchers starting out with cancer miRNA research, but is also intended for the experienced researcher who wants to incorporate concepts and tools from systems biology and bioinformatics into his work. Bioinformaticians and modellers are provided with a general perspective on miRNA biology in cancer, and the state of the art in computational miRNA biology.

Structure

Chapters 1–4. The first four chapters of this book provide an overview about miRNA genomics, biogenesis and mechanisms of target regulation and discuss state-of-the-art methods for the study of miRNAs in cancer, from transcriptomic profiling to applications in clinical practice. Computational tools, developed specifically for miRNA research, are introduced, including algorithms for target prediction, bioinformatics and model-based systems biology approaches to investigate miRNA-regulated networks.

Chapters 5–8. This part of the book presents case studies that discuss the role of miRNAs in various cancer types and their interaction with tumour suppressors and oncogenes. Cancer specific expression signatures and miRNA regulated systems, relevant for tumorigenesis, metastisation and resistance to treatment, are discussed. These chapters nicely illustrate the interdisciplinary approach that combines computational approaches with wet lab experiments, which nowadays forms an integral part of miRNA research.

Chapters 9–11. These three chapters deal with advanced computational approaches to investigate regulatory networks, composed of miRNAs, transcription factors and target genes with a focus on regulatory motifs, like feedback and feed-forwards loops. Moreover, approaches for model generation and simulation studying dynamical aspects of miRNA-target regulation are introduced.

Chapters 12–13. Chapter 12 provides an overview and classification of microRNA web resources dealing with miRNAs, their expression profiles, target genes, function and diseases associations. Chapter 13 introduces a web-based platform that integrates CLIP-Seq and Degradome-Seq data, for the detection of miRNA-target pairs and the identification of regulatory modules.

Chapters 15–16. These two chapters deal with methods for the analysis and integration of miRNA and mRNA expression profiles for the data driven prediction of miRNA-target relationships and the inference of phenotypic associations. Furthermore, methods for the analysis of combinatorial regulation of targets and pathways by miRNAs are discussed.

How to Read This Book

Biologists and Clinicians

Readers with prior knowledge of miRNA cancer biology can skip *Chap. 1* and can directly proceed to the second, third and fourth chapter (*Chaps. 2, 3 and 4*) in order to get a comprehensive overview about computational tools.

Depending on the reader's interest, one can proceed with one or more of the specialized chapters that discuss the interplay between miRNAs and p53 respectively E2F1 (*Chaps. 5* and *8*) and the role of miRNAs in melanoma and in the lung (*Chaps. 6 and 7*).

What follows are chapters that require the reader to have a basic understanding of principles in computational biology. Nevertheless, by reading these chapters one can get an idea of the value that bioinformatics and systems biology approaches have in unraveling underlying mechanisms of miRNA regulation in cancer.

Finally, web resources and tools that can be used to enrich the knowledge about all aspects of miRNA regulation are introduced in *Chap. 12*. *Chapter 13* deals with a particularly interesting resource for the identification of miRNA-target interactions.

Computational Scientists and Modellers

For computer scientists and modellers it is in any case wise to read the first two chapters to get a comprehensive introduction into the field of miRNAs in human cancers and an overview about established applications of computational algorithms in miRNA biology (*Chaps. 1 and* 2). Those, who are already familiar with miRNA-target prediction and validation methods might want to skip the third chapter. In any case we recommend reading the fourth chapter, introducing systems biology approaches that have only recently entered the scene of computational miRNA biology (*Chap. 4*). Further, state-of-the-art developments are covered after the chapters that describe some case scenarios of miRNA biology and cancer. Thus, we would recommend readers with more interest in computational approaches to proceed with *Chaps. 9, 10, 11, 14, 15 and 16.*

Acknowledgements

First of all we would like to thank all authors for making the effort and contributing to this book with their highly engaging chapters.

We would like to thank Xin Lai for offering his help whenever needed and to all the other colleagues in the Department of Systems Biology and Bioinformatics at the University of Rostock who create a unique working environment.

Rostock, Germany

Ulf Schmitz
Olaf Wolkenhauer
Julio Vera

Contents

MicroRNAs in Human Cancer

1

Thalia A. Farazi, Jessica I. Hoell, Pavel Morozov, and Thomas Tuschl

Abstract

Mature microRNAs (miRNAs) are single-stranded RNA molecules of 20–23-nucleotide (nt) length that control gene expression in many cellular processes. These molecules typically reduce the translation and stability of mRNAs, including those of genes that mediate processes in tumorigenesis, such as inflammation, cell cycle regulation, stress response, differentiation, apoptosis, and invasion. miRNA targeting is initiated through specific base-pairing interactions between the 5′ end ("seed" region) of the miRNA and sites within coding and untranslated regions (UTRs) of mRNAs; target sites in the 3′ UTR lead to more effective mRNA destabilization. Since miRNAs frequently target hundreds of mRNAs, miRNA regulatory pathways are complex. To provide a critical overview of miRNA dysregulation in cancer, we first discuss the methods currently available for studying the role of miRNAs in cancer and then review miRNA genomic organization, biogenesis, and mechanism of target recognition, examining how these processes are altered in tumorigenesis. Given the critical role miRNAs play in tumorigenesis processes and their disease specific expression, they hold potential as therapeutic targets and novel biomarkers.

Keywords

microRNA • Cancer • mRNA destabilization • 3′ UTR • Genomics • Deep sequencing • Post-transcriptional gene regulation

1.1 miRNA Overview

miRNAs were originally shown to be important in timing of larval development in *C. elegans*, leading to the identification of the miRNAs *lin-4* and *let-7* [1, 2]. Our initial understanding of miRNA-mRNA target recognition came from

*T. T. is cofounder of and scientific advisor to Alnylam Pharmaceuticals and scientific advisor to Regulus Therapeutics.

T.A. Farazi • J.I. Hoell • P. Morozov • T. Tuschl (✉)
Howard Hughes Medical Institute, Laboratory of RNA Molecular Biology, The Rockefeller University, New York, NY, 10065, USA
e-mail: ttuschl@mail.rockefeller.edu

U. Schmitz et al. (eds.), *MicroRNA Cancer Regulation: Advanced Concepts, Bioinformatics and Systems Biology Tools*, Advances in Experimental Medicine and Biology 774,
DOI 10.1007/978-94-007-5590-1_1, © Springer Science+Business Media Dordrecht 2013

observations of sequence complementarity of the *lin-4* RNA to multiple conserved sites within the *lin-14* 3′ UTR [1, 3]; molecular genetic analysis had shown that this complementarity was required for the repression of *lin-14* by *lin-4* [4]. Homologues of *let-7* or *lin-4/mir-125* were thereafter shown to have temporal expression patterns in other organisms, including mammals, and to regulate mammalian development [5–8]. Given their integral role in development, it was no surprise that miRNAs were soon found to be important in tumorigenesis, and since their discovery close to 5,000 publications associate miRNAs to cancer, including over 1,000 reviews (recent examples include [9–11]). miRNAs were initially linked to tumorigenesis due to their apparent proximity to chromosomal breakpoints [12] and their dysregulated expression levels in many malignancies [13, 14].

Given the wealth of rapidly accumulating information implicating miRNAs in cancer, to allow the reader to critically assess the reports exploring the function of miRNAs in malignancies, we first review the methods used to study the expression and role of miRNAs in tumors, and then review the evidence that relates miRNA genomic organization, biogenesis, target recognition and function to tumorigenesis. An overview of miRNA cistronic expression and sequence similarity allows a better understanding of the regulation of miRNA expression and the factors contributing to technical limitations in accuracy of miRNA detection. Understanding the regulatory potential of miRNAs based on sequence similarity families and miRNA abundance allows evaluation of which miRNAs are important regulators of tumorigenesis pathways.

1.2 Methods for Studying miRNA Genetics and Expression

1.2.1 miRNA Profiling

The main methods currently used for miRNA profiling are sequencing, microarray and real-time RT-PCR based approaches (reviewed in [15–17]). The input material initially used for these studies comprised high quality preserved fresh frozen samples, but recently it has been possible to obtain reproducible and comparable profiles using formalin-fixed paraffin-embedded tissues (FFPE), making these archived tumor collections accessible for study [18–20]. Microarrays generally provide fold-changes in miRNA expression between samples, with members of miRNA sequence families prone to cross-hybridization [21–24]. More recently, calibration cocktails of synthetic miRNAs were used in array experiments to derive absolute abundance of miRNAs [25]. RT-PCR methods are lower throughput and require normalization (i.e. candidate reference genes including other small noncoding RNAs [26, 27]). Mean expression normalization has been suggested as an alternative RT-PCR normalization method for reduction of technical variation to allow appreciation of biological changes [28]. If external miRNA standards are used for quantification (i.e. [29, 30]), the most abundant miRNA, which may vary in length due to 3′ end heterogeneity, should be used as a calibration standard. Sequencing methods, besides their obvious potential to identify new miRNAs, editing and mutation events, estimate miRNA abundance based on frequency of sequence reads (e.g. [5, 7, 8, 31–34]). Given the dramatic increase in sequencing power, bar-coding samples can allow multiple specimens to be processed at the same time, reducing the cost and effort of profiling, and paving the way for large specimen studies [34–36]. Ligation biases between miRNAs and 5′ and 3′ adapters for RT-PCR amplification exist in sequencing methods, and miRNA read frequencies may not always reflect the absolute expression levels, but these variations are irrelevant when monitoring fold-changes between samples. A study with a synthetic pool of 770 miRNA sequences showed that overall, these biases do not prevent identification of miRNAs, and allowed estimation of these biases [36]. For example, certain miRNAs could be over-represented due to higher ligation efficiency (such as miR-21, which was ~2-fold over-represented), while other miRNAs could be under-represented (such as miR-31, which was >5-fold under-represented). However, given the increasing depth of sequencing, most under-represented miRNAs are identified with sufficient sequence reads to allow for a statistically significant comparison across parallel processed samples.

Recent studies have compared the results obtained using multiple platforms [37]. A study of miRNA expression in liposarcoma revealed excellent agreement between bar-coded next generation sequencing and microarray profiles [38], while another study of miRNA expression in breast cancer showed good agreement between bar-coded sequencing and another hybridization-based method, Northern blotting [39].

Finally, choosing the appropriate statistical analysis to evaluate the data depends on the methodology used to obtain the profiles, ranging from established SAM analysis for microarray data [40], to newly developed techniques for sequencing data [34, 41, 42]. Recent in situ hybridization (ISH) advances allowed sensitive detection of miRNAs in heterogeneous tissues, defining miRNA cellular localization [43–45]. The potential of miRNA localization to suggest function for a subpopulation of cells was demonstrated early on, as in the case of *lsy-6* expressed in less than ten neurons in *C. elegans* controlling left/right asymmetry [46].

1.2.2 miRNA Databases and Validation

It is critical to know which miRNAs are validated and have the potential to regulate cellular functions, especially given the frequent revisions of the miRNA database, miRBase (www.mirbase.org) [47], and the dramatic increase in the number of novel and re-annotated miRNAs through the use of deep-sequencing technologies. It is extremely challenging to establish the validity of novel miRNAs, particularly when their definition is based on a handful of sequence reads. The latest release of miRBase (version 17) includes 1,424 human miRNA precursors. Compared to version 16, version 17 includes 385 novel human miRNA precursors, 45 name changes, 1 sequence revision, and the removal of 2 precursors. Given the recent explosion in acquisition of next generation sequencing profiles, miRBase has now added features to allow evaluation of microRNA annotation [48]. The database mapped reads from short RNA deep-sequencing experiments to miRNAs and developed web interfaces to view these mappings. This is an important step in characterizing the newly identified miRNAs as prototypical miRNAs (consisting of a hairpin structure and processing sites consistent with RNase III cleavage steps).

The challenge of constantly revising and curating existing databases based on newly acquired sequencing data is illustrated in two recent studies re-evaluating mouse and human miRNAs. A recent study of 60 million small RNA sequence reads generated from a variety of adult and embryonic mouse tissues confirmed 398 annotated miRNA genes and identified 108 novel miRNA genes but was unable to find sequencing evidence for 150 previously annotated mouse miRNAs. Ectopic expression of the confirmed and newly identified miRNA hairpin sequences yielded small RNAs with the classical miRNA features but failed to support other previously annotated sequences (of the 17 tested miRNAs with no read evidence, only one yielded a single sequence read, while of 28 tested miRNAs with insufficient number of reads, only 4 were verified) [49]. A more recent study has re-annotated human miRNAs based on read evidence from over 1,000 human samples [39]. miRNAs were curated both on the basis of read counts, as well as patterns compatible with traditional miRNA processing, re-defining prototypical miRNAs (557 precursors, corresponding to 1,112 mature and star sequences (miRNA*, described in the following section), miR-451 and miR-618 being the only miRNAs without a star sequence). 269 not yet reported star sequences were added (compared to miRBase 16), putative miRNAs from miRBase, for which read evidence was not obtained, were ignored, and specific miRNAs were renamed according to the read ratio between mature and star sequences. The importance of curated miRNA databases is especially evident in assessing the statistical significance of differentially expressed miRNAs to identify potential biomarkers based on microarray studies. Including miRNAs without strong read evidence in such comparisons could skew the results.

1.3 Mechanisms of Alteration of miRNA Levels in Malignancy

We review miRNA biogenesis (Fig. 1.1) and illustrate which steps of the biogenesis pathway are linked to malignancy, starting from miRNA genomic localization, transcriptional regulation, processing steps and post-transcriptional modification. There is evidence supporting the association of the first three processes and/or the factors that control them with tumorigenesis, whereas evidence relating post-transcriptional miRNA modifications to cancer is not clear-cut.

1.3.1 General Principles of miRNA Genomic Organization

miRNAs are frequently expressed as polycistronic transcripts. To date, 1,424 human miRNA precursor sequences have been deposited in miRBase [47]. Approximately one-third (497) of these miRNAs are located in 156 clusters, each measuring ≤51 kb in the human genome (51 kb being the longest distance between miRNAs belonging to the same cluster, Fig. 1.2). These miRNA clusters are co-expressed based on evidence from miRNA profiling data from a variety of tissues and cell lines [22, 33, 34, 49]. The genomic organization of representative oncogenic (miR-17 and miR-21) and tumor suppressor (let-7 and miR-141) sequence families (described in following section) is illustrated in Fig. 1.2. Presentation of miRNA profiles in the form of expression clusters provides a readily interpretable summary of expression data and stresses the importance of cistronic expression regulation; dysregulation of one member of the cluster should be accompanied by similar dysregulation of other cluster members [39]. Since miRNA genes are frequently multi-copy, determining the relative contribution of each genomic location to mature miRNA expression is challenging.

1.3.2 Alterations in Genomic miRNA Copy Numbers and Location

Changes in miRNA expression between normal and tumor specimens are often attributed to the location of miRNAs in regions of chromosomal instability (amplification, translocation or deletion), or nearby chromosomal breakpoints, initially locating 52.5% of miRNA genes in cancer-associated regions or fragile sites [12]. The miRNA cluster *mir-15a/16-1* is located in a frequently deleted genomic locus containing a putative tumor suppressor containing region in chronic B-cell lymphocytic leukemia (B-CLL) [50]. Other examples include deletion of *let-7g/mir-135-1* in a variety of human malignancies [12], amplification of *mir-17-92* cluster in lymphoma [51], translocation of *mir-17-92* in T-cell acute lymphoblastic leukemia (T-ALL) [52] and amplification of *mir-26a* in glioblastoma [53].

1.3.3 Alterations in miRNA Transcriptional Regulation

Some autonomously expressed miRNA genes have promoter regions that allow miRNAs to be highly expressed in a cell-type-specific manner, and can even drive high levels of oncogenes in cases of chromosomal translocation. The *mir-142* gene, strongly expressed in hematopoietic cells, is located on chromosome 17 and was found at the breakpoint junction of a t(8;17) translocation to *MYC*, which causes an aggressive B-cell leukemia [54]. The translocated *MYC* gene, which was also truncated at the first exon, was located only four nucleotides from the 3′ end of the *mir-142* precursor, placing it under the control of the upstream *mir-142* promoter. In an animal model for Hepatocellular Carcinoma (HCC), a similar event placed *C-MYC* downstream of the *mir-122a* promoter, which is active only in hepatocytes [55].

Many transcription factors regulate miRNA expression in a tissue-specific and disease state-specific fashion, and some miRNAs are regulated by well-established tumor suppressor or oncogene pathways such as TP53, MYC, and RAS (reviewed in [56]). The miRNA and its transcriptional

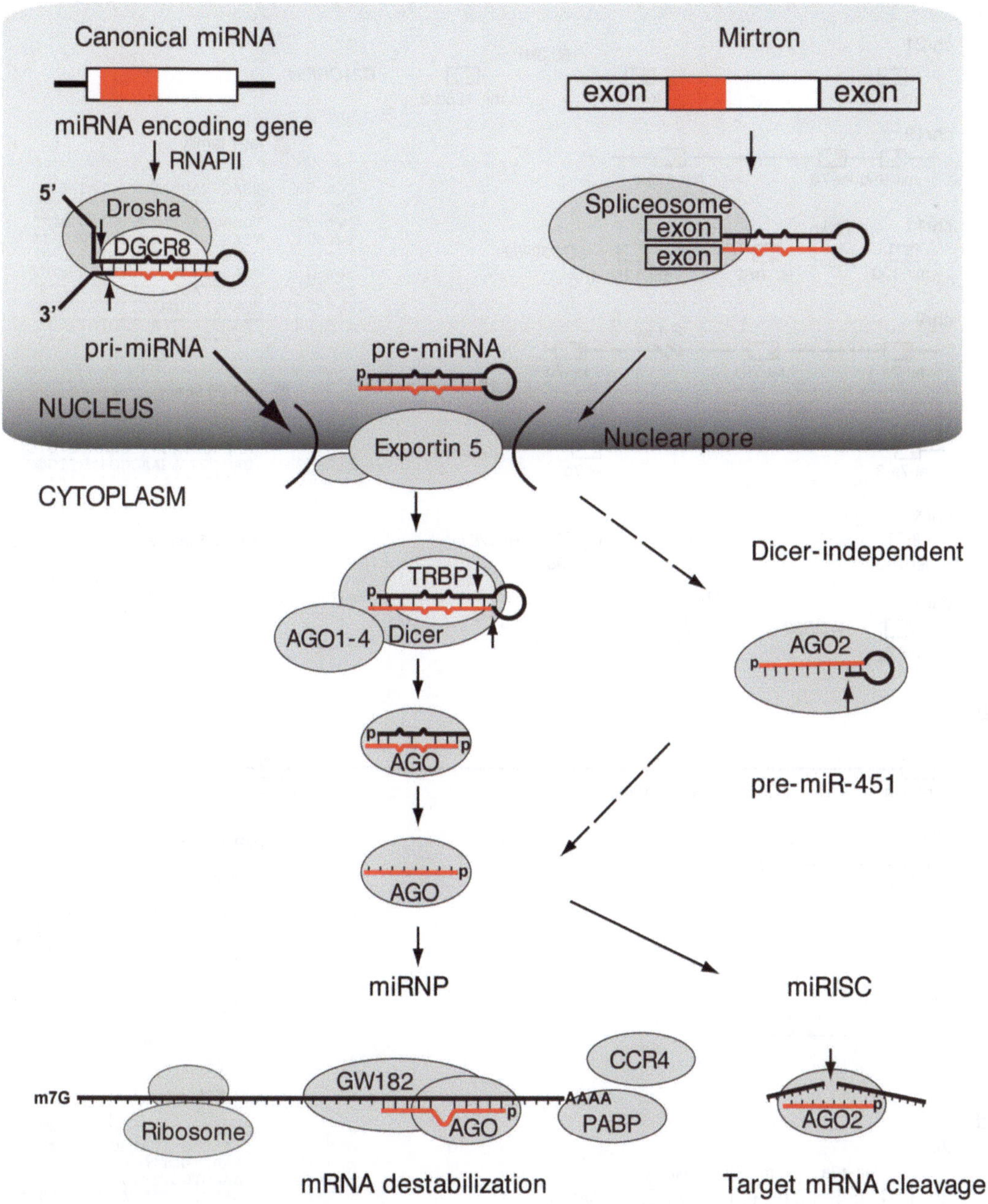

Fig. 1.1 miRNA biogenesis pathway. miRNAs are transcribed by RNAPII to produce pri-miRNAs. Canonical miRNAs are processed by the endoribonuclease Drosha in partnership with its RBP partner DGCR8; mirtrons are instead processed by the spliceosome. The processed pre-miRNA is transported to the cytoplasm through an export complex consisting of exportin 5. The pre-miRNA is subsequently processed in the cytoplasm by another endoribonuclease Dicer in partnership with its RBP partner TRBP to form the final 21–23 nucleotide miRNA product. miR-451 is not processed by Dicer, but is rather cleaved by AGO2. Mature miRNAs (indicated in *red*) are then incorporated into AGO 1 through 4, forming miRNPs, also known as miRISC. miRNPs also incorporate other proteins, such as GW182. miRNPs are thought to direct miRNA mediated destabilization (i.e. through interaction with CCR4) or miRNA mediated translational repression (i.e. through interaction with ribosomes) of miRNAs without perfectly complementary mRNA targets. miRISC is thought to direct AGO2 catalyzed target mRNA cleavage of miRNA fully or nearly fully complementary mRNA targets

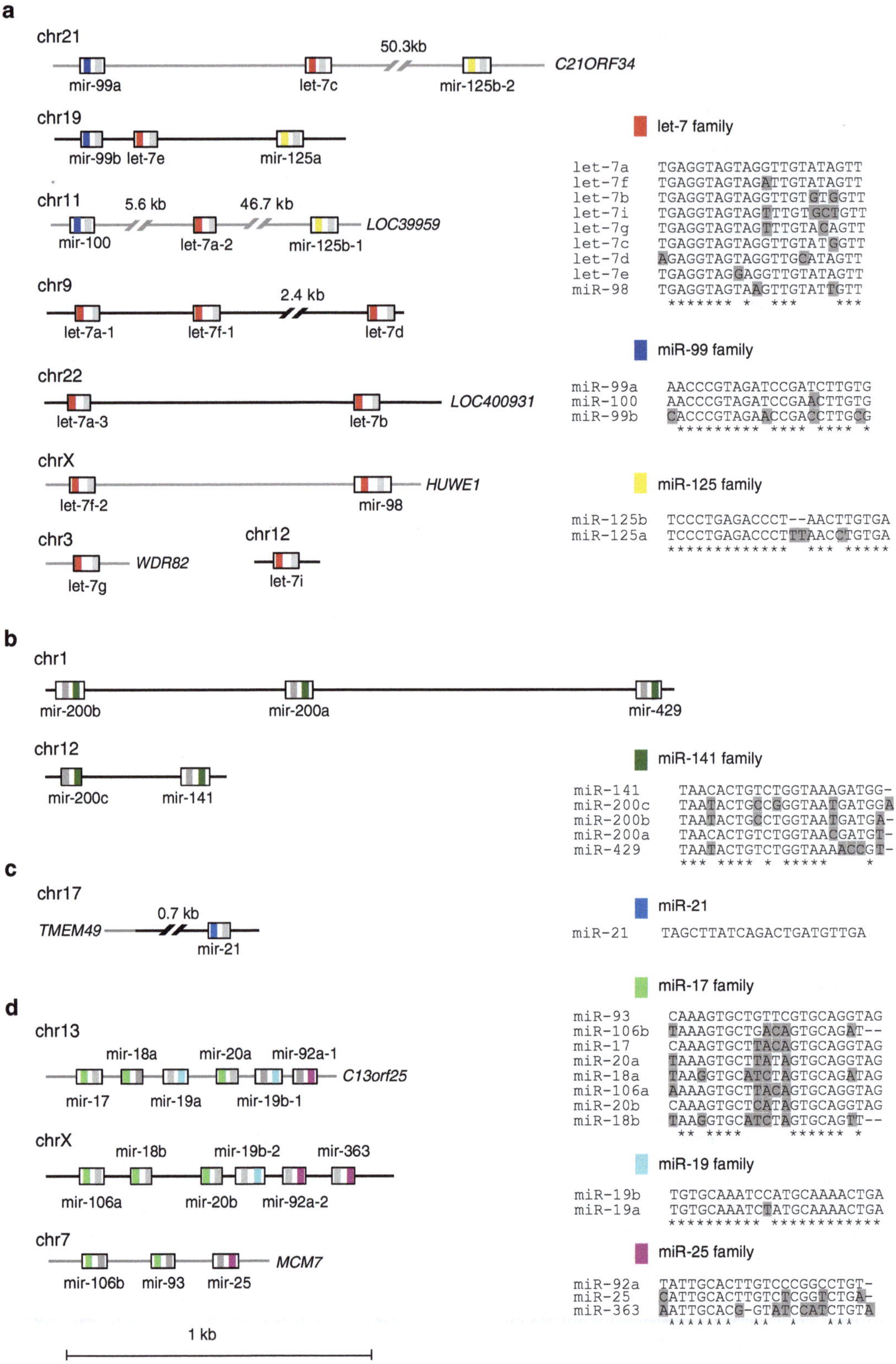
a
chr21
50.3kb
C21ORF34
mir-99a
let-7c
mir-125b-2
chr19
mir-99b let-7e
mir-125a
chr11
5.6 kb
46.7 kb
LOC39959
mir-100
let-7a-2
mir-125b-1
chr9
2.4 kb
let-7a-1
let-7f-1
let-7d
chr22
LOC400931
let-7a-3
let-7b
chrX
HUWE1
let-7f-2
mir-98
chr3
WDR82
let-7g
chr12
let-7i
let-7 family
let-7a TGAGGTAGTAGGTTGTATAGTT
let-7f TGAGGTAGTAGATTGTATAGTT
let-7b TGAGGTAGTAGGTTGTGTGGTT
let-7i TGAGGTAGTAGTTTGTGCTGTT
let-7g TGAGGTAGTAGTTTGTACAGTT
let-7c TGAGGTAGTAGGTTGTATGGTT
let-7d AGAGGTAGTAGGTTGCATAGTT
let-7e TGAGGTAGGAGGTTGTATAGTT
miR-98 TGAGGTAGTAAGTTGTATTGTT
miR-99 family
miR-99a AACCCGTAGATCCGATCTTGTG
miR-100 AACCCGTAGATCCGAACTTGTG
miR-99b CACCCGTAGAACCGACCTTGCG
miR-125 family
miR-125b TCCCTGAGACCCT--AACTTGTGA
miR-125a TCCCTGAGACCCTTTAACCTGTGA
b
chr1
mir-200b
mir-200a
mir-429
chr12
mir-200c
mir-141
miR-141 family
miR-141 TAACACTGTCTGGTAAAGATGG-
miR-200c TAATACTGCCGGGTAATGATGGA
miR-200b TAATACTGCCTGGTAATGATGA-
miR-200a TAACACTGTCTGGTAACGATGT-
miR-429 TAATACTGTCTGGTAAAACCGT-
c
chr17
0.7 kb
TMEM49
mir-21
miR-21
miR-21 TAGCTTATCAGACTGATGTTGA
d
chr13
mir-18a
mir-20a
mir-92a-1
C13orf25
mir-17
mir-19a
mir-19b-1
chrX
mir-18b
mir-19b-2
mir-363
mir-106a
mir-20b
mir-92a-2
chr7
MCM7
mir-106b
mir-93
mir-25
1 kb
miR-17 family
miR-93 CAAAGTGCTGTTCGTGCAGGTAG
miR-106b TAAAGTGCTGACAGTGCAGAT--
miR-17 CAAAGTGCTTACAGTGCAGGTAG
miR-20a TAAAGTGCTTATAGTGCAGGTAG
miR-18a TAAGGTGCATCTAGTGCAGATAG
miR-106a AAAAGTGCTTACAGTGCAGGTAG
miR-20b CAAAGTGCTCATAGTGCAGGTAG
miR-18b TAAGGTGCATCTAGTGCAGTT--
miR-19 family
miR-19b TGTGCAAATCCATGCAAAACTGA
miR-19a TGTGCAAATCTATGCAAAACTGA
miR-25 family
miR-92a TATTGCACTTGTCCCGGCCTGT-
miR-25 CATTGCACTTGTCTCGGTCTGA-
miR-363 AATTGCACG-GTATCCATCTGTA

regulators can participate in complex feedback regulation loops. Examples include the TP53 regulated *mir-34a* [57, 58], the RAS regulated *mir-21* [33, 59, 60] and the MYC regulated *mir-17-92* gene cluster [61, 62].

miRNA dysregulation has also been linked to changes in epigenetic regulation, such as the methylation status of miRNA genes, which results in alterations in their expression levels [63, 64]. Examples of methylated miRNA genes include *mir-127* in bladder cancer cells [65] and *mir-9-1* in breast cancer [66].

1.3.4 miRNA Biogenesis Pathway in Tumorigenesis

miRNA biogenesis has been reviewed extensively [56, 67–73] (Fig. 1.1). miRNA pathway components could either be mis-expressed in tumors or mutated (reviewed in [74, 75]). Post-transcriptional regulation of miRNAs themselves through RNA editing or terminal modifications was shown to alter miRNA targeting, processing and stability, but connection of these modifications to tumorigenesis has not yet been definitive (reviewed in [56, 75, 76]).

1.3.4.1 miRNA Biogenesis

Briefly, the mature 20–23-nt miRNA molecules are excised in a multi-step process from primary transcripts (pri-miRNAs) that contain one or more 70-nt hairpin miRNA precursors (pre-miRNA) and have their own promoters or share promoters with coding genes. These hairpin structures are recognized in the nucleus by DGCR8, a double-stranded RNA-binding protein (dsRBP), and RNASEN, also known as RNase III Drosha, and excised to yield pre-miRNAs. These molecules are subsequently transported by XPO5 (exportin 5) to the cytoplasm where they are further processed by DICER1 (Dicer) in complex with the dsRBPs TARBP2 (TRBP) and/or PRKRA to yield an RNA duplex processing intermediate composed of mature miRNA and miRNA* sequences. Some miRNAs bypass the general miRNA processing and their maturation can be independent of DGCR8 and RNASEN, such as miR-320 or miR-484 [77], or are DICER1 independent, such as erythropoiesis-related miR-451 [78, 79]. DGCR8 and RNASEN independent miRNAs include mirtrons and tailed mirtrons, which release their pre-miRNA by splicing and exonuclase trimming [80, 81]. A recent review describes alternative processing pathways and enumerates settings in which alternative miRNA pathways contribute to distinct phenotypes among miRNA biogenesis mutants [82].

While the mature miRNA is loaded into the Argonaute/EIF2C (AGO) proteins that are at the core of the miRNA-containing ribonucleoprotein complex (miRNP), sometimes also referred to as RNA-induced silencing complex (miRISC), the miRNA* is released and degraded. miR-451 is generated from an unusual hairpin structure that is processed by AGO2 instead of DICER1 [78, 79]. The miRNPs contain a member of the AGO family (1–4), which binds the miRNA and mediates target mRNA recognition. Several other RBPs have been implicated in miRNA biogenesis, including DHX9, DDX6, MOV10, DDX5, DDX17, LIN28A, HNRNPA1 and KSRP [56, 83]. Following transcription, miRNAs can be modified by several enzymes, including deaminases, resulting in miRNA editing, and terminal uridylyl transferases (TUTases), leading to pre-miRNA uridylylation, potentially affecting the

Fig. 1.2 miRNA genomic and functional organization. The genomic and functional organization of four miRNA clusters is clarified: (**a**) *let-7/mir-98* cluster, (**b**) *mir-141/mir-200a* cluster, (**c**) *mir-21* cluster and (**d**) *mir-17-92* cluster. The genomic locations for each of the miRNA members are defined. *Grey* lines denote intronic regions. miRNA mature sequences are color coded according to the sequence family they belong to (i.e. in the *let-7/mir-98* cluster *red* signifies the let-7 sequence family). The star sequence is defined with a *grey* bar. The sequence families are depicted as sequence alignments compared to the most highly expressed miRNA family member shown on *top*, based on profiles of over 1,000 human specimens [39]. *Shaded residues* denote differences from the most highly expressed miRNA family member amount and ratio of miRNA and miRNA* (e.g. [84]), or their sequences (e.g. [85]).

1.3.4.2 Alterations in RNASEN/DGCR8 and DICER1/TARBP2

Inhibition of the miRNA biogenesis pathway leads to severe developmental defects and is lethal in many organisms (reviewed earlier in [86], recent examples include [77, 78]), and perturbations of this pathway predispose to tumorigenesis [87]. Initial miRNA expression profiling experiments suggested that miRNAs are less abundant in tumors compared to their normal tissue counterparts [14], leading to the proposal that miRNAs are predominantly tumor suppressors rather than oncogenes. Quantification of absolute miRNA levels, not only relative abundance, in miRNA profiling methods is necessary to clarify these observations. 27% of various tumors are found to have a hemizygous deletion of the gene that encodes DICER1 [88]. Global knockdown of mature miRNAs by targeting DICER1, RNASEN and its cofactor DGCR8 increases the oncogenic potential of already transformed cancer cell lines and accelerates tumor formation [87]. Reductions in the amount of DICER1 resulting in impaired miRNA processing have also been shown to increase the rate of tumor formation in two different cancer mouse models, a K-RAS-driven lung cancer [88] and an Rb-driven retinoblastoma [89]. DICER1 is therefore considered a haploinsufficient tumor suppressor, requiring partial deletion for its associated tumorigenesis phenotype [89]. The phosphorylation of the DICER1 cofactor TARBP2 by the mitogen-activated protein kinase Erk enhances pre-miRNA processing of oncogenic miRNAs, such as miR-21, and decreases production of tumor suppressor let-7a [90]. Moreover, TARBP2 is mutated in some colon and gastric cancers with microsatellite instability, and TARBP2 frameshift mutations correlate with DICER1 destabilization; in cell lines and xenografts with TARBP2 mutations, reintroduction of wild type TARBP2/DICER1 slowed tumor growth [91, 92]. Finally, DICER1 was also recently implicated as a metastasis suppressor (reviewed in [93]).

1.3.4.3 Alterations in Other Pathway-Related RBPs

Firstly, expression of LIN28A blocks processing of tumor suppressor pri- and pre-let-7 [94–98], thus maintaining expression of genes that drive self-renewal and proliferation (reviewed in [99]); tumors that express LIN28A were indeed shown to be poorly differentiated and more aggressive than LIN28A-negative tumors. Secondly, the helicases DDX5 and DDX17 are thought to stimulate processing of one third of all murine miRNAs by acting as a scaffold and recruiting factors to the RNASEN complex and thereby promoting pri-miRNA processing [100]. Association of DDX17 and DDX5 RNA helicases through interactions mediated by the tumor suppressor TP53 with the RNASEN/DGCR8 complex facilitates the conversion of pri- to pre-miRNAs [101]. Specifically, the DDX5-mediated interaction of the RNASEN complex with the tumor suppressor TP53 was shown to have a stimulatory effect on the tumor suppressor pri-miR-16-1, pri-miR-143 and pri-miR-145 processing in response to DNA damage in cancer cells [101]. Thus, TP53 mutations, often observed in malignancies, led to a decrease in pre-miRNA production. Thirdly, oncogenic SMADs, downstream effectors of the TGF-β superfamily pathways, have been shown to control RNASEN-mediated miRNA maturation through interaction with DDX5, promoting expression of oncogenic miR-21 [102]. KSRP promotes the biogenesis of a subset of miRNAs, including let-7a, by serving as a component of both DICER1 and RNASEN complexes affecting proliferation, apoptosis and differentiation [103]. In a final example, inactivating mutations of XPO5 in tumors with microsatellite instability result in the nuclear retention of miRNAs [104]. Restoration of XPO5 function reverses the impaired export of pre-miRNAs and has tumor suppressor features.

1.4 Dysregulation of miRNA-mRNA Target Recognition

1.4.1 miRNA Function/Mechanism

As described above, miRNAs function through the AGO proteins, containing both RNA-binding domains and RNase H domains (reviewed in [105]). The four human *Ago* genes are co-expressed and bind to miRNAs irrespective of their sequence. AGO2, in contrast to the other members, retains an active RNase H domain and thus is able to directly cleave target RNAs with extensive complementarity to the bound miRNAs. The assembly of the miRNP complex involves multiple AGO conformational transitions captured in a series of crystal structures (reviewed in [106]). The mRNA target is recognized by pairing of the miRNA seed region (position 2–8) to complementary sequences located mainly in the target 3′ UTR, but also in the coding regions. Target mRNA recognition and regulation involves members of the GW182/TNRC6 family. TNRC6 proteins act at the effector step of silencing, downstream of AGO proteins, and play a crucial role in miRNA silencing in animals (reviewed in [107]). Proteomic approaches identified additional AGO-interacting proteins, some of which likely represent mRNA-interacting partners that co-purified with miRNA-targeted mRNPs; their function in RNA silencing processes and potentially tumorigenesis remains to be established.

In mammalian cells under steady state conditions, miRNAs have been shown to destabilize targeted transcripts [108–111] through a variety of mechanisms, including de-capping and de-adenylation; target mRNA and protein abundance changes track closely [108, 109, 112, 113]. These studies also showed that miRNAs destabilize mRNAs preferably through binding sites located in their 3′ UTRs [114–118]. Ribosome profiling studies demonstrated that the ribosome density of miRNA targets was unaltered, while changes in miRNA levels were inversely correlated to mRNA and protein abundance, emphasizing the role of miRNAs in regulation of mRNA stability but not translation [119]. Translational regulation by miRNA targeting is considered to predominantly act at the level of translation initiation. Identification of miRNA/mRNA ribonucleoprotein components in processing bodies (P-bodies) also implies their role in mRNA storage and RNA turnover. An excellent recent review describes the different mechanisms implicated in miRNA function, highlighting the different experiments supporting translational repression versus mRNA decay and the evolution in our current thinking [107].

1.4.2 Organization of miRNAs into Sequence Families

Certain miRNAs share sequence similarity in regions that are critical for mRNA target recognition, specifically the seed region, and are best viewed as a family when considering mRNA target regulation and functional consequences of altered miRNA expression. miRNAs can be grouped in sequence families, based not only on their seed sequence similarity but also overall sequence similarity given that the miRNA 3′ end also contributes to miRNA targeting, although to a lesser extent (reviewed in [68]) (Fig. 1.2). Changes in the overall abundance of miRNA sequence families relate directly to target regulation. In a MYC-driven B-cell lymphoma mouse model, a conditional knockout of the oncogenic *miR-17-92* gene cluster induces apoptosis, which can be reduced by reintroduction of only one of the four sequence families produced from the cluster [120].

1.4.3 miRNA-mRNA Stoichiometry

The majority of miRNA profiling studies do not provide an estimate of miRNA abundance, which is critical in our understanding of the role of miRNA-mRNA mediated regulation in tumorigenesis. Only the most abundantly expressed miRNAs occupy a substantial fraction of their available mRNA target sites and affect target mRNA stability [118]. Abundant miRNAs that behave as "switches", turned on or off during the tumorigenesis process, as shown in developmental processes, have the most significant regulatory potential, given that miRNAs usually only lead to modest 1.5- to 4-fold regulation of their target expression [112, 113, 115]. However, given that specific mRNAs are subject to regulation by multiple miRNAs of unrelated families, cumulative effects of lower expressed miRNAs may be relevant [67, 121, 122]. Furthermore, in the rare circumstance that miRNAs share near perfect complementarity to mRNAs, they may act in a siRNA-like catalytic mode, cleaving mRNA targets even at low miRNA abundance. To conclude, the interplay between miRNAs expressed in particular tissues, the levels of their respective expressed targets, as well as other post-transcriptional gene regulatory mechanisms (such as regulation by RBPs or other competing interactions – see below) is likely responsible for balancing miRNA conferred regulation.

1.4.4 Changes in the miRNA Targets

The binding sites of miRNAs in mRNAs can be altered through a variety of mechanisms, such as point mutations, translocations, shortening of the 3′ UTR, competition with other RBPs or decoy molecules for mRNA binding. Point mutations in miRNA targets can both create or destroy a miRNA binding site [123–125]. Chromosomal translocations can remove miRNA binding sites from their regulated oncogenes, such as in the case of let-7 targeting of the 3′ UTR of the *Hmga2* gene [126]. Shortening of the 3′ UTR through alternative polyadenylation can relax miRNA mediated regulation of known oncogenes, such as *IGF2BP1/IMP1*, and lead to oncogenic transformation [127], as does use of decoy pseudogenes, as in the case of *PTEN*, by saturating miRNA binding sites [128]. Finally, cooperativity or competition of miRNAs for mRNA target site binding with other RBPs, such as ELAVL1 (HuR), DND1 and PUM1, can also de-repress target expression [129–132]. This topic is discussed in a recent review [83].

1.5 Cancer Tissues Have Distinct miRNA Profiles

We will first discuss the state of current miRNA profile databases, and then explore the issue of tissue heterogeneity in the tissue profiles before summarizing the role of miRNA dysregulation in malignancies.

1.5.1 miRNA Cancer Database

The development of miRNA microarrays, RT-PCR platforms and deep sequencing methodologies has resulted in an exponential acquisition of miRNA profiles. Some of the published miRNA profiles are available in the NCBI Gene Expression Omnibus, similarly to mRNA profiles (other resources include www.microrna.org, http://www.mirz.unibas.ch). Larger cancer and blood-borne disease collections have recently been published using various platforms [133–135]. However, there is no database or viewer that allows for cross-platform comparison of existing data.

1.5.2 Tissue Heterogeneity

Tissues are generally composed of multiple cell types, each with their distinct gene expression program. Disease not only alters the expression programs of the affected cell type, but often also its cell type composition. To best separate these effects in the profiling of heterogeneous tumor samples, it may be useful to profile tumor cell lines and individual cell types that may be present in a tumor sample, or define miRNA cellular localization by performing RNA ISH. Figure 1.3 compares miRNA abundance profiles of normal breast, an estrogen receptor positive invasive ductal breast carcinoma, the estrogen receptor positive ductal cell line MCF7, human fat and blood [38, 39]. Strikingly, we can model the profile of a human cancer by simply combining tumor cell line and human fat profiles at equal ratio. This demonstrates that the MCF7 tumor cell line may be a good disease model for deciphering miRNA regulatory networks, as it expresses many of the miRNAs present in the predominant tumor derived cell type and highlights the need for individual cell type miRNA profiles.

1.5.3 miRNAs as Tumor Suppressors and Oncogenes

miRNA dysregulation could be used as a diagnostic tool even if the particular miRNAs do not serve any regulatory function. Alternatively,

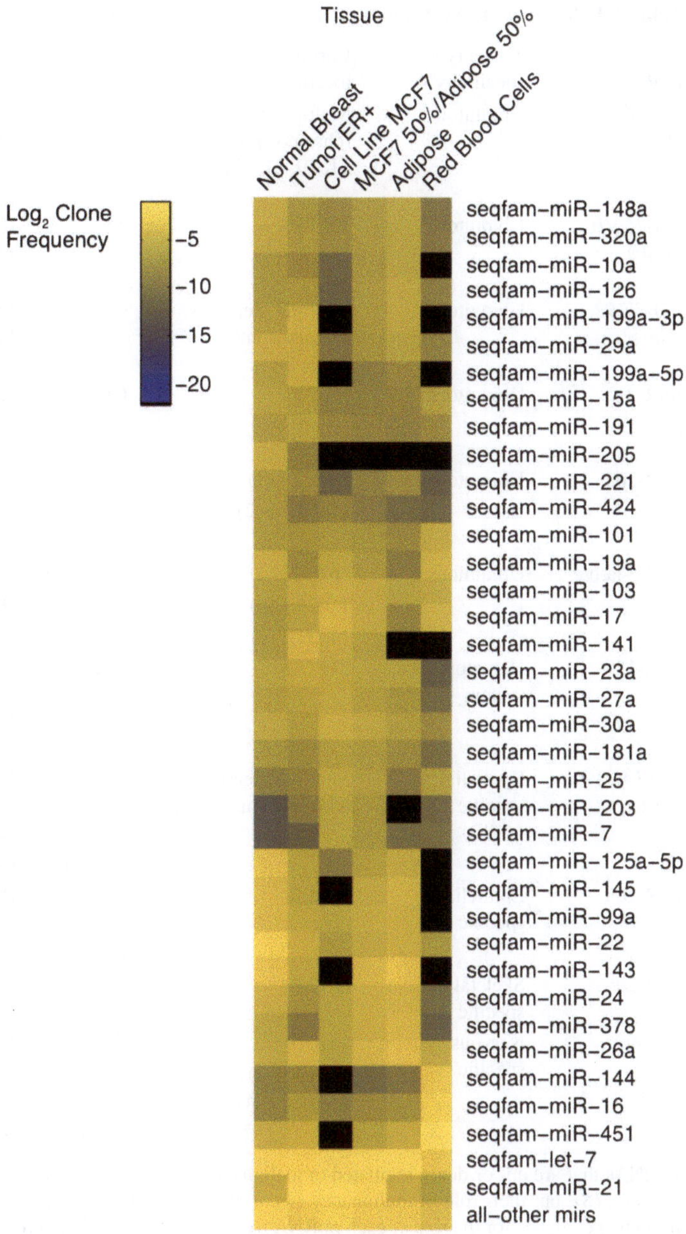

Fig. 1.3 miRNA breast tumor and cell line profiles. Comparison of abundance profiles of the *top* expressed miRNA sequence families of normal breast, an estrogen receptor positive invasive ductal carcinoma breast tumor (ER+), the MCF7 ductal derived cell line, human subcutaneous adipose tissue and red blood cells

miRNA dysregulation could drive tumorigenesis through the roles miRNAs can adopt as tumor suppressors or oncogenes. miRNAs that are up- or down-regulated in malignancies are respectively referred to as oncogenic or tumor-suppressor miRNAs, sometimes even if there is no evidence for their causative role in tumorigenesis. Some of the most commonly dysregulated miRNAs are summarized in Table 1.1 (reviewed in [11]).

Functional studies performed in cancer cell lines or mouse models of various malignancies through over-expression or knockdown of miRNAs have supported a role for some of these miRNAs in tumorigenesis. Over-expression of tumor suppressor miRNAs, such as let-7g, reduced tumor burden in a K-RAS murine lung cancer model [172]. Over-expression of the oncogenic *mir-17-92* gene cluster led to a lymphoproliferative

Table 1.1 Some of the most common cancer-associated miRNAs

miRNA	Tissue type specificity	Chromosomal location	Property	Malignancy
let-7/98 cluster	Ubiquitous	Multiple members (chromosomes 3, 9, 11, 19, 21, 22, X)	TS	CLL [136], lymphoma [137], gastric [138], lung [139], prostate [9], breast [140], ovarian [138], colon [138], leiomyoma [138], melanoma [138]
mir-15a/16-1 cluster	Ubiquitous	13q14.2	TS	CLL [141], lymphoma [9], multiple myeloma [9], pituitary adenoma [142], prostate [142], pancreatic [142]
mir-17-92 cluster	Ubiquitous	Multiple members (chromosomes 7, 13, X)	OG	Lymphoma [143], multiple myeloma [9], lung [139], colon [143], medulloblastoma [144], breast [140], prostate [145]
miR-21	Ubiquitous	17q23.1	OG	Lymphoma, breast, lung, prostate, gastric, cervical, head and neck, colorectal, glioblastoma (for all: [146])
miR-26a	Ubiquitous	3p22.2 (–1) 12q14.1 (–2)	TS	Lymphoma [147], hepatocellular carcinoma [148], thyroid carcinoma [149]
			OG	Glioblastoma [53, 150]
miR-34a/b/c	Ubiquitous	1p36.22 (a) 11q23.1 (b) 11q23.1 (c)	TS	CLL [136], lymphoma [9], pancreatic [9], colon [9], neuroblastoma [151], glioblastoma [152]
miR-155	Hematopoietic system	21q21.3	OG	Lymphoma (i.e. Burkitt's, Hodgkin's, non-Hodgkin's) [9], CLL ([9], [18]), breast [140], lung [9], colon [9], pancreatic [9]
mir-141/200a cluster	Epithelial specific	Multiple members (chromosomes 1, 12)	TS	Breast [140, 153], renal clear cell carcinoma [154], gastric [155], bladder [156]
			OG/TS	Ovarian [157–159]
miR-205	Epithelial specific	1q32.2	TS	Prostate [160, 161], bladder [162], breast [153, 163, 164], esophageal [165]
			OG	Ovarian [166]
miR-206	Skeletal muscle specific	6p12.2	TS	Rhabdomyosarcoma [30], breast [167]
miR-9	Nervous system specific	1q22 (–1) 5q14.3 (–2) 15q26.1 (–3)	TS	Medulloblastoma [168], ovarian [169]
			OG/TS	Breast [66, 170, 171]

miRNAs that are up- or down-regulated in malignancies are respectively referred to as oncogenic (*OG*) or tumor-suppressor (*TS*), but their role in malignancy is not always experimentally validated. Given the number of manuscripts providing evidence for the role of each miRNA based on patient, cell culture or animal model studies, reviews are often cited instead of original reports to limit the number of references, and only a few selected reports are presented if no review is presented

disorder, and higher level expression of the cluster in MYC-driven B-cell lymphomas dramatically increased tumorigenicity [62, 173]. Overexpression of another oncogene, miR-21, frequently highly expressed in solid and hematologic malignancies, resulted in a pre-B malignant lymphoid like phenotype whereas subsequent miR-21 inactivation in the same model led to apoptosis and tumor regression [174]. Transgenic mice models with loss and gain of function of miR-21 combined with a model of lung cancer confirmed the role of miR-21 as an enhancer of tumorigenesis when over-expressed, or a partial protector when genetically deleted [59]. Ectopic expression of miR-155 in bone marrow induced polyclonal pre-B cell proliferation progressing to B-cell leukemia or myeloproliferation in mice [175, 176].

Metastasis-related miRNAs have been identified in various malignancies mainly from cell line and xenograft experiments (reviewed in [177]). Examples include breast cancer-related miR-10b, miR-9, miR-31 and miR-335 among others. The interesting regulatory roles of these miRNAs cannot easily be validated in large clinical studies. Two clinical studies with long-term follow-up data instead identified miR-210 to be associated with tumor aggressiveness [178, 179], pointing to difficulties reconciling cell line, xenograft model and patient materials, due to tissue heterogeneity discussed earlier, the heterogeneous nature of the malignancy and timing of clinical specimen acquisition. Tumor miRNA profiles cannot dissect contributions from subpopulations of cells that may be important for tumor characteristics such as metastasis, while cell line miRNA profiles cannot capture the cellular interactions from supporting cell types in the tumor microenvironment. Patient samples are often collected at time of diagnosis, by which time a tumor is already well established and cannot unravel early changes that may be critical in tumor initiation or later changes important in metastasis.

1.5.4 miRNA-Regulated Pathways

The observed effects of miRNA mis-expression on tumor initiation, maintenance or metastasis can be explained by the mRNA targets and pathways they regulate, which include known tumor suppressors and oncogenes (reviewed in [11]). miRNAs regulate a large number of genes, some estimates reporting miRNA regulation of up to 60% of the human genome, making it challenging to attribute a phenotype after mis-expression of a particular miRNA through its action on only a subset of targets [111, 180]. If a few of these targets control rate-limiting steps in the studied tumorigenesis processes within the specified tissues and cell types, such as metastasis, then miRNA regulation of a handful of targets could potentially explain the phenotype resulting from miRNA mis-expression [181].

Examples of miRNA regulated cancer pathways include differentiation, apoptosis, proliferation and stem cell maintenance, a process important for disease relapse and/or metastasis. The skeletal muscle-specific miR-206 blocks human rhabdomyosarcoma growth in mouse xenograft models by inducing myogenic differentiation [30], while the mir-141/200a cluster is critical in the epithelial to mesenchymal transition (EMT) in various malignancies (reviewed in [182]). Sustained expression of endogenous *mir-17-92* cluster is required to suppress apoptosis in Myc-driven B-cell lymphomas in a conditional knockout allele of *mir-17-92* cluster [120]. TP53-regulated, ectopically expressed miR-34 induced cell cycle arrest in both primary and tumor derived cell lines, downregulating genes promoting cell cycle progression (reviewed in [58]). In a final example of miRNA regulated cancer pathways, isolation of a subset of highly tumorigenic breast cancer cells that were thought to have stemness properties showed that these cells do not express let-7 family members and that expression of let-7 or its known target RAS leads to loss of self renewal [183].

1.6 Alterations of miRNA Sequence

miRNA dysregulation could be a result of mutations in miRNA genes in well-conserved regions in their mature sequence affecting mRNA targeting, or the remainder of the miRNA precursor potentially affecting processing and stability of the mature miRNA (reviewed in [75]). For example, a mutation in the seed region of *mir-96* was shown to lead to hearing loss in a mouse model [184] and was identified in families with non-syndromic progressive sensorineural hearing loss [185], while a point mutation in the viral *mir-K5* precursor stem loop was shown to interfere with its processing and reduce mature miR-K5 accumulation [186]. Germline deletion of the *mir-17-92* gene cluster was another recent example causing skeletal growth defects in humans [187]. If miRNAs are drivers of oncogenic and tumor suppressor pathways we would expect to find miRNA mutations that can also be causative of the disease. So far the only mutation identified in a miRNA that could lead to malignancy is miR-16, where a germline mutation potentially affects miR-16 biogenesis and abundance in a kindred with familial CLL [188] and New Zealand black mice that naturally develop CLL-like disease [189]. Single nucleotide polymorphisms (SNPs), located both in precursor and mature miRNA sequences, have been examined in the context of disease risk for various malignancies but have not been validated as causative (reviewed in [75]).

1.7 miRNA Target Identification

The currently available target prediction databases (reviewed in [68]) do not easily allow prioritizing the involvement of reported targets in certain phenotypes, thus necessitating the selection of a few targets from a list of hundreds for further study and validation, based on a priori knowledge of potentially involved biological pathways. Since the prediction algorithms do not always produce identical target lists, use of multiple algorithms and comparison or intersection of their results narrows the list to higher confidence targets. Targets are only relevant to a specific phenotype if they are expressed in the studied tissue, an issue not addressed by most computational prediction algorithms. Recently, new algorithms are trying to prioritize computationally predicted targets using integrated miRNA and mRNA profiles [134]. Biochemical identification methods in cell lines and tissues are being established and further refine our understanding of miRNA-mRNA target binding recognition. These methods involve two approaches: over-expression or down-regulation of studied miRNAs followed by assessment of transcriptome-wide mRNA levels by mRNA microarray analysis (e.g. [118]) or deep sequencing technology after immunoprecipitation of miRNAs and mRNAs complexed with AGO, the main component of the miRNA effector complex, to not only identify mRNA targets, but also localize their precise binding sites [190, 191].

1.8 miRNAs as Diagnostics

miRNAs demonstrated their potential as diagnostic tumor markers early on when their profiles were shown to correlate with the tumor embryonic origin, thus defining tumors of unknown origin indistinguishable by histology and assigned based on clinical information [14]. miRNA expression patterns have been linked to clinical outcomes given that miRNAs modulate tumor behavior such as tumor progression and metastasis. Expression of let-7 is downregulated in non-small cell lung cancer patients [192] and is associated with poor prognosis [125, 193], whereas a miRNA signature was identified to be associated with prognosis in CLL [188]. Advances in miRNA detection, such as ISH or RT-PCR, may allow miRNAs to be used as diagnostic and prognostic markers in the clinic.

1.9 miRNAs as Therapeutics

Because miRNAs affect the expression of multiple genes and thereby tune multiple points in disease pathways, miRNAs and their regulated genes represent interesting drug targets. Antisense oligonucleotide targeting experiments in human cell lines, mice [117, 194–197] and non-human primates [198] have demonstrated the feasibility of manipulating miRNA levels. miR-143 was initially shown to promote adipocyte differentiation and could be a target for therapies in obesity and metabolic diseases [194]. Alternatively, "miRNA sponges" have been exploited to reduce miRNA expression in mammalian cells and mouse models by using RNA transcripts expressed from strong promoters containing miRNA-complementary binding sites (reviewed in [199]). Systemic administration of antisense oligonucleotide therapeutics to miR-122, a liver-enriched miRNA, in mice and primates was shown to alter lipid metabolism and hepatitis C viral load, resulting in reduced liver damage [117, 195–197, 200, 201]. At the same time, systemic delivery of a miRNA mimic for miR-26a in a murine model of HCC reduced tumor size [148]. The new and exciting advances in delivery of miRNA inhibitors and mimics hold the promise of quickly translating our knowledge of miRNAs into treating disease.

Acknowledgements We thank Iddo Ben-Dov for sharing his unpublished data and Miguel Brown and Aleksandra Mihailovic for assistance with figure generation. We thank Markus Hafner, Kemal Akat, and Neil Renwick for their help with editing the manuscript. T.F. is supported by Grant #UL1 TR000043 from the National Center for Research Resources and the National Center for Advancing Translational Sciences (NCATS), NIH. J.I.H. is supported by the Deutsche Forschungsgemeinschaft. T.T. is an HHMI investigator, and work in his laboratory was supported by NIH grant MH08442, RC1CA145442 and the Starr Cancer Foundation. We apologize to those investigators whose work we could not cite due to space constraints.

References

1. Lee RC, Feinbaum RL, Ambros V (1993) The C. elegans heterochronic gene lin-4 encodes small RNAs with antisense complementarity to lin-14. Cell 75:843–854
2. Reinhart BJ, Slack FJ, Basson M et al (2000) The 21-nucleotide let-7 RNA regulates developmental timing in Caenorhabditis elegans. Nature 403:901–906
3. Wightman B, Ha I, Ruvkun G (1993) Posttranscriptional regulation of the heterochronic gene lin-14 by lin-4 mediates temporal pattern formation in C. elegans. Cell 75:855–862
4. Wightman B, Burglin TR, Gatto J et al (1991) Negative regulatory sequences in the lin-14 3′-untranslated region are necessary to generate a temporal switch during Caenorhabditis elegans development. Genes Dev 5:1813–1824
5. Lagos-Quintana M, Rauhut R, Lendeckel W et al (2001) Identification of novel genes coding for small expressed RNAs. Science 294:853–858
6. Lagos-Quintana M, Rauhut R, Yalcin A et al (2002) Identification of tissue-specific microRNAs from mouse. Curr Biol 12:735–739
7. Lau NC, Lim LP, Weinstein EG et al (2001) An abundant class of tiny RNAs with probable regulatory roles in Caenorhabditis elegans. Science 294:858–862
8. Lee RC, Ambros V (2001) An extensive class of small RNAs in Caenorhabditis elegans. Science 294:862–864
9. Garofalo M, Croce CM (2010) microRNAs: master regulators as potential therapeutics in cancer. Annu Rev Pharmacol Toxicol 51:25–43
10. Medina PP, Slack FJ (2008) microRNAs and cancer: an overview. Cell Cycle 7:2485–2492
11. Ventura A, Jacks T (2009) MicroRNAs and cancer: short RNAs go a long way. Cell 136:586–591
12. Calin GA, Sevignani C, Dumitru CD et al (2004) Human microRNA genes are frequently located at fragile sites and genomic regions involved in cancers. Proc Natl Acad Sci U S A 101:2999–3004
13. Calin GA, Liu CG, Sevignani C et al (2004) MicroRNA profiling reveals distinct signatures in B cell chronic lymphocytic leukemias. Proc Natl Acad Sci U S A 101:11755–11760
14. Lu J, Getz G, Miska EA et al (2005) MicroRNA expression profiles classify human cancers. Nature 435:834–838
15. Aravin A, Tuschl T (2005) Identification and characterization of small RNAs involved in RNA silencing. FEBS Lett 579:5830–5840
16. Creighton CJ, Reid JG, Gunaratne PH (2009) Expression profiling of microRNAs by deep sequencing. Brief Bioinform 10:490–497
17. Meyer SU, Pfaffl MW, Ulbrich SE (2010) Normalization strategies for microRNA profiling experiments: a 'normal' way to a hidden layer of complexity? Biotechnol Lett 32(12):1777–1788
18. Lawrie CH, Soneji S, Marafioti T et al (2007) MicroRNA expression distinguishes between germinal center B cell-like and activated B cell-like subtypes of diffuse large B cell lymphoma. Int J Cancer 121:1156–1161
19. Weng L, Wu X, Gao H et al (2010) MicroRNA profiling of clear cell renal cell carcinoma by whole-genome small RNA deep sequencing of paired frozen and formalin-fixed, paraffin-embedded tissue specimens. J Pathol 222:41–51
20. Xi Y, Nakajima G, Gavin E et al (2007) Systematic analysis of microRNA expression of RNA extracted from fresh frozen and formalin-fixed paraffin-embedded samples. RNA 13:1668–1674
21. Barad O, Meiri E, Avniel A et al (2004) MicroRNA expression detected by oligonucleotide microarrays: system establishment and expression profiling in human tissues. Genome Res 14:2486–2494
22. Baskerville S, Bartel DP (2005) Microarray profiling of microRNAs reveals frequent coexpression with neighboring miRNAs and host genes. RNA 11:241–247
23. Thomson JM, Parker JS, Hammond SM (2007) Microarray analysis of miRNA gene expression. Methods Enzymol 427:107–122
24. Nelson PT, Baldwin DA, Scearce LM et al (2004) Microarray-based, high-throughput gene expression profiling of microRNAs. Nat Methods 1:155–161
25. Bissels U, Wild S, Tomiuk S et al (2009) Absolute quantification of microRNAs by using a universal reference. RNA 15:2375–2384
26. Peltier HJ, Latham GJ (2008) Normalization of microRNA expression levels in quantitative RT-PCR assays: identification of suitable reference RNA targets in normal and cancerous human solid tissues. RNA 14:844–852
27. Fiedler SD, Carletti MZ, Christenson LK (2010) Quantitative RT-PCR methods for mature microRNA expression analysis. Methods Mol Biol 630:49–64
28. Mestdagh P, Van Vlierberghe P, De Weer A et al (2009) A novel and universal method for microRNA RT-qPCR data normalization. Genome Biol 10:R64

29. Smith RD, Brown B, Ikonomi P et al (2003) Exogenous reference RNA for normalization of real-time quantitative PCR. Biotechniques 34:88–91
30. Taulli R, Bersani F, Foglizzo V et al (2009) The muscle-specific microRNA miR-206 blocks human rhabdomyosarcoma growth in xenotransplanted mice by promoting myogenic differentiation. J Clin Invest 119:2366–2378
31. Berezikov E, Thuemmler F, van Laake LW et al (2006) Diversity of microRNAs in human and chimpanzee brain. Nat Genet 38:1375–1377
32. Houbaviy HB, Murray MF, Sharp PA (2003) Embryonic stem cell-specific microRNAs. Dev Cell 5:351–358
33. Landgraf P, Rusu M, Sheridan R et al (2007) A mammalian microRNA expression atlas based on small RNA library sequencing. Cell 129:1401–1414
34. Witten D, Tibshirani R, Gu SG et al (2010) Ultra-high throughput sequencing-based small RNA discovery and discrete statistical biomarker analysis in a collection of cervical tumours and matched controls. BMC Biol 8:58
35. Vigneault F, Sismour AM, Church GM (2008) Efficient microRNA capture and bar-coding via enzymatic oligonucleotide adenylation. Nat Methods 5:777–779
36. Hafner M, Renwick N, Brown M et al (2011) RNA-ligase-dependent biases in miRNA representation in deep-sequenced small RNA cDNA libraries. RNA 17:1697–1712
37. Git A, Dvinge H, Salmon-Divon M et al (2010) Systematic comparison of microarray profiling, real-time PCR, and next-generation sequencing technologies for measuring differential microRNA expression. RNA 16:991–1006
38. Ugras S, Brill E, Jacobsen A et al (2011) Small RNA sequencing and functional characterization reveals microRNA-143 tumor suppressor activity in liposarcoma. Cancer Res 71:5659–5669
39. Farazi TA, Horlings HM, Ten Hoeve JJ et al (2011) MicroRNA sequence and expression analysis in breast tumors by deep sequencing. Cancer Res 71:4443–4453
40. Tusher VG, Tibshirani R, Chu G (2001) Significance analysis of microarrays applied to the ionizing radiation response. Proc Natl Acad Sci U S A 98:5116–5121
41. Berninger P, Gaidatzis D, van Nimwegen E et al (2008) Computational analysis of small RNA cloning data. Methods 44:13–21
42. Robinson MD, McCarthy DJ, Smyth GK (2010) EdgeR: a Bioconductor package for differential expression analysis of digital gene expression data. Bioinformatics 26:139–140
43. Nelson PT, Baldwin DA, Kloosterman WP et al (2006) RAKE and LNA-ISH reveal microRNA expression and localization in archival human brain. RNA 12:187–191
44. Pena JT, Sohn-Lee C, Rouhanifard SH et al (2009) miRNA in situ hybridization in formaldehyde and EDC-fixed tissues. Nat Methods 6:139–141
45. Sempere LF, Christensen M, Silahtaroglu A et al (2007) Altered MicroRNA expression confined to specific epithelial cell subpopulations in breast cancer. Cancer Res 67:11612–11620
46. Johnston RJ, Hobert O (2003) A microRNA controlling left/right neuronal asymmetry in Caenorhabditis elegans. Nature 426:845–849
47. Griffiths-Jones S, Saini HK, van Dongen S et al (2008) miRBase: tools for microRNA genomics. Nucleic Acids Res 36:D154–D158
48. Kozomara A, Griffiths-Jones S (2010) miRBase: integrating microRNA annotation and deep-sequencing data. Nucleic Acids Res 39:D152–D157
49. Chiang HR, Schoenfeld LW, Ruby JG et al (2010) Mammalian microRNAs: experimental evaluation of novel and previously annotated genes. Genes Dev 24:992–1009
50. Calin GA, Dumitru CD, Shimizu M et al (2002) Frequent deletions and down-regulation of micro- RNA genes miR15 and miR16 at 13q14 in chronic lymphocytic leukemia. Proc Natl Acad Sci U S A 99:15524–15529
51. Tagawa H, Seto M (2005) A microRNA cluster as a target of genomic amplification in malignant lymphoma. Leukemia 19:2013–2016
52. Mavrakis KJ, Wolfe AL, Oricchio E et al (2010) Genome-wide RNA-mediated interference screen identifies miR-19 targets in Notch-induced T-cell acute lymphoblastic leukaemia. Nat Cell Biol 12:372–379
53. Huse JT, Brennan C, Hambardzumyan D et al (2009) The PTEN-regulating microRNA miR-26a is amplified in high-grade glioma and facilitates gliomagenesis in vivo. Genes Dev 23:1327–1337
54. Gauwerky CE, Huebner K, Isobe M et al (1989) Activation of MYC in a masked t(8;17) translocation results in an aggressive B-cell leukemia. Proc Natl Acad Sci U S A 86:8867–8871
55. Etiemble J, Moroy T, Jacquemin E et al (1989) Fused transcripts of c-myc and a new cellular locus, hcr in a primary liver tumor. Oncogene 4:51–57
56. Krol J, Loedige I, Filipowicz W (2010) The widespread regulation of microRNA biogenesis, function and decay. Nat Rev Genet 11:597–610
57. Chang TC, Wentzel EA, Kent OA et al (2007) Transactivation of miR-34a by p53 broadly influences gene expression and promotes apoptosis. Mol Cell 26:745–752
58. He L, He X, Lowe SW et al (2007) microRNAs join the p53 network – another piece in the tumour-suppression puzzle. Nat Rev Cancer 7:819–822
59. Hatley ME, Patrick DM, Garcia MR et al (2010) Modulation of K-Ras-dependent lung tumorigenesis by MicroRNA-21. Cancer Cell 18:282–293
60. Huang TH, Wu F, Loeb GB et al (2009) Up-regulation of miR-21 by HER2/neu signaling promotes cell invasion. J Biol Chem 284:18515–18524

61. O'Donnell KA, Wentzel EA, Zeller KI et al (2005) c-Myc-regulated microRNAs modulate E2F1 expression. Nature 435:839–843
62. He L, Thomson JM, Hemann MT et al (2005) A microRNA polycistron as a potential human oncogene. Nature 435:828–833
63. Han L, Witmer PD, Casey E et al (2007) DNA methylation regulates MicroRNA expression. Cancer Biol Ther 6:1284–1288
64. Saito Y, Jones PA (2006) Epigenetic activation of tumor suppressor microRNAs in human cancer cells. Cell Cycle 5:2220–2222
65. Saito Y, Liang G, Egger G et al (2006) Specific activation of microRNA-127 with downregulation of the proto-oncogene BCL6 by chromatin-modifying drugs in human cancer cells. Cancer Cell 9:435–443
66. Lehmann U, Hasemeier B, Christgen M et al (2008) Epigenetic inactivation of microRNA gene hsa-mir-9-1 in human breast cancer. J Pathol 214:17–24
67. Bartel DP (2004) MicroRNAs: genomics, biogenesis, mechanism, and function. Cell 116:281–297
68. Bartel DP (2009) MicroRNAs: target recognition and regulatory functions. Cell 136:215–233
69. Brodersen P, Voinnet O (2009) Revisiting the principles of microRNA target recognition and mode of action. Nat Rev Mol Cell Biol 10:141–148
70. Carthew RW, Sontheimer EJ (2009) Origins and mechanisms of miRNAs and siRNAs. Cell 136:642–655
71. Ghildiyal M, Zamore PD (2009) Small silencing RNAs: an expanding universe. Nat Rev Genet 10:94–108
72. Kim VN, Han J, Siomi MC (2009) Biogenesis of small RNAs in animals. Nat Rev Mol Cell Biol 10:126–139
73. Winter J, Jung S, Keller S et al (2009) Many roads to maturity: microRNA biogenesis pathways and their regulation. Nat Cell Biol 11:228–234
74. Kwak PB, Iwasaki S, Tomari Y (2010) The microRNA pathway and cancer. Cancer Sci 101(11):2309–2315
75. Ryan BM, Robles AI, Harris CC (2010) Genetic variation in microRNA networks: the implications for cancer research. Nat Rev Cancer 10:389–402
76. Nishikura K (2010) Functions and regulation of RNA editing by ADAR deaminases. Annu Rev Biochem 79:321–349
77. Yi R, Pasolli HA, Landthaler M et al (2009) DGCR8-dependent microRNA biogenesis is essential for skin development. Proc Natl Acad Sci U S A 106:498–502
78. Cheloufi S, Dos Santos CO, Chong MM et al (2010) A dicer-independent miRNA biogenesis pathway that requires Ago catalysis. Nature 465:584–589
79. Yang JS, Maurin T, Robine N et al (2010) Conserved vertebrate mir-451 provides a platform for Dicer-independent, Ago2-mediated microRNA biogenesis. Proc Natl Acad Sci U S A 107:15163–15168
80. Babiarz JE, Ruby JG, Wang Y et al (2008) Mouse ES cells express endogenous shRNAs, siRNAs, and other Microprocessor-independent, Dicer-dependent small RNAs. Genes Dev 22:2773–2785
81. Berezikov E, Chung WJ, Willis J et al (2007) Mammalian mirtron genes. Mol Cell 28:328–336
82. Yang JS, Lai EC (2011) Alternative miRNA biogenesis pathways and the interpretation of core miRNA pathway mutants. Mol Cell 43:892–903
83. van Kouwenhove M, Kedde M, Agami R (2011) MicroRNA regulation by RNA-binding proteins and its implications for cancer. Nat Rev Cancer 11:644–656
84. Hagan JP, Piskounova E, Gregory RI (2009) Lin28 recruits the TUTase Zcchc11 to inhibit let-7 maturation in mouse embryonic stem cells. Nat Struct Mol Biol 16:1021–1025
85. Kawahara Y, Zinshteyn B, Chendrimada TP et al (2007) RNA editing of the microRNA-151 precursor blocks cleavage by the Dicer-TRBP complex. EMBO Rep 8:763–769
86. Stefani G, Slack FJ (2008) Small non-coding RNAs in animal development. Nat Rev Mol Cell Biol 9:219–230
87. Kumar MS, Lu J, Mercer KL et al (2007) Impaired microRNA processing enhances cellular transformation and tumorigenesis. Nat Genet 39:673–677
88. Kumar MS, Pester RE, Chen CY et al (2009) Dicer1 functions as a haploinsufficient tumor suppressor. Genes Dev 23:2700–2704
89. Lambertz I, Nittner D, Mestdagh P et al (2010) Monoallelic but not biallelic loss of Dicer1 promotes tumorigenesis in vivo. Cell Death Differ 17:633–641
90. Paroo Z, Ye X, Chen S et al (2009) Phosphorylation of the human microRNA-generating complex mediates MAPK/Erk signaling. Cell 139:112–122
91. Melo SA, Ropero S, Moutinho C et al (2009) A TARBP2 mutation in human cancer impairs microRNA processing and DICER1 function. Nat Genet 41:365–370
92. Garre P, Perez-Segura P, Diaz-Rubio E et al (2010) Reassessing the TARBP2 mutation rate in hereditary nonpolyposis colorectal cancer. Nat Genet 42:817–818; author reply 818
93. Valastyan S, Weinberg RA (2010) Metastasis suppression: a role of the Dice(r). Genome Biol 11:141
94. Newman MA, Thomson JM, Hammond SM (2008) Lin-28 interaction with the Let-7 precursor loop mediates regulated microRNA processing. RNA 14:1539–1549
95. Piskounova E, Viswanathan SR, Janas M et al (2008) Determinants of microRNA processing inhibition by the developmentally regulated RNA-binding protein Lin28. J Biol Chem 283:21310–21314
96. Rybak A, Fuchs H, Smirnova L et al (2008) A feedback loop comprising lin-28 and let-7 controls pre-let-7 maturation during neural stem-cell commitment. Nat Cell Biol 10:987–993

97. Viswanathan SR, Daley GQ, Gregory RI (2008) Selective blockade of microRNA processing by Lin28. Science 320:97–100
98. Viswanathan SR, Powers JT, Einhorn W et al (2009) Lin28 promotes transformation and is associated with advanced human malignancies. Nat Genet 41:843–848
99. Viswanathan SR, Daley GQ (2010) Lin28: a microRNA regulator with a macro role. Cell 140:445–449
100. Fukuda T, Yamagata K, Fujiyama S et al (2007) DEAD-box RNA helicase subunits of the Drosha complex are required for processing of rRNA and a subset of microRNAs. Nat Cell Biol 9:604–611
101. Suzuki HI, Yamagata K, Sugimoto K et al (2009) Modulation of microRNA processing by p53. Nature 460:529–533
102. Davis BN, Hilyard AC, Lagna G et al (2008) SMAD proteins control DROSHA-mediated microRNA maturation. Nature 454(7200):56–61
103. Trabucchi M, Briata P, Garcia-Mayoral M et al (2009) The RNA-binding protein KSRP promotes the biogenesis of a subset of microRNAs. Nature 459:1010–1014
104. Melo SA, Moutinho C, Ropero S et al (2010) A genetic defect in exportin-5 traps precursor microRNAs in the nucleus of cancer cells. Cancer Cell 18:303–315
105. Ender C, Meister G (2010) Argonaute proteins at a glance. J Cell Sci 123:1819–1823
106. Parker JS (2010) How to slice: snapshots of Argonaute in action. Silence 1:3
107. Huntzinger E, Izaurralde E (2011) Gene silencing by microRNAs: contributions of translational repression and mRNA decay. Nat Rev Genet 12: 99–110
108. Linsley PS, Schelter J, Burchard J et al (2007) Transcripts targeted by the microRNA-16 family cooperatively regulate cell cycle progression. Mol Cell Biol 27:2240–2252
109. Zhao Y, Ransom JF, Li A et al (2007) Dysregulation of cardiogenesis, cardiac conduction, and cell cycle in mice lacking miRNA-1-2. Cell 129:303–317
110. Bagga S, Bracht J, Hunter S et al (2005) Regulation by let-7 and lin-4 miRNAs results in target mRNA degradation. Cell 122:553–563
111. Lim LP, Lau NC, Garrett-Engele P et al (2005) Microarray analysis shows that some microRNAs downregulate large numbers of target mRNAs. Nature 433:769–773
112. Baek D, Villen J, Shin C et al (2008) The impact of microRNAs on protein output. Nature 455(7209): 64–71
113. Selbach M, Schwanhausser B, Thierfelder N et al (2008) Widespread changes in protein synthesis induced by microRNAs. Nature 455(7209):58–63
114. Grimson A, Farh KK, Johnston WK et al (2007) MicroRNA targeting specificity in mammals: determinants beyond seed pairing. Mol Cell 27:91–105
115. Hausser J, Landthaler M, Jaskiewicz L et al (2009) Relative contribution of sequence and structure features to the mRNA binding of Argonaute/EIF2C-miRNA complexes and the degradation of miRNA targets. Genome Res 19:2009–2020
116. Karginov FV, Conaco C, Xuan Z et al (2007) A biochemical approach to identifying microRNA targets. Proc Natl Acad Sci U S A 104:19291–19296
117. Krützfeldt J, Rajewsky N, Braich R et al (2005) Silencing of microRNAs in vivo with 'antagomirs'. Nature 438:685–689
118. Landthaler M, Gaidatzis D, Rothballer A et al (2008) Molecular characterization of human Argonaute-containing ribonucleoprotein complexes and their bound target mRNAs. RNA 14:2580–2596
119. Guo H, Ingolia NT, Weissman JS et al (2010) Mammalian microRNAs predominantly act to decrease target mRNA levels. Nature 466:835–840
120. Mu P, Han YC, Betel D et al (2009) Genetic dissection of the miR-17 92 cluster of microRNAs in Myc-induced B-cell lymphomas. Genes Dev 23:2806–2811
121. Wu S, Huang S, Ding J et al (2010) Multiple microRNAs modulate p21Cip1/Waf1 expression by directly targeting its 3′ untranslated region. Oncogene 29:2302–2308
122. Krek A, Grun D, Poy MN et al (2005) Combinatorial microRNA target predictions. Nat Genet 37:495–500
123. Chin LJ, Ratner E, Leng S et al (2008) A SNP in a let-7 microRNA complementary site in the KRAS 3′ untranslated region increases non-small cell lung cancer risk. Cancer Res 68:8535–8540
124. Jiang S, Zhang HW, Lu MH et al (2010) MicroRNA-155 functions as an OncomiR in breast cancer by targeting the suppressor of cytokine signaling 1 gene. Cancer Res 70:3119–3127
125. Takamizawa J, Konishi H, Yanagisawa K et al (2004) Reduced expression of the let-7 microRNAs in human lung cancers in association with shortened postoperative survival. Cancer Res 64:3753–3756
126. Mayr C, Hemann MT, Bartel DP (2007) Disrupting the pairing between let-7 and Hmga2 enhances oncogenic transformation. Science 315:1576–1579
127. Mayr C, Bartel DP (2009) Widespread shortening of 3′ UTRs by alternative cleavage and polyadenylation activates oncogenes in cancer cells. Cell 138:673–684
128. Poliseno L, Salmena L, Zhang J et al (2010) A coding-independent function of gene and pseudogene mRNAs regulates tumour biology. Nature 465:1033–1038
129. Bhattacharyya SN, Habermacher R, Martine U et al (2006) Relief of microRNA-mediated translational repression in human cells subjected to stress. Cell 125:1111–1124
130. Kim HH, Kuwano Y, Srikantan S et al (2009) HuR recruits let-7/RISC to repress c-Myc expression. Genes Dev 23:1743–1748
131. Kedde M, Strasser MJ, Boldajipour B et al (2007) RNA-binding protein Dnd1 inhibits microRNA access to target mRNA. Cell 131:1273–1286
132. Kedde M, van Kouwenhove M, Zwart W et al (2010) A Pumilio-induced RNA structure switch in p27-3′

UTR controls miR-221 and miR-222 accessibility. Nat Cell Biol 12:1014–1020
133. Volinia S, Galasso M, Costinean S et al (2010) Reprogramming of miRNA networks in cancer and leukemia. Genome Res 20:589–599
134. Mestdagh P, Lefever S, Pattyn F et al (2011) The microRNA body map: dissecting microRNA function through integrative genomics. Nucleic Acids Res 39(20):e136
135. Keller A, Leidinger P, Bauer A et al (2011) Toward the blood-borne miRNome of human diseases. Nat Methods 8(10):841–843
136. Calin GA, Cimmino A, Fabbri M et al (2008) MiR-15a and miR-16-1 cluster functions in human leukemia. Proc Natl Acad Sci U S A 105:5166–5171
137. Roush S, Slack FJ (2008) The let-7 family of microRNAs. Trends Cell Biol 18:505–516
138. Peter ME (2009) Let-7 and miR-200 microRNAs: guardians against pluripotency and cancer progression. Cell Cycle 8:843–852
139. Osada H, Takahashi T (2011) let-7 and miR-17-92: small-sized major players in lung cancer development. Cancer Sci 102:9–17
140. O'Day E, Lal A (2010) MicroRNAs and their target gene networks in breast cancer. Breast Cancer Res 12:201
141. Aqeilan RI, Calin GA, Croce CM (2010) miR-15a and miR-16-1 in cancer: discovery, function and future perspectives. Cell Death Differ 17:215–220
142. Finnerty JR, Wang WX, Hebert SS et al (2010) The miR-15/107 group of microRNA genes: evolutionary biology, cellular functions, and roles in human diseases. J Mol Biol 402:491–509
143. Mendell JT (2008) miRiad roles for the miR-17-92 cluster in development and disease. Cell 133:217–222
144. Uziel T, Karginov FV, Xie S et al (2009) The miR-17 92 cluster collaborates with the Sonic Hedgehog pathway in medulloblastoma. Proc Natl Acad Sci U S A 106:2812–2817
145. Poliseno L, Salmena L, Riccardi L et al (2010) Identification of the miR-106b~25 microRNA cluster as a proto-oncogenic PTEN-targeting intron that cooperates with its host gene MCM7 in transformation. Sci Signal 3:ra29
146. Jazbutyte V, Thum T (2010) MicroRNA-21: from cancer to cardiovascular disease. Curr Drug Targets 11:926–935
147. Sander S, Bullinger L, Klapproth K et al (2008) MYC stimulates EZH2 expression by repression of its negative regulator miR-26a. Blood 112:4202–4212
148. Kota J, Chivukula RR, O'Donnell KA et al (2009) Therapeutic microRNA delivery suppresses tumorigenesis in a murine liver cancer model. Cell 137:1005–1017
149. Visone R, Pallante P, Vecchione A et al (2007) Specific microRNAs are downregulated in human thyroid anaplastic carcinomas. Oncogene 26:7590–7595
150. Kim H, Huang W, Jiang X et al (2010) Integrative genome analysis reveals an oncomir/oncogene cluster regulating glioblastoma survivorship. Proc Natl Acad Sci U S A 107:2183–2188
151. Cole KA, Attiyeh EF, Mosse YP et al (2008) A functional screen identifies miR-34a as a candidate neuroblastoma tumor suppressor gene. Mol Cancer Res 6:735–742
152. Li N, Fu H, Tie Y et al (2009) miR-34a inhibits migration and invasion by down-regulation of c-Met expression in human hepatocellular carcinoma cells. Cancer Lett 275:44–53
153. Gregory PA, Bracken CP, Bert AG et al (2008) MicroRNAs as regulators of epithelial-mesenchymal transition. Cell Cycle 7:3112–3118
154. Nakada C, Matsuura K, Tsukamoto Y et al (2008) Genome-wide microRNA expression profiling in renal cell carcinoma: significant down-regulation of miR-141 and miR-200c. J Pathol 216:418–427
155. Du Y, Xu Y, Ding L et al (2009) Down-regulation of miR-141 in gastric cancer and its involvement in cell growth. J Gastroenterol 44:556–561
156. Adam L, Zhong M, Choi W et al (2009) miR-200 expression regulates epithelial-to-mesenchymal transition in bladder cancer cells and reverses resistance to epidermal growth factor receptor therapy. Clin Cancer Res 15:5060–5072
157. Bendoraite A, Knouf EC, Garg KS et al (2009) Regulation of miR-200 family microRNAs and ZEB transcription factors in ovarian cancer: evidence supporting a mesothelial-to-epithelial transition. Gynecol Oncol 116:117–125
158. Park SM, Gaur AB, Lengyel E et al (2008) The miR-200 family determines the epithelial phenotype of cancer cells by targeting the E-cadherin repressors ZEB1 and ZEB2. Genes Dev 22:894–907
159. Hu X, Macdonald DM, Huettner PC et al (2009) A miR-200 microRNA cluster as prognostic marker in advanced ovarian cancer. Gynecol Oncol 114:457–464
160. Gandellini P, Folini M, Longoni N (2009) miR-205 Exerts tumor-suppressive functions in human prostate through down-regulation of protein kinase Cepsilon. Cancer Res 69:2287–2295
161. Schaefer A, Jung M, Mollenkopf HJ et al (2010) Diagnostic and prognostic implications of microRNA profiling in prostate carcinoma. Int J Cancer 126:1166–1176
162. Wiklund ED, Bramsen JB, Hulf T et al (2011) Coordinated epigenetic repression of the miR-200 family and miR-205 in invasive bladder cancer. Int J Cancer 128:1327–1334
163. Iorio MV, Ferracin M, Liu CG et al (2005) MicroRNA gene expression deregulation in human breast cancer. Cancer Res 65:7065–7070
164. Wu H, Zhu S, Mo YY (2009) Suppression of cell growth and invasion by miR-205 in breast cancer. Cell Res 19:439–448
165. Feber A, Xi L, Luketich JD et al (2008) MicroRNA expression profiles of esophageal cancer. J Thorac Cardiovasc Surg 135:255–260; discussion 260
166. Iorio MV, Visone R, Di Leva G et al (2007) MicroRNA signatures in human ovarian cancer. Cancer Res 67:8699–8707

167. Negrini M, Calin GA (2008) Breast cancer metastasis: a microRNA story. Breast Cancer Res 10:203
168. Ferretti E, De Smaele E, Po A et al (2009) MicroRNA profiling in human medulloblastoma. Int J Cancer 124:568–577
169. Laios A, O'Toole S, Flavin R et al (2008) Potential role of miR-9 and miR-223 in recurrent ovarian cancer. Mol Cancer 7:35
170. Ma L, Young J, Prabhala H et al (2010) miR-9, a MYC/MYCN-activated microRNA, regulates E-cadherin and cancer metastasis. Nat Cell Biol 12: 247–256
171. Sun Y, Wu J, Wu SH et al (2009) Expression profile of microRNAs in c-Myc induced mouse mammary tumors. Breast Cancer Res Treat 118:185–196
172. Kumar MS, Erkeland SJ, Pester RE et al (2008) Suppression of non-small cell lung tumor development by the let-7 microRNA family. Proc Natl Acad Sci 105:3903–3908
173. Xiao C, Srinivasan L, Calado DP et al (2008) Lymphoproliferative disease and autoimmunity in mice with increased miR-17-92 expression in lymphocytes. Nat Immunol 9:405–414
174. Medina PP, Nolde M, Slack FJ (2010) OncomiR addiction in an in vivo model of microRNA-21-induced pre-B-cell lymphoma. Nature 467:86–90
175. Costinean S, Zanesi N, Pekarsky Y et al (2006) Pre-B cell proliferation and lymphoblastic leukemia/high-grade lymphoma in E(mu)-miR155 transgenic mice. Proc Natl Acad Sci U S A 103:7024–7029
176. O'Connell RM, Rao DS, Chaudhuri AA et al (2008) Sustained expression of microRNA-155 in hematopoietic stem cells causes a myeloproliferative disorder. J Exp Med 205:585–594
177. Hurst DR, Edmonds MD, Welch DR (2009) Metastamir: the field of metastasis-regulatory microRNA is spreading. Cancer Res 69:7495–7498
178. Camps C, Buffa FM, Colella S et al (2008) hsa-miR-210 is induced by hypoxia and is an independent prognostic factor in breast cancer. Clin Cancer Res 14:1340–1348
179. Foekens JA, Sieuwerts AM, Smid M et al (2008) Four miRNAs associated with aggressiveness of lymph node-negative, estrogen receptor-positive human breast cancer. Proc Natl Acad Sci 105:13021–13026
180. Friedman RC, Farh KK, Burge CB et al (2009) Most mammalian mRNAs are conserved targets of microRNAs. Genome Res 19:92–105
181. Valastyan S, Benaich N, Chang A et al (2009) Concomitant suppression of three target genes can explain the impact of a microRNA on metastasis. Genes Dev 23:2592–2597
182. Cano A, Nieto MA (2008) Non-coding RNAs take centre stage in epithelial-to-mesenchymal transition. Trends Cell Biol 18:357–359
183. Yu F, Yao H, Zhu P et al (2007) let-7 regulates self renewal and tumorigenicity of breast cancer cells. Cell 131:1109–1123
184. Lewis MA, Quint E, Glazier AM et al (2009) An ENU-induced mutation of miR-96 associated with progressive hearing loss in mice. Nat Genet 41:614–618
185. Mencia A, Modamio-Hoybjor S, Redshaw N et al (2009) Mutations in the seed region of human miR-96 are responsible for nonsyndromic progressive hearing loss. Nat Genet 41:609–613
186. Gottwein E, Cai X, Cullen BR (2006) Expression and function of microRNAs encoded by Kaposi's sarcoma-associated herpesvirus. Cold Spring Harb Symp Quant Biol 71:357–364
187. de Pontual L, Yao E, Callier P (2011) Germline deletion of the miR-17 approximately 92 cluster causes skeletal and growth defects in humans. Nat Genet 43(10):1026–1030
188. Calin GA, Ferracin M, Cimmino A et al (2005) A MicroRNA signature associated with prognosis and progression in chronic lymphocytic leukemia. N Engl J Med 353:1793–1801
189. Raveche ES, Salerno E, Scaglione BJ et al (2007) Abnormal microRNA-16 locus with synteny to human 13q14 linked to CLL in NZB mice. Blood 109:5079–5086
190. Chi SW, Zang JB, Mele A et al (2009) Argonaute HITS-CLIP decodes microRNA-mRNA interaction maps. Nature 460:479–486
191. Hafner M, Landthaler M, Burger L et al (2010) Transcriptome-wide identification of RNA-binding protein and microRNA target sites by PAR-CLIP. Cell 141:129–141
192. Johnson CD, Esquela-Kerscher A, Stefani G et al (2007) The let-7 MicroRNA represses cell proliferation pathways in human cells. Cancer Res 67: 7713–7722
193. Yanaihara N, Caplen N, Bowman E et al (2006) Unique microRNA molecular profiles in lung cancer diagnosis and prognosis. Cancer Cell 9:189–198
194. Esau C, Kang X, Peralta E et al (2004) MicroRNA-143 regulates adipocyte differentiation. J Biol Chem 279:52361–52365
195. Krutzfeldt J, Kuwajima S, Braich R et al (2007) Specificity, duplex degradation and subcellular localization of antagomirs. Nucleic Acids Res 35:2885–2892
196. Elmen J, Lindow M, Silahtaroglu A et al (2008) Antagonism of microRNA-122 in mice by systemically administered LNA-antimiR leads to up-regulation of a large set of predicted target mRNAs in the liver. Nucleic Acids Res 36:1153–1162
197. Esau C, Davis S, Murray SF et al (2006) miR-122 regulation of lipid metabolism revealed by in vivo antisense targeting. Cell Metab 3:87–98
198. Elmen J, Lindow M, Schutz S et al (2008) LNA-mediated microRNA silencing in non-human primates. Nature 452:896–899
199. Ebert MS, Sharp PA (2010) MicroRNA sponges: progress and possibilities. RNA 16:2043–2050
200. Lanford RE, Hildebrandt-Eriksen ES, Petri A et al (2010) Therapeutic silencing of microRNA-122 in primates with chronic hepatitis C virus infection. Science 327:198–201
201. Meister G, Tuschl T (2004) Mechanisms of gene silencing by double-stranded RNA. Nature 431: 343–349

Bioinformatics, Non-coding RNAs and Its Possible Application in Personalized Medicine

2

Simona Rossi and George A. Calin

Abstract

Non-coding RNAs are important actors in human biology. A massive amount of data has been created and manipulated, and important findings have been extracted thanks in part to bioinformatics approaches and consequent experimental validation; many of these results are for a specific class of non-coding RNAs, the microRNAs (miRNAs), that are important regulators of gene expression although their transcriptional regulation is not yet well understood. Their involvement in cancer development and progression makes the related research field an integrated one, composed of bioinformaticians, clinicians, statisticians and biologists, as well as informaticians and data miners that cure data manipulation and storage especially due to the output of the latest technologies, like the Next Generation Sequencers.

In this chapter we report the main miRNA findings of the last 10 years in terms of identification and prediction techniques, data generation and manipulation methods, as well as possible use in clinical practice.

Keywords

Non-coding RNA • microRNAs • Cancer • Biomarker • Prognosis • Prediction • Personalized medicine • Target prediction • Deep sequencing and network

S. Rossi, Ph.D. (✉)
SIB Swiss Institute of Bioinformatics, Bâtiment Génopode, Quartier Sorge,
CH-1015 Lausanne, Switzerland
e-mail: simona.rossi@isb-sib.ch

G.A. Calin, M.D., Ph.D.
Department of Experimental Therapeutics, Unit 1950, The University of Texas MD Anderson Cancer Center, 1515 Holcombe Blvd, Houston, TX 77030, USA
e-mail: gcalin@mdanderson.org

2.1 Introduction

A non-coding RNA (ncRNA) is a functional RNA molecule that is not translated into a protein and has important biological functions. Research in the past decade has shown that several types of ncRNAs, including long ncRNAs, ultraconserved genes, and, in particular, microRNAs (miRNAs), are involved in cancer development and progression.

U. Schmitz et al. (eds.), *MicroRNA Cancer Regulation: Advanced Concepts, Bioinformatics and Systems Biology Tools*, Advances in Experimental Medicine and Biology 774,
DOI 10.1007/978-94-007-5590-1_2, © Springer Science+Business Media Dordrecht 2013

Table 2.1 Types of non-coding RNAs (ncRNAs)

ncRNA	Description	Approximate size (nucleotides)	Functional characteristics
lncRNA	Long non-coding RNA	>200	Regulates gene expression at the level of chromatin remodeling, transcription, and post-transcriptional processing
miRNA	MicroRNA	19–25	Uses post-transcriptional repression/activation of gene expression
siRNA	Small interfering/ silencing RNA	20–25	Silences gene expression through the RNA interference pathway
piRNA	PIWI-interacting RNA	24–30	Guides PIWI proteins to direct chromatin modification and transposon silencing
snoRNA	Small nucleolar RNA	70–240	Performs site-specific modification of RNAs
smRNA	Small modulatory RNA	~20	Performs transcriptional activation of neuronal differentiation by converting the NRSF/REST transcription factor from a repressor to an activator of neuronal genes

As Mattick and Makunin wrote 7 years ago:

> RNA regulatory networks may determine most of our complex characteristics, play a significant role in disease…. We predict that mutations/ variations in many if not most ncRNA sequences, especially those that are involved in regulatory networks, will lead to a variety of milder phenotypes than the usually severe consequences of mutations in proteins, and will have a major influence on quantitative trait variation, developmental differences and abnormalities, cancer and other complex diseases such as neurological disorders [1].

Indeed, alterations in protein-coding genes and ncRNAs play a fundamental role in cancer. However, although the full coding component of the human genome has been sequenced for various cancers in recent years, such information related to ncRNAs is still fragmentary. Current methods for high-throughput sequencing, such as next-generation sequencing (NGS), offer the opportunity to investigate the entire ncRNA genome. However, no standardized method is currently available that allows a convenient analysis of these huge data sets, in part owing to the difficulty of managing short reads.

In this chapter, we will first discuss current technologies and then how they can be used in clinical practice, by focusing on advances in miRNAs, the ncRNAs whose importance in disease development and progression is most evident.

2.2 Background

2.2.1 Types of ncRNAs

As previously mentioned, several types of ncRNAs are associated with cancer (Table 2.1):

1. miRNAs are involved in cancer predisposition, development, and progression, and they can be used for diagnostic purposes and prognostic evaluation [2]
2. Long ncRNAs have been found to be associated with metastasis [3]
3. Ultraconserved genes are deregulated in cancer [4], and miRNAs may regulate the expression of ultraconserved regions in various cancers, including colorectal cancer and chronic lymphocytic leukemia [5].

2.2.2 miRNA Function

Structurally, miRNAs are short (19–25-nucleotide) RNAs processed from hairpin loop structures (pre-miRNAs; 60–110 nucleotides in length) that regulate the expression of protein-coding genes through imperfect complementarity with target messenger RNAs. Much of the current research on miRNAs is focused on the elucidation of miRNA function, typically using the gene expression profiling approach. Each miRNA has been studied for its singular contribution to differential expression between two classes of samples or for its ability to extract predictive signatures. However, the effects of miRNAs on cell pathology and physiology are likely to be complex for two reasons: (1) each miRNA can control translation of tens or even hundreds of different coding messengers, and (2) translation of a single messenger can be controlled by more than one miRNA.

A large amount of data has been analyzed worldwide during the past decade. In fact, the amount of published literature related to miRNAs, and ncRNAs in general, has been growing exponentially (Fig. 2.1) allowing the discovery of several miRNA properties. We will summarize the most important of them in this report:

- miRNAs can act as oncogenes or tumor-suppressor genes.
- miRNAs can be either up-regulated or down-regulated in tumor progression and metastasis.
- miRNAs affect a number of pathways that contribute to metastasis, including migration, invasion, cell proliferation, epithelial-to-mesenchymal transition, angiogenesis, and apoptosis. One miRNA can affect multiple metastasis-contributing pathways, and its expression can be regulated by multiple proteins.
- miRNAs can be used as prognostic markers for survival or predictive markers for treatment efficacy, which will be increasingly important in the era of personalized medicine.

2.2.3 miRNA Networks in Cancer

As the number of miRNAs discovered in the human genome has grown, it has become increasingly clear that miRNA regulation of a specific target gene or protein, influence on cellular behavior, and ultimately contribution to the development of disease is highly complex. Although the reductionist approach of investigating single miRNA–single target gene relationships is of value, tumor initiation and progression is likely dependent on multiple miRNAs and genes. Therefore, the integration of both miRNA and target gene patterns of expression to identify "network" deregulation, as Volinia and colleagues [6] have recently done, is critical to our understanding of the role that miRNAs play both as biomarkers and as therapeutic targets.

Volinia and colleagues examined miRNA networks in cancer and leukemia in a collection of more than 4,000 samples, of which about 3,000 were neoplastic. One aim was to identify benchmarks: they found that the miR-17-92 family was amplified in cancer and the miR-143-145 cluster was deleted. In addition, they experimentally validated the miR-155 chronic lymphocytic leukemia network in transgenic mice. Analysis of miRNA tissue specificity in 50 different normal tissues grouped by 17 systems, corresponding to 1,107 human samples, revealed that a small set of miRNAs were tissue-specific and many others were broadly expressed. Moreover, they studied 51 oncologic or hemato-oncologic disorders and identified cancer-type-specific miRNAs. They were able to infer genetic networks for miRNAs in normal tissues and in their pathologic counterparts. Overall, miRNA networks in cancer cells defined independently regulated miRNAs, and target genes of uncoordinated miRNAs were involved in cancer-specific pathways.

Volinia and colleagues found that the most common pathways identified by Kyoto Encyclopedia of Genes and Genomes (KEGG) analysis of deregulated miRNAs in cancers included the Wnt, phosphatidylinositol, focal adhesion, and vascular endothelial growth factor pathways. After directly comparing miRNA networks of normal tissues with those of solid cancers, they identified differences in the representation of miRNA hubs. Applying their approach to specific cancers (e.g., lung cancer and acute myeloid leukemia), they detected differences not only in the distribution of miRNAs but also in the number of networks. For example, normal lung tissue was characterized by one interconnected network of miRNAs, whereas adenocarcinoma of the lung was characterized by eight separate networks in addition to the large network.

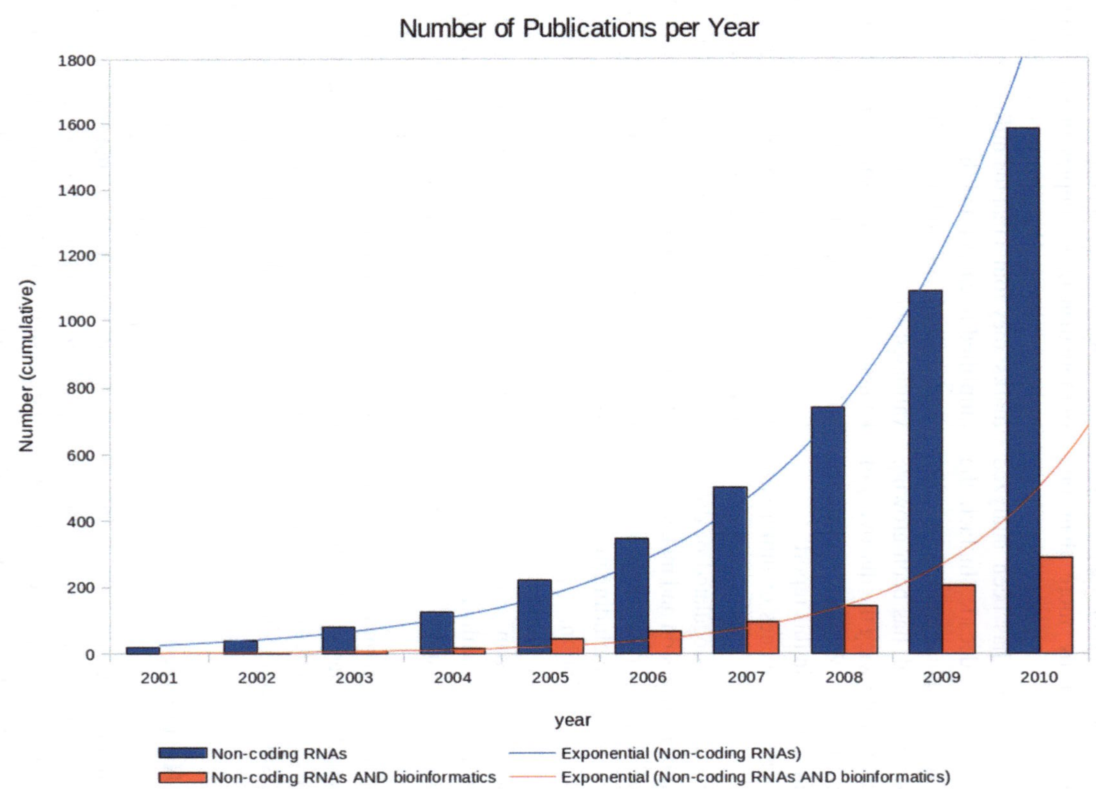

Fig. 2.1 In the past decade, the amount of published literature on non-coding RNAs has increased exponentially. The number of articles on non-coding RNAs and bioinformatics has also progressively increased (graph is based on published articles listed in PubMed; http://www.ncbi.nlm.nih.gov/pubmed/)

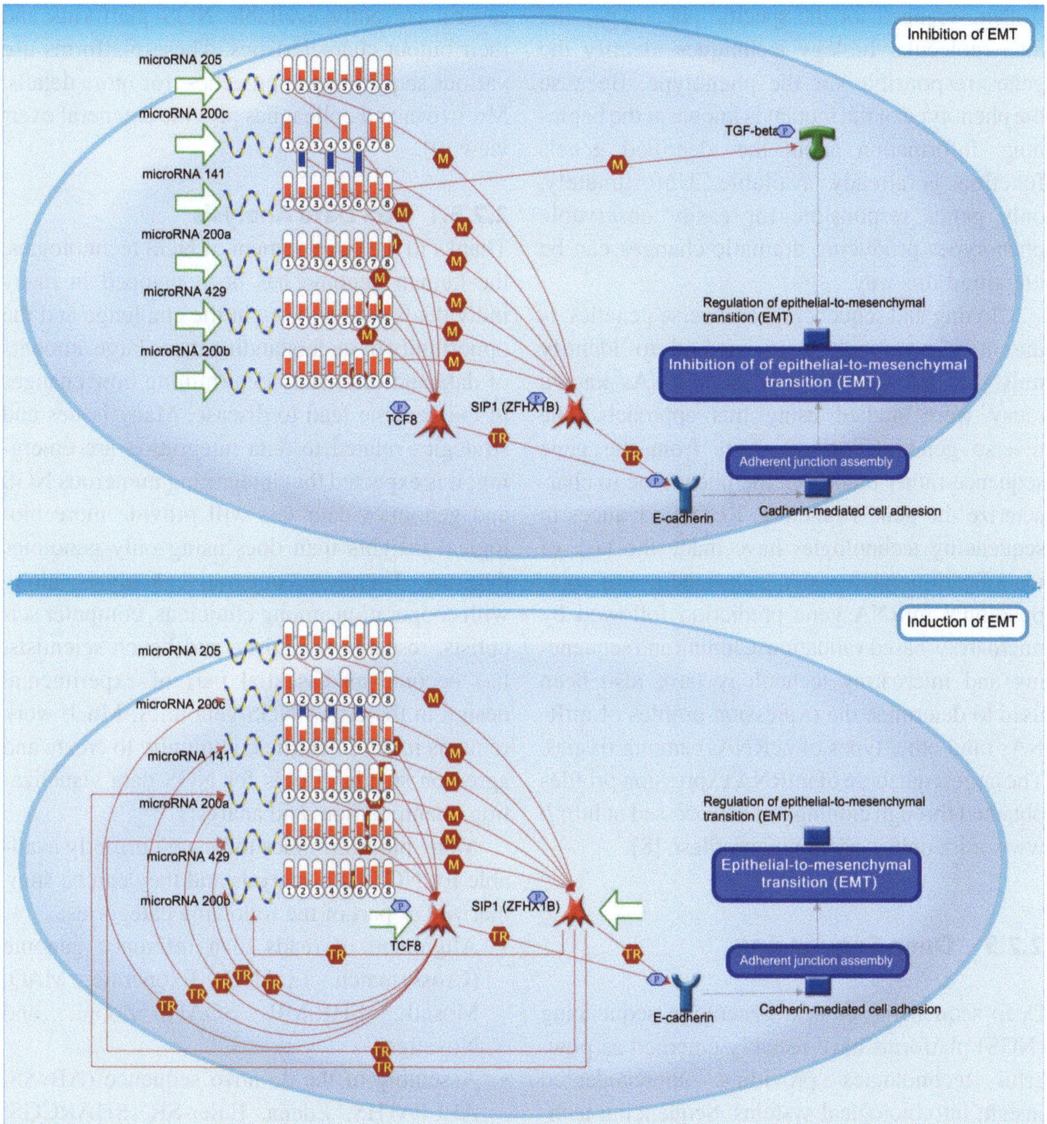

Fig. 2.2 Sample Metacore map, obtained from data in E-TABM-1135 (http://www.ebi.ac.uk/arrayexpress/browse.html?keywords=E-TABM-1135&expandefo=on). Ferracin and colleagues used these arrays [7]

As research on this topic continues, several applications are available to help scientists study the interactions between genomics data and the pathways in which they are involved. These applications include MetaCore (http://www.genego.com/metacore.php; see an example in Fig. 2.2), Ingenuity (http://www.ingenuity.com/), and Cytoscape (http://www.cytoscape.org/).

2.2.4 miRNA Identification

miRNAs can be identified using one of two methods: forward or reverse genetics. Phenotyping, or forward genetics, was the first technique used to identify miRNAs. This technique identifies mutations that produce specific phenotypes. Mutagens are used to generate a panel of mutants that are further screened for the specific phenotype, and then molecular biology techniques identify the genes responsible for the phenotype. Because the phenotype of the mutant is known at the beginning, information about the identified gene's function is already available. Unfortunately, only genes responsible for easily observable phenotypes producing dramatic changes can be identified this way.

Cloning and sequencing, or reverse genetics, is the most commonly used method to identify miRNAs; the vast majority of miRNAs known today were studied using this approach. The reverse genetics method starts from the gene sequence rather than from the phenotype to characterize the gene's function. Recent advances in sequencing technologies have made the task of identification much easier, as has the use of computational miRNA gene prediction followed by microarray-based validation. Cloning and sequencing and microarray technology have also been used to determine the expression profiles of miRNAs (and other types of ncRNAs) among tissues. The largest database of miRNA expression profiles obtained through cloning can be accessed at http://www.mirz.unibas.ch/cloningprofiles/, [8].

2.2.5 Deep Sequencing

Deep sequencing or next-generation sequencing (NGS) platforms have recently emerged as powerful technologies providing unprecedented insight into biological systems. Sequencing technology has come a long way since Sanger first introduced pyrosequencing and assembly as a means of pyrosequencing entire genomes (www.sanger.ac.uk). Initially, this technology was applicable only to small genomic sequences, such as the genomes of the bacteriophages, viruses, and bacterial artificial chromosomes, as sequencing was expensive and required a great deal of manual labor to assemble the reads into the underlying sequence. Today, sequencing and assembly methodologies can be applied to entire mammalian genomes and most of the labor is automated.

NGS, or third generation sequencers, were first introduced in the early 2000s. Table 2.2 lists several currently available NGS platforms and their output characteristics. These platforms use various sequencing approaches; for more details, Morozova and colleagues provide a general overview [9].

2.2.5.1 NGS Data Analysis

Thanks to the development of NGS technologies, the human genome has been mapped in many individuals, which brought the challenge and the opportunity of understanding these large amounts of data and ultimately determining how changes in the genome lead to disease. Many issues and strategies related to data integration are emerging; it is expected that integrating numerous NGS and genomics data sets will provide more biological insights than does using only genomics data set. Therefore, integrative analysis, along with cooperation among clinicians, computer scientists, research scientists, and bench scientists, has become an essential part of experimental design in the era of NGS genomics. Much work remains in the scientific community to create and agree on standard tools for NGS data visualization, manipulation, and analysis.

Many bioinformatics tools are currently available for NGS data analysis and they can be summarized as part of the following categories:

- Alignment of reads to a reference genome (Cross_match, ELAND, Exonerate, MAQ, Mosaik, SHRiMP, SOAP, Zoom, and Novalign)
- Assembly of the de novo sequence (ABySS, ALLPATHS, Edena, Euler-SR, SHARCGS, SHRAP, SSAKE, and Velvet)
- Quality control and base calling
- Polymorphism detection
- Genome browsing and annotation

2.2.5.2 NGS and ncRNAs

With NGS techniques, it has become possible to study ncRNAs in a high-throughput manner. Using specialized algorithms, ncRNA classes such as miRNAs can be detected in deep sequencing data. Unfortunately, none of the existing software provides a method to rigorously address the problem of reads mapped multiple times [10], which usu ally happens with the short reads (shorter than 36

Table 2.2 Common used next generation sequencing platforms and some of their features

Sequencing Platform	Sequencing chemistry	Template amplification method	Read length (base pairs)	Sequencing throughput (megabits per hour)
ABI3730xl Genome Analyzer	Automated sanger sequencing	In vivo amplification via cloning	700–900	0.03–0.07
Roche (454) FLX	Pyrosequencing on solid support	Emulsion polymerase chain reaction	200–300	13
Illumina Genome Analyzer	Sequencing-by-synthesis with reversible terminators	Bridge polymerase chain reaction	32–40	25
ABI SOLiD	Sequencing by ligation	Emulsion polymerase chain reaction	35	21–28

nucleotides) seen in small RNAs. Most current software packages for the analysis of small RNA sequencing data remove such reads from the analysis, thus sacrificing accuracy and any expression profile generated for that short read.

Nonetheless, many NGS tools are available to study ncRNAs using various approaches. These tools include the following:

- miRNAkey, a software package designed to be used as a base station for the analysis of miRNA sequencing data. It can be locally run on any Unix/Linux or Mac computer with 64-bit architecture via the graphical user interface or on a computer-cluster via the command line. The software is freely available for download at http://ibis.tau.ac.il/miRNAkey.
- ALPS (alignment of pattern matrices score), which is used to classify ncRNAs using position and size information in deep sequencing data (http://www.bio.ifi.lmu.de/ALPS). ALPS is a scoring system that uses only primary information from a deep sequencing experiment (i.e., the relative positions and lengths of reads) to classify ncRNAs. ALPS makes no further assumptions about, for example, common structural properties in the ncRNA class, but it is nevertheless able to identify ncRNA classes with high accuracy. ALPS can recognize multiple classes of ncRNAs, as well as novel ncRNA classes, provided these unknown ncRNAs have a characteristic pattern of deep sequencing read lengths and positions.
- MIReNA, a genome-wide search algorithm that looks for miRNA sequences by exploring a multidimensional space defined by only five (physical and combinatorial) parameters characterizing acceptable pre-miRNAs. MIReNA validates pre-miRNAs with high sensitivity and specificity and detects new miRNAs by homology from known miRNAs or from deep sequencing data.

Comparison of MIReNA with miRDeep using deep sequencing data to predict miRNAs highlights the highly specific predictive power of MIReNA [11]. Other available tools, such as MiPred [12], miRabela [13], microPred [14], and MiRDeep [15], do not provide a differential expression analysis among the known miRNAs in the input samples and do not have a graphic interface. In addition, the UEA sRNA toolkit [16], miRanalyzer, [17], SeqBuster [18], DSAP [19], and mirTools [20] are also available, but these require many processing steps and are mostly web-based; thus they are either limited in file size or add a long upload stage to the analysis process.

2.3 miRNA Gene Prediction

2.3.1 Prediction Factors

The first miRNA strongly conserved in animals, let-7, was discovered 13 years ago [21]. Since then, many methods for genome-wide prediction of miRNA genes have been developed and used.

Although a standard miRNA prediction procedure is not yet available, researchers have discovered and incorporated into their methods several important factors for effective and accurate prediction:

- Presence of stable hairpin precursors. Lee and colleagues indicated that mature miRNAs are processed from stem-loop precursor structures [22].
- Relative symmetry of the internal loops of stem-loop precursor structures. Although the presence of a miRNA in a sequence implies the existence of a hairpin, it would appear that the presence of a hairpin does not imply the existence of a miRNA. Secondary structures were considered by miRNA prediction programs starting in 2003 [23].
- Position-dependent selection strength. The pattern of evolutionary conservation along the miRNA precursor was one of the first factors to be incorporated into miRNA prediction methods [24]. The majority of miRNAs are likely conserved; thus it would make sense to consider sequence conservation in the prediction of miRNAs. However, several independent researchers have identified species-specific miRNAs [25, 26]. Thus, most of the current miRNA prediction programs that rely on cross-species sequence conservation as a major defining characteristic risk missing non-conserved miRNAs. Although most miRNAs may be conserved, it is estimated that 7% of human and 11% of mouse miRNAs are species-specific [24], indicating that, when a program requires cross-species conservation, it is worth considering which species are being compared. The more distant two species are in phylogeny, the less likely conservation-based definitions are to facilitate the identification of new miRNAs [26].
- Free-energy estimations. These are widely used by miRNA prediction programs as defining criteria [27].
- Complementarity of the conforming sequences. Hofacker and colleagues used the Vienna library to predict likely paired sequences by creating base pairing probability matrices that were based on the likelihood of the existence of a particular base pair [28].
- Machine learning. Hidden Markov Models (HMMs), an example of machine learning, have recently been used for the first time in predicting miRNA sequences [29]. An HMM attempts to build a statistical model of a system by using examples provided to it. It is possible to train an HMM to evaluate whether candidate sequences of RNA are miRNAs using sequences of known miRNAs. However, an HMM can also be overtrained to the point where it is capable of recognizing only its training data as valid miRNAs.

Many of the preconditions or assumptions used in miRNA prediction are based on the belief that currently known miRNAs are representative of all existing and not yet discovered miRNAs. Of course, we will not be able to evaluate the extent of this problem until after we have complete knowledge of all miRNAs [30].

2.3.2 Prediction Methods

Many methods have been used for miRNA gene prediction (see Table 2.3); these methods are each based on one or more of a relatively small number of parameters:

- Information present in miRNA genes
- Information about the miRNA-target interaction
- Number of miRNAs in a given genome
- Evolutionary selection of miRNA precursors
- Features used to distinguish between miRNA precursors and other types of stem loops (these features can be used either to filter out unlikely miRNA precursors or to compute scores indicative of the likelihood with which a stem-loop will be recognized and processed as a miRNA precursor)

By virtue of their structure and mechanism of action, computational methods have been devised to investigate the encoding of miRNA genes and the targets of miRNA action. A number of factors (e.g., as mentioned above, sequence conservation, secondary structure, folding energetics) were considered in the implementation of these computational solutions. Most of these programs rely on both sequence conservation and secondary structure estimation (Fig. 2.3). Two exceptions are PalGrade, which appears to consider only secondary structure, and ProMiR, which combines conservation with an HMM. The success of these

Table 2.3 Commonly used microRNA (miRNA) prediction software

Category	Program	Main feature(s)	Website or Reference
Homology-based	miRAlign	Genome-wide approach that detects homologous and distant homologous miRNAs using sequence and structure alignments	http://bioinfo.au.tsinghua.edu.cn/miralign/
	Phylogenetic shadowing	Scans pairwise alignments for patterns of conservation reflecting pre-miRNA characteristics	[31]
Gene-based	miRScan	Scans genome alignments for conserved stem-loop regions, taking into account conservation of the surrounding sequence	http://genes.mit.edu/mirscan/
	miRSeeker	Scans genome alignments for conserved miRNA genes	[32]
Mixed approach	PalGrade	Scans genomes for potential miRNA hairpins and uses miRNA microarrays for further filtering to determine the most probable conserved and non-conserved miRNAs, then performs cloning and sequencing of candidate miRNAs	[33]
Machine learning	BayesMiRNAfinder	Predicts mature miRNAs on the basis of secondary structure and sequence features via the Naïve Bayes classifier, then uses comparative analysis to reduce the number of false positive predictions	http://wotan.wistar.upenn.edu/miRNA/register.php
	microPred	Support vector machine-based algorithm that considers pre-miRNA, pseudo pre-miRNA, and non-coding RNA hairpins in predicting conserved and non-conserved pre-miRNA candidates	http://www.comlab.ox.ac.uk/microPred/microPred-server.html
	MiPred	Uses a hybrid feature to identify pre-miRNA-like hairpins and applies the random forest classifier to predict real pre-miRNAs	http://www.bioinf.seu.edu.cn/miRNA/
	miR-KDE	Predicts species-specific pre-miRNAs from hairpins based on similar stem-loop features using the relaxed variable kernel density estimator	[34]
	miRFinder	Scans pairwise alignments, selecting for candidate hairpins using 18 parameters and employing the support vector machine program to distinguish pre-miRNAs from non-miRNA hairpins	http://www.bioinformatics.org/mirfinder/
	MiRPred	Focuses only on the secondary structure of miRNA precursors by employing linear genetic programming and 16 motif-bases classifiers	[35]
	ProMiR II	Probabilistic co-learning model that employs the hidden Markov model-based algorithm to predict clustered, non-clustered, conserved, and non-conserved miRNAs	http://cbit.snu.ac.kr/~ProMiR2/examples.html
	RNAmicro	Classifies raw results from other large-scale non-coding RNA surveys to determine putative RNAs that are conserved in both sequence and secondary structure and uses the support vector machine classifier plus additional filters to identify real miRNA hairpins	http://www.tbi.univie.ac.at/~jana/software/RNAmicro.html
	Triplet-SVM-classifier	Uses the triplet elements to represent local contiguous structure-sequence information and uses a support vector machine classifier for miRNA prediction	http://bioinfo.au.tsinghua.edu.cn/mirnasvm/

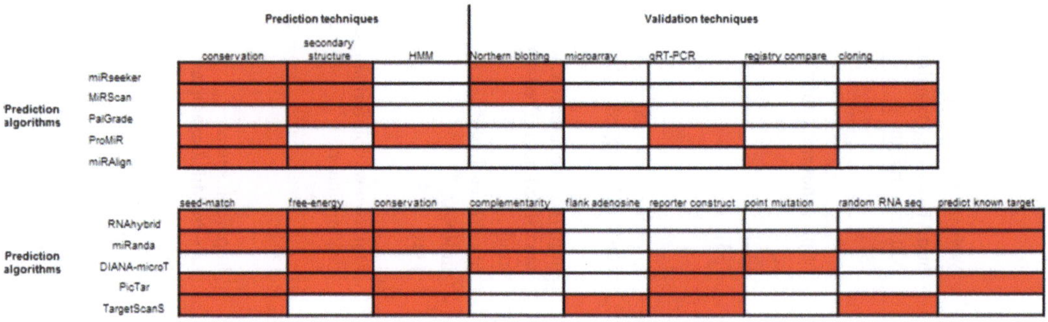

Fig. 2.3 Prediction techniques, validation techniques, and algorithms used in microRNA and microRNA target prediction programs. *Red* indicates the presence of the feature. HMM indicates hidden Markov model; qRT-PCR, quantitative real-time polymerase chain reaction. Figure obtained as modified and merged version of Tables 1,2, and 3 previously published in [33]

computational solutions has been evaluated based on the program's ability to both elucidate new miRNAs and deduce the targets of miRNA action. The programs miRseeker, miRScan, PalGrade, ProMiR, and miRAlign have also been evaluated for their ease of implementation [33].

Regardless of the prediction strategy used in the program, error checking must be installed to differentiate between true positives and false positives. The most robust method of error checking involves verification of miRNA identity using outside criteria. Currently, the most sensitive and discriminant method for detecting miRNA expression is quantitative real-time polymerase chain reaction. One advantage of this method is that it can verify miRNAs that are expressed at low levels. However, no standard verification method exists, and each miRNA prediction program has sought to verify its computational conclusions differently, for example, MiRseeker and MiRScan use northern blotting (Fig. 2.3).

2.4 miRNA Target Prediction

The search for miRNA targets is actually the search for knowledge of miRNA function. miRNAs are believed to play an important role in regulating developmental gene expression, and the current number of experimentally verified targets is likely a small percentage of the true number. Furthermore, each miRNA can target more than one messenger RNA [36], and more than one miRNA can regulate a single target site. miRNA targets are generally believed to be regions of messenger RNA to which the miRNA binds, repressing gene expression. It is also known that miRNAs bind with partial complementarity.

As Vatolin and colleagues pointed out, "there is no simple method to validate targets reproducibly, efficiently and inexpensively" [37]. *In silico* evaluations are often used in target prediction algorithms, likely because they are faster and less expensive than *in vivo* evaluations. Many software tools for target prediction have been deployed (see Table 2.4); the most frequently used and reported in literature include TargetScanS, PicTar, DIANA-microT, miRanda, and RNAhybrid, all of which have been compared by Doran and Strauss [33]. Although the most common evaluation method compares predictions against the list of current experimentally validated targets, some current programs are also able to predict targets that have not yet been verified. PicTar and TargetScanS have been described as the best performers, both with approximately 20–30% false-positive rates. Although they use different search criteria, 80–90% of the human targets they predict are identical. The false-positive rate for miRanda is about 24%, but miRanda has lower sensitivity than either PicTar or TargetScanS [38]. As good as these programs are, Sood and colleagues believe that 50% of the miRNA targets are missed by programs such as PicTar and miRanda owing to misalignments between the mouse and human genomes [39].

Table 2.4 Commonly used microRNA (miRNA) target prediction software

Category	Program	Main feature(s)	Website or Reference
Seed-based	DIANA-microT 3.0	Identifies conserved and non-conserved miRNA recognition elements, providing confidence scores for each prediction that correlate with protein expression levels	http://diana.cslab.ece.ntua.gr/microT/
	ElMMo	Analyzes three different types of seed sequences and incorporates site conservation statistics by taking into account phylogenetic relationships between species	http://www.mirz.unibas.ch/ElMMo2/
	MicroInspector	Species-specific target prediction that reports strong and weak interactions and provides a secondary structure of the miRNA–messenger RNA (mRNA) duplex	http://bioinfo.uni-plovdiv.bg/microinspector/
	miRanda	Identifies maximum local complementarity between the 3′ untranslated region and the miRNA, which is further filtered using the binding energy of the duplex and evolutionary conservation in multiple genomes	http://www.microrna.org/microrna/home.do
	PicTar	Identifies targets based on sequence complementarity between single or multiple miRNAs and the 3′ untranslated region	http://pictar.mdc-berlin.de/
	TargetScan	Incorporates the "offset 6mer" miRNA binding site, probability of conserved targeting, and multiple miRNA "context scores" reflecting various features of miRNA–mRNA binding	http://www.targetscan.org/
Pattern-based	Rna22	Identifies putative target islands based on patterns that are conserved within the miRNA sequences of the same or different organisms, then examines whether the miRNA query sequence can form heteroduplexes with the identified target island that satisfy the user's parameter settings	http://cbcsrv.watson.ibm.com/rna22.html
Machine learning	NBmiRTar	Uses both "seed" and "out-seed" segments of the miRNA–mRNA duplex and can be used to further filter output files from miRanda	http://wotan.wistar.upenn.edu/NBmiRTar/login.php
	miRTif	Support vector machine algorithm, trained on all experimentally validated miRNA–mRNA pairs (both positive and negative datasets), that can be applied as post-processing software for miRanda, TargetScan, or PicTar	http://mirtif.bii.a-star.edu.sg/
	miTarget2	Support vector machine classifier, trained on multiple microarray datasets, that can identify miRNA–mRNA-degraded targets	http://mirdb.org
	TargetBoost	Considers duplex stability and complementarity for predictions	https://demo1.interagon.com/targetboost/
	TargetMiner	Uses an experimentally validated (by protein and mRNA levels) negative training dataset to train the support vector machine classifier and identify unique miRNA–mRNA duplex features to use for target prediction	http://www.isical.ac.in/~bioinfo_miu/targetminer20.htm

(continued)

Table 2.4 (continued)

Category	Program	Main feature(s)	Website or Reference
Targets secondary structures	PITA	Focuses on target-site accessibility by calculating the difference between energy gained from the formation of the miRNA–mRNA duplex and energy required to unfold the miRNA binding site on the mRNA target	http://genie.weizmann.ac.il/pubs/mir07/
Thermodynamic model	FASTH	Calculates minimum free energy of the duplex structure between miRNA and mRNA and then filters the dataset using miRNA–mRNA associations not related to energy	[40]
	RNA-Hybrid	Focuses on the energetically favorable hybridization sites for predicting miRNA target pairs	http://bibiserv.techfak.uni-bielefeld.de/rnahybrid/welcome.html
Mixed approach	HOCTAR	Uses miRNA-expressing host genes as proxies for miRNA expression by exhibiting the inverse relationship of the expression	http://hoctar.tigem.it/
	mirWIP	Applies characteristics of miRNA-mRNA duplexes that were enriched by immunoprecipitating the RISC complex	http://146.189.76.171/query.php

Most of these programs require some degree of sequence complementarity (commonly in the form of a seed of 6–8 nucleotides) as well as favorable free energy in the miRNA-target duplex, although it has been shown that free energy calculations do not have much impact on the accuracy of the program's predictions [41]. The program generally assumes that target site occupancy depends on the strength of the base pairing, although TargetScanS is a notable exception to this. Berezikov and colleagues state that 7% of human and 11% of novel mouse miRNAs are species–specific [24]. The opposite is probably true as well: conservation does not necessarily indicate that the target site is functional, or that it has the same function as the parent gene [42].

A current and well maintained list of miRNA-target interactions with experimental support is essential for thoroughly elucidating miRNA function under different conditions and in different species. The miRTarBase database (http://miRTarBase.mbc.nctu.edu.tw/) accumulates miRNA-target interactions by manually surveying pertinent literature and using text mining to filter research articles related to functional studies of miRNAs. MiRTarBase currently contains 3,969 experimentally verified miRNA-target interactions between 625 miRNAs and 2,433 target genes among 14 species. miRTarBase provides the largest and most updated collection of miRNA-target interactions as it compares its content with that of similar, previously developed databases.

2.5 miRNAs in Clinical Practice

miRNAs are involved in cancer predisposition, development, and progression through gene deregulation and/or single-nucleotide polymorphism. MiRNAs in a cancer setting can be classified on the basis of their main functions:

- Oncomir: a miRNA that can function as a tumor suppressor or as an oncogene depending on its target
- Metastamir: a miRNA that can have either a prometastatic or an antimetastatic effect
- Apoptomir: a miRNA that is involved in apoptosis
- Hypoxamir: a miRNA that is up-regulated by hypoxia
- Angiomir: a miRNA that can regulate angiogenesis; proangiomirs promote angiogenesis and antiangiomirs inhibit angiogenesis

miRNAs are unique candidates for targeted therapy because they have the ability to affect multiple molecules simultaneously along the same pathway. Thus, there is speculation that miRNAs may one day play a role in the development of new therapeutic applications [43]. Given the role that miRNAs have in driving disease, the opportunity to modulate their regulation may represent a means of help patients live longer and maintain better quality of life.

2.5.1 Cancer Predisposition and Development

MiRNA involvement in cancer was first found for mir-15a and miR-16 in chronic lymphocytic leukemia [44]; this has been confirmed by subsequent research [45, 46]. Since then, many other miRNA alterations have been observed in virtually every type of cancer (for a complete review, see [2]); part of these alterations are summarized in Fig. 2.4. For example, miR-155, when overexpressed in chronic lymphocytic leukemia, is associated with a faster rate of carcinogenesis [47], and miR-17-92 was found to lead to cancer growth when overexpressed in lung and breast cancers [48]. On the other hand, the presence of some miRNAs has been shown to decrease the incidence and growth rate of cancer. For example, miR-126 was shown to decrease the growth of cancer cells when it was introduced into cancer cells in which its expression was decreased [49]. Sequence variations that cause miRNAs to be abnormally expressed may provide a new means of determining cancer predisposition [46]. For example, miR-16 has been shown to be involved in predisposition for chronic lymphocytic leukemia [4]. Venkatachalam and colleagues found that germline variations help pinpoint which genes are involved in familial predisposition to cancer [50].

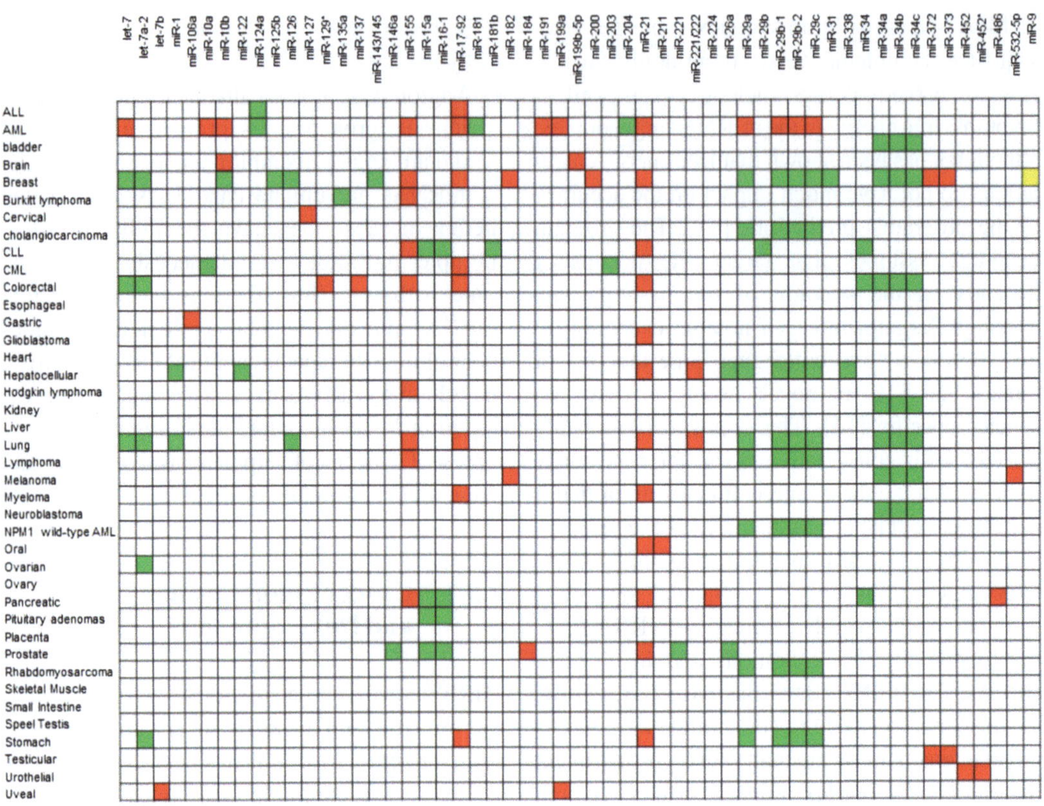

Fig. 2.4 MicroRNAs that are deregulated in various types of cancer. A *green square* represents down-regulation, a *red square* represents up-regulation and a *yellow square* indicates that both regulations have been highlighted

2.5.1.1 Diagnosis, Prognosis and Metastasis

miRNAs have several properties that would make them effective diagnostic markers for cancer, for example, checking a patient's expression levels of specific miRNAs would help a clinician decide whether the patient is at risk for developing cancer or whether the patient's tumor has metastasized. The deregulation of miRNA expression can be used as a diagnostic tool: Fig. 2.4 lists the miRNAs most often deregulated in cancer. The stability of miRNAs in formalin-fixed, paraffin-embedded tissues and body fluids is advantageous for biomarker discovery and validation, plus miRNAs can be extracted from small biopsy specimens. In addition, miRNAs are potential therapeutic agents for personalized cancer management.

Several miRNAs are also involved in disease progression and prognosis:

- miR-135a is associated with poor prognosis in Hodgkin lymphoma [51]
- miR-191 and miR-199a are associated with poor prognosis in acute myeloid leukemia [52]
- miR-10b is associated with progression high-grade gliomas [53]
- miR-21 expression levels increase with advanced tumor stages in oral and colorectal cancer [54], and miR-21 expression is associated with poor prognosis in chronic lymphocytic leukemia patients with 17p deletion [55]
- miR-146a expression decreases as prostate carcinomas become more advanced [56]
- miR-182 expression increases with progression of melanoma [57]

The first evidence of miRNA involvement in metastasis was found for miR-10b as a promoter of breast cancer metastasis [58], and later miR-335 was shown to reduce the progression of metastatic breast cancer when it was introduced into cancer cells in which its expression was decreased [49]. After these initial findings, many associations between miRNAs and metastasis have been found in several types of cancer, including miR-21 up-regulation in breast, oral, and colorectal cancer [59, 60]; miR-126 down-regulation in metastatic relapse of breast cancer [61]; and miR-122 down-regulation in intrahepatic metastases [62]. The involvement of miR-21 in several cancers suggests that it may have a general role in tumor progression and that there may be metastatic pathways common to multiple cancers, (Fig. 2.4).

2.6 Conclusion

miRNAs, and ncRNAs in general, are involved in the development and progression of various cancers. New technology has elucidated the function of specific miRNAs and their potential use in targeted therapies for these cancers; however, standardized methods for predicting and sequencing miRNAs and miRNA targets remain to be developed. The biomedical and bioinformatics research community must work to fill in the gaps in miRNA research in order to be able to translate the findings into clinical practice.

References

1. Mattick JS, Makunin IV (2006) Non-coding RNA. Hum Mol Genet 15(Spec No 1):R17–R29, PMID: 16651366
2. Garzon R, Marcucci G, Croce CM (2010) Targeting microRNAs in cancer: rationale, strategies and challenges. Nat Rev Drug Discov 9(10):775–789
3. Gupta RA, Shah N, Wang KC, Kim J, Horlings HM, Wong DJ, Tsai MC, Hung T, Argani P, Rinn JL, Wang Y, Brzoska P, Kong B, Li R, West RB, van de Vijver MJ, Sukumar S, Chang HY (2010) Long non-coding RNA HOTAIR reprograms chromatin state to promote cancer metastasis. Nature 464(7291):1071–1076
4. Wojcik SE, Rossi S, Shimizu M, Nicoloso MS, Cimmino A, Alder H, Herlea V, Rassenti LZ, Rai KR, Kipps TJ, Keating MJ, Croce CM, Calin GA (2010) Non-codingRNA sequence variations in human chronic lymphocytic leukemia and colorectal cancer. Carcinogenesis 31(2):208–215
5. Rossi S, Kopetz S, Davuluri R, Hamilton SR, Calin GA (2010) MicroRNAs, ultraconserved genes and colorectal cancers. Int J Biochem Cell Biol 42(8): 1291–1297
6. Volinia S, Galasso M, Costinean S, Tagliavini L, Gamberoni G, Drusco A, Marchesini J, Mascellani N, Sana ME, Abu Jarour R, Desponts C, Teitell M, Baffa R, Aqeilan R, Iorio MV, Taccioli C, Garzon R, Di Leva G, Fabbri M, Catozzi M, Previati M, Ambs S, Palumbo T, Garofalo M, Veronese A, Bottoni A, Gasparini P, Harris CC, Visone R, Pekarsky Y, de la Chapelle A, Bloomston M, Dillhoff M, Rassenti LZ, Kipps TJ, Huebner K, Pichiorri F, Lenze D, Cairo S, Buendia MA, Pineau P, Dejean A, Zanesi N, Rossi S, Calin GA, Liu CG, Palatini J, Negrini M, Vecchione A, Rosenberg A, Croce CM (2010) Reprogramming of miRNA networks in cancer and leukemia. Genome Res 20(5):589–599
7. Ferracin M, Pedriali M, Veronese A, Zagatti B, Gafà R, Magri E, Lunardi M, Munerato G, Querzoli G, Maestri I, Ulazzi L, Nenci I, Croce CM, Lanza G, Querzoli P, Negrini M (2011) MicroRNA profiling for the identification of cancers with unknown primary tissue-of-origin. J Pathol 225(1):43–53. doi:10.1002/path.2915
8. Mihaela Z, Jaskiewicz L (2009) Computational biology of small regulatory RNAs (chapter 5). In: Appel RD, Ernest F, Appel RD (eds) Bioinformatics: a Swiss perspective, 1st edn. World Scientific Publishing Company, Singapore. ISBN 9812838775
9. Morozova O, Hirst M, Marra MA (2009) Applications of new sequencing technologies for transcriptome analysis. Annu Rev Genomics Hum Genet 10:135–151, PMID: 19715439
10. Li R, Yu C, Li Y, Lam TW, Yiu SM, Kristiansen K, Wang J (2009) SOAP2: an improved ultrafast tool for short read alignment. Bioinformatics 25(15): 1966–1967
11. Mathelier A, Carbone A (2010) MIReNA: finding microRNAs with high accuracy and no learning at genome scale and from deep sequencing data. Bioinformatics 26(18):2226–2234
12. Jiang P, Wu H, Wang W, Ma W, Sun X, Lu Z (2007) MiPred: classification of real and pseudo microRNA precursors using random forest prediction model with combined features. Nucleic Acids Res 35(Web Server issue):W339–W344, PMID: 17553836
13. Sewer A, Paul N, Landgraf P, Aravin A, Pfeffer S, Brownstein MJ, Tuschl T, van Nimwegen E, Zavolan M (2005) Identification of clustered microRNAs using an ab initio prediction method. BMC Bioinformatics 6:267, PMID:16274478
14. Batuwita R, Palade V (2009) microPred: effective classification of pre-miRNAs for human miRNA gene prediction. Bioinformatics 25(8):989–995
15. Friedlander MR et al (2008) Discovering microRNAs from deep sequencing data using miRDeep. Nat Biotechnol 26:407–415
16. Moxon S, Moulton V, Kim JT (2008) A scoring matrix approach to detecting miRNA target sites. Algorithms Mol Biol 3:3, PMID: 18377655
17. Hackenberg M, Sturm M, Langenberger D, Falcón-Pérez JM, Aransay AM (2009) miRanalyzer: a micro RNA detection and analysis tool for next-generation sequencing experiments. Nucleic Acids Res 37(Web Server issue):W68–W76, PMID: 19433510

18. Pantano L, Estivill X, Martí E (2010) SeqBuster, a bioinformatic tool for the processing and analysis of small RNAs datasets, reveals ubiquitous miRNA modifications in human embryonic cells. Nucleic Acids Res 38(5):e34
19. Huang PJ, Liu YC, Lee CC, Lin WC, Gan RR, Lyu PC, Tang P (2010) DSAP: deep-sequencing small RNA analysis pipeline. Nucleic Acids Res 38(Web Server issue):W385–W391, PMID: 20478825
20. Zhu E, Zhao F, Xu G, Hou H, Zhou L, Li X, Sun Z, Wu J (2010) mirTools: microRNA profiling and discovery based on high-throughput sequencing. Nucleic Acids Res 38(Web Server issue):W392–W397
21. Pasquinelli AE, Reinhart BJ, Slack F, Martindale MQ, Kuroda MI, Maller B, Hayward DC, Ball EE, Degnan B, Müller P, Spring J, Srinivasan A, Fishman M, Finnerty J, Corbo J, Levine M, Leahy P, Davidson E, Ruvkun G (2000) Conservation of the sequence and temporal expression of let-7 heterochronic regulatory RNA. Nature 408(6808):86–89
22. Lee RC, Ambros V (2001) An extensive class of small RNAs in Caenorhabditis elegans. Science 294 (5543):862–864
23. Lai EC, Tomancak P, Williams RW, Rubin GM (2003) Computational identification of Drosophila microRNA genes. Genome Biology 4:R42. doi:10.1186/gb-2003-4-7-r42
24. Berezikov E, van Tetering G, Verheul M, van de Belt J, van Laake L, Vos J, Verloop R, van de Wetering M, Guryev V, Takada S, van Zonneveld A, Mano H, Plasterk R, Cuppen E (2006) Many novel mammalian microRNA candidates identified by extensive cloning and RAKE analysis. Genome Res 16:1289–1298
25. Pfeffer S, Zavolan M, Grasser F, Chien M, Russo J, Ju J, John B, Enright A, Marks D, Sander C, Tuschl T (2004) Identification of virus encoded microRNAs. Science 304:734–736
26. Berezikov E, Guryev V, van de Belt J, Wienholds E, Plasterk RH, Cuppen E (2005) Phylogenetic shadowing and computational identification of human microRNA genes. Cell 120(1):21–24
27. Mathews DH, Turner DH, Zuker M (2007) RNA secondary structure prediction. Curr Protoc Nucleic Acid Chem; Chapter 11:Unit 11.2. PMID: 18428968
28. Hofacker I, Fekete M, Flamm C, Huynen M, Rauscher S, Stolorz P, Stadler P (1998) Automated detection of conserved RNA structure elements in complete RNA virus genomes. Nucleic Acids Res 26:3825–3826
29. Nam J, Shin K, Han J, Lee Y, Kim V, Zhang B (2005) Human microRNA prediction through a probabilistic colearning model of sequence and structure. Nucleic Acids Res 33:3570–3581
30. Mattick JS, Makunin IV (2005) Small regulatory RNAs in mammals. Hum Mol Genet 14:R121–R132
31. Berezikov E, Guryev V, van de Belt J, Wienholds E, Plasterk RH, Cuppen E (2005) Phylogenetic shadowing and computational identification of human microRNA genes. Cell 120(1):21–24
32. Lai EC, Tomancak P, WilliamsRW, Rubin GM (2003) Computational identification of Drosophila microRNA gene. Genome Biology 4:R42. doi:10.1186/gb-2003-4-7-r42
33. Doran J, Strauss WM (2007) Bio-informatic trends for the determination of miRNA-target interactions in mammals. DNA Cell Biol 26(5):353–360
34. Chang DT, Wang CC, Chen JW (2008) Using a kernel density estimation based classifier to predict species-specific microRNA precursors. BMC Bioinformatics 9 Suppl 12:S2
35. Brameier M, Wiuf C (2007) Ab initio identification of human microRNAs based on structure motifs. BMC Bioinformatics 8:478
36. John B, Enright A, Aravin A, Tuschl T, Sander C, Marks D (2004) Human microRNA targets. PLoS Biol 2:e363
37. Vatolin S, Navaratne K, Weil RJ (2006) A novel method to detect functional microRNA targets. J Mol Biol 358(4):983–996
38. Rajewsky N (2006) MicroRNA target predictions in animals. Nat Genet 38:S8–S13
39. Sood P, Krek A, Zavolan M, Macino G, Rajewsky N (2006) Cell-type-specific signatures of microRNAs on target mRNA expression. Proc Natl Acad Sci U S A 103(8):2746–2751
40. Ragan C, Cloonan N, Grimmond SM, Zuker M, Ragan MA (2009) Transcriptome-wide prediction of miRNA targets in human and mouse using FASTH. PLoS One 4(5):e5745
41. Lewis B, Burge C, Bartel D (2005) Conserved seed pairing, often flanked by adenosines, indicates that thousands of human genes are microRNA targets. Cell 120:15–20
42. Brennecke J, Stark A, Russell R, Cohen S (2005) Principles of microRNA-target recognition. PLoS Biol 3:e85
43. Krützfeldt J, Poy MN, Stoffel M (2006) Strategies to determine the biological function of microRNAs. Nat Genet 38(Suppl):S14–S19
44. Calin GA, Dumitru CD, Shimizu M, Bichi R, Zupo S, Noch E, Aldler H, Rattan S, Keating M, Rai K, Rassenti L, Kipps T, Negrini M, Bullrich F, Croce CM (2002) Frequent deletions and down-regulation of micro- RNA genes miR15 and miR16 at 13q14 in chronic lymphocytic leukemia. Proc Natl Acad Sci U S A 99(24):15524–15529
45. Calin GA, Liu CG, Sevignani C, Ferracin M, Felli N, Dumitru CD, Shimizu M, Cimmino A, Zupo S, Dono M, Dell'Aquila ML, Alder H, Rassenti L, Kipps TJ, Bullrich F, Negrini M, Croce CM (2004) MicroRNA profiling reveals distinct signatures in B cell chronic lymphocytic leukemias. Proc Natl Acad Sci U S A 101(32):11755–11760
46. Calin GA, Croce CM (2009) Chronic lymphocytic leukemia: interplay between noncoding RNAs and protein-coding genes. Blood 114(23):4761–4770
47. Fabbri M, Garzon R, Andreeff M, Kantarjian HM, Garcia-Manero G, Calin GA (2008) MicroRNAs and noncoding RNAs in hematological malignancies:

molecular, clinical and therapeutic implications. Leukemia 22(6):1095–1105
48. Taby R, Issa JP (2010) Cancer epigenetics. CA Cancer J Clin 60(6):376–392
49. Edwards JK, Pasqualini R, Arap W, Calin GA (2010) MicroRNAs and ultraconserved genes as diagnostic markers and therapeutic targets in cancer and cardiovascular diseases. J Cardiovasc Transl Res 3(3):271–279
50. Venkatachalam R, Ligtenberg MJ, Hoogerbrugge N, Schackert HK, Görgens H, Hahn MM, Kamping EJ, Vreede L, Hoenselaar E, van der Looij E, Goossens M, Churchman M, Carvajal-Carmona L, Tomlinson IP, de Bruijn DR, Van Kessel AG, Kuiper RP (2010) Germline epigenetic silencing of the tumor suppressor gene PTPRJ in early-onset familial colorectal cancer. Gastroenterology 139(6):2221–2224
51. Navarro A, Diaz T, Martinez A, Gaya A, Pons A, Gel B, Codony C, Ferrer G, Martinez C, Montserrat E, Monzo M Regulation of JAK2 by miR-135a: prognostic impact in classic Hodgkin lymphoma. Blood 114(14):2945–2951
52. Garzon R, Volinia S, Liu CG, Fernandez-Cymering C, Palumbo T, Pichiorri F, Fabbri M, Coombes K, Alder H, Nakamura T, Flomenberg N, Marcucci G, Calin GA, Kornblau SM, Kantarjian H, Bloomfield CD, Andreeff M, Croce CM (2008) MicroRNA signatures associated with cytogenetics and prognosis in acute myeloid leukemia. Blood 111(6):3183–3189
53. Gabriely G, Yi M, Narayan RS, Niers JM, Wurdinger T, Imitola J, Ligon KL, Kesari S, Esau C, Stephens RM, Tannous BA, Krichevsky AM (2011). Human glioma growth is controlled by microRNA-10b. Cancer Res 71(10):3563–3572
54. Schetter AJ, Okayama H, Harris CC (2012) The role of microRNAs in colorectal cancer. Cancer J 18(3):244–252
55. Rossi S, Shimizu M, Barbarotto E, Nicoloso MS, Dimitri F, Sampath D, Fabbri M, Lerner S, Barron LL, Rassenti LZ, Jiang L, Xiao L, Hu J, Secchiero P, Zauli G, Volinia S, Negrini M, Wierda W, Kipps TJ, Plunkett W, Coombes KR, Abruzzo LV, Keating MJ, Calin GA (2010) microRNA fingerprinting of CLL patients with chromosome 17p deletion identify a miR-21 score that stratifies early survival. Blood 116(6):945–952
56. Leite KR, Tomiyama A, Reis ST, Sousa-Canavez JM, Sañudo A, Camara-Lopes LH, Srougi M (2011) MicroRNA expression profiles in the progression of prostate cancer-from high-grade prostate intraepithelial neoplasia to metastasis. Urol Oncol 2011 Aug 29. [Epub ahead of print]
57. Segura MF, Hanniford D, Menendez S, Reavie L, Zou X, Alvarez-Diaz S, Zakrzewski J, Blochin E, Rose A, Bogunovic D, Polsky D, Wei J, Lee P, Belitskaya-Levy I, Bhardwaj N, Osman I, Hernando E (2009) Aberrant miR-182 expression promotes melanoma metastasis by repressing FOXO3 and microphthalmia-associated transcription factor. Proc Natl Acad Sci U S A 106(6):1814–1819
58. Ma L, Teruya-Feldstein J, Weinberg RA (2007) Tumour invasion and metastasis initiated by microRNA-10b in breast cancer. Nature 449(7163):682–688
59. Slaby O, Svoboda M, Fabian P, Smerdova T, Knoflickova D, Bednarikova M, Nenutil R, Vyzula R (2007) Altered expression of miR-21, miR-31, miR-143 and miR-145 is related to clinicopathologic features of colorectal cancer. Oncology 72(5–6):397–402
60. Li J, Huang H, Sun L, Yang M, Pan C, Chen W, Wu D, Lin Z, Zeng C, Yao Y, Zhang P, Song E (2009) MiR-21 indicates poor prognosis in tongue squamous cell carcinomas as an apoptosis inhibitor. Clin Cancer Res 15(12):3998–4008
61. Hurst DR, Edmonds MD, Welch DR (2009) Metastamir: the field of metastasis-regulatory microRNA is spreading. Cancer Res 69(19):7495–7498
62. Tsai WC, Hsu PW, Lai TC, Chau GY, Lin CW, Chen CM, Lin CD, Liao YL, Wang JL, Chau YP, Hsu MT, Hsiao M, Huang HD, Tsou AP (2009) MicroRNA-122, a tumor suppressor microRNA that regulates intrahepatic metastasis of hepatocellular carcinoma. Hepatology 49(5):1571–1582

MicroRNA Target Prediction and Validation

3

William Ritchie, John E.J. Rasko, and Stéphane Flamant

Abstract

The accurate prediction and validation of microRNA targets is essential to understanding the function of microRNAs. Computational predictions indicate that all human genes may be regulated by microRNAs, with each microRNA possibly targeting thousands of genes. Here we discuss computational and experimental methods for identifying mammalian microRNA targets. We describe microRNA target prediction resources and procedures that are suitable for experiments where more accurate prediction of microRNA targets is more important than detecting all putative targets. We then discuss experimental methods for identifying and validating microRNA target genes, with an emphasis on the target reporter assay as the method of choice for specifically testing functional microRNA target sites.

Keywords

microRNA • Target genes • microRNA expression • Gene regulation • Experimental validation • Target prediction algorithms • Bioinformatics

W. Ritchie (✉)
Gene and Stem Cell Therapy Program, Centenary Institute, University of Sydney,
Locked Bag No 6, Newtown, NSW 2042, Australia
e-mail: w.ritchie@centenary.usyd.edu.au

J.E.J. Rasko
Gene and Stem Cell Therapy Program, Centenary Institute, University of Sydney,
Locked Bag No 6, Newtown, NSW 2042, Australia

Cell and Molecular Therapies, RPA Hospital,
Locked Bag No 6, Newtown, NSW 2042, Australia
e-mail: j.rasko@centenary.usyd.edu.au

S. Flamant
INSERM UMR-935, Campus CNRS, Bâtiment A,
7 rue Guy Môquet, 94802 Villejuif, France
e-mail: stephane.flamant@inserm.fr

U. Schmitz et al. (eds.), *MicroRNA Cancer Regulation: Advanced Concepts, Bioinformatics and Systems Biology Tools*, Advances in Experimental Medicine and Biology 774,
DOI 10.1007/978-94-007-5590-1_3, © Springer Science+Business Media Dordrecht 2013

3.1 Introduction

In 2002, Eric Lai [1] compared the sequences of 11 microRNAs to the K box and Brd Box motifs that were known to mediate post-transcriptional regulation in *Drosophila*. He demonstrated that the first eight nucleotides, now called the seed region, of microRNAs (miRNAs), were perfectly complementary to these motifs and concluded that this complementarity may be essential in post-transcriptional regulation by microRNAs. This simple bioinformatics analysis established one of the strongest predictive features used in target prediction to date. Since then, the microRNA repertoire has grown exponentially and numerous experimental methods have been developed to confirm microRNA targets. None of these advances has produced a unique feature of microRNA targeting that is more telling than the seed region. They have instead led to the conclusion that microRNA regulation is very intricate and diverse. For this reason, the computational and experimental methods that have been developed generally focus on specific aspects of microRNA regulation and are used to either investigate the physical interaction between microRNAs and their putative targets or the functional outcome of microRNA targeting. Here we describe these computational and experimental methods and explain which specific aspects of microRNA regulation they focus on.

3.2 Computational Methods to Identify microRNA Targets

Despite a plethora of different algorithms and methods to predict microRNA targets, most rely on similar sequence-based approaches for their starting point. These algorithms initially search for some degree of sequence complementarity between the miRNA of interest and the 3′ untranslated region (3′UTR) of mRNAs with emphasis on the miRNA seed region (nt 2–8). Because the miRNA:mRNA duplex can contain mismatches, gaps and G:U pairs, the number of possible targets based uniquely on this alignment is too large to be informative. Additional steps are therefore required to refine target predictions and rank them according to statistical confidence. Here we describe the most commonly used methods for detecting miRNA targets, classified according to the criteria used to refine the initial sequence analysis (Fig. 3.1). For each approach we provide examples of commonly used algorithms and discuss their limitations.

3.2.1 Thermodynamic Stability of the microRNA:mRNA Duplex

miRanda [2], the first freely-available prediction program measures the thermodynamic stability between a miRNA and its putative target to increase prediction accuracy. Different scores for the C:G, A:U, and G:U pairs are used to measure stability with a requirement for more stable energy scores at the 5′ end of the miRNA. A user-defined threshold can then be set to eliminate unstable duplexes. Since miRanda became available, more complete models to calculate the stability of RNA duplexes have been published and successfully used to predict miRNA targets. The standalone algorithm RNAhybrid [3], for example, calculates the most stable hybridization site between two sequences and can easily be incorporated into existing prediction algorithms. The PITA algorithm [4] also uses thermodynamic stability of a miRNA:mRNA duplex but compares it to the stability of local structures within the 3′UTR of the target mRNA. If the duplex is predicted to occur within a region of the 3′UTR that is already involved in a stable structure, the miRNA is less likely to bind to its target. This approach is limited by the accurate prediction of stable secondary structures, which becomes unreliable when considering long distance interactions and therefore larger RNA structures.

3.2.2 Sequence Conservation of the Target Site Between Multiple Species

Evaluating sequence conservation of predicted targets between distantly related species efficiently reduces the number of false positive

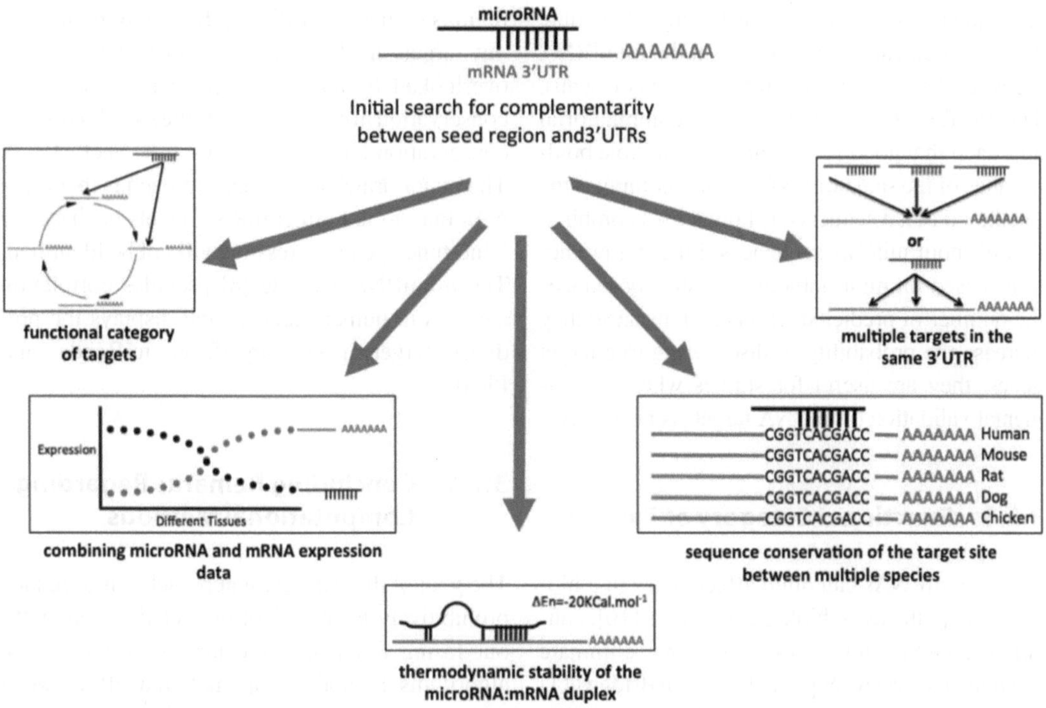

Fig. 3.1 Computational methods to identify miRNA targets. After the initial search for sequence complementarity between the seed region of the miRNA (nt 2–8) and the putative mRNA target, most algorithms will use additional criteria to refine predictions. The **functional category of targets** can be used to search for targets that belong to the same biological pathway or process. **Combining microRNA and mRNA expression data** and searching for negative correlations between them can efficiently predict miRNA targets regulated through mRNA destabilization. The **thermodynamic stability of the microRNA:mRNA duplex** searches for stronger physical interaction between the miRNA and its targets. Investigating **sequence conservation of the target site between multiple species** or **multiple target sites in the same 3′UTR** can be used to rank putative targets according to their statistical likelihood

predictions. Most algorithms will require that the predicted miRNA target site be located in homologous regions of the 3′UTR, and that the seed binding region be in a highly conserved region. TargetScan [5] initially searches for conserved seed pairing regions in 3′UTR alignments between 28 vertebrate species. This set of putative targets is then refined using a context score based on the target position in the 3′UTR and surrounding sequence composition and further refined by considering 3′ pairing of the miRNA within [6]. This approach is of little use in detecting species-specific binding sites or binding sites of species-specific miRNAs. TargetScan also provides non-conserved targets on their website.

3.2.3 Multiple Targets in the Same 3′UTR

Recent analysis demonstrates that numerous mRNAs are targeted by the same miRNA at different sites within their 3′UTR. This multi-targeting occurs at a significantly higher rate than expected. Focusing therefore on mRNAs that have more than one predicted site for the same miRNA in the 3′UTR can increase the signal to noise ratio for different algorithms [7, 8]. Although this approach will eliminate numerous true target sites it has the advantage of producing a list of high confidence gene targets. This method requires the user to first select one or more target prediction programs and subsequently refine their results for multi-targeting. This last step can be performed on the mimiRNA website [8] (http://mimirna.centenary.org.au). The PicTar [9] algorithm uses a combinatorial approach that not only accounts for multiple binding sites of the same miRNA but also computes the likelihood that a sequence is bound by a combination of input miRNA sequences. Filtering predictions based on multi-targeting drastically reduces the number of predicted targets and, because they increase the probability of discovering true target genes, they are useful for studies where experimental validation of miRNA targets is necessary.

3.2.4 Functional Category of Targets

Because miRNAs can often affect genes in a biochemical pathway or biological process [10], considering the function of target genes may eliminate biologically irrelevant predictions. mirBridge [11] starts with a set of genes with a known function and searches for enrichment of putative targets based on sequence analysis amongst this gene set. This approach is useful for experiments where a specific function or pathway is being dissected but may prove limiting in studies where a specific miRNA or mRNA is being analysed with no prior knowledge of its function.

3.2.5 Combining microRNA and mRNA Expression Data

Numerous miRNAs inhibit gene expression by destabilizing mRNAs [12]. As a consequence, mRNA targets should be expressed at lower levels in tissues where the miRNA is expressed. Correlating mRNA and miRNA expression across multiple tissues and selecting those pairs that are negatively correlated can successfully detect target genes [13]. Because this method is independent of any sequence analysis, it can be used to filter predictions made by any of the aforementioned algorithms. Another advantage of this approach is that it is not restricted to targets located in the 3′UTR. Although there are fewer published examples of miRNA targets in other regions of mature mRNAs, there may be numerous targets in the coding region that have been overlooked because the high level of sequence conservation in exons prohibits the use of sequence conservation-based techniques (see Sect. 3.3.3). The major drawback of this approach is that miRNAs that do not affect mRNA levels or that only "fine-tune" gene expression will not be identified. The mimiRNA website [8] provides correlation analysis in human samples and displays the predicted targets from TargetScan, miRanda, and PicTar.

3.2.6 Concluding Remarks Regarding Computational Methods

The goal of these different approaches is to reduce prohibitively large lists of predicted targets without losing too many true targets. Tuning these algorithms to find an optimal tradeoff between accuracy and sensitivity is currently impossible because relatively few targets have been validated experimentally. As a result, the efficiency of these algorithms is often tested by measuring the enrichment for predicted targets amongst a set of mRNAs or proteins for which the expression is subject to perturbation of miRNA expression. A recent study based on protein expression following both miRNA overexpression and knockdown found that TargetScanS and Pictar gave the best results [14]. However, this type of benchmark does not account for off-target effects which may be prevalent considering that miRNAs often target transcription enhancers and repressors [13]. One commonly used approach to enhance the quality of target predictions is to consider the overlap between multiple programs. We do not recommend this as there is no proof that this will increase prediction quality and it will systematically reduce the number of candidates [7].

3.2.7 Future Directions

The degree of sequence conservation of a target or its involvement in a pathway for which other targets are predicted (described in Sects. 3.2.2,

3.2.3 and 3.2.4 above) does not imply the biological mechanism through which a specific miRNA binds to its targets. Binding of miRNA:mRNA pairs is affected by spatial and temporal co-expression of the miRNA:mRNA pair, target site availability, and the formation of a stable duplex at the target site. Future algorithms will be required to investigate these three criteria to discover the whole repertoire of miRNA targets.

Co-expression of miRNA:mRNA pairs is often evaluated by simultaneous sequencing of mRNA enriched libraries and small RNA libraries from the same cells. As more of these experiments are performed on different cell types and even subcellular localizations, prediction tools will be able to integrate co-expression data with increasing efficiency.

Target site availability is currently evaluated by folding a small sequence of RNA around the putative target. As discussed above, this does not take into account long distance interactions between different regions of the same RNA molecule. Such interactions are currently impossible to predict because there is insufficient biochemical data on the stability of large RNA structures and because the number of possible suboptimal structures that could be predicted is prohibitively large. Moreover, target site accessibility should take into account RNA binding proteins, the prediction of which suffers the same limitations as miRNA targets.

The stability of the miRNA:mRNA duplex has been thoroughly investigated through machine learning models and *in vivo* mutagenesis assays [15]. The results of these studies show that there is no clear-cut rule on the amount of sequence complementarity required between the miRNA and its target or at what position complementarity should occur. These most likely depend on the region of the mature miRNA that is exposed in the active site of Argonaute proteins and are therefore available to interact with its target. Understanding the different conformations of the Argonaute proteins should therefore allow for more accurate target predictions.

3.3 Experimental Identification and Validation of microRNA Targets

The identification of microRNAs and their target genes was originally conducted through classic genetic studies in the worm *Caenorhabditis elegans*, whereby a miRNA mutant displayed an opposite phenotype to that shown by the corresponding target gene null mutant [16]. Although this method was appropriate for small organisms such as nematodes [17] or the fruit fly *Drosophila melanogaster* [18], it remains limited for larger animals like mammals. Therefore artificial systems are needed to identify and validate miRNA target genes. Validation of a putative miRNA target site requires that a physical interaction between a miRNA and its target mRNA will lead to decreased production of the corresponding protein. Such physical interaction implies the spatiotemporal co-expression of the regulating miRNA and its target gene. On this basis, modulating miRNA expression levels should result in changes in the amount of a reporter protein such as luciferase or GFP, which are quantified in comparison to controls. Several methods have been designed to experimentally identify targeted mRNAs at various steps along the miRNA regulatory pathway (Fig. 3.2). Since the net result of miRNA-mediated gene regulation is a decrease in the amount of target protein being produced, methods measuring changes in protein output resulting from variations of miRNA expression have become a standard approach to identifying and validating miRNA targets. In addition, a number of biochemical methods have been developed in order to experimentally identify miRNA:mRNA pairs isolated from immunopurified ribonucleoprotein complexes or enriched miRNA:mRNA duplexes. Here we describe some of the methods used to experimentally identify and validate miRNA target genes (see also refs 19–21 for review).

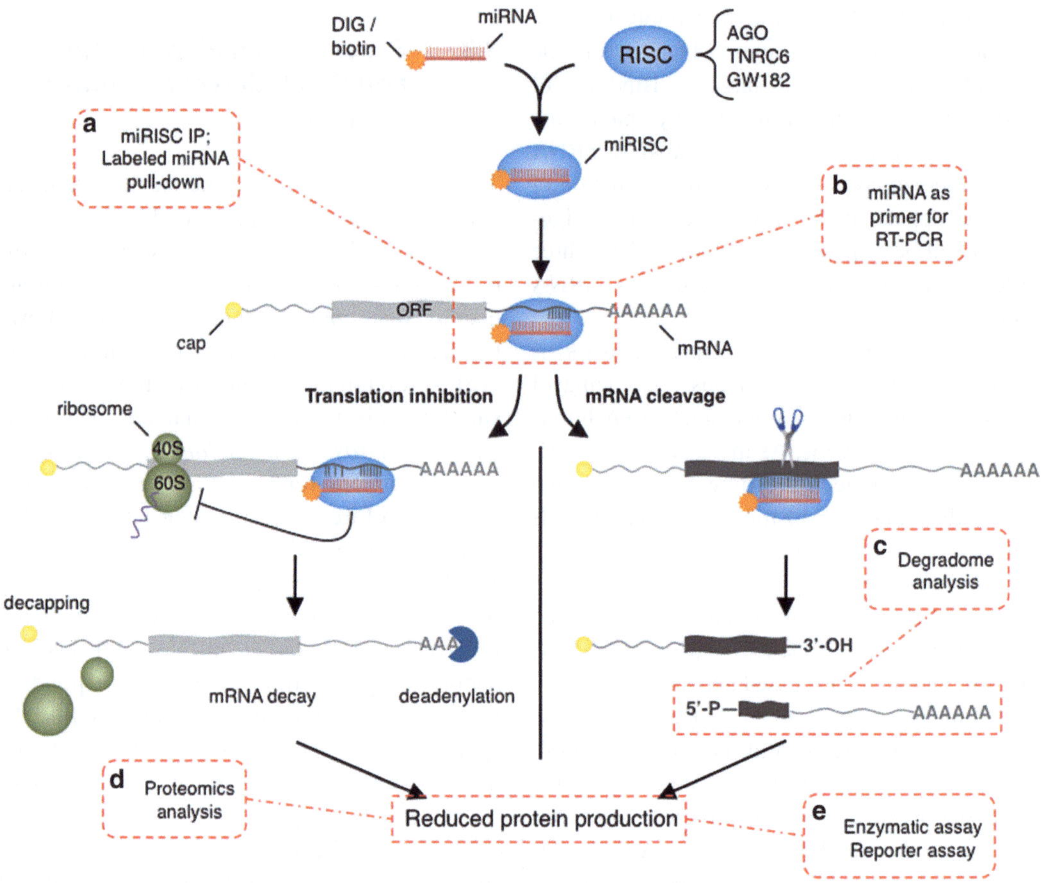

Fig. 3.2 Experimental methods designed to identify and validate targeted mRNA based on the relevant part of the miRNA regulatory pathway. Once loaded into the RNA-induced silencing complex (miRISC), miRNA drives miRISC to the targeted mRNA. Depending on the level of complementarity between the miRNA and the mRNA target site, miRISC follows two different routes to inhibit protein production. Partial base pairing between miRNA and mRNA (*left*) leads to translation inhibition and mRNA decay. High complementarity between miRNA and targeted transcript (*right*) results in mRNA cleavage by Argonaute slicing activity. (**a**) Biochemical methods have been designed in order to purify miRNA:mRNA complexes by immunoprecipitation (IP) or pull-down of labeled miRNA from miRISC components (Sect. 3.3.3). (**b–c**) Molecular approaches are used to identify target genes through miRNA-primed reverse transcription of targeted mRNA template, or by analysis of cleavage products (Sect. 3.3.4). (**d**) Proteomics analysis identifies changes in protein output upon miRNA expression variations (Sect. 3.3.2). (**e**) Target genes are ultimately validated by reporter assay (Sect. 3.3.1). *DIG* digoxigenin, *ORF* open reading frame

3.3.1 Reporter Assays

In vitro reporter assays have been designed to confirm the interaction between a given miRNA and a putative target mRNA. The rationale is that upon binding to its target site(s) a given miRNA will inhibit reporter protein production, thereby leading to reduced protein amount or activity which can be measured compared to relevant controls [22–25]. Typically the putative miRNA target site is cloned downstream of the open reading frame of a reporter gene, *e.g.* luciferase (*Renilla* or firefly) or GFP, and the recombinant plasmid is transfected into mammalian cells. Depending on the size of the 3′UTR to be tested, the full-length UTR or a fragment containing the predicted binding site is used. However, a partial UTR sequence may give erroneous positive results due to higher accessibility of the miRNA consequent to loss of secondary structures in the UTR. The recombinant reporter plasmid and a vector overexpressing the miRNA of interest, or a synthetic double-stranded oligonucleotide (miRNA mimic), are then transiently transfected into mammalian cells, usually HeLa or HEK293 cells, and luciferase activity or fluorescence intensity is measured 24–48 h later. It is important to assess endogenous miRNA expression levels in the cell system used for the assay, as the endogenous expression of miRNAs is not the same from one cell type to another, and some miRNAs display tissue-specific expression (*e.g.* hematopoietic-, brain-, embryonic stem cell-restricted miRNAs). Alternatively, cells can be transfected with the reporter vector alone if they express suitable endogenous levels of the candidate miRNA. Reduction of miRNA expression can be achieved using miRNA inhibitors such as modified antisense oligonucleotides [26] or sponge vectors [27], which constitute an elegant option when cells have high endogenous miRNA levels.

Importantly, transfection controls must be chosen carefully. These controls include reporter vectors without the UTR sequence, or with a UTR cloned in the antisense orientation. Also, cells must be co-transfected with a control luciferase reporter vector to normalize for variations in transfection efficiencies. Alternatively, dual luciferase reporter systems can be used, in which UTR sequences are cloned downstream to one luciferase gene (*Renilla*), while the other luciferase reporter (firefly) remains unaltered and is used for normalization. Specificity of miRNA regulation is assessed by co-transfection of an irrelevant miRNA or scrambled RNA duplexes. In these conditions, only transfection with the relevant miRNA should result in a decrease of reporter activity/expression. However, this result could be due to some off-target effect of the miRNA, which is provided in supra-physiological amounts to the cell when overexpressed, or indirect regulation by targeting genes that, in return, affect expression of the reporter. To confirm the specific inhibition of a miRNA on a target gene, it is therefore essential that the predicted binding sites be disrupted and that modified UTR sequences be tested in the reporter assay as well. This strategy not only definitively validates the miRNA:mRNA interaction and regulation, but also identifies which site(s) is/are true functional binding site(s) in the case of multiple predicted miRNA target sites. Last, a modified miRNA mimic harboring the complementary sequence to the mutated UTR can be used to rescue target regulation of the mutated UTR reporter constructs. In summary, a valid reporter assay should be carried out by co-transfecting (1) a reporter plasmid containing the full 3′UTR sequence, and (2) the same reporter construct with a disrupted target site, together with a miRNA overexpressing vector *vs.* scramble sequence.

The reporter assay described above indicates that, when a given miRNA and target gene are expressed simultaneously in the same cell, they are likely to interact and this interaction might result in miRNA-mediated reduced expression of the target gene. It remains, however, an artificial system in which both the miRNA and the targeted UTR are overexpressed in a heterologous system. It is thus recommended to confirm, when possible, that such regulation does occur on the endogenous gene. Changes in protein amounts upon miRNA overexpression/inhibition can be measured by Western blot, flow cytometry, or immunocytochemistry experiments. If antibodies are not available, other validation methods can be used, for example, based on enzymatic activity, ligand binding, etc. Another indication of miRNA-induced gene regulation can be provided by target transcript quantification. Although miRNAs were originally shown to regulate gene expression by repressing mRNA translation without affecting transcript level, it is now widely accepted that miRNA-mediated regulation is frequently accompanied by mRNA destabilization, essentially due to increased deadenylation [28, 29]. Transcripts displaying reduced levels upon miRNA ectopic expression are subsequently analysed for the presence of miRNA target sites in their 3′UTR using prediction algorithms (see Sect. 3.2) in order to identify putative miRNA target genes [12, 26, 30].

3.3.2 Proteomics Methods

Several proteomics studies have been designed to identify miRNA target genes. Vinther et al. [31] used stable isotope labeling by amino acids in cell culture (SILAC), in which proteins are metabolically labeled by cells growing in medium containing heavy isotopes of essential amino acids. Differences in protein synthesis are determined by mass spectrometry as the ratio of peptide peak intensities from light and heavy isotopes [32]. Of 504 proteins investigated by SILAC, they identified a set of 12 proteins with reduced expression in HeLa cells overexpressing miR-1 and grown in medium containing heavy isotopes, as compared to control cells grown with light isotopes. Seed region complementary sites were found in the 3′UTR of corresponding genes for 8 of these proteins, which was a significant enrichment for miR-1 seed motif when compared with entire 3′UTR sequence databases. These investigators used the luciferase reporter assay to confirm miR-1 regulation for 6 out of 11 target genes tested [31].

The SILAC method was subsequently used in two large-scale proteomics studies to identify target genes of several miRNAs [14, 33]. In both cases, HeLa cells were transfected with different miRNA duplexes, and protein output was measured 48 h post-transfection. Selbach et al. used a modified version of SILAC in which cells were pulse-labeled (pSILAC) so that heavy isotopes were primarily incorporated into newly synthesized proteins [14]. In addition, SILAC was used to study the impact of miR-223 deficiency in mouse neutrophils [33] and let-7b knockdown in HeLa cells [14]. The authors concluded that each miRNA regulates hundreds of target proteins, though to a relatively modest degree. Motif analysis revealed a significant enrichment for corresponding miRNA seed complementary sites in the 3′UTR of repressed genes, as compared to an unmodified protein set. While Baek et al. found that most repressed targets displayed detectable mRNA destabilization [33], Selbach et al. identified substantial direct regulation by translation inhibition [14]. Overall, these studies suggested that miRNAs act primarily by fine tuning expression of a large number of target genes.

Zhu et al. used two-dimensional differentiation in-gel electrophoresis (2D-DIGE) to identify miR-21 targets in a mouse breast cancer model [34]. Proteins were extracted from tumors derived from human MCF7 cells treated with anti-miR-21 antisense or control oligonucleotide. After labeling with two different fluorescent dyes, both protein samples were separated by 2D-polyacrylamide gel electrophoresis (PAGE) in the same gel. Fluorescence intensity was measured by gel imaging, and differentially expressed proteins were purified from the gel prior to identification by mass spectrometry. This method identified seven proteins that were up-regulated in anti-miR-21 treated tumors, including tropomyosin (TPM) 1, which was further validated by reporter assay and Western blot [34]. Of note, several proteins were also found to be down-regulated upon anti-miR-21 treatment in this study, which suggests an indirect effect of miR-21.

Another approach for target identification combined miRNA and protein expression analysis with computational predictions. miRNA profiling was performed to identify differentially expressed miRNAs between two samples, which were compared to proteomics data generated by 2D-PAGE associated to mass spectrometry [35] or reverse-phase protein arrays [36]. Reciprocally expressed miRNAs and proteins were then compared to miRNA target predictions to identify relevant target genes. This analysis resulted in the identification of 52 and 17 miRNA:gene target pairs in rat kidney [35] and human cartilage [36], respectively. More recently, a targeted proteomics approach was designed to identify let-7 miRNA target genes in *C. elegans* [37]. The method combined isotope-coded affinity tag (ICAT) protein labeling [38] and detection by selected reaction monitoring mass spectrometry [39] to quantify protein levels between wild type and *let-7* mutant whole worms. By definition, the ICAT labeling is restricted to proteins harbouring mass spectrometry-detectable peptides that contain cysteine residues [38]. This limitation implied working on a predefined set of proteins predicted as let-7 targets that met these requirements, leading to consequent reduced proteome coverage. Of 161 proteins analysed, 29 were significantly altered in mutant worms, including ten that were downregulated, suggesting an indirect effect of miRNA regulation [37]. Ten of the identified targets were further validated by genetic analysis and, for one of them, by reporter assay. The authors then used a modified method based on metabolic labeling of worms using heavy isotopes [40], to facilitate full coverage of the *C. elegans* peptide repertoire. Of 27 predicted miR-58 targets, four were identified as significantly upregulated in a *miR-58* mutant using this modified method [37].

3.3.3 Biochemical Approaches

miRNA-mediated gene silencing in mammals requires a functional miRNA-loaded RNA-induced silencing complex (miRISC) machinery (Fig. 3.2). Several studies identified miRNA target transcripts by virtue of their association with miRISC components by co-immunoprecipitation of human or *Drosophila* Argonaute (AGO) proteins [35, 41–46], human TNRC6 proteins [45], or nematode GW182 protein family AIN1-2 [47]. This strategy was originally used by Mourelatos et al. to identify new miRNAs that were co-immunoprecipitated with AGO2/EIF2C2-containing complex in HeLa cells [48]. Immunoprecipitated mRNAs were then identified by cloning, microarray analysis, or deep sequencing. A first strategy consists in the purification of all miRISC-associated mRNA species in a given cell type, in order to identify the global "targetome" of that cell type, without preliminary knowledge of the presence of any specific miRNA. Sequence motif analysis is then performed to identify miRNA complementary sites enriched in miRISC-bound mRNAs compared to whole cell mRNAs, thus inferring which miRNAs are co-expressed. Easow et al. used this approach in *Drosophila* S2 cells stably expressing FLAG/HA-Ago1 [42]. Microarray analysis revealed significant enrichment of transcripts containing complementary sites for miR-184, miR-7 and miR-314, in anti-HA pulled down mRNAs. Similarly, Beitzinger et al. pulled down AGO1- and AGO2-associated transcripts from HEK293 cells and identified immunoprecipitated mRNAs by complementary DNA (cDNA) library preparation and sequencing [41]. Another approach consists in comparing miRISC-associated mRNAs of cells transfected with, or deprived of, a given miRNA to mock-transfected or unmodified control cells. Easow et al. found a significant overrepresentation of miR-1 complementary sequences in Ago1 co-purified transcripts from miR-1 transfected S2 cells compared to untransfected cells [42]. Several studies using this method, also called RIP-Chip (ribonu cleoprotein immunoprecipitation-gene chip), reported identification of miRNA targets in 293 cells [43–45], Hodgkin lymphoma cell lines [49], human H4 glioneuronal cells [46], and *C. elegans* [47]. In this latter study, high-throughput sequencing was used to identify co-immunoprecipitated miRNAs as well. Notably, this experimental procedure allowed the identification of miRNA target genes with stable mRNA levels that are likely to be primarily regulated by translational repression [43].

Recently, the HITS-CLIP method (high-throughput sequencing by crosslinking and immunoprecipitation) was developed to identify direct protein/RNA interactions [50]. This approach uses UV irradiation to crosslink nucleic acids and proteins in close proximity, which are then immunopurified using an antibody to a miRISC component. Partial RNA digestion leaves miRISC-protected RNA fragments, which are then identified by high throughput sequencing. Chi et al. used HITS-CLIP to purify Ago2-bound mRNA and miRNA species from mouse brain as well as miR-124 transfected HeLa cells [51]. As in other studies, bound mRNAs were enriched for complementary sites to miRNAs that were either highly endogenously expressed or over-expressed following transfection. This approach, also called CLIP-Seq, was used to isolate Argonaute protein ALG-1-bound mRNAs in *C. elegans* [52] and Ago2-purified transcripts in wild type versus *dicer*$^{-/-}$ mouse ES cells [53]. An improvement of the method, named photoactivatable-ribonucleoside-enhanced (PAR)-CLIP, was recently described, in which crosslinking efficiency was enhanced by incorporation of the photoactivatable nucleoside analog 4-thiouridine into transcripts of cultured cells [54]. Upon UV crosslinking at 365 nm, thymidine located at the crosslinking sites are converted to cytidine, which allows for the precise identification of RNA-protein binding site. PAR-CLIP method was used to identify miRNA target sites of mRNAs associated to AGO and TNRC6 family proteins in 293 cells. Deep sequencing of bound RNAs revealed enrichment of complementary sites for the most highly expressed miRNAs [54].

Interestingly, these high-throughput studies revealed that a high proportion (25–50 %) of the binding sites were located within the coding sequence (CDS) region of bound mRNAs [51, 52, 54]. This observation suggests that functional miRNA target sites may not only be located in 3′UTRs as previously thought, in agreement with a number of recent reports identifying miRNA target sites in CDS [55–58]. Furthermore, Schnall-Levin et al. recently demonstrated frequent CDS targeting through repeated miRNA binding sites, of paralogous families of the C_2H_2 zinc-finger genes, which typically contain many tandem repeats of the finger motif [59]. Similarly, building on previously published microarray data in mammalian cells transfected with, or deprived of, specific miRNA [14, 33], Fang and Rajewsky showed that CDS target sites act synergistically with 3′UTR sites for miRNA-mediated regulation of gene expression [60]. Of importance, most prediction algorithms could not identify this class of miRNA target sites because of the "3′UTR-only" rule. However, the PITA algorithm [4], which mainly identifies target site accessibility, and the rna22 program [57, 61], which identifies over-represented sequence patterns, can be used to detect miRNA binding sites located outside the UTR. In addition, the mimiRNA algorithm [8], which identifies miRNA:mRNA pairs that display conserved negative correlation of expression across several tissues, can be used to select candidate target genes prior to searching for putative binding sites. Of note, CDS target site validation requires a modified reporter assay, whereby the target-site-containing sequence is fused in frame with a reporter CDS [42, 59]. Alternatively, co-transfection of wild type and mutated versions of the targeted CDS associated with two different epitope tags, *e.g.* Myc and FLAG, has been used to monitor by Western blot the level of protein down-regulation upon miRNA co-expression [58].

An alternative strategy to the aforementioned protein pull-down methods was proposed by Orom and Lund, who developed an affinity-based target gene identification procedure [62]. In this case, transfection of a biotinylated synthetic miRNA allows the purification of miRNA:mRNP complexes using streptavidin-agarose beads. This strategy is attractive since it allows target gene identification of a specific miRNA, whereas other methods seek to isolate virtually all miRNA-regulated transcripts. By purifying a biotin-tagged bantam miRNA in *Drosophila* S2 cells, the endogenous target gene *Hid* was efficiently identified [62]. The same group subsequently used this technique to isolate mRNAs bound to biotinylated miR-10a in mouse ES cells. Surprisingly, microarray analysis revealed that 55 of the 100 most enriched mRNAs corresponded to ribosomal protein genes, with no enrichment for known miR-10a targets or transcripts with miR-10a complementary sites [63]. They further showed that miR-10a bound conserved sites in the 5′ UTR of these genes, leading to upregulation of ribosomal protein translation and ribosome formation, resulting in a ~30 % increase of global protein synthesis [63]. Combined to 4-thiouridine modified nucleotides and UV crosslinking, biotin-tagged miRNA 'pullout' was used to demonstrate direct interaction between miR-34a and MYC transcript in human fibroblasts. Similarly the LAMP (labeled miRNA pull-down) assay was developed [64, 65], in which synthetic miRNAs were labeled with digoxigenin (DIG), and binding RNAs were isolated using anti-DIG agarose beads. The LAMP method was used to isolate known targets of *C. elegans* let-7 and lin-4, and zebrafish let-7 and miR-1. Specifically, 302 transcripts enriched using DIG-tagged miR-1 pull down (compared to mutated miR-1 control) were identified, including the known miR-1 target Hand2 [66]. An improvement of the method, called TAP-Tar (tandem affinity precipitation target identification) was recently described, which combined HA-tagged AGO1-2 immunoprecipitation followed by biotinylated miRNA pull down using streptavidin beads in HeLa cells [67]. This two-step procedure was shown to recover the known miR-20a target E2F1 more efficiently than each pull down method used separately.

3.3.4 Molecular Methods

Vatolin et al. reported the use of endogenous miRNAs as primers for cDNA synthesis by reverse transcriptase on the targeted mRNA template [68]. Although pairing of the target mRNA to the miRNA 3′ end is usually weaker than to the 5′ end (the seed region), the hypothesis underpinning this work was that the miRNA 3′ end could form a temporary stable duplex with the target mRNA to initiate cDNA synthesis (Fig. 3.2). Using cytoplasmic extracts, a first round of reverse transcription elongates the miRNA sequence to generate cDNA-miRNA molecules, which are purified and used as secondary primers to drive a second round of reverse transcription, thereby increasing the specificity of the reaction. After ligation of an adapter sequence at the 5′ end, cDNAs are PCR amplified using a primer from the adapter and a gene-specific primer corresponding to a target RNA of interest. PCR products are then cloned and sequenced to identify the regulatory miRNA based on homology searches of the appropriate databases. Vatolin et al. recovered partial sequences of miRNAs associated to β-actin, N-Ras and K-Ras mRNAs from human hTERT-RPE1 epithelial cells, and confirmed their functional regulation by Western blot and luciferase assay upon miRNA overexpression [68].

Andachi modified the method by ligating an adapter sequence to the 3′ end of the cDNA and by using a biotinylated, miRNA-specific primer together with an adapter-specific primer for PCR amplification [69]. The amplification product was purified using avidin beads, and further PCR amplified with adapter-specific and nested miRNA-specific primers. When applied to *C. elegans*, this method isolated the known lin-4 target gene lin-14, and identified the K10C3.4 gene as a new target for let-7, which was further validated through reporter assay and genetic complementation analysis [69]. The two methods described above allow identification of miRNA:mRNA pairs by either target gene- or miRNA-specific analysis, and are not suitable for high-throughput identification of miRNA targets.

In the specific context of miRNA-mediated cleavage of a target gene (Fig. 3.1), several studies identified mRNA cleavage products by RNA ligase mediated-5′ rapid amplification of cDNA ends (RLM-RACE) [70–77]. In the original method, an RNA adapter was ligated to the 5′ phosphate of cleaved, uncapped poly-A^+ RNAs. After reverse transcription with oligo-(dT), cDNAs were amplified using adapter- and gene-specific primers, before cloning and sequencing. This approach was used to validate miR-171-mediated cleavage of several transcripts of the SCL family of transcription factors in *Arabidopsis thaliana* [70], as well as Hoxb8 mRNA cleavage by miR-196 in mouse embryos [25]. In addition, the 5′ end of the cloned mRNA was shown to map to the nucleotide pairing with the tenth nucleotide of the miRNA.

An improved method, named PARE (parallel analysis of RNA ends), was developed for genome-wide identification of miRNA-induced cleavage products [71, 72]. In this modified protocol, the 5′ RNA adapter was engineered to contain an MmeI restriction site, and after reverse transcription and second strand cDNA synthesis, double-stranded molecules were digested with MmeI, generating 20–21 nt tag sequences attached to the adapter. A DNA adapter was then ligated at the 3′ end of the tag, which was PCR amplified using 5′adapter- and 3′adapter-specific primers. Tags were analysed by high-throughput sequencing and matched to the *Arabidopsis* genome to identify corresponding target genes and infer regulatory miRNAs. This 'degradome' tag analysis identified a large proportion of known *Arabidopsis* miRNA and *trans*-acting siRNA (ta-siRNA) target genes, although most of the tags represented mRNA degradation products unrelated to these small RNAs [71, 72]. PARE was also used to identify miRNA and ta-siRNA target genes in rice [75]. A modified RLM-RACE methodology was also developed, in which

Arabidopsis cleaved transcripts were linearly amplified by *in vitro* transcription using a T7 promoter, prior to microarray analysis [73, 78]. Of the 228 candidate targets identified, 14 corresponded to previously known miRNA targets [73].

Although this approach is most suited to plants, in which extensive base pairing between miRNA and mRNA leads to miRISC-mediated cleavage of targeted mRNA, several studies reported PARE analysis of the degradome in mammalian cells [74, 76, 77]. Karginov et al. compared degradome tags from wild type versus *Ago2*$^{-/-}$ mouse ES cells, in order to identify miRNA-specific cleavage products. Tag abundance peaked at nucleotide position 10 of the miRNA in wild type cells, whereas no peak was identified in *Ago2*$^{-/-}$ cells [74]. This study also identified a number of target genes subjected to direct Drosha-mediated endonucleolytic cleavage, as well as Ago2- and Drosha-independent cleavage sites that were conserved in human 293 cells. In another study, Shin et al. defined a class of metazoan target sites named 'centered sites', which lack perfect seed pairing and 3′-compensatory pairing, but instead harbour 11–12 contiguous nucleotides that pair with miRNA nt 4–15 [76]. Using RLM-RACE degradome sequencing, they identified a set of genes targeted for miRNA-mediated cleavage in HeLa cells and human brain, though of low abundance. Although most of the putative target genes were attributed to three highly expressed miRNAs (miR-196a, -28, -151-5p), a total of 18 additional miRNA target genes were identified [76]. Likewise, Bracken et al. performed degradome analysis on six adult mouse tissues and d16.5 whole mouse embryo, resulting in the identification of 23 putative miRNA-mediated cleavage sites, most of which displayed low read frequency [77]. Although these studies revealed the existence of miRNA-guided cleavage of target mRNAs in mammals, such targeting remains restricted to a limited number of genes. In addition, degradome analyses showed that a substantial proportion of transcripts were subjected to endonucleolytic cleavage, though most of them were not related to miRNA regulation [76, 77].

3.3.5 Concluding Remarks

Here we have considered a diverse array of computational and experimental methods used for genome-wide identification of miRNA target genes, each of which exhibits its own strengths and weaknesses. Yet, high-throughput approaches require formal validation to discriminate direct from indirect targeting, and identify functional miRNA target sites among the plethora of predictions. In this regard, reporter assays can provide such information, although they should be supported by other validation analysis, notably showing miRNA and mRNA co-expression and targeted protein output variations upon miRNA expression modulation.

The experimental methods described above highlight the existence of a large number of miRNA binding sites outside the 3′UTR of interacting mRNAs, particularly in the CDS. Although target sites in the CDS do not appear to be as effective in regulating protein output as those present in the 3′UTR [59], their contribution to the fine-tuning of gene expression has been essentially ignored so far. In addition, most of the widely used target prediction algorithms consider sites solely located within the 3′UTR, which renders CDS target site analysis even more difficult. Implemented computational methods will undoubtedly be developed in the future in order to investigate CDS target sites, together with the recently identified centered sites [76].

New models to explore miRNA function are regularly described, among which miRNA loss- and gain-of-function approaches will play an increasing role. Such models have proved useful for functional analyses of miRNA activity and target gene identification in nematode and *Drosophila*, and to a lesser extent in mouse ([79–83], see ref [84] for review). The mirKO resource [85] that was recently made available for the scientific community should aid in deciphering new miRNA functions and targets in the mouse. Likewise, the generation of miRNA/mRNA targeting networks through computational analysis of putative target gene function [86] should provide additional hints towards functional miRNA target gene identification.

Acknowledgments The authors thank DIM Biothérapies, Stem Pole Ile-de-France, Cure The Future (Cell and Gene Trust), the Rebecca L Cooper Medical Research Foundation and the Cancer Council NSW [Project Grant 1006260] and the Australian National Health and Medical Research Council [Training Fellowship 571156] for support.

References

1. Lai EC (2002) Micro RNAs are complementary to 3′UTR sequence motifs that mediate negative post-transcriptional regulation. Nat Genet 30:363–364
2. John B, Enright AJ, Aravin A et al (2004) Human MicroRNA targets. PLoS Biol 2:e363
3. Kruger J, Rehmsmeier M (2006) RNAhybrid: microRNA target prediction easy, fast and flexible. Nucleic Acids Res 34:W451–W454
4. Kertesz M, Iovino N, Unnerstall U et al (2007) The role of site accessibility in microRNA target recognition. Nat Genet 39:1278–1284
5. Lewis BP, Burge CB, Bartel DP (2005) Conserved seed pairing, often flanked by adenosines, indicates that thousands of human genes are microRNA targets. Cell 120:15–20
6. Grimson A, Farh KK, Johnston WK et al (2007) MicroRNA targeting specificity in mammals: determinants beyond seed pairing. Mol Cell 27:91–105
7. Ritchie W, Flamant S, Rasko JE (2009) MicroRNA target prediction: traps for the unwary. Nat Methods 6:397–398
8. Ritchie W, Flamant S, Rasko JE (2010) MimiRNA: a microRNA expression profiler and classification resource designed to identify functional correlations between microRNAs and their targets. Bioinformatics 26:223–227
9. Krek A, Grun D, Poy MN et al (2005) Combinatorial microRNA target predictions. Nat Genet 37:495–500
10. Xiao C, Rajewsky K (2009) MicroRNA control in the immune system: basic principles. Cell 136:26–36
11. Tsang JS, Ebert MS, van Oudenaarden A (2010) Genome-wide dissection of microRNA functions and cotargeting networks using gene set signatures. Mol Cell 38:140–153
12. Lim LP, Lau NC, Garrett-Engele P et al (2005) Microarray analysis shows that some microRNAs downregulate large numbers of target mRNAs. Nature 433:769–773
13. Ritchie W, Rajasekhar M, Flamant S et al (2009) Conserved expression patterns predict microRNA targets. PLoS Comput Biol 5:e1000513
14. Selbach M, Schwanhausser B, Thierfelder N et al (2008) Widespread changes in protein synthesis induced by microRNAs. Nature 455:58–63
15. Brennecke J, Stark A, Russell RB et al (2005) Principles of microRNA-target recognition. PLoS Biol 3:e85
16. Lee RC, Feinbaum RL, Ambros V (1993) The C. elegans heterochronic gene lin-4 encodes small RNAs with antisense complementarity to lin-14. Cell 75:843–854
17. Reinhart BJ, Slack FJ, Basson M et al (2000) The 21-nucleotide let-7 RNA regulates developmental timing in Caenorhabditis elegans. Nature 403:901–906
18. Brennecke J, Hipfner DR, Stark A et al (2003) Bantam encodes a developmentally regulated microRNA that controls cell proliferation and regulates the proapoptotic gene hid in Drosophila. Cell 113:25–36
19. Orom UA, Lund AH (2010) Experimental identification of microRNA targets. Gene 451:1–5
20. Thomas M, Lieberman J, Lal A (2010) Desperately seeking microRNA targets. Nat Struct Mol Biol 17:1169–1174
21. Thomson DW, Bracken CP, Goodall GJ (2011) Experimental strategies for microRNA target identification. Nucleic Acids Res 39:6845–6853
22. Doench JG, Petersen CP, Sharp PA (2003) SiRNAs can function as miRNAs. Genes Dev 17:438–442
23. Zeng Y, Yi R, Cullen BR (2003) MicroRNAs and small interfering RNAs can inhibit mRNA expression by similar mechanisms. Proc Natl Acad Sci U S A 100:9779–9784
24. Kiriakidou M, Nelson PT, Kouranov A et al (2004) A combined computational-experimental approach predicts human microRNA targets. Genes Dev 18:1165–1178
25. Yekta S, Shih IH, Bartel DP (2004) MicroRNA-directed cleavage of HOXB8 mRNA. Science 304:594–596
26. Krutzfeldt J, Rajewsky N, Braich R et al (2005) Silencing of microRNAs in vivo with 'antagomirs'. Nature 438:685–689
27. Ebert MS, Neilson JR, Sharp PA (2007) MicroRNA sponges: competitive inhibitors of small RNAs in mammalian cells. Nat Methods 4:721–726
28. Standart N, Jackson RJ (2007) MicroRNAs repress translation of m7Gppp-capped target mRNAs *in vitro* by inhibiting initiation and promoting deadenylation. Genes Dev 21:1975–1982
29. Guo H, Ingolia NT, Weissman JS et al (2010) Mammalian microRNAs predominantly act to decrease target mRNA levels. Nature 466:835–840
30. Farh KK, Grimson A, Jan C et al (2005) The widespread impact of mammalian MicroRNAs on mRNA repression and evolution. Science 310:1817–1821
31. Vinther J, Hedegaard MM, Gardner PP et al (2006) Identification of miRNA targets with stable isotope labeling by amino acids in cell culture. Nucleic Acids Res 34:e107
32. Ong SE, Blagoev B, Kratchmarova I et al (2002) Stable isotope labeling by amino acids in cell culture, SILAC, as a simple and accurate approach to expression proteomics. Mol Cell Proteomics 1:376–386
33. Baek D, Villen J, Shin C et al (2008) The impact of microRNAs on protein output. Nature 455:64–71
34. Zhu S, Si ML, Wu H et al (2007) MicroRNA-21 targets the tumor suppressor gene tropomyosin 1 (TPM1). J Biol Chem 282:14328–14336
35. Tian Z, Greene AS, Pietrusz JL et al (2008) MicroRNA-target pairs in the rat kidney identified by microRNA

microarray, proteomic, and bioinformatic analysis. Genome Res 18:404–411
36. Iliopoulos D, Malizos KN, Oikonomou P et al (2008) Integrative microRNA and proteomic approaches identify novel osteoarthritis genes and their collaborative metabolic and inflammatory networks. PLoS One 3:e3740
37. Jovanovic M, Reiter L, Picotti P et al (2010) A quantitative targeted proteomics approach to validate predicted microRNA targets in C. elegans. Nat Methods 7:837–842
38. Gygi SP, Rist B, Gerber SA et al (1999) Quantitative analysis of complex protein mixtures using isotope-coded affinity tags. Nat Biotechnol 17:994–999
39. Anderson L, Hunter CL (2006) Quantitative mass spectrometric multiple reaction monitoring assays for major plasma proteins. Mol Cell Proteomics 5:573–588
40. Krijgsveld J, Ketting RF, Mahmoudi T et al (2003) Metabolic labeling of C. elegans and D. melanogaster for quantitative proteomics. Nat Biotechnol 21:927–931
41. Beitzinger M, Peters L, Zhu JY et al (2007) Identification of human microRNA targets from isolated argonaute protein complexes. RNA Biol 4:76–84
42. Easow G, Teleman AA, Cohen SM (2007) Isolation of microRNA targets by miRNP immunopurification. RNA 13:1198–1204
43. Karginov FV, Conaco C, Xuan Z et al (2007) A biochemical approach to identifying microRNA targets. Proc Natl Acad Sci U S A 104:19291–19296
44. Hendrickson DG, Hogan DJ, Herschlag D et al (2008) Systematic identification of mRNAs recruited to argonaute 2 by specific microRNAs and corresponding changes in transcript abundance. PLoS One 3:e2126
45. Landthaler M, Gaidatzis D, Rothballer A et al (2008) Molecular characterization of human Argonaute-containing ribonucleoprotein complexes and their bound target mRNAs. RNA 14:2580–2596
46. Wang WX, Wilfred BR, Hu Y et al (2010) Anti-Argonaute RIP-Chip shows that miRNA transfections alter global patterns of mRNA recruitment to microribonucleoprotein complexes. RNA 16:394–404
47. Zhang L, Ding L, Cheung TH et al (2007) Systematic identification of C. elegans miRISC proteins, miRNAs, and mRNA targets by their interactions with GW182 proteins AIN-1 and AIN-2. Mol Cell 28:598–613
48. Mourelatos Z, Dostie J, Paushkin S et al (2002) miRNPs: a novel class of ribonucleoproteins containing numerous microRNAs. Genes Dev 16:720–728
49. Tan LP, Seinen E, Duns G et al (2009) A high throughput experimental approach to identify miRNA targets in human cells. Nucleic Acids Res 37:e137
50. Licatalosi DD, Mele A, Fak JJ et al (2008) HITS-CLIP yields genome-wide insights into brain alternative RNA processing. Nature 456:464–469
51. Chi SW, Zang JB, Mele A et al (2009) Argonaute HITS-CLIP decodes microRNA-mRNA interaction maps. Nature 460:479–486
52. Zisoulis DG, Lovci MT, Wilbert ML et al (2010) Comprehensive discovery of endogenous Argonaute binding sites in Caenorhabditis elegans. Nat Struct Mol Biol 17:173–179
53. Leung AK, Young AG, Bhutkar A et al (2011) Genome-wide identification of Ago2 binding sites from mouse embryonic stem cells with and without mature microRNAs. Nat Struct Mol Biol 18:237–244
54. Hafner M, Landthaler M, Burger L et al (2010) Transcriptome-wide identification of RNA-binding protein and microRNA target sites by PAR-CLIP. Cell 141:129–141
55. Duursma AM, Kedde M, Schrier M et al (2008) miR-148 targets human DNMT3b protein coding region. RNA 14:872–877
56. Forman JJ, Legesse-Miller A, Coller HA (2008) A search for conserved sequences in coding regions reveals that the let-7 microRNA targets Dicer within its coding sequence. Proc Natl Acad Sci U S A 105:14879–14884
57. Tay Y, Zhang J, Thomson AM et al (2008) MicroRNAs to Nanog, Oct4 and Sox2 coding regions modulate embryonic stem cell differentiation. Nature 455:1124–1128
58. Schnall-Levin M, Zhao Y, Perrimon N et al (2010) Conserved microRNA targeting in Drosophila is as widespread in coding regions as in 3′UTRs. Proc Natl Acad Sci U S A 107:15751–15756
59. Schnall-Levin M, Rissland OS, Johnston WK et al (2011) Unusually effective microRNA targeting within repeat-rich coding regions of mammalian mRNAs. Genome Res 21:1395–1403
60. Fang Z, Rajewsky N (2011) The impact of miRNA target sites in coding sequences and in 3′UTRs. PLoS One 6:e18067
61. Miranda KC, Huynh T, Tay Y et al (2006) A pattern-based method for the identification of MicroRNA binding sites and their corresponding heteroduplexes. Cell 126:1203–1217
62. Orom UA, Lund AH (2007) Isolation of microRNA targets using biotinylated synthetic microRNAs. Methods 43:162–165
63. Orom UA, Nielsen FC, Lund AH (2008) MicroRNA-10a binds the 5′UTR of ribosomal protein mRNAs and enhances their translation. Mol Cell 30:460–471
64. Christoffersen NR, Shalgi R, Frankel LB et al (2010) p53-independent upregulation of miR-34a during oncogene-induced senescence represses MYC. Cell Death Differ 17:236–245
65. Hsu RJ, Yang HJ, Tsai HJ (2009) Labeled microRNA pull-down assay system: an experimental approach for high-throughput identification of microRNA-target mRNAs. Nucleic Acids Res 37:e77
66. Zhao Y, Samal E, Srivastava D (2005) Serum response factor regulates a muscle-specific microRNA that targets Hand2 during cardiogenesis. Nature 436:214–220
67. Nonne N, Ameyar-Zazoua M, Souidi M et al (2010) Tandem affinity purification of miRNA target mRNAs (TAP-Tar). Nucleic Acids Res 38:e20
68. Vatolin S, Navaratne K, Weil RJ (2006) A novel method to detect functional microRNA targets. J Mol Biol 358:983–996

69. Andachi Y (2008) A novel biochemical method to identify target genes of individual microRNAs: identification of a new Caenorhabditis elegans let-7 target. RNA 14:2440–2451
70. Llave C, Xie Z, Kasschau KD et al (2002) Cleavage of Scarecrow-like mRNA targets directed by a class of Arabidopsis miRNA. Science 297:2053–2056
71. Addo-Quaye C, Eshoo TW, Bartel DP et al (2008) Endogenous siRNA and miRNA targets identified by sequencing of the Arabidopsis degradome. Curr Biol 18:758–762
72. German MA, Pillay M, Jeong DH et al (2008) Global identification of microRNA-target RNA pairs by parallel analysis of RNA ends. Nat Biotechnol 26:941–946
73. Franco-Zorrilla JM, Del Toro FJ, Godoy M et al (2009) Genome-wide identification of small RNA targets based on target enrichment and microarray hybridizations. Plant J 59:840–850
74. Karginov FV, Cheloufi S, Chong MM et al (2010) Diverse endonucleolytic cleavage sites in the mammalian transcriptome depend upon microRNAs, Drosha, and additional nucleases. Mol Cell 38:781–788
75. Li YF, Zheng Y, Addo-Quaye C et al (2010) Transcriptome-wide identification of microRNA targets in rice. Plant J 62:742–759
76. Shin C, Nam JW, Farh KK et al (2010) Expanding the microRNA targeting code: functional sites with centered pairing. Mol Cell 38:789–802
77. Bracken CP, Szubert JM, Mercer TR et al (2011) Global analysis of the mammalian RNA degradome reveals widespread miRNA-dependent and miRNA-independent endonucleolytic cleavage. Nucleic Acids Res 39:5658–5668
78. Jiao Y, Riechmann JL, Meyerowitz EM (2008) Transcriptome-wide analysis of uncapped mRNAs in Arabidopsis reveals regulation of mRNA degradation. Plant Cell 20:2571–2585
79. Xiao C, Calado DP, Galler G et al (2007) MiR-150 controls B cell differentiation by targeting the transcription factor c-Myb. Cell 131:146–159
80. Zhao Y, Ransom JF, Li A et al (2007) Dysregulation of cardiogenesis, cardiac conduction, and cell cycle in mice lacking miRNA-1-2. Cell 129:303–317
81. Johnnidis JB, Harris MH, Wheeler RT et al (2008) Regulation of progenitor cell proliferation and granulocyte function by microRNA-223. Nature 451:1125–1129
82. Ventura A, Young AG, Winslow MM et al (2008) Targeted deletion reveals essential and overlapping functions of the miR-17 through 92 family of miRNA clusters. Cell 132:875–886
83. Patrick DM, Zhang CC, Tao Y et al (2010) Defective erythroid differentiation in miR-451 mutant mice mediated by 14-3-3zeta. Genes Dev 24:1614–1619
84. Park CY, Choi YS, McManus MT (2010) Analysis of microRNA knockouts in mice. Hum Mol Genet 19:R169–R175
85. Prosser HM, Koike-Yusa H, Cooper JD et al (2011) A resource of vectors and ES cells for targeted deletion of microRNAs in mice. Nat Biotechnol 29:840–845
86. Tsang JS, Ebert MS, van Oudenaarden A (2010) Genome-wide dissection of microRNA functions and cotargeting networks using gene set signatures. Mol Cell 38:140–153

MicroRNA-Regulated Networks: The *Perfect Storm* for Classical Molecular Biology, the *Ideal Scenario* for Systems Biology

4

Julio Vera, Xin Lai, Ulf Schmitz, and Olaf Wolkenhauer

Abstract

MicroRNAs (miRNAs) are involved in many regulatory pathways some of which are complex networks enriched in regulatory motifs like positive or negative feedback loops or coherent and incoherent feedforward loops. Their complexity makes the understanding of their regulation difficult and the interpretation of experimental data cumbersome. In this book chapter we claim that systems biology is the appropriate approach to investigate the regulation of these miRNA-regulated networks. Systems biology is an interdisciplinary approach by which biomedical questions on biochemical networks are addressed by integrating experiments with mathematical modelling and simulation. We here introduce the foundations of the systems biology approach, the basic theoretical and computational tools used to perform model-based analyses of miRNA-regulated networks and review the scientific literature in systems biology of miRNA regulation, with a focus on cancer.

Keywords

miRNA regulated networks • miRNA target hub • miRNA cluster • Feedback loop • Feedforward loop • Post-transcriptional regulation • miRNA network motifs • Kinetic models • Bistability • Ultrasensitivity

J. Vera (✉) • X. Lai • U. Schmitz
Department of Systems Biology and Bioinformatics, Institute of Computer Science, University of Rostock, 18051 Rostock, Germany
e-mail: julio.vera@uni-rostock.de

O. Wolkenhauer
Department of Systems Biology and Bioinformatics, Institute of Computer Science, University of Rostock, 18051 Rostock, Germany

Stellenbosch Institute for Advanced Study (STIAS), Wallenberg Research Centre, Stellenbosch University, Stellenbosch, South Africa

4.1 Introduction

MiRNAs, a large class of small ncRNAs (typically 20–23nt in length) play an important role in the regulation of gene expression. In the last decade it has been established that miRNA-mediated translational repression is crucial in the regulation of core processes with remarkable time-, cell-, and tissue-specificity, such as proliferation, differentiation, apoptosis, response to environmental

U. Schmitz et al. (eds.), *MicroRNA Cancer Regulation: Advanced Concepts, Bioinformatics and Systems Biology Tools*, Advances in Experimental Medicine and Biology 774,
DOI 10.1007/978-94-007-5590-1_4, © Springer Science+Business Media Dordrecht 2013

stresses and organ development [1, 2]. Dysregulation of these precise processes, alone or in combination, plays an important role in the emergence and progression of cancer, and therefore it is not surprising that mounting *in vitro, in vivo* and clinical evidences assign a non-trivial role to miRNAs in oncogenesis [3].

What makes miRNAs extremely intriguing biomolecules is that typically one miRNA is capable of regulating the expression of multiple genes. Computational methods commonly assign dozens to hundreds of target gene candidates to individual miRNAs, while experimentation often confirms dozens of them. On the other hand, many genes are targeted for repression by a high number of miRNAs, which seem to regulate those genes individually or cooperatively [4]. To make the understanding of miRNA-mediated regulation more cumbersome, some transcription factors (TF) promote the parallel expression of many miRNAs, some of which target the same genes or several genes involved in the same signalling pathways [5]. Thus, pathways involving miRNAs are often complex regulatory networks, enriched in motifs like feedback loops and feedforward loops, whose regulation is difficult to understand and make the direct interpretation of experimental data elaborate. Thus, one could say that the investigation of miRNA regulation is the perfect storm for conventional molecular biology, far from the idyllic image of "one pathway, few components, simple interactions" that used to work.

In this book chapter we claim that systems biology is the appropriate approach to investigate the regulation of those complex miRNA-regulated networks. Systems biology is an interdisciplinary approach, focussing on the investigation of spatiotemporal processes in biochemical networks, by which biomedical questions are addressed by integrating experiments in iterative cycles with mathematical modelling and simulation. Mathematical modelling is a tool to formulate hypotheses, develop more directed and better designed experiments, and which will allow to make predictions [6]. It is actually *the approach* when dealing with high-throughput data of highly interconnected biochemical networks, composed of dozens to hundreds of proteins, genes and miRNAs. But also when dealing with networks enriched in non-linear motifs like feedback and feedforward loops, which induce non-intuitive regulatory patterns like ultrasensitivity, bistability, or oscillations.

In this book chapter we first introduce the foundations of the systems biology approach. Secondly, we show that miRNAs are embedded in complex network motifs, some of them common to other biomolecules like feedback loops, others specific to miRNAs, like miRNA clusters or target hubs. Our opinion is that the involvement of miRNAs in those complex motifs makes the use of systems biology necessary to understand miRNA regulation. Thirdly, we present and discuss some basic computational and theoretical tools used to perform model-based predictions in miRNA-regulated networks, whose understanding is required to interpret the results shown in systems biology-based papers and some of the chapters in this book. Finally, we make an extensive review of the scientific papers published to date in which systems biology has been used to investigate miRNA regulation.

4.2 Systems Biology in a Nutshell

To investigate a biological system following the systems biology approach, an iterative process is conducted that includes the following four key steps: (a) the construction of a regulatory map including the interactions among the molecular entities already experimentally proved and/or those under investigation; (b) the construction of a mathematical model of ordinary differential equations or other modelling formalisms, based on the regulatory map; (c) the calibration and validation of the model (process in which the model is characterized and validated using experimental data); and (d) the analysis of the properties of the biological network using the mathematical model and computational tools. A final step always includes the validation of the new insights generated with the mathematical model by means of new biological experiments. In the following, those steps are discussed in further detail.

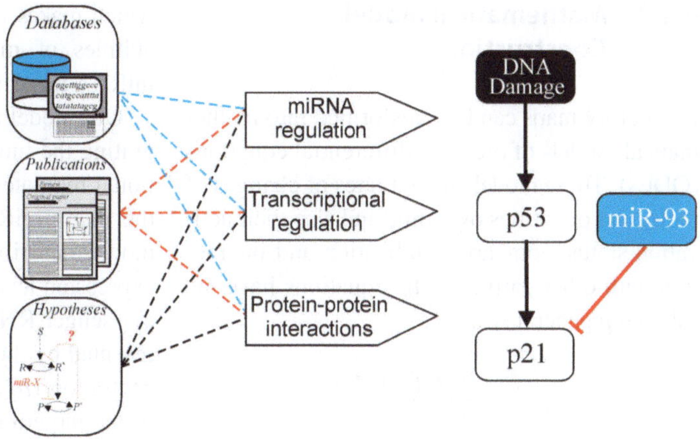

Fig. 4.1 Construction of a regulatory map. Relevant information about to the molecules (genes, proteins, miRNAs,...) and biochemical processes of interest is retrieved from relevant biomedical publications and databases, and organised into a graphical scheme called regulatory map

4.2.1 Network Construction

The first step towards the construction of a mathematical model is to retrieve the existing knowledge about the biochemical system under investigation. Relevant information about the molecules (genes, proteins, miRNAs, transcription factors. etc., Fig. 4.1) and biochemical processes of interest is retrieved from biomedical publications and databases, and further processed, annotated and organised into a so-called regulatory map. The regulatory map is a visualization of the state-of-the-art of the biomedical knowledge about the biochemical network. It contains as well some *a priori* hypotheses on the role of some of the molecules involved in the network that can drive the design of the mathematical model and experiments.

There are many resources available for this purpose. For example, protein-protein interactions can be extracted from databases like the Human Protein Reference Database HPRD and STRING [7, 8]. Information about miRNA:target regulations can be extracted from databases of experimentally validated interactions like miRecords [9] and Tarbase [10]. In addition, a number of online resources contain collections of predicted miRNA:target interactions (e.g., miRWalk [11] and miRGen 2.0 [12]). Information about TFs controlling the expression of the miRNA under investigation can be obtained either from databases of experimentally proved TFs of miRNAs (see TransmiR [13]) or from online resources of predicted putative TFs like PuTmiR [14]. MiRNA web resources are discussed in detail in the Chap. 12.

In many cases these regulatory maps are hand drawn and customized for the investigated pathway. However, they can also be implemented using a well established protocol, the Systems Biology Graphical Notation (SBGN; [15]), which contains a standardised set of symbols to represent biochemical processes and molecules. The advantage is that standardised regulatory maps, generated and curated with specific software tools, can be shared on public repositories and can be further annotated with additional information. For example, they can be used to display information concerning critical ontology terms associated with each entity in the network. These standard regulatory maps also facilitate the dissemination of the results of simulation or wet lab experiments.

A well constructed regulatory map is *per se* an excellent tool to investigate some features of complex regulatory networks involving TFs, signalling proteins and miRNAs. For example, they can be used to detect putative regulatory motifs, including feedback and feedforward-loops or miRNA target hubs (for further details on those motifs, see the coming section in this chapter and Chaps. 9 and 10). The detection of these putative regulatory motifs can be a good starting point for the design of experiments to analyse their features and regulation.

4.2.2 Mathematical Model Construction

Regulatory maps can be transformed into mathematical models of ordinary differential equations (ODEs). These models reflect rates of changes of molecular quantities over time and also their activation status, compartmentalization and interaction with other partners. The equations have the following general structure:

$$\frac{d}{dt}P = \sum_i F_i(S,k,P)$$

In this kind of mathematical equations, the left-hand side of the equation accounts for the variation in time of *P*, which represents a set of time-dependent variables that account for the interacting proteins, miRNAs and other molecules. F_i are the rate equations of the biochemical processes considered in the network which influence the values of *P* (transcription, miRNA-mediated regulation, protein-protein interactions…). *S* is a set of model variables accounting for the input signals of the network, biological signals that are external and not regulated by the biochemical system analysed, but affecting it. *k* are the kinetic rates, fixed numbers associated with given biochemical properties, that characterize numerically the rate equations. In the following, we illustrate the basic features of ODE models with a small example that describes a sub-module of the network accounting for the miRNA-mediated regulation of p21 [16]. The cell cycle regulator and tumour suppressor p21 (a.k.a. CDKN1A or Cip1/Waf1) is a well-known case of a signalling protein repressed via miRNA regulation. p21 is a cell cycle regulator vital in both the G_1/S and G_2/M cell cycle arrest after DNA damage [17]. The expression of p21 is widely dependent on environmental conditions and it can be transcriptionally regulated through p53-dependent and -independent mechanisms [18]. p21 is an important cancer-related protein, with well-proved tumour suppressor activity, and also tumour-promoting activity [19]. Recently, it has been shown that p21 expression undergoes regulation by numerous miRNAs, and it is integrated in a complex regulatory network involving multiple TF/miRNA regulatory motifs [16, 20], which makes it an excellent candidate to test the abilities of mathematical modelling to dissect miRNA regulation.

Our model accounts for the sub-network integrating the tumour suppressor p53, its transcriptional regulated protein p21 and miR-93, one of the many miRNAs proved to repress p21. The model describes the evolution in time of the expression levels for three chemical species: p21 messenger RNA (mRNA) (in the equations represented by the variable *mp21*); p21 protein concentration (*p21*); and the free cytosolic fraction of the targeting miRNA-93 (*miR93*; see Fig. 4.2 for complete explanation).

The extent of the miRNA induced post-transcriptional repression and other basal and dynamic features of the regulatory network are tightly controlled by the efficiency of the molecular events here described (e.g. p53-mediated transcription, mRNA and miRNA basal turnover, efficiency in the association of miRNA and mRNA…), which are characterized in the model by the numerical values of the rate constants. Hence, the values of those constants are critical to define the properties of the biochemical network and their determination is a critical process in the construction of a mathematical model which is described in detail in the following subsection.

We note that there is a multiplicity of mathematical modelling frameworks that can be used to analyse biochemical models, from which ODE models are the most commonly used but not the unique possible choice. The right choice of the modelling framework is a trade-off between several issues, including the precision in the current knowledge about the biochemical network investigated, the quality and amount of the experimental data available and the nature of the network properties that one tries to analyse by means of modelling. Alternative modelling frameworks include many variations of ODE models like mass-action kinetics, models containing Michaelis-Menten and Hill equations and power-law models, and also Boolean models, cellular automata and partial differential equations. For further details, we refer the reader to papers, which are devoted to this topic [23, 24].

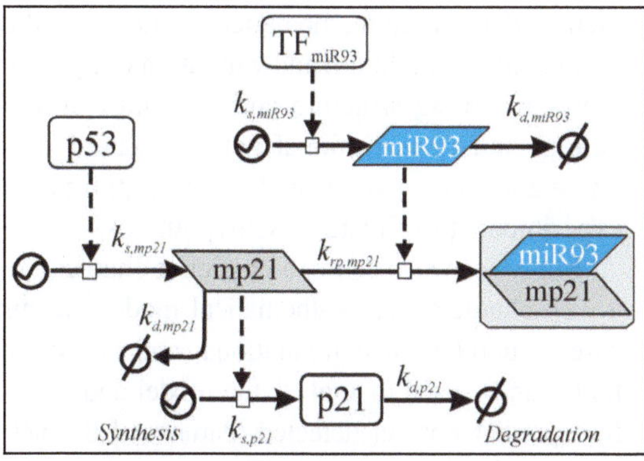

$$\frac{d}{dt}mp21 = k_{s,mp21} \cdot F_{TFact}(p53) - mp21 \cdot \left[k_{d,mp21} + k_{rp,mp21} \cdot miR93\right]$$

$$\frac{d}{dt}miR93 = k_{s,miR93} \cdot F_{TFact}(TF_{miR93}) - miR93 \cdot \left[k_{d,miR93} + k_{rp,mp21} \cdot mp21\right]$$

$$\frac{d}{dt}p21 = k_{s,p21} \cdot mp21 - k_{d,p21} \cdot p21 + \sum_j \sigma_j \cdot k_{ppi,p21} \cdot p21^{gij} \cdot Prot\bar{X}_j^{gj}$$

Fig. 4.2 **Mathematical model construction.** The model describes the evolution in time of the expression levels for three chemical species: p21 mRNA (*mp21*), p21 protein concentration (*p21*), and the free cytosolic fraction of the targeting miRNA-93 (*miR93*). *Left*: Scheme of the model, with all the biochemical processes considered. *Right*: Mathematical model in ODEs, in which each biochemical process considered has been modelled using conventional mass action rates or power-law terms [21, 22]. The following processes are modelled: For p21 mRNA (*mp21*): (i) p53-mediated synthesis (whose efficiency is represented by the kinetic constant $k_{s,mp21}$); (ii) basal degradation ($k_{d,mp21}$); and (iii) miRNA-93 repression of mp21 translation ($k_{rp,mp21}$). For miRNA-93 (*miR93*): (i) basal synthesis ($k_{s,miR93}$), mediated by its TF (TF_{miR93}); (ii) basal degradation ($k_{d,miR93}$); and (iii) association with the mRNA into the repression complex [*mp21* | *miR93*] ($k_{rp,mp21}$). For the p21 protein levels (*p21*): (i) mRNA mediated synthesis of protein ($k_{s,p21}$); and (ii) basal degradation ($k_{d,p21}$); (iii) further protein-protein interactions between p21 and signalling proteins X_j (σ_j is the stochiometric coefficient and $k_{ppl,p21}$ the rate constant)

4.2.3 Model Calibration

In the process of model calibration, quantitative data are integrated with the mathematical model to assign values to the model parameters, in a manner such that the model, once characterized, mimics the behaviour of the system represented by the experimental data available. The process use to follow a sequence of iterative cycles of data-driven parameter estimation, model quality assessment, model structure modification (when required) and re-estimation of model parameters. The critical element of the process is the design of a set of suitable perturbation experiments that help to characterise the dynamics of the investigated network. Those experiments must be performed in a quantitative fashion, such that quantitative data, describing the reduction of mRNA and protein levels induced by the miRNAs and signalling events, are produced and subsequently processed and analysed (see [25] for a detailed list of quantitative experimental techniques).

The process of parameter estimation, the step in which reliable approximations are assigned to the model parameters, can be performed following different strategies. In some rare cases, a bibliographic search is enough, and information in publications where a similar biological model and similar experimental conditions are considered suffices to assign values to the parameters. Another option is to manually train the model, by manually tuning the parameter values until the model simulation matches with the available data. The most reliable option is to perform a quantitative data fitting, in which the values of the model parameters are iteratively modified until the differences between the experimental data and the model simulations for identical biological conditions are minimised (Fig. 4.3). There is a plethora of computational methods, based on the mathematical principle of the "maximum likelihood", that can be used for this purpose, while a similar number of software tools is available (e.g. PottersWheel [26] or COPASI [27]).

Parameter estimation is not always necessary to make a model useful for the analysis of a regulatory network. In some cases, computational methods scanning the whole range of feasible values for the model parameters allow investigating the structural and dynamical features of the system and detect the so-called design principles. Design principles are general patterns in the performance

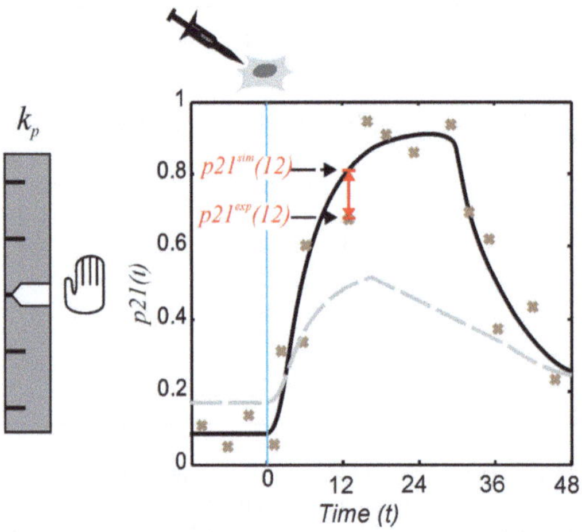

Fig. 4.3 Parameter estimation. Computational techniques are used to estimate the optimal parameter values that can minimise the distance (*red arrow*) between the model predictions, $p21^{sim}(t)$, and the available experimental data, $p21^{exp}(t)$, for the considered experimental conditions. The distance is calculated by the following cost function:

$$Min\ F_{error} = \frac{1}{n_{exp} \cdot n_{var} \cdot n_{tp}} \sum_{k=1}^{n_{exp}} \sum_{j=1}^{n_{var}} \sum_{i=1}^{n_{tp}} \left(X_{k,j}(t_i) - X_{k,j}^{exp}(t_i) \right)^2$$

In which, n_{exp} is the number of experiments, n_{var} is the number of measured quantities (observables, e.g. mRNA, protein or miRNA expression levels), and n_{tp} is the number of time points where each observable was measured. $X_{k,j}(t_i)$ is the value of the jth observable at the ith time point obtained after numerical simulation of the model for the kth experiment and $X_{k,j}^{exp}(t_i)$ is the corresponding value of the jth observable at the ith time point measured in the kth experiment. When the model parameters are properly estimated, there is good agreement between the model simulations (*solid black line*) and the experimental data (*brown crosses*). With non-optimal parameter values, model simulations are not able to reproduce the dynamics of the data (*dashed grey line*)

of the biochemical network that are associated with well defined dynamical regimes and parameter value intervals in the network [28].

4.2.4 Model Analysis

In some cases, to establish the structure of the mathematical model in a reliable manner using this iterative cycle of mathematical modelling and experimentation is already a valuable achievement: a data-based mathematical model is a valid tool for the formulation and validation of hypotheses concerning structure and dynamics of signalling and transcriptional networks. And it is also a good strategy to help designing appropriate experiments to validate these hypotheses.

But we can also confer a predictive character to a well-characterised mathematical model. In this case, a number of computational and analytical tools can be used to analyse the model and therefore predict not yet detected features of the network under investigation. Those are especially useful for the detection and analysis of regulatory motifs, which are associated with an inherent complexity and an unforeseeable non-linear behaviour. The combination of simple regulatory motifs like positive or negative feedback loops and coherent or incoherent feedforward loops can induce a complex non-intuitive dynamical behaviour. To comprehend the behaviour of those complex networks it is necessary to use mathematical models in combination with computational methods and analytical tools. Interestingly, miRNAs are deeply involved in networks enriched in that kind of regulatory motifs [29]. In the following sections we first introduce the definition and main properties of those regulatory motifs and secondly, discuss the tools that allow the analysis of them.

4.3 MicroRNAs as Components of Complex Network Motifs

The biological networks in which miRNAs are embedded are enriched in regulatory motifs displaying complex steady-state and transient patterns. Among them, we find positive and negative feedback loops, coherent and incoherent feedforward loops, miRNA target hubs and miRNA clusters. In the following sections, we give a detailed definition of these motifs and their main structural and dynamical properties.

4.3.1 Feedback Loops

We can distinguish two types of feedback loops involving miRNAs, positive and negative feedback loops. In positive feedback loops the activation of a biochemical event positively regulates a biochemical process upstream the system. In the simplest setup of a miRNA-mediated positive feedback loop, the expression of the miRNA is inhibited by one of its target proteins (Fig. 4.4). This is for example the case for the transcription factor E2F1 and some of its promoted miRNAs [30]. However, the structure can be more complex and involve one or more signal mediators. One such example is the network integrated by p53, its negative regulator NAD-dependent deacetylase sirtuin-1 (Sirtl) and the p53-promoted miR-34a, which represses Sirt1 (Sirt1⊣p53→miR-34a⊣Sirt1; see [31]). Positive feedback loops can induce the emergence of bistable expression of both the miRNA and the TF involved. There, perturbations driven by external regulatory signals can induce an irreversible transition from a situation of miRNA-induced TF repression to a new state, in which miRNA is silenced and TF overcomes permanent high expression. In this manner, positive feedback loops can induce the conversion of a transient signal into a longer lasting cellular response [32], but also induce signal amplification and in some cases instability [33].

A negative feedback loop appears when a molecule positively regulates the expression or activation of its own inhibitor. In the simplest configuration of a miRNA-mediated negative feedback loop, the TF which is targeted by the miRNA (miR) acts as an activator of miRNA expression (Fig. 4.5). Negative feedback loops

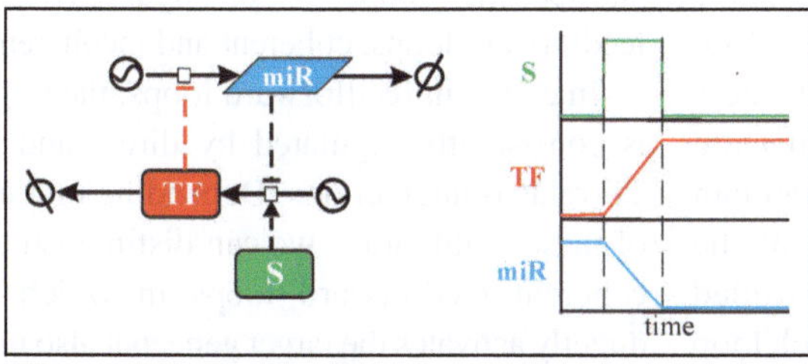

$$\frac{d}{dt}TF = k_{s,TF} \cdot \frac{F_{TF}(S)}{\left(k_{rp,TF} + miR\right)^{gi}} - k_{d,TF} \cdot TF$$

$$\frac{d}{dt}miR = \frac{k_{s,miR}}{\left(k_{rp,miR} + TF\right)^{gj}} - k_{d,miR} \cdot miR$$

Fig. 4.4 miRNA-mediated positive feedback loop. This positive feedback loop is composed of a double negative loop (*TF*⊣miR and miR⊣*TF*). *Left*: The expression of a miRNA is inhibited by the one of its target proteins (*TF*). In positive feedback loops, transient external signals (*S*) can trigger a permanent transition from a scenario of miRNA-induced *TF* repression to that of a permanent *TF*-mediated miRNA repression and high *TF* expression. *Right*: A simple mathematical model accounting for this positive feedback loop composed of a miRNA (*miR*), and its target and negative transcriptional regulator (*TF*). The following processes are modelled: For the miRNA (*miR*), TF-repressed synthesis ($k_{s,miR}$) and basal degradation ($k_{d,miR93}$) are modelled. For the *TF* expression levels (*TF*), S-mediated and miR-repressed protein synthesis ($k_{s,TF}$) and basal degradation ($k_{d,TF}$) are modelled. For the negative regulation, we used a modified power-law term [21, 34]

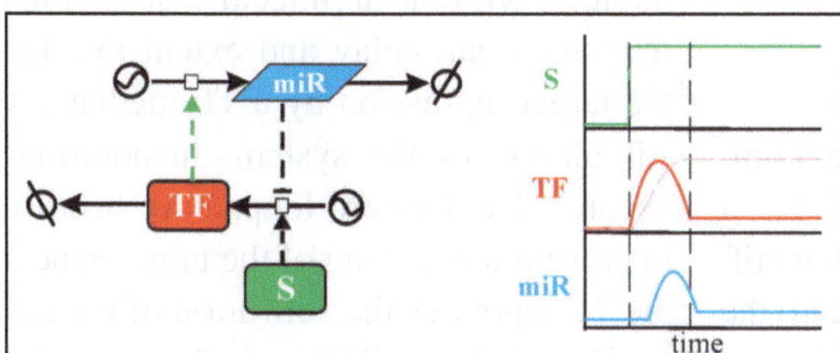

$$\frac{d}{dt}TF = k_{s,TF} \cdot \frac{F_{TF}(S)}{\left(k_{rp,TF} + miR\right)^{gi}} - k_{d,TF} \cdot TF$$

$$\frac{d}{dt}miR = k_{s,miR} \cdot F_{miR}(TF) - k_{d,miR} \cdot miR$$

Fig. 4.5 miRNA-mediated negative feedback loop. *Left*: The expression of a miRNA is promoted by one of its target proteins (*TF*). In negative feedback loops, the system maintains homeostasis and an increase of the external signal (*S*), which promotes the expression of *TF* is compensated by the increased expression of the miRNA such that levels of *TF* remain stable for a wide range of values for *S*. *Right*: A simple mathematical model accounting for this negative feedback loop composed of a miRNA (*miR*), and its target and positive transcriptional regulator (*TF*). The following processes are modelled: For the miRNA (*miR*), TF-mediated synthesis ($k_{s,miR}$) and basal degradation ($k_{d,miR93}$) are modelled. For the TF expression levels (*TF*), S-mediated and miR-repressed protein synthesis ($k_{s,TF}$) and basal degradation ($k_{d,TF}$). For the negative regulation, we used a modified power-law term [21, 34]

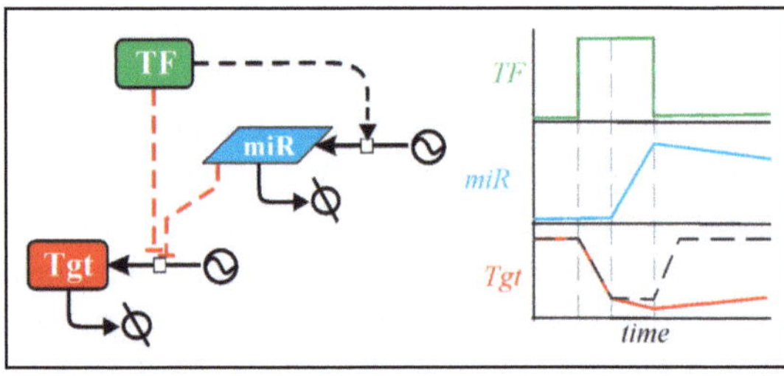

$$\frac{d}{dt}miR = k_{s,miR} \cdot F_{miR}(TF) - k_{d,miR} \cdot miR$$

$$\frac{d}{dt}Tgt = \frac{k_{s,Tgt}}{(k_{mR} + miR)^{gi} \cdot (k_{TF} + TF)^{gj}} - k_{d,Tgt} \cdot Tgt$$

Fig. 4.6 miRNA-mediated coherent feedforward loop. *Left*: The expression of a target is negatively regulated by a *TF* in two ways: directly, by repressing the transcription of the target gene, and indirectly, by promoting the expression of a miRNA which represses the target gene. *Right*: A simple mathematical model accounting for this coherent feedforward loop. The following processes are modelled: For *miR*, the *TF*-mediated synthesis ($k_{s,miR}$) and basal degradation ($k_{d,miR}$) are modelled. For the target (*Tgt*), the *TF*- and *miR*-repressed protein synthesis ($k_{s,Tgt}$) and basal degradation ($k_{d,tgt}$) are modelled. For the negative regulation, we used a modified power-law term. This particular feedforward loop, in which two inhibitors independently repress the target, is associated with a delay or extension in the target repression termination for transient stimuli (*centre*, *red line*) compared to the situation without feedforward loop (*centre*, *dashed line*)

can induce signal termination, and in some cases oscillatory expression of the loop components. Moreover, negative feedback loops are a simple class of homeostatic systems, that are able to fine-tune gene expression and maintain steady-state levels of both components of the system against noise and uncontrolled fluctuations [35]. Such a negative feedback loop can maintain homeostasis and prevent uncontrolled growth and proliferation, phenomena typically observable in cancer development. Other more complex miRNA feedback loop systems are possible, in which the regulation is mediated by protein mediators. For further details and discussion on this kind of systems we refer to the Chap. 9.

4.3.2 Feedforward Loops

MiRNA-mediated feedforward loops are composed of three molecular components: a TF, a miRNA and a target gene. In this regulatory motif, the TF regulates both the target gene and the miRNA, and the miRNA represses the target gene. Thus, the target gene is a downstream component whose expression is simultaneously regulated by the TF and the TF-regulated miRNA (Fig. 4.6). In some cases, the target gene can also regulate the TF and/or the miRNA, and in that case the three components system could result in both feedforward and feedback loops. We distinguish two kinds of feedforward loops: coherent and incoherent.

In coherent feedforward loops, the target gene is consistently regulated by direct and indirect TF-related interactions. Depending on the nature of these regulations, we can distinguish between coherent feedforward loops in which the TF directly activates the target gene, but also indirectly activates it through repressing the miRNA (double positive regulation), and those in which the TF directly represses the target gene, and indirectly represses it through activating the miRNA (double negative regulation). Coherent miRNA-mediated feedforward loops are supposed to serve as sign-sensitive delay elements, delaying the response of the target gene to the double regulation exerted by the TF. In Fig. 4.6 we see that modules with the architecture of double negative regulation can delay and extent the duration of the target repression by a TF-mediated transient stimulation of the system. In addition, double negative feedforward loops have been suggested to prevent the leaking of the target gene, in a way the TF represses the activation of the target gene itself and the miRNA further destabilizes its mRNA at the posttranscriptional level to achieve the complete function blocking of the targeted gene [35].

In incoherent feedforward loops, the target gene gets oppositely regulated by the TF and the TF-modulated miRNA. We can distinguish

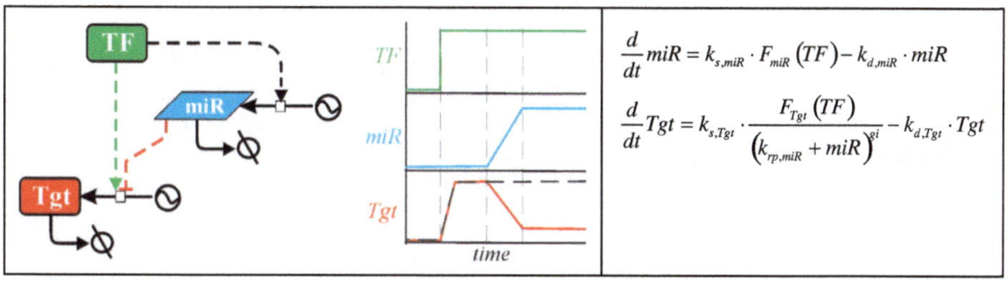

Fig. 4.7 miRNA-mediated incoherent feedforward loop. *Left*: The expression of a target is regulated by a *TF* in two ways: positively, by direct activation of the target gene expression, and negatively, by promoting the expression of a miRNA which represses the target gene. *Right*: A simple mathematical model accounting for this coherent feedforward loop. Processes modelled and variables are identical to those in Fig. 4.6, excepting the *TF*-activation and miR-repression of target synthesis ($k_{s,Tgt}$). For the case of a step-like *TF* activation, the system can generate a pulse response in the protein levels (*centre*, *red line*) compared to the situation without feedforward loop (*centre*, *dashed line*)

between those feedforward loops in which the target gene is directly activated by TF and indirectly repressed via miRNA regulation (incoherent type A) and those in which the target gene is directly repressed by TF and indirectly activated via TF-mediated activation of the miRNA (incoherent type B). Incoherent miRNA-mediated feedforward loops can display accelerated response to the stimulus signal. The accelerated response occurs when the stimulus signal is turned on in the incoherent feedforward loop type A, while in the incoherent feedforward loop type B it occurs when the stimulus signal is turned off. Other dynamical properties are associated with these modules; for example, in incoherent feedforward loops type A a pulse-like response of the target protein levels can be triggered in response to a step-like TF activation in case of Fig. 4.7. These systems have been proved as noise buffers, which is in favour of a fine-tuning function of the miRNA regulation [36].

Recent publications, in which bioinformatics methodologies were used, have revealed that miRNA-mediated feedforward loops are prevalent mechanisms of gene expression at the genome-scale level. For a detailed discussion of other types of coherent and incoherent feedforward loops not discussed here, the interested readers are referred to the publication of Mangan and Alon [37]. In addition, many biological examples of feedforward loops can be found in the paper published by Re [38].

4.3.3 MicroRNA Target Hubs

In the context of miRNA regulation, target hubs are genes that are regulated by a large number of different miRNAs (many miRNAs targeting a unique gene, Fig. 4.8). In a computational analysis, Shalgi and collaborators found 470 genes in the human genome, each potentially regulated by at least 15 different miRNAs [29]. It is well established that in most of the cases single miRNAs induce a rather mild regulation of the target. However, it is claimed that concerted expression of several of those miRNAs can induce a much stronger repression of a target hub [39, 40]. Moreover, they found that these genes use to be further connected to many other proteins through protein-protein interactions and/or transcriptional regulations, integrating regulatory networks enriched with miRNA-mediated feedforward and feedback loops. In this manner, the multiplicity of miRNAs regulating target hubs may lead to non-linear features like cross-talk and cooperativity and generation of tissue-specific regulatory patterns [16].

The first instance of an experimentally proved miRNA target hub is the Cyclin-dependent kinase inhibitor 1A (CDKN1A), also known as p21, which is widely involved in cell cycle regulation and described as tumour suppressor [41]. In a recent publication, Wu and collaborators [20] subjected 266 miRNAs which were predicted as putative post-transcriptional regulators of p21 to experimental

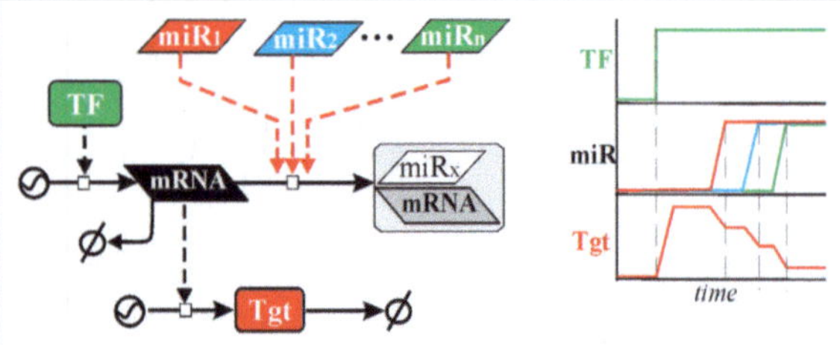

$$\frac{d}{dt}mRNA = k_s \cdot F\,(TF) - k_d \cdot mRNA - \sum_i k_i \cdot miR_i \cdot mRNA$$

$$\frac{d}{dt}Tgt = k_{s,Tgt} \cdot mRNA - k_{d,Tgt} \cdot Tgt$$

Fig. 4.8 **miRNA target hubs**. *Left*: miRNA target hubs are genes that are negatively regulated by at least 15 different miRNAs. In the scheme, we depict a system that integrates the mRNA and protein of a target hub (respectively *mRNA* and *Tgt*), n different miRNAs (miR_i), and the *TF* promoting the expression of the target hub. *Right*: A simple mathematical model accounting for the regulation of the target hub. The following processes are modelled: For the mRNA (*mRNA*), TF-mediated synthesis (k_s), basal degradation (k_d) and the sum of the n-independent miRNA-mediated target repression processes. For the protein expression levels (*Tgt*), mRNA-mediated synthesis ($k_{s,tgt}$) and basal degradation ($k_{d,tgt}$). It is well established that in most of the cases single miRNAs induce a rather mild regulation of the target. However, it is claimed that concerted expression of several of those miRNAs can induce a much stronger repression of a target hub (*centre*)

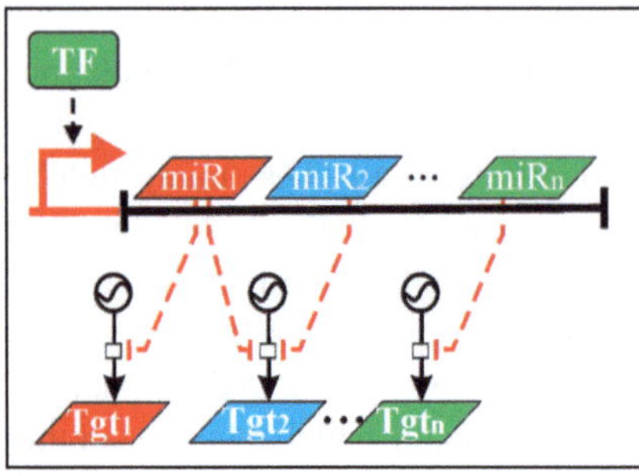

$$\frac{d}{dt}miR_i = k_{s,miRi} \cdot F_{miRi}\,(TF) - k_{d,miRi} \cdot miR_i \quad i = 1,n$$

$$\frac{d}{dt}Tgt_i = \frac{k_{s,Tgti}}{\left(k_{rp,miRij} + miR_j\right)^{gij}} - k_{d,Tgti} \cdot Tgt_i \quad i = 1,n$$

Fig. 4.9 **miRNA clusters**. *Left*: A miRNA cluster is a set of two or more miRNAs that are transcribed from physically adjacent miRNA genes. miRNAs in the same cluster can be transcriptionally activated by the same factors or signals. *Right*: A simple mathematical model accounting for a miRNA cluster integrating n-different miRNAs (miR_i), regulated by the same TF (*TF*). The miRNAs can repress alone or in cooperation a number of target genes (Tgt_i). The following processes are modelled: For the miRNAs (miR_i), *TF*-mediated synthesis ($k_{s,miRi}$) and basal degradation ($k_{d,miRi}$) are modelled. For the targets, expression levels (Tgt_i), *miR*-repression of protein synthesis ($k_{s,Tgti}$) and basal degradation ($k_{d,Tgti}$) are modelled

validation. For a subset of 28 miRNAs the ability to repress p21 could be proved. What links their finding to cancer is that fact that eight out of the 28 p21-regulating miRNAs originate from the chromosome 19 miRNA cluster, which is supposed to promote cell-cycle progression and proliferation in cancer-related cell lines [42].

4.3.4 MicroRNA Clusters

A miRNA cluster is a set of two or more miRNAs that are transcribed from physically adjacent miRNA genes and fulfil three additional criteria: they are transcribed in the same orientation, not separated by a transcription unit and not separated by any miRNA gene in the opposite orientation. Under these conditions, miRNAs in the same cluster can be transcriptionally activated by the same factors or signals (Fig. 4.9). Although the usual size for miRNA clusters is 2–3 miRNAs, larger clusters exist. Using bioinformatics methods, Weber [43] identified 37 putative human miRNA clusters which are conserved in the mouse genome and 19 of them are located in characterized transcription units.

The miR-17-92 cluster, found on human chromosome 13 and composed of six miRNAs (miR-17, miR-18a, miR-19a, miR-20a, miR-19b-1 and miR-92-1) is one of the most studied miRNA clusters. It is involved in tumour formation and development. Interestingly, this cluster can be activated by the TF c-Myc. Subsequently the protein E2F1, which is also c-Myc activated, is downregulated by two of the clustered miRNAs, miR-17-5p and miR-20a. This is a good example of a network that integrates a miRNA cluster and an incoherent feedforward loop ([44] and Chap. 9).

4.4 Tools to Perform Model-Based Predictions for MicroRNA-Regulated Networks

4.4.1 Predictive Simulations

A well calibrated mathematical model is a powerful tool to dissect the features of networks involving miRNAs. The underlying idea is that after calibration a model becomes predictive and then simulations can be used to extrapolate the behaviour of the network under experimental conditions not yet tested. It is the same approach as used in modern weather forecasting, in which sophisticated simulations of mathematical models are used to actually foresee the coming weather.

A "predictive" mathematical model can be useful in many ways. Firstly, it can be used for a faster and more accurate design of new experiments aiming for the validation of hypotheses (e.g. about the regulation and structure of the network) which have been encoded in the model. In this case the model is a tool to cut short the process of network elucidation and saves experimental effort.

Secondly, a mathematical model becomes necessary when the system investigated is a large and highly interconnected biochemical network, or we intend to integrate diverse information sources like databases, high-throughput data, and patient data (Fig. 4.10). In this case, a well constructed and calibrated mathematical model becomes a computational extension of the scientist's intuition, which allows expanding the rational thinking far beyond the maximum six linked factors/associations that an average brain can simultaneously handle. Numerous evidences show that miRNA-related regulatory networks involved in cancer are massive systems composed of hundreds of interacting proteins, miRNAs and TFs. Consequently, it is quite likely that models will become an inevitable strategy towards the understanding of miRNAs in cancer.

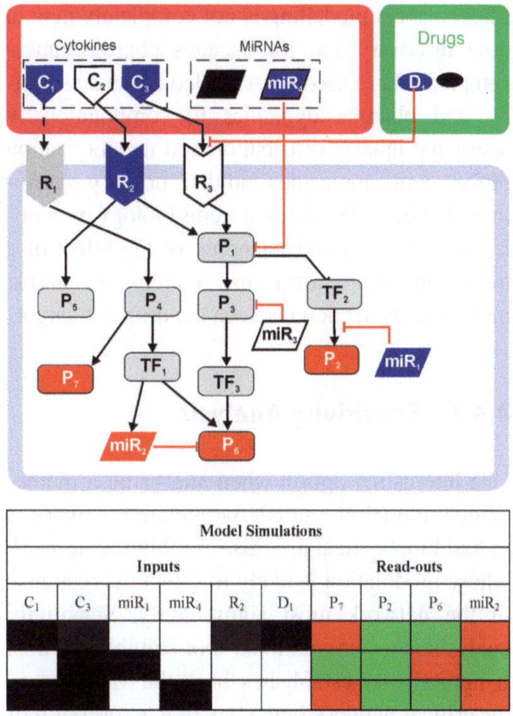

Fig. 4.10 Predictive simulations for large and highly interconnected biochemical networks. For networks involving dozens to hundreds of interacting proteins, drugs, miRNAs and genes (*top*), scientist's intuition does not suffice to interpret the outcome of complex experiments. In this case, a mathematical model allows expanding the scientist's rational thinking to integrate and analyse large amounts of data and making predictions about the network response under complex biological scenarios. These models can be used to simulate complex experimental scenarios involving the overlapping effect of many input signals (cytokines, miRNAs, drugs...) and compute the expected values for a number of critical read-outs of the network (*bottom*)

Finally, data-based mathematical models can be used to test and prototype strategies for the therapeutic manipulation of biochemical networks, including those based on miRNAs and siRNAs, which can boost the design of novel therapies.

Mathematical modelling is not completely new in drug discovery and for decades pharmaceutical companies have used the so-called pharmacokinetics and pharmacodynamics to computationally assess the uptake, distribution and metabolization of therapeutic molecules and their primary physiological effects [45]. What systems biology provides is a system-level understanding of the effect of a therapeutic drug rather than merely the simple molecular interaction between the drug and target.

4.4.2 Sensitivity Analysis

One interesting option when analyzing a mathematical model of a miRNA-regulated network is to find biochemical processes within the network whose modulation is likely to affect the response of the network most significantly. Sensitivity analysis is a computational tool suitable for this. It numerically establishes how the variation in the critical network outcomes (e.g. expression of critical targeted genes) can be categorized and assigned to different sources of variation in the system in an either qualitative or quantitative manner [46]. These sources of variation include the distinctive regulation of processes integrated in the network like the intensity of input signals, the abundance of TFs, the availability of miRNAs and proteins regulating gene targets in a transcriptional or post-transcriptional manner and more. In the context of a mathematical model, we typically associate this with changes in the values of the associated model parameters; by the use of sensitivity analysis, one can look for those parameters for which a numerical variation significantly affects critical responses of the system (Fig. 4.11).

We can categorize sensitivity analysis into two types [46]: (i) local sensitivity analysis, in which small variations around the nominal values of the parameters are applied (one-by-one) to determine which parameters (and their associated processes) have a significant influence on the critical

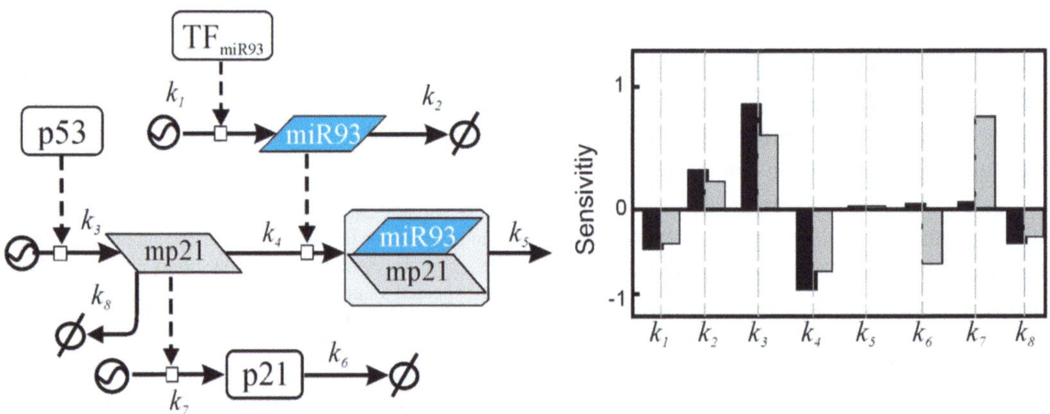

Fig. 4.11 Sensitivity analysis. This computational tool provides information about model parameters (and associated regulatory processes) for which a variation in their value significantly affects the input–output behaviour of the system. In the common methods for estimating local sensitivities, the value of a given parameter is slightly modified (1–5% up or down) and the model is used to calculate the resultant change in the critical response of interest, e.g. the expression levels of any protein, mRNA or miRNA of interest. The absolute sensitivity index is calculated with the equation:

$$S_k^{p21} = \frac{\Delta p21}{\Delta k} = \frac{\Delta p21(k+\Delta k) - p21(k)}{\Delta k}$$

where p21 represents the analysed response variable; k the parameter whose value is modified, and Δk accounts for perturbation of the parameter. In our figure, we analysed the sensitivity of p21 expression with respect to the parameter values of the p21 regulating processes. Sensitivities can be positive (an increase in the model parameter values induces an increase in p21 expression level) or negative values (an increase in the model parameter values induces a decrease in p21 expression level). The absolute value of the sensitivity accounts for the strength that a model parameter value change has on p21 expression levels: values nearly zero indicate that the representative process has no influence on p21 expression levels responses of a network; and (ii) global sensitivity analysis, in which one considers large ranges of feasible values for the parameters when performing the computational analysis and investigates the simultaneous variations of many model parameters (instead of a one-by-one analysis).

Biochemical processes in the network are later ranked according to the impact of their modulation on the input–output behaviour of the system. This information, combined with predictive simulations, can be used to suggest key biochemical processes that might be deregulated in pathological conditions and thus these processes become potential therapeutic targets.

4.4.3 Stability and Bifurcation Analysis

As we mentioned in the previous section, the biochemical networks in which miRNAs participate are enriched in complex regulatory motifs like positive and negative feedback loops. It has been theoretically and experimentally proved that this kind of network motifs can result in highly non-linear behaviour, which challenges the common sense of even well-trained biologists. Behaviour like self-sustained oscillations, ultrasensitivity or bistability can emerge from these network motifs (Fig. 4.12). Biochemical networks containing one or several of these motifs can be analysed by means of more sophisticated mathematical tools, which can detect the emergence of those non-linearities and their consequences in the regulation of the system.

In many cases, the emergence of this non-linear behaviour can be linked to changes of the values of some model parameters in a critical interval. Computational stability and bifurcation analysis are sophisticated mathematical methodologies that allow assessing the emergence of these patterns and identifying which processes in the network are linked to them.

In bifurcation analysis, theoretical and computational tools are used to study qualitative

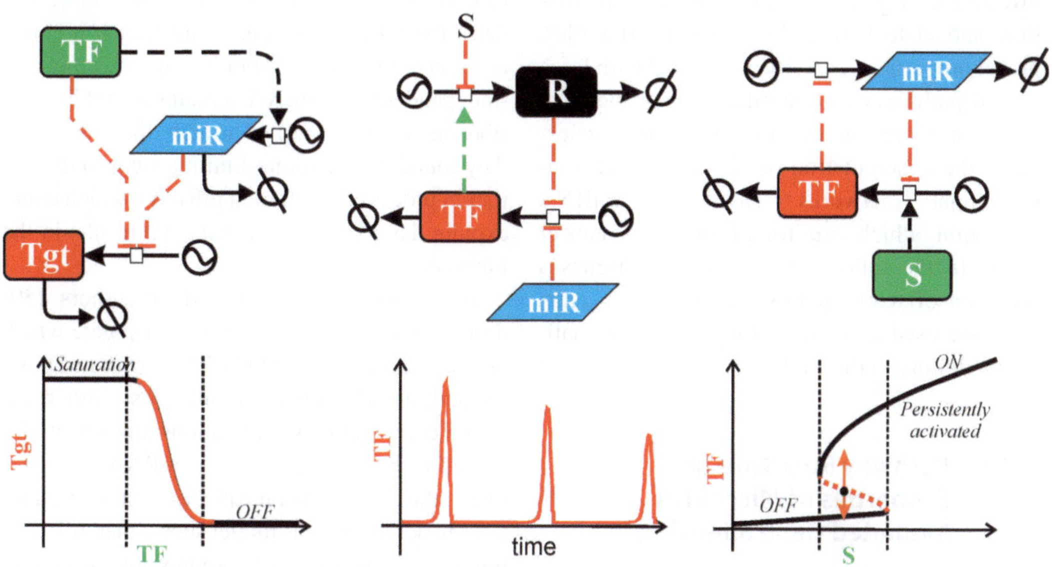

Fig. 4.12 **Ultrasensitivity** makes a biochemical network able to transform graded input signals into discrete all-or-none outputs, creating real thresholds for the input signal that determine the activation of the system. **Sustained oscillations.** In network motifs that include negative feedback loops, under some biological conditions the concentration of one or more of the compounds of the network oscillates over time, even with constant external stimulation. **Bistability (or multistability)**. In networks with positive feedback loops, small perturbations for a given experimental scenario can shift the system between totally different fates, for example, inducing a transition between a scenario in which transient stimulation induces quick signal termination to another in which it provokes persistent activation

changes in the behaviour of nonlinear dynamical systems under variations of system's parameters. By qualitative changes we understand the appearance or extinction of solutions to the model equations (transition from single solutions to multistability or *vice versa*), or a change in the stability of the current state (transitions from steady-state behaviour to oscillations or *vice versa*). Those transitions between qualitatively different regulatory patterns in the network happen in the so-called bifurcation points, critical values for some model parameters in which a continuous change in the parameter value induces a discontinuous and possibly irreversible response of the network. Chap. 9 offers a nice introduction to the basics of these tools.

4.5 Real Life Examples of Systems Biology Used to Investigate MicroRNA Regulation

Mathematical modelling can actually be used to investigate basic properties of miRNA regulation and detect so-called design principles, global properties generally associated with large sets of pathways with similar but not identical structure. Moreover, it is a suitable tool to investigate the dysregulation of signalling and transcriptional networks involving miRNA regulation which are found in many cancer types. In the following section, we discuss a selection of works published in the last years that have used systems biology and mathematical modelling to this end.

4.5.1 Pathway Structure and Properties of MicroRNA-Mediated Gene Silencing

Recently, mathematical modelling was introduced to understand the mechanism by which the gene silencing is accomplished. Levine et al. [47] developed a quantitative model to show how local and global properties of the miRNA-mediated silencing mechanism could affect target mRNA and protein levels. Their analyses suggest that the target specific properties (e.g. the number of binding sites on target mRNAs) can result in different effects on target mRNA levels; likewise, different cellular conditions can cause different behaviour of the same miRNA-target pair.

MiRNAs are reported to repress protein translation by a variety of mechanisms, including: (1) initiation block: repressing cap recognition or 60S subunit joining; (2) post-initiation block: inhibiting elongation, triggering ribosome drop-off, or facilitating proteolysis of nascent polypeptides; (3) affecting the mRNA: deadenylation of mRNA to induce quick mRNA decay, or translocation of mRNA to P-bodies [48].

Due to the complication of the translation initiation process, it is possible that, when different steps are affected by miRNA or RISC components, the regulation of protein production might be divergent. To investigate this problem, Nissan and Parker [49] developed computational models of translation initiation considering miRNA repression. The models were used to predict the effect of miRNA repression on different target mRNAs whose translation might be cap-dependent (the mRNA contains a functional m^7G cap or a non-functional ApppN cap structure) or -independent (the mRNA contains a viral internal ribosomal entry site). Through their analysis, they found different rate-limiting steps in the initiation process for different mRNAs, which could explain the divergent experimental results in the literature.

In parallel, Whichard and co-authors [50] derived a model and used it to investigate which biochemical events in the miRNA-mediated post-transcriptional regulation are most important. Using computational simulations and sensitivity analysis, they found that the miRNA synthesis rate is the dominant controller of protein production. In addition, their model shows that miRNAs might exert potent target repression in certain conditions. MiRNA-mediated mechanisms for the repression of protein translation were also investigated in a recent paper by Zinovyev et al. [51] and are subject of a complete chapter in this book by the same authors (see Chap. 11 for further details).

4.5.2 Modeling of MicroRNA Regulation at Single Cell Level

To investigate the effects of miRNAs on target gene expression in single cells, Mukherji et al. [52] adopted a two-colour fluorescent reporter system which allows them to measure the change of gene expression when miRNA binding sites are present or absent within reporters. Through their single-cell analysis, two intriguing features were found: (1) although the average level of protein repression by miRNAs is modest, in agreement with previous population-based measurements, the repression among individual cells varies dramatically; (2) regulation by miRNAs establishes a threshold level of target mRNA which determines the degree of the repression of protein production. Below the threshold the protein production is highly repressed, however, near this threshold protein expression responds sensitively to target mRNA input. This result is also consistent with the mathematical model which described the effect of molecular titration on the sensitive response of transcription above the threshold. The model showed that with the increasing abundance of the mRNA targets, the availability of miRNA for repression is diluted; the strength of the interaction between miRNA and its target and their relative abundance decides the sharpness of the switch from full repression to escape from miRNA repression.

4.5.3 Identification and Modelling of MicroRNA-Mediated Network Motifs

miRNAs are important components embedded in gene regulatory networks and are found to establish different kinds of network motifs with their TFs and targets. The most common miRNA-mediated network motifs found in gene regulatory networks are feedback and feedforward loops. Such loops can be further differentiated into several subtypes: feedback loops can be either positive or negative if a miRNA and its TF are positively or negatively co-regulated by each other; the feedforward loops can be either coherent or incoherent, if a miRNA and its TF consistently or oppositely regulate their common target.

Bioinformatics algorithms are a powerful tool for detecting miRNA-mediated motifs in gene regulatory networks. By using a bioinformatics approach, Tsang et al. [35] found that the miRNA-mediated motifs, which are frequently observed in regulatory networks of bacteria and yeast, are also prevalent in mouse and human. Particularly, for miRNAs upregulated in neuronal cells, the coherent feedforward and negative feedback loops involving these miRNAs are identified to be more prevalent in mature neurons than other types of motifs. The recurrences of these miRNA-mediated motifs suggest that miRNAs have important biological functions and enhance the robustness of gene regulation in mammals. Re et al. [38] computationally identified a total of 638 putative miRNA-mediated feedforward loops in human gene regulatory networks. The authors filtered the results for motifs with cancer relevant features and demonstrated for some of them experimentally that miRNA-mediated feedforward loops are involved in various aspects of organism development and differentiation, suggesting a crucial role of miRNA-mediated feedforward loops in gene regulatory networks.

Other groups focus on searching for gene-specific network motifs containing miRNA regulation. For example, p53-specific feedback and feedforward loops containing miRNA regulation were identified and studied by Sinha et al. [53]. The authors used a bioinformatics based integrative approach to reveal the miRNA-mediated network motifs underlying the p53 regulatory network, which suggests the important contribution of miRNAs for p53 to control signalling pathways involved in tumour suppression. Moreover, Martinez and Walhout [54] summarized a bunch of experimentally verified feedback and feedforward loops, which are composed of miRNA and TF regulations. In this review paper, the authors demonstrated that the existence of these network motifs reveals not only the reciprocal regulation between miRNAs and TFs but also their coordination in regulating shared target genes at genome-scale level.

Apart from bioinformatics approaches, the employment of mathematical modelling becomes a popular strategy to help scientists unravel interesting dynamic properties of miRNA-mediated network motifs, which are thought to be essential for maintaining the operation of gene regulatory networks. Some mathematical models show that the existence of miRNA-mediated network motifs leads to more effective noise buffering in gene expression. In support of this, Xu et al. [55] set up both deterministic and stochastic models to study and analyze four kinds of miRNA-mediated motifs, which were categorized by four types of external input signals: (1) the same signal acts on two miRNAs and their common target gene; (2) different signals act on one miRNA and its target gene; (3) the signal acts on the target gene only; (4) the signal acts on the miRNA only. Their numerical simulations indicated that all the four miRNA-mediated motifs exhibit strong robustness to external random perturbations in the target gene expression. Similarly, Osella et al. [36] used the same two modelling methods to investigate the role of miRNA-mediated feedforward loops in buffering noise in target gene expression. They demonstrated that compared to the simple gene activation by a TF, the system containing miRNA-mediated repression shows greater ability to dampen fluctuations in the target gene expression. In addition, the oscillatory behaviour of genes was mathematically proved to be affected by miRNA regulation. Xie et al. [56] incorporated miRNA regulation into a gene network containing delayed negative feedback loops to elucidate the possible effect of miRNAs on oscillatory gene expression. They showed that the effect of miRNAs on mRNA stability can decide whether the gene expression oscillates or not.

In our own work, we proposed to combine both bioinformatics and mathematical modelling for studying miRNA network motifs. Our approach relies on the construction of computational models of miRNA regulatory networks by integrating published knowledge and the analysis of these networks via computer simulation of biomedically relevant scenarios and bifurcation analysis [57]. We followed this line in a recent paper, in which we propose an integrative systems biology approach that combines bioinformatics tools, used to set up the structure of a miRNA-regulated network, and mathematical modelling, which is employed to investigate its regulatory features.

We applied this method to investigate the regulation of miRNA target hubs, using p21 as example [16]. A recent study by Wu et al. [20] identified more than 20 miRNAs that can regulate the expression of p21 post-transcriptionally, which made p21 the first experimentally confirmed case of a miRNA target hub. Furthermore, there are evidences suggesting that pairs of miRNAs can cooperate if their binding sites reside in close proximity [29, 39]. We adopted a systems biology approach to investigate mechanisms of collective miRNA repression of the cell cycle regulator p21. In our study, we first integrated miRNA target site and TF predictions with data from the literature and from web resources to generate a regulatory map for the target hub p21 (Fig. 4.13a, b). Our

Fig. 4.13 Computational analysis of target hub p21 repression regulated by multiple and cooperative miRNAs. **(a)** Illustrative representation of the p21 regulatory map. The system is composed of TFs, miRNAs involved in the regulation of p21 as well as p21 interacting proteins. The repression of p21 by miRNAs is conducted through either translation repression or mRNA degradation. Network motifs, such as feedback and feedforward loops could be detected in this network. **(b)** Putative miRNA target sites in the p21 3′ UTR. We used the results of two established target prediction algorithms, miRanda and RNA22 [58, 59]. **(c)** Cooperative miRNA regulation of p21 expression in different cellular functions. Our model indicates that, for different cellular functions, the distinctive combination of miRNA states (on/off; *left*) leads to different p21 steady state levels. We also illustrated the effect of cooperativity among miRNAs against non-cooperativity (*right*). **(d)** The response of p21 to DNA damage (DD) with and without feedforward loop. In the simulation here depicted, we assumed a step-like activation of DNA damage response (*top*). The bottom plot describes the response of p21 to DNA damage under two conditions: feedforward loop and no feedforward loop. The simulation suggests that the suppression of the feedforward loop mediated by a miRNA can favour cancer progression by delaying the initiation of p21-triggered cell cycle arrest in response to DNA damage

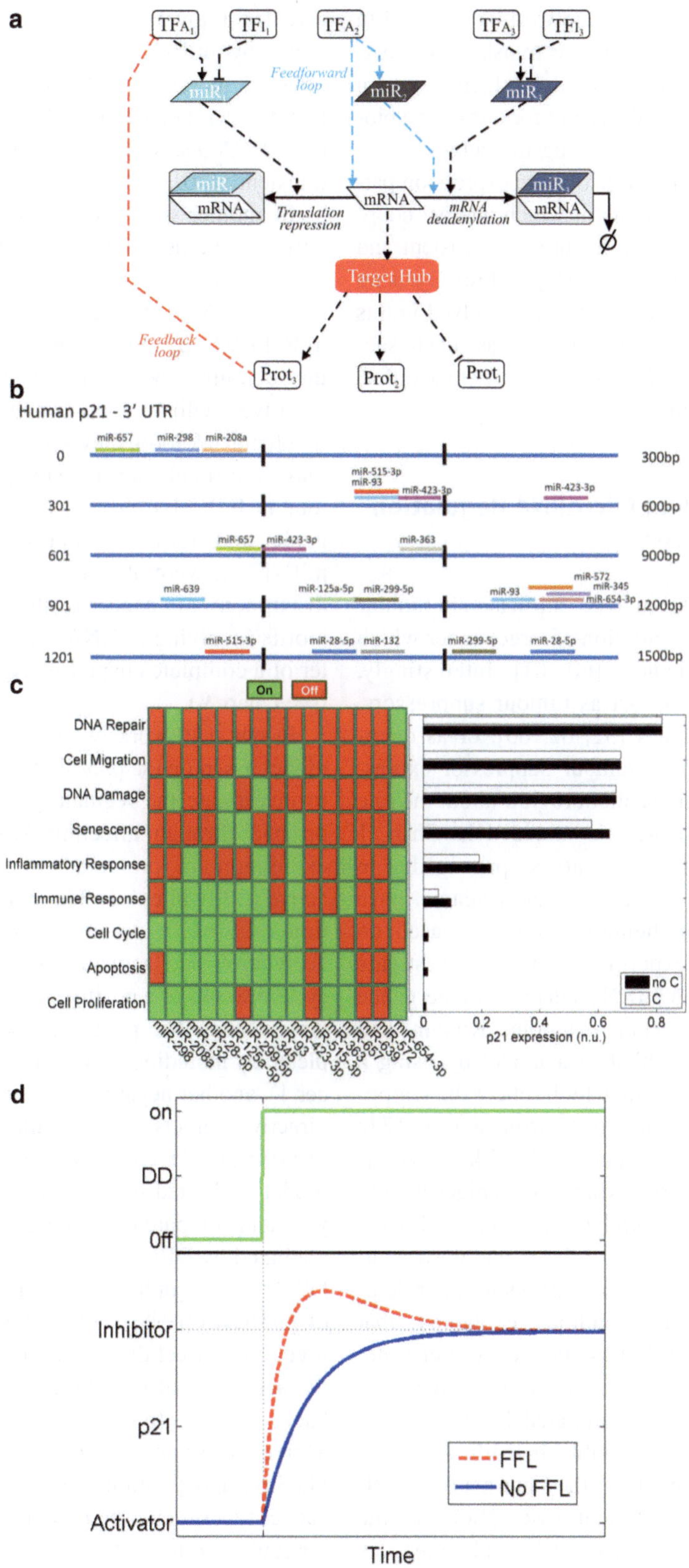
a
TFA1
TFI1
TFA2
TFA3
TFI3
Feedforward loop
miR1
miR2
miR3
mRNA
Translation repression
mRNA deadenylation
Target Hub
Feedback loop
Prot3
Prot2
Prot1
b
Human p21 - 3' UTR
0
300bp
301
600bp
601
900bp
901
1200bp
1201
1500bp
c
On
Off
DNA Repair
Cell Migration
DNA Damage
Senescence
Inflammatory Response
Immune Response
Cell Cycle
Apoptosis
Cell Proliferation
p21 expression (n.u.)
no C
C
d
on
DD
Off
Inhibitor
p21
Activator
FFL
No FFL
Time

analysis showed that this network is enriched in feedforward loops. Then, we translated the network into a data-driven model, which was then used for a kinetic analysis and for testing the biological hypotheses concerning the network. Our analysis indicated that distinctive expression patterns for miRNAs, some of which interact cooperatively, fine-tune the features of transient and long-term regulation of p21 (Fig. 4.13c). By using GO term associations of the TFs involved in this network our mathematical model was able to successfully predict p21 protein levels for nine different cellular functions [16].

4.5.4 Modeling MicroRNA Regulation in Cancer

Many miRNAs seem to be implicated in tumour progression via regulation of target genes which play a role in cancer [60, 61]. Interestingly, these miRNAs can act as tumour suppressors, silencing oncogenes [32, 62, 63], or as oncogenes by inhibiting tumour suppressor expression [64, 65]. An excellent review about the role of miRNAs in cancer is included in Chap. 1 of this book. Thus, it is not surprising that in the last years several research teams have started using mathematical models based on the clinical and experimental data to investigate the role of certain miRNAs in cancer emergence and progression. For example, Khanin and Vinciotti [66] established a model by using a set of microarray data, which reflect the temporal gene expression profile from a miR-124a transfection experiment. miR-124a acts as tumour suppressor which is epigenetically silenced in hepatocellular carcinoma and acute lymphoblastic leukemia [67, 68] and downregulated in glioblastoma [69]. By using their data-driven model, the authors explained that upregulation of miR-124a targets at later time-points is primarily due to the decay of free miRNAs, which they substantiated by a predicted and experimental verifiable half-life for miR-124a. Similarly, by using the same set of experimental data but different assumptions in the mathematical model, Vohradsky and colleagues revealed a novel mechanism of mRNA accumulation by miR-124a, by which miR-124a can organize its target mRNA response in a switch-like manner, i.e. the miRNA has big influence on the mRNA decay and the influence drops to zero transiently[70].

In another study, Aguda and colleagues [30] derived a mathematical model for a specific cancer network, which includes a feedback loop formed by Myc and E2F, and the miRNA cluster miR-17-92. By analyzing the consequence of the coupling between the miRNA-mediated negative feedback loop and the E2F/Myc positive feedback loop, they showed that miR-17-92 plays a critical role in shaping bistable behaviour in E2F/Myc protein levels, and the oncogenic and tumour suppressor properties of miR-17-92 were also demonstrated. A detailed discussion of cancer-related feedback loop motifs involving miRNA regulation is the matter of a complete chapter by Aguda in this book (see Chap. 9).

In our own work, we set up a mathematical model based on the p53/Sirt1 signalling pathway to investigate the regulatory effect of miR-34a on p53 through affecting Sirt1[31]. miR-34a expression is promoted by transcriptionally active p53, while miR-34a represses the translation of Sirt1, a protein known to inhibit the acetylation and activation of p53. The overall system works as a positive feedback loop system (Fig. 4.14a). The picture of the system is completed by including DBC1 (deleted in breast cancer 1), another negative regulator of Sirt1, which directly interacts with and inhibits Sirt1 activity *in vitro* and *in vivo*. By using a mathematical model calibrated with quantitative data we analysed and compared the strength of p53 activation mediated by DBC1 and miR-34a. We found that DBC1, which indirectly enhances the activation of p53 (DBC1—|Sirt1—|p53*), shows a stronger regulatory effect than miR-34a, which enhances the activation of p53 through the positive feedback loop (p53* → miR-34a—|Sirt1—|p53*). Moreover, we show that the loss of p53 activity via Sirt1 upregulation, that can be found in some cancer types, could be partially compensated by upregulating miR-34a expression (Fig. 4.14b).

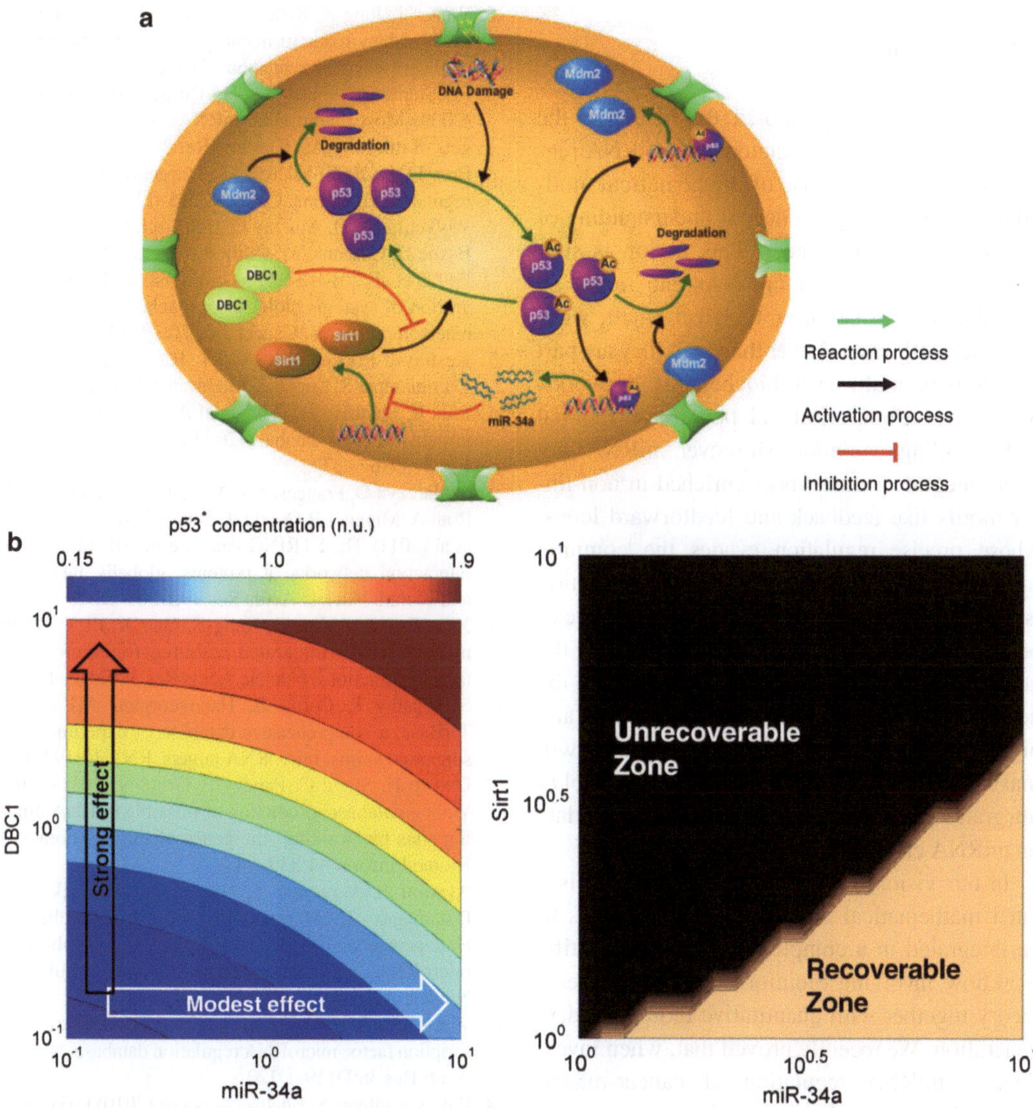

Fig. 4.14 Modelling miR-34a regulation in the p53/Sirt1 signaling pathway. (**a**). Scheme of the p53/Sirt1 signalling pathway. In this pathway, we included the new-found miR-34 mediated feedback loop in which miR-34a is transcriptionally promoted by p53 and represses the production of Sirt1. Other than miR-34a, DBC1 (deleted in breast cancer 1), another negative regulator of Sirtl, directly interacts with Sirt1 and inhibits Sirt1 activity *in vitro* and *in vivo*, and therefore it can activate p53 acetylation and subsequently upregulate p53-mediated pathways. (**b**). We perturbed the parameters accounting for the concentrations of DBC1 and miR-34a in the interval [10^{-1} 10^{1}] and computed the steady-state levels of active p53 (*p53**; *left*). The simulations suggest that although the concentration of DBC1 and miR-34a are perturbed in the same normalised interval, changes in the concentration of DBC1 induce bigger variations in the steady-state levels of *p53** than the modulation of miR-34a. Moreover, we ran the simulations concerning the loss of *p53** due to the upregulation of Sirt1, which is commonly found in cancerous cells (*right*). The results show that in case of the intermediate upregulation (~$10^{0.5}$) of Sirt1 the loss of *p53** can be compensated by upregulating miR-34a expression (Recoverable Zone). This counterbalance, however, does not work for strongly upregulated Sirt1 (10^{1}) (Unrecoverable Zone)

4.6 Summary

In this chapter we promoted the idea that the inherent complexity associated with miRNA regulation will make the use of mathematical modelling necessary to get a deeper understanding of miRNA biology in cancer. This notion is supported by the fact that miRNAs can rarely be considered in isolation or to affect small, well-delimited pathways. Rather than that, they are part of complex multi-level biochemical networks, which involve transcriptional, post-transcriptional and signalling regulation. Moreover, miRNAs are often integrated in networks enriched in non-linear motifs like feedback and feedforward loops, whose precise regulation evades the common rational thinking. Thus, miRNA cancer regulation is a complex phenomenon from the perspective of both the scale and the structural complexity of the networks in which it is involved. In addition, the development of more sophisticated, accurate and high-throughput experimental techniques will make mathematical modelling a necessary tool to integrate massive, multiple-type quantitative data on miRNA cancer regulation.

In our vision, rather than an exercise of isolated mathematical abstraction, modelling is to be integrated in a complex, concerted scientific workflow involving bioinformatics and systems theory together with quantitative biological data generation. We recently proved that, when investigating miRNA regulation of cancer-related genes, this approach is possible, desirable and probably necessary [16]. This complex workflow is what we consider genuine in systems biology.

Acknowledgements This work is supported by the German Federal Ministry of Education and Research (BMBF) as part of the projects e: Bio-miRSys [C.N. 0316175A] and e: Bio-Metsys [C.N. 0316171].

References

1. Bartel DP (2004) MicroRNAs: genomics, biogenesis, mechanism, and function. Cell 116:281–297
2. Hudder A, Novak RF (2008) miRNAs: effectors of environmental influences on gene expression and disease. Toxicol Sci 103:228–240
3. Winter J, Jung S, Keller S, Gregory RI, Diederichs S (2009) Many roads to maturity: microRNA biogenesis pathways and their regulation. Nat Cell Biol 11:228–234
4. Friedman RC, Farh KK-H, Burge CB, Bartel DP (2009) Most mammalian mRNAs are conserved targets of microRNAs. Genome Res 19:92–105
5. Bartel DP (2009) MicroRNAs: target recognition and regulatory functions. Cell 136:215–233
6. Wolkenhauer O, Auffray C, Baltrusch S, Blüthgen N, Byrne H, Cascante M, Ciliberto A, Dale T, Drasdo D, Fell D et al (2010) Systems biologists seek fuller integration of systems biology approaches in new cancer research programs. Cancer Res 70:12–13
7. Keshava Prasad TS, Goel R, Kandasamy K, Keerthikumar S, Kumar S, Mathivanan S, Telikicherla D, Raju R, Shafreen B, Venugopal A et al (2009) Human protein reference database–2009 update. Nucleic Acids Res 37:D767–D772
8. Szklarczyk D, Franceschini A, Kuhn M, Simonovic M, Roth A, Minguez P, Doerks T, Stark M, Muller J, Bork P et al (2011) The STRING database in 2011: functional interaction networks of proteins, globally integrated and scored. Nucleic Acids Res 39:D561–D568
9. Xiao F, Zuo Z, Cai G, Kang S, Gao X, Li T (2009) miRecords: an integrated resource for microRNA-target interactions. Nucleic Acids Res 37:D105–D110
10. Sethupathy P, Corda B, Hatzigeorgiou AG (2006) TarBase: a comprehensive database of experimentally supported animal microRNA targets. RNA 12:192–197
11. Dweep H, Sticht C, Pandey P, Gretz N (2011) miR-Walk – database: prediction of possible miRNA binding sites by 'walking' the genes of three genomes. J Biomed Inform 44:839–847
12. Alexiou P, Vergoulis T, Gleditzsch M, Prekas G, Dalamagas T, Megraw M, Grosse I, Sellis T, Hatzigeorgiou AG (2010) miRGen 2.0: a database of microRNA genomic information and regulation. Nucleic Acids Res 38:D137–D141
13. Wang J, Lu M, Qiu C, Cui Q (2010) TransmiR: a transcription factor-microRNA regulation database. Nucleic Acids Res 38:D119–D122
14. Bandyopadhyay S, Bhattacharyya M (2010) PuTmiR: a database for extracting neighboring transcription factors of human microRNAs. BMC Bioinformatics 11:190
15. Le Novère N, Hucka M, Mi H, Moodie S, Schreiber F, Sorokin A, Demir E, Wegner K, Aladjem MI, Wimalaratne SM et al (2009) The systems biology graphical notation. Nat Biotechnol 27:735–741
16. Lai X, Schmitz U, Gupta S, Bhattacharya A, Kunz M, Wolkenhauer O, Vera J (2012) Computational analysis of target hub gene repression regulated by multiple and cooperative miRNAs. Nucleic Acids Res 40:8818–8834
17. Jung Y-S, Qian Y, Chen X (2010) Examination of the expanding pathways for the regulation of p21 expression and activity. Cell Signal 22:1003–1012
18. Gartel AL, Tyner AL (1999) Transcriptional regulation of the p21((WAF1/CIP1)) gene. Exp Cell Res 246:280–289
19. Abbas T, Dutta A (2009) p21 in cancer: intricate networks and multiple activities. Nat Rev Cancer 9:400–414

20. Wu S, Huang S, Ding J, Zhao Y, Liang L, Liu T, Zhan R, He X (2010) Multiple microRNAs modulate p21Cip1/Waf1 expression by directly targeting its 3′ untranslated region. Oncogene 29:2302–2308
21. Vera J, Balsa-Canto E, Wellstead P, Banga JR, Wolkenhauer O (2007) Power-law models of signal transduction pathways. Cell Signal 19:1531–1541
22. Vera J, Rath O, Balsa-Canto E, Banga JR, Kolch W, Wolkenhauer O (2010) Investigating dynamics of inhibitory and feedback loops in ERK signalling using power-law models. Mol Biosyst 6:2174–2191
23. Kholodenko B, Yaffe MB, Kolch W (2012) Comput ational approaches for analyzing information flow in biological networks. Sci Signal 5:re1
24. Saez-Rodriguez J, Alexopoulos LG, Epperlein J, Samaga R, Lauffenburger DA, Klamt S, Sorger PK (2009) Discrete logic modelling as a means to link protein signalling networks with functional analysis of mammalian signal transduction. Mol Syst Biol 5:331
25. Vera J, Wolkenhauer O (2008) A system biology approach to understand functional activity of cell communication systems. Methods Cell Biol 90:399–415
26. Maiwald T, Timmer J (2008) Dynamical modeling and multi-experiment fitting with PottersWheel. Bioinformatics 24:2037–2043
27. Mendes P, Hoops S, Sahle S, Gauges R, Dada J, Kummer U (2009) Computational modeling of biochemical networks using COPASI. Methods Mol Biol 500:17–59
28. Savageau MA, Coelho PMBM, Fasani RA, Tolla DA, Salvador A (2009) Phenotypes and tolerances in the design space of biochemical systems. Proc Natl Acad Sci U S A 106:6435–6440
29. Shalgi R, Lieber D, Oren M, Pilpel Y (2007) Global and local architecture of the mammalian microRNA-transcription factor regulatory network. PLoS Comput Biol 3:e131
30. Aguda BD, Kim Y, Piper-Hunter MG, Friedman A, Marsh CB (2008) MicroRNA regulation of a cancer network: consequences of the feedback loops involving miR-17-92, E2F, and Myc. Proc Natl Acad Sci U S A 105:19678–19683
31. Lai X, Wolkenhauer O, Vera J (2012) Modelling miRNA regulation in cancer signalling systems: miR-34a regulation of the p53/Sirt1 signalling module. In: Computational modelling of signalling networks, Liu X, Betterton M (eds.), Humana press 2012, ISBN: 9781617798230
32. Johnson SM, Grosshans H, Shingara J, Byrom M, Jarvis R, Cheng A, Labourier E, Reinert KL, Brown D, Slack FJ (2005) RAS is regulated by the let-7 microRNA family. Cell 120:635–647
33. Nikolov S, Yankulova E, Wolkenhauer O, Petrov V (2007) Principal difference between stability and structural stability (robustness) as used in systems biology. Nonlinear Dynamics Psychol Life Sci 11: 413–433
34. Sorribas A, Hernández-Bermejo B, Vilaprinyo E, Alves R (2007) Cooperativity and saturation in biochemical networks: a saturable formalism using Taylor series approximations. Biotechnol Bioeng 97:1259–1277
35. Tsang J, Zhu J, van Oudenaarden A (2007) MicroRNA-mediated feedback and feedforward loops are recurrent network motifs in mammals. Mol Cell 26:753–767
36. Osella M, Bosia C, Corá D, Caselle M (2011) The role of incoherent microRNA-mediated feedforward loops in noise buffering. PLoS Comput Biol 7:e1001101
37. Mangan S, Alon U (2003) Structure and function of the feed-forward loop network motif. Proc Natl Acad Sci U S A 100:11980–11985
38. Re A, Corá D, Taverna D, Caselle M (2009) Genome-wide survey of microRNA-transcription factor feed-forward regulatory circuits in human. Mol Biosyst 5:854–867
39. Doench JG, Sharp PA (2004) Specificity of microRNA target selection in translational repression. Genes Dev 18:504–511
40. Saetrom P, Heale BS, Snøve O Jr, Aagaard L, Alluin J, Rossi JJ (2007) Distance constraints between microRNA target sites dictate efficacy and cooperativity. Nucleic Acids Res 35:2333–2342
41. Harper JW, Adami GR, Wei N, Keyomarsi K, Elledge SJ (1993) The p21 Cdk-interacting protein Cip1 is a potent inhibitor of G1 cyclin-dependent kinases. Cell 75:805–816
42. Rippe V, Dittberner L, Lorenz VN, Drieschner N, Nimzyk R, Sendt W, Junker K, Belge G, Bullerdiek J (2010) The two stem cell microRNA gene clusters C19MC and miR-371-3 are activated by specific chromosomal rearrangements in a subgroup of thyroid adenomas. PLoS One 5:e9485
43. Weber MJ (2005) New human and mouse microRNA genes found by homology search. FEBS J 272:59–73
44. O'Donnell KA, Wentzel EA, Zeller KI, Dang CV, Mendell JT (2005) c-Myc-regulated microRNAs modulate E2F1 expression. Nature 435:839–843
45. Rowland M, Tozer TN (1995) Clinical pharmacokinetics: concepts and applications. Williams & Wilkins, Baltimore
46. Saltelli A, Tarantola S, Campolongo F (2000) Sensitivity analysis as an ingredient of modeling. Stat Sci 15:377–395
47. Levine E, Ben Jacob E, Levine H (2007) Target-specific and global effectors in gene regulation by MicroRNA. Biophys J 93:L52–L54
48. Fabian MR, Sonenberg N, Filipowicz W (2010) Regulation of mRNA translation and stability by microRNAs. Annu Rev Biochem 79:351–379
49. Nissan T, Parker R (2008) Computational analysis of miRNA-mediated repression of translation: implications for models of translation initiation inhibition. RNA 14:1480–1491
50. Whichard ZL, Motter AE, Stein PJ, Corey SJ (2011) Slowly produced microRNAs control protein levels. J Biol Chem 286:4742–4748
51. Zinovyev A, Morozova N, Nonne N, Barillot E, Harel-Bellan A, Gorban AN (2010) Dynamical modeling of microRNA action on the protein translation process. BMC Syst Biol 4:13

52. Mukherji S, Ebert MS, Zheng GXY, Tsang JS, Sharp PA, van Oudenaarden A (2011) MicroRNAs can generate thresholds in target gene expression. Nat Genet 43:854–859
53. Sinha AU, Kaimal V, Chen J, Jegga AG (2008) Dissecting microregulation of a master regulatory network. BMC Genomics 9:88
54. Martinez NJ, Walhout AJM (2009) The interplay between transcription factors and microRNAs in genome-scale regulatory networks. Bioessays 31:435–445
55. Xu F, Liu Z, Shen J, Wang R (2009) Dynamics of microRNA-mediated motifs. IET Syst Biol 3:496–504
56. Xie Z-R, Yang H-T, Liu W-C, Hwang M-J (2007) The role of microRNA in the delayed negative feedback regulation of gene expression. Biochem Biophys Res Commun 358:722–726
57. Nikolov S, Vera J, Schmitz U, Wolkenhauer O (2011) A model-based strategy to investigate the role of microRNA regulation in cancer signalling networks. Theory Biosci=Theorie in Den Biowissenschaften 130:55–69
58. John B, Enright AJ, Aravin A, Tuschl T, Sander C, Marks DS (2004) Human MicroRNA targets. PLoS Biol 2:e363
59. Miranda KC, Huynh T, Tay Y, Ang Y-S, Tam W-L, Thomson AM, Lim B, Rigoutsos I (2006) A pattern-based method for the identification of MicroRNA binding sites and their corresponding heteroduplexes. Cell 126:1203–1217
60. Schultz J, Lorenz P, Gross G, Ibrahim S, Kunz M (2008) MicroRNA let-7b targets important cell cycle molecules in malignant melanoma cells and interferes with anchorage-independent growth. Cell Res 18:549–557
61. Garzon R, Calin GA, Croce CM (2009) MicroRNAs in cancer. Annu Rev Med 60:167–179
62. Takamizawa J, Konishi H, Yanagisawa K, Tomida S, Osada H, Endoh H, Harano T, Yatabe Y, Nagino M, Nimura Y et al (2004) Reduced expression of the let-7 microRNAs in human lung cancers in association with shortened postoperative survival. Cancer Res 64:3753–3756
63. Akao Y, Nakagawa Y, Naoe T (2006) let-7 microRNA functions as a potential growth suppressor in human colon cancer cells. Biol Pharm Bull 29:903–906
64. Iorio MV, Ferracin M, Liu C-G, Veronese A, Spizzo R, Sabbioni S, Magri E, Pedriali M, Fabbri M, Campiglio M et al (2005) MicroRNA gene expression deregulation in human breast cancer. Cancer Res 65:7065–7070
65. Chan JA, Krichevsky AM, Kosik KS (2005) MicroRNA-21 is an antiapoptotic factor in human glioblastoma cells. Cancer Res 65:6029–6033
66. Khanin R, Vinciotti V (2008) Computational modeling of post-transcriptional gene regulation by microRNAs. J Comput Biol 15:305–316
67. Agirre X, Vilas-Zornoza A, Jiménez-Velasco A, Martin-Subero JI, Cordeu L, Gárate L, San José-Eneriz E, Abizanda G, Rodríguez-Otero P, Fortes P et al (2009) Epigenetic silencing of the tumour suppressor microRNA Hsa-miR-124a regulates CDK6 expression and confers a poor prognosis in acute lymphoblastic leukemia. Cancer Res 69:4443–4453
68. Furuta M, Kozaki K, Tanaka S, Arii S, Imoto I, Inazawa J (2010) miR-124 and miR-203 are epigenetically silenced tumour-suppressive microRNAs in hepatocellular carcinoma. Carcinogenesis 31:766–776
69. Fowler A, Thomson D, Giles K, Maleki S, Mreich E, Wheeler H, Leedman P, Biggs M, Cook R, Little N et al (2011) miR-124a is frequently down-regulated in glioblastoma and is involved in migration and invasion. Eur J Cancer 47:953–963
70. Vohradsky J, Panek J, Vomastek T (2010) Numerical modelling of microRNA-mediated mRNA decay identifies novel mechanism of microRNA controlled mRNA downregulation. Nucleic Acids Res 38:4579–4585

The p53/microRNA Network in Cancer: Experimental and Bioinformatics Approaches

5

Sabine Hünten, Helge Siemens, Markus Kaller, and Heiko Hermeking

Abstract/Summary

In the recent years, microRNAs (miRNAs) were identified as important components of the signaling cascades that mediate and regulate tumor suppression exerted by p53. This review illustrates some of the main principles that underlie the mechanisms by which miRNAs participate in p53's function and how they were identified. Furthermore, the current status of the research on the connection between p53 and miRNAs, as well as alterations in the p53/miRNA pathways found in cancer will be summarized and discussed. In addition, experimental and bioinformatics approaches, which can be applied to study the connection between p53 and miRNAs are described. Although, some of the central miRNA-encoding genes that mediate the effects of p53, such as the *miR-34* and *miR-200* families, have been identified, many additional analyses remain to be performed to fully elucidate the connections between p53 and miRNAs.

Keywords

p53 • microRNA • miRNA • Tumor suppression • SILAC • Next generation sequencing • Genome-wide analysis • miR-34 • miR-34a • miR-34b/c

Sabine Hünten, Helge Siemens, and Markus Kaller are equally contributing authors.

S. Hünten • H. Siemens • M. Kaller •
H. Hermeking, Ph.D. (✉)
Experimental and Molecular Pathology, Institute of Pathology, Ludwig-Maximilians-Universität München, Thalkirchner Strasse 36, D-80337 Munich, Germany
e-mail: heiko.hermeking@med.uni-muenchen.de

5.1 Introduction to p53 Biology

The p53 transcription factor is encoded by a tumor suppressor gene, which is presumably the most commonly mutated gene in human cancer [1]. In addition, many of the cancers without *p53* mutation may harbor alterations up- or down-stream of

U. Schmitz et al. (eds.), *MicroRNA Cancer Regulation: Advanced Concepts, Bioinformatics and Systems Biology Tools*, Advances in Experimental Medicine and Biology 774,
DOI 10.1007/978-94-007-5590-1_5, © Springer Science+Business Media Dordrecht 2013

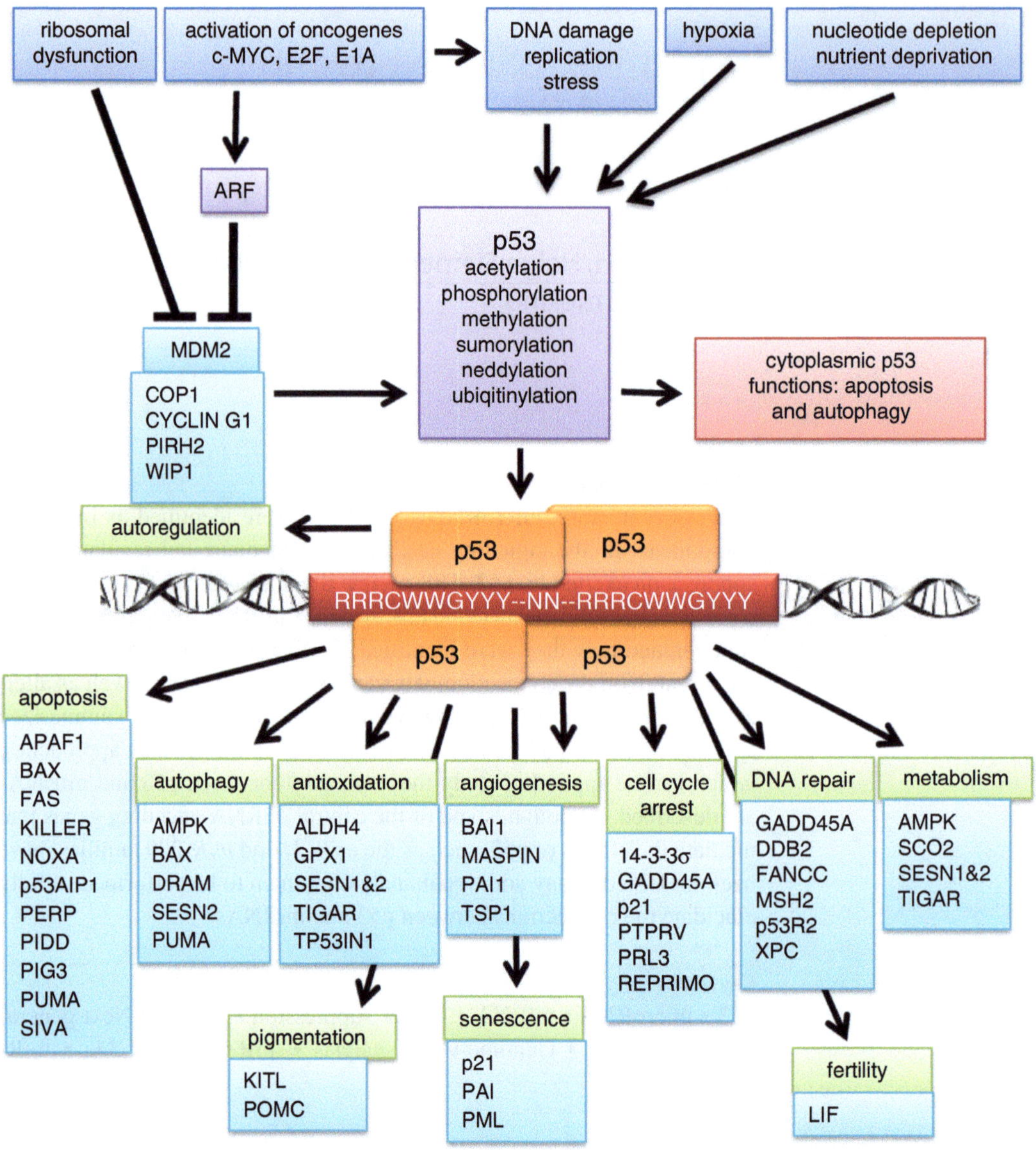

Fig. 5.1 p53 as a central mediator of stress responses. In this model the types of stress and cellular events (*dark blue*) leading to activation of p53 and the protein encoding genes activated by p53 are depicted. p53 is shown as a symbolic tetramer occupying a p53 binding motif (in *red*) containing two palindromic DNA sequences (*white letters* with R = purines (A or G), Y = pyrimidines (C or T), W = A or T and N = bases representing spacers between the two palindromic half-sites). Processes (*green*) regulated by p53 and the respective p53 target genes (*light blue*) implicated are indicated

p53, which also impede the ability of p53 to suppress tumor cell growth. p53 and its loss may represent attractive targets for tumor therapy [2]. Most *p53* mutations target the DNA binding domains of p53, suggesting that the regulation of specific target genes is central for the tumor suppression mediated by p53. However, alternative functions of p53 in the cytoplasm and in mitochondria have also been described [3]. p53's transcriptional activity is induced by various forms of cellular stress that cause diverse post-translational modifications of p53, which are thought to allow a fine-tuning of the cellular response to the type and extent of stress experienced by the respective cell ([4] summarized in Fig. 5.1). For example repairable DNA damage may cause a transient cell cycle arrest, whereas extensive damage may induce apoptosis via generating different levels of p53 activity. DNA damage in the form of double-strand DNA breaks was one of the first inducers of p53 to be discovered. Subsequently, ribosomal, replication, metabolic, oxidative and transcriptional stress, as well as hypoxia were found to cause an increase in p53's transcriptional activity. These alterations stimulate distinct signaling cascades, which activate enzymes that modify p53 or regulate co-factors binding to p53. For example DNA double strand breaks lead to activation of the ATM kinase, which phosphorylates p53 at multiple N-terminal residues [5] and thereby increases its transactivation activity. Furthermore, p53 may be activated by inhibition of the MDM2 protein, which represents an E3-ubiquitin ligase that marks p53 for proteasomal degradation. p53 forms tetramers, that bind to palindromic recognition sites often organized in tandem repeats with spacers of varying length between them (Fig. 5.1). Promoters display gradual responsiveness to p53 either due to different numbers of p53 binding motifs or due to the presence of high affinity versus low affinity sites [6]. For example, the *p21* gene has a high affinity p53-binding site and mediates cell cycle arrest, whereas genes that mediate cell death harbor low affinity p53 binding sites. Therefore, apoptosis is presumably only induced when p53 is strongly activated. p53 directly activates a large set of genes, which mediate numerous cellular functions that contribute to tumor suppression. Many, but not all of these protein coding target genes are depicted in Fig. 5.1. The activation of p53 target genes is either caused by an increase in p53 abundance after p53 stabilization, anti-repression of specific genes after removal of repressive MDM2/MDMX from p53 by acetylation and/or phosphorylation, as well as by formation of promoter-specific transcriptional complexes [4]. Furthermore, p53 may mediate the specific repression of genes. However, the mechanisms of gene repression by p53 are less well understood and may be indirect to some extent [6]. In the recent years, miRNAs were shown to represent important mediators of gene repression caused by p53.

5.2 p53 and the miRNA World: Current State of the Art

miRNAs have presumably evolved to allow organisms to effectively deal with stress [7, 8]. In line with this notion the p53 stress-response pathway is heavily interconnected with miRNAs not only by regulating their expression and processing, but also since p53 itself represents a down-stream target of miRNAs (see Figs. 5.2, 5.3 and 5.4). The protein-coding genes regulated by p53 elicit several cellular phenotypes/processes, which contribute to tumor suppression, as for example induction of cell cycle arrest, senescence and apoptosis, as well as inhibition of metastasis, angiogenesis and glycolysis [9–15]. Interestingly, these processes are also regulated and in some cases induced by p53-regulated miRNAs [10, 12]. In the last 5 years the characterization of a number of miRNAs directly regulated by p53 and the cellular effects of these connections have been reported. For an overview see Fig. 5.3.

5.2.1 The *miR-34* Genes

In 2007 the miR-34 genes, *miR-34a* and *miR-34b/c*, were reported to be directly regulated by p53 by a number of laboratories using diverse approaches [16–21]. For example, we determined the abundance of miRNAs in libraries representing small RNAs generated after p53 activation using a next generation sequencing approach [16]: we found that miR-34a showed the most pronounced increase among all detected miRNAs after p53 activation, which is mediated by p53 binding sites in the promoter region of its host gene. When ectopically expressed, miR-34a and miR-34b/c displayed tumor suppressive activities, i.e. they caused induction of apoptosis and senescence, inhibition of cell cycle progression, and a decrease of angiogenesis (reviewed in [10, 12, 22]). These effects were mediated by direct down-regulation of the expression of numerous key regulators and effectors of these processes as BCL-2, Cyclin E, CDK4 and CDK6. Meanwhile, a large number of additional miR-34 targets have

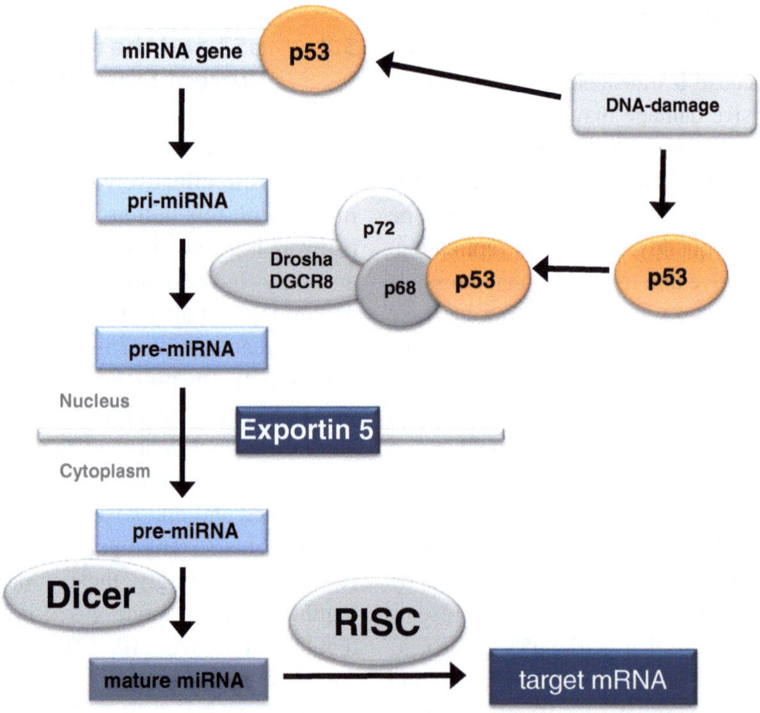

Fig. 5.2 Effects of p53 on the miRNA processing pathway. The synthesis of miRNAs in mammalian cells and the known modes of regulation by p53 are depicted

been identified using a variety of approaches (reviewed in [12]; see also [23, 24] and references therein). Among the miR-34 targets SIRT1, c-MET, Axl, c-/N-MYC, LDH-A and SNAIL seem to be especially relevant for the suppression of cancer. In fact their common up-regulation in tumors could be due to the frequent inactivation of the p53/miR-34 axis during tumor development ([25, 26]; see also next sub-chapter). These targets contribute to the suppression of migration and invasion (SNAIL, c-MET, Axl) and metabolism (LDH-A). In the case of c-MET it was recently shown that p53 down-regulates c-*MET* expression via SP1-mediated occupancy and repression of the c-*MET* promoter and by inducing miR-34a/b/c, which directly target the 3′-UTR of the c-*MET* mRNA [27]. p53 may suppress metastasis by antagonizing epithelial-mesenchymal transitions, which have been implicated in the early, invasive stages of metastasis. Instead, p53 activation promotes mesenchymal-epithelial transition (MET) and favors the epithelial state of cells [28]. We recently found that p53-induced MET is mediated by induction of *miR-34a* and *miR-34b/c* in colorectal cancer cell lines [29]. miR-34a and miR-34b/c achieve this effect by negatively regulating a master-regulator of EMT, the SNAIL transcription factor [29, 30]. In addition, we found that the *miR-34a* and the *miR-34b/c* genes are directly repressed by SNAIL [29]. Therefore, miR-34a/b/c and SNAIL form a double-negative feed-back loop (summarized in [31]). Stemness represents another important oncogenic trait of cancer cells which is suppressed by miR-34. It was shown that miR-34 directly suppresses CD44, which blocks the expansion of cancer-initiating tumor stem cells in a mouse model of prostate cancer [32]. When miR-34a is ectopically expressed, stemness markers as CD133, CD44 and BMI-1 are down-regulated in colorectal cancer cells [29]. Furthermore, it was recently reported that, similar to p53, the miR-34 miRNAs provide a barrier for somatic cell reprogramming and the generation of IPS

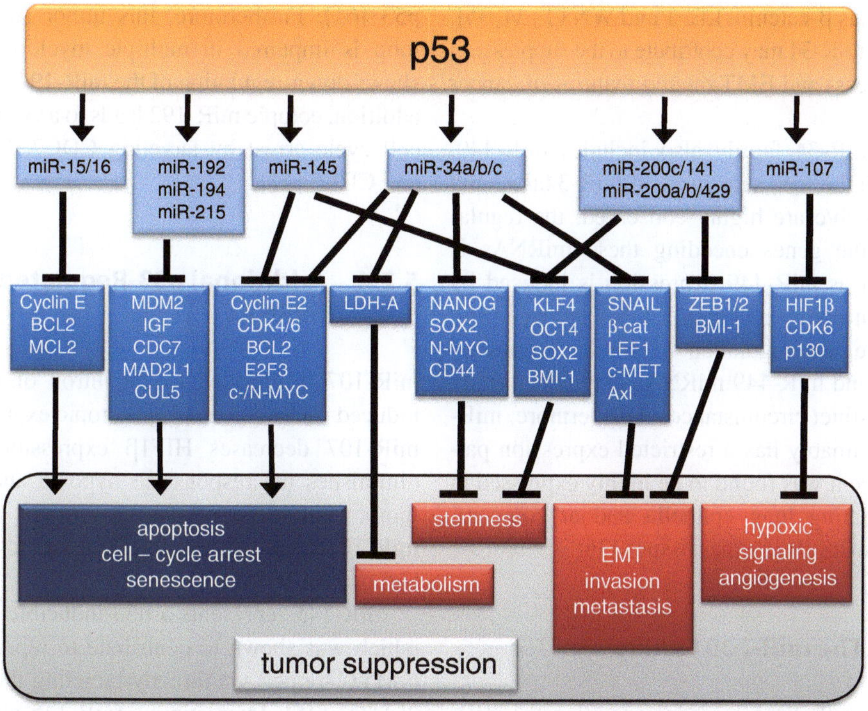

Fig. 5.3 Regulation of miRNA expression by p53. Model summarizing direct transcriptional activation of miRNA-encoding genes, the affected miRNA targets and the reported cellular effects, which collectively contribute to tumor suppression by p53. The *arrows* or inhibition symbols pointing to the cellular processes represent the summation of the regulations resulting from the activity of the indicated p53-induced miRNAs. For details see the main text

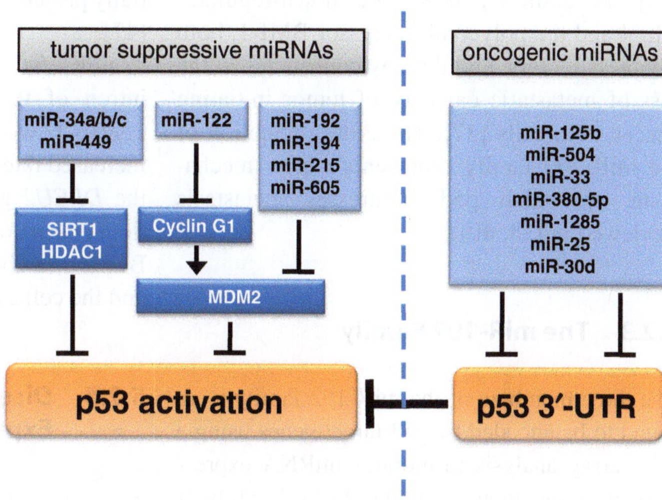

Fig. 5.4 Regulation of p53 by miRNAs. Model summarizing the regulation of the 3′-UTR of *p53*, as well as the down-regulation of p53-modifying enzymes by miRNAs. Oncogenic miRNAs with seed-matching sequences in the 3′-UTR of *p53* or tumor suppressive miRNAs directly down-regulating the indicated proteins are depicted. For details see the main text

(induced pluripotent stem) cells from mouse embryo fibroblast [33]. miR-34 mediated this effect by direct down-regulation of NANOG, SOX2 and N-MYC. Therefore, cancer cells with loss of miR-34 expression may also be more prone to become tumor initiation cells, which exhibit features of stem cells. Furthermore, miR-34 inhibits components of the wnt/β-catenin/TCF pathway, as β-catenin, LEF1 and WNT1 [34, 35]. Thereby, miR-34 may contribute to the suppression of stemness- and EMT-related features of cancer cells.

The miR-34 family also includes miR-449. Although the seed sequences of miR-34a/b/c and miR-449a/b/c are highly conserved, the regulation of the genes encoding these miRNAs is divergent as miR-449 expression is induced by E2F1, but not by p53 and/or DNA damage [36]. Therefore, the regulation of similar targets by miR-34 and miR-449 miRNAs may occur under rather distinct circumstances. Furthermore, miR-449 presumably has a restricted expression pattern, since it was found to be highly expressed in differentiating lung epithelia and at comparatively low levels in other tissues [36].

5.2.2 The miR-200 Family

More recently the two genes encoding the miR-200 family, which give rise to the miR-200c/141 and the 200a/200b/429 miRNAs, were identified as direct p53 target genes that enforce mesenchymal-epithelial transitions (MET) [37, 38] by targeting the EMT-regulators ZEB1 and ZEB2 [39]. In addition, miR-200c down-regulates KLF4 and the polycomb repressor BMI-1, both stemness factors, and thereby contributes to the loss of metastatic capacity of tumor initiating cancer stem cells [37]. Therefore, induction of the miR-200 family represents a new mechanism by which p53 suppresses metastasis (reviewed in [28, 40]).

5.2.3 The miR-192 Family

The three members of the miR-192 family were found to be encoded by p53 target genes using a microarray analysis to monitor miRNA expression after treatment with the MDM2 inhibitor Nutlin-3a [41]. These authors also found that ectopic miR-192 expression induces p21 in a p53-dependent manner. Later it was shown that the miR 192 family targets the IGF pathway and also MDMD2, which results in the activation of p53 [61]. Furthermore, this tumor suppressive loop is impaired in multiple myeloma, which shows down-regulation of the miR-192 family. In addition, ectopic miR-192 leads to a G_1 and G_2/M cell cycle arrest by targeting CDC7, MAD2L1 and CUL5 [42].

5.2.4 Additional p53-Regulated miRNAs

miR-107 is encoded by an intron of the p53-induced *PANK1* gene [43]. Ectopic expression of miR-107 decreases HIF1β expression, which diminishes the response to hypoxia and blocks tumor angiogenesis and growth. In addition, miR-107 targets the cell cycle regulators CDK6 and p130/pRBL2 [44].

miR-145 represents a p53-inducible miRNA, which was shown to contribute to repression of c-MYC by p53 via directly targeting the c-*MYC* 3′-UTR [45]. Interestingly, miR-145 also negatively regulates OCT4, SOX2 and KLF4, and thereby represses pluripotency in human embryonic stem cells [46]. Therefore, miR-145 may at least in part explain why deletion of p53 strongly enhances the generation of IPS cells and potentially promotes the expansion of cancer stem cells [47].

miR-15a and miR-16-1 are encoded by an intron of the *DLEU2* mRNA. Initially, miR-15a/16-1 were shown to be processed at an increased rate after p53 activation [16, 48]. Later, the *DLEU2* gene was shown to be a transcriptional target of p53 [49]. Since miR-15/16 target BCL2 and Cyclin E, they affect both, apoptosis and the cell cycle.

5.2.5 Direct Regulation of p53 Expression by miRNAs

Seversal recent publications demonstrated that miRNAs contribute to the tight control under which p53 is placed in the cell by directly interacting with the 3′-UTR of *p53* mRNA (summarized in Fig. 5.4). By computational analysis of putative miRNA binding sites using TargetScan and mirBase prediction software a binding site of miR-125b was identified in the 3′-UTR of *p53* [50]. MiR-125b is expressed at high levels in the brain and conserved between human, zebrafish and other vertebrates. Ectopic expression of miR-125b decreased p53 protein levels and apoptosis in human cells, whereas inhibition of miR-125b had the opposite effect in lung fibroblasts and zebrafish brain. When zebrafish were treated with DNA damaging agents miR-125b expression was down-regulated, presumably allowing the observed increase in p53 protein. Analysis of 89 colorectal cancer samples revealed that elevated expression of miR-125b is associated with increased tumor size and invasion, and also correlates with poor prognosis and decreased survival [51]. These results are in accordance with a negative regulation of p53 by miR-125b.

By an *in silico* search two miR-504 seed-matching sequences were identified in the 3′-UTR of p53 [52]. Accordingly, ectopic expression of miR-504 down-regulated p53 protein levels, reduced p53-dependent apoptosis and cell cycle arrest, and resulted in increased tumor formation *in vivo*.

miR-33 also targets p53 by binding to two seed-matching motifs in the 3′-UTR of p53 [53]. Interestingly, miR-33 is down-regulated in hematopoietic stem cells (HSC) and up-regulated in more differentiated progenitor cells in super-p53 mice, which are endowed with an extra copy of p53. Ectopic expression of miR-33 in HSC results in increased stemness and decreased recipient survival. In mouse embryonic fibroblasts miR-33 promotes neoplastic transformation presumably via down-regulation of p53.

miR-380-5p was found to down-regulate p53 in neuroblastomas, which commonly express wild-type p53 [54]. Neuroblastomas with elevated expression of miR-380-5p showed a decreased patient survival. Furthermore, miR-380-5p was highly expressed in mouse embryonic stem cells and its ectopic expression cooperated with HRAS in transformation, abrogation of oncogene-induced senescence and promoted tumor formation in mice. Finally, *in vivo* delivery of a miR-380-5p antagonist decreased tumor size in an orthotopic mouse model of neuroblastoma.

In a systematic, bioinformatics screen 107 potential p53-targeting miRNAs were identified using TargetScan [55]. When these candidates were experimentally tested in a dual-reporter assay, miR-1285 turned out to be the most effective repressor of p53's 3′-UTR reporter activity. In line with these results, miR-1285 decreased p53 mRNA- and protein-levels by directly binding to the 3′-UTR of p53 via two seed-matching sequences.

In a similar bioinformatics screen using less stringent criteria and four different miRNA target prediction methods (Miranda, TargetScan, PicTar and RNA22) 67 candidate miRNAs with the potential to directly inhibit p53 expression were identified [56]. In a subsequent, experimental screen only eight of these had an inhibitory effect on p53-mediated transactivation. Of these only three were effective in a dual reporter assay employing the *p53* 3′-UTR: miR-200a, -30d and -25. By mutation of the respective corresponding seed-matching sequences in reporter constructs only miR-30d and miR-25 were validated as direct regulators of the *p53* 3′-UTR. In contrast, miR-200a presumably affects the *p53* 3′-UTR by indirect regulation, e.g. via modulation of transcription factors that regulate miRNAs, which directly target p53. In a cellular assay ectopic miR-30d and miR-25 decreased p53 levels, p53 target expression and downstream effects of p53 as apoptosis, cell cycle arrest and senescence. The opposite was observed, when both miRNAs were inhibited by antagomirs. In line with these observations, miR-25 and miR-30d were found to be up-regulated in multiple myeloma cells, which showed a concomitant down-regulation of p53 mRNA expression. Furthermore, inhibition of miR-25 and miR-30d induced p53 and apoptosis in a multiple myeloma cell line. Therefore, miR-25 and miR-30d presumably represent oncogenic miRNAs. These examples show that the bioinformatics identification of miRNA/target mRNA interactions has to be validated experimentally as it currently generates mainly false predictions.

5.2.6 Indirect Regulation of p53 by miRNAs

Several examples of p53 being subject to indirect regulation by miRNAs via down-regulation of up-stream regulators of p53 have been documented. One of the first cases was the regulation of SIRT1 by miR-34a [57]. An *in silico* search for miR-34a targets, which might affect p53 resulted in the analysis and experimental confirmation of SIRT1 as a miR-34a target. As a consequence of SIRT1 down-regulation by miR-34a an increase in p53 activity and enhanced expression of its targets p21 and PUMA, as well as increased apoptosis was observed. Since miR-34a itself is induced by p53 the regulations connecting miR-34a, SIRT1 and p53 constitute a positive feed-back loop. In tumors this self activating loop may be disrupted by the silencing of *miR-34* genes by CpG methylation [12, 25, 26] and mutation/inactivation of p53.

As mentioned above, miR-449 is similar to miR-34, but regulated by other factors, as for example E2F1. Interestingly, when miR-449 was expressed ectopically it also indirectly activated p53 via directly suppressing the expression of SIRT1 [58]. This may allow additional pathways to increase p53 activity.

Also miR-122 leads to an up-regulation of p53 [59]. However, this is accomplished even more indirectly, since the miR-122-mediated down-regulation of Cyclin G1 presumably inhibits recruitment of PP2A phosphatase to MDM2 resulting in decreased MDM2 activity and increased p53 levels/activity. In line with this scenario ectopic miR-122 expression increased the sensitivity of hepatocellular carcinoma derived cell-lines to doxorubicin.

More recently, miR-885-5p was shown to activate p53 and the expression of p53 target genes [60]. However, although miR-885-5p was shown to target CDK2 and MCM5, the mechanism of the miR-885-5p effect on p53 remained unclear.

miR-192/194/215 are transcriptionally induced by p53 and negatively modulate MDM2 activity [61]. Interestingly, their ectopic expression enhanced the therapeutic effectiveness of MDM2 inhibitors against multiple myeloma (MM), an incurable B cell neoplasm, in experimental settings. A similar feedback loop was recently described for miR-605, which is also induced by p53 and negatively regulates MDM2 expression [62].

5.2.7 Direct Involvement of p53 in miRNA Processing and Maturation

Since the levels of certain processed miRNAs were increased after p53 activation even in the absence of an induction of the corresponding primary miRNAs (pri-miRNA), the possibility that p53 may directly affect the processing of miRNAs was analyzed [48]. Indeed, these authors found that p53 interacts with the miRNA processing complex DROSHA through association with the DEAD-box RNA helicase p68 (indicated in Fig. 5.2). Thereby, p53 enhances processing of specific pri-miRNAs with growth suppressive function (e.g. miR-16-1, miR-143, miR-145) to precursor miRNAs (pre-miRNAs) resulting in a significant increase in the corresponding miRNAs. Therefore, direct transcriptional regulation of any miRNA-encoding gene by p53 should not be deduced from the detection of an increase in mature miRNA levels by techniques like miR-Seq. Such analysis should be complemented by quantifications of the pri-miRNA levels and detection of p53 occupancy at the promoter of the respective pri-miRNA encoding gene.

Another link between p53 and miRNA-processing has been observed in conditional DICER knock-out mice [63]. DICER deficiency and therefore incomplete miRNA maturation induces p53 and p19/ARF, which leads to reduced proliferation and premature senescence. Interestingly, deletion of *Ink4/Arf* or *p53* prevents premature senescence induced by deletion of DICER. Therefore, a p53-dependent checkpoint seems to monitor proper miRNA processing.

5.2.8 The p53-Relatives p63 and p73 in the Regulation of miRNAs

The p53 family members p63 and p73 have also been implicated in the regulation of miRNA expression and processing. TAp63 was shown to coordinately regulate DICER and miR-130b to suppress metastasis [64]. In contrast to *p53*, the *p63* and *p73* genes are not affected by mutations in tumors. p73 promotes genome stability and mediates chemosensitivity, whereas p63 largely lacks these p53-like functions and instead promotes proliferation and cell survival. p63 and p73 were shown to be connected via miRNA regulations: p63 represses the expression of miR-193-5p, which targets p73, thereby causing an increase in p73 expression, whereas p73 induces miR-193-5p [65]. Interestingly, therapeutic inhibition of miR-193-5p effectively blocked tumor progression when combined with an otherwise ineffective chemotherapy in an orthotopic tumor model.

5.3 Alterations of the p53/miRNA Network in Human Cancer

Similar to protein coding genes miRNA-encoding genes may harbor oncogenic or tumor suppressive functions. As discussed above, p53-induced miRNAs promote tumor suppressive processes, as cell cycle arrest, senescence, inhibition of EMT and metastasis. During cancer initiation or progression cells with inactivation of miRNA-encoding genes may have a selective advantage, since they presumably display a weakened or missing induction of these tumor suppressive mechanisms. In tumors miRNA-encoding genes may be inactivated by a number of different mechanisms. The p53-inducible miRNAs discussed above are likely to be down-regulated in at least half of all tumors due to the mutational inactivation of p53. However, in tumors retaining wild-type p53 the p53-regulated miRNA-encoding genes represent good candidates for being subject to inactivating events. These include loss by deletion or other structural changes as translocations. In addition, down-regulation of miRNA expression by epigenetic silencing via CpG methylation and/or deacetylation of promoter regions has been described. Furthermore, indirect down-regulation due to mutations of other up-stream regulatory transcription factors and alterations in the miRNA processing machinery has been observed. Another mode of inactivation may be due to the aberrant expression of a seed-match containing RNA, a so-called competing endogenous RNA (ceRNA), which sequesters the respective miRNA [66]. This mechanism was originally observed in plant cells [67]. The existence of cancer-relevant ceRNAs in human cells was documented by the identification of RNAs, which regulate expression of the PTEN tumor suppressor via this route [68]. A further possibility of miRNA inactivation was suggested to occur by mutation of seed sequences or altered processing of miRNAs. However, such alterations were only rarely observed until now [69, 70]. Furthermore, an escape from miRNA action by deletion or mutation of seed matching sequences in the respective target mRNA is conceivable. Indeed, such alterations have been observed in mRNAs encoding oncogenic factors [71, 72]. For an overview of reported alterations in the p53/miRNA network detected in cancer see Table 5.1.

5.3.1 Cancer-Specific Alteration of the miR-15/16 Encoding *dLEU2* Gene

The first report of a genetic inactivation of a miRNA was the observation that the *dLEU2* gene, which is located on chromosome 13q14 and encodes the miR-15a and miR-16-1 miRNAs, is commonly deleted in chronic lymphocytic leukemia (CLL)[75]. More recently, it was shown that experimental deletion of *miR-15a/16-1* or of the entire *dLEU2* gene predisposes mice to CLL [123]. Therefore, *dLEU2* is presumably the tumor suppressor gene located in the 13q14 region. Importantly, this study provided the first proof for a *bona fide* tumor suppressor gene function of an miRNA.

5.3.2 Cancer-Specific Alterations of the *miR-34* Family

The *miR-34a* and *miR-34b/c* genes are frequently silenced by CpG methylation in a variety of tumor types [12, 25, 26, 94]. *MiR-34a* methylation was initially shown to occur in numerous cell lines

Table 5.1 Alterations of p53-regulated miRNAs in human tumors

miRNA-gene	Tumor type	Mechanism	Frequency [%]	n=	References
miR-15a/16-1	Prostate cancer	Deletion	80	20+15	[73]
	Chronic Lymphocytic Leukemia/CLL	Germline Mutation in the primary precursor	15	75 cancer (+ control: 160 normal)	[74]
	Chronic Lymphocytic Leukemia/CLL	Deletion	68, 51	60, 322	[75, 76]
	Mantle Cell Lymphoma/MCL	Deletion	55	53	[77]
	Mantle Cell Lymphoma/MCL	Downregulation	71	30	[76]
	Non Small Cell Lung Cancer/NSCLC	Deletion or down-regulated	74	23	[78, 79]
	Pituitary tumors [Cushing's Disease / CD]	n.d.	–	14 (+7controls)	[80]
	Ovarian	Downregulation		38	[81]
	Non-Hodgkin's Lymphoma/NHL	Deletion	43	43	[82]
	Hodgkin's disease/HD	Deletion	29	7	[82]
	Multiple Myeloma/ MM	Downregulation	54	37	[83]
	Pituitary adenoma	Deletion	n.d.	20	[84]
	Pancreatic cancer	Downregulation	70	10	[85]
	Prostate cancer	Downregulation	100	23	[86]
	Prostate cancer cell lines	Downregulation	–	50	[87]
miR-16-2	Progression of prostate carcinogenesis	Downregulation	–	63	[88]
miR-34a	Non Small Cell Lung Cancer/NSCLC	Downregulation	91	23	[79]
	Acute Myeloid Leukemia/AML	Hypermethylation	0	20	[89]
	Non-Hodgkin's Lymphoma/NHL	Hypermethylation	18.8	32	[89]
	Acute Lymphoblastic Leukemia/ALL	Hypermethylation	0	20	[89]
	Chronic Lymphocytic Leukemia/CLL	Hypermethylation	4	50	[89]
	Chronic Myeloid Leukemia/CML	Hypermethylation	0	11	[89]
	Multiple Myeloma/ MM	Hypermethylation	5, 5	55	[89]
	Colorectal cancer cell lines	Hypermethylation, p53 mutation	23	13	[25]
	Prostate cancer	Hypermethylation	79	24	[25]
	Breast cancer cell lines	Hypermethylation	25	24	[25]
	Kidney cancer cell lines	Hypermethylation	21	14	[25]

(continued)

Table 5.1 (continued)

miRNA-gene	Tumor type	Mechanism	Frequency [%]	n=	References
	Bladder cancer cell lines	Hypermethylation	33	6	[25]
	Lung cancer cell lines	Hypermethylation	29	24	[25]
	Melanoma	Hypermethylation	63	32	[25]
miR-34b/c	Gastric cancer	Hypermethylation	70	118	[90]
	Non Small Cell Lung Cancer/NSCLC	Hypermethylation		161	[91]
	Primary melanoma cell lines	Downregulation		2	[92]
	Head and Neck Cancer/H&N	Downregulation	n.d.	10	[93]
	Colon cancer	Hypermethylation	90	111	[94]
	MYC translocation-negative Burkitt Lymphoma	Downregulation	100	5	[95]
miR-34a/b/c	Malignant Pleural Mesothelioma/MPM	Hypermethylation	28(a)/85(b/c)	47	[96]
	Ovarian cancer	p53 Mutation, Hypermethylation	100(a)/72(b/c), 62(a)/69(b/c)	89, 13	[26, 97]
	Colorectal Cancer	Hypermethylation	74(a)/99(b/c)	114	[26]
	Pancreatic cancer	Hypermethylation	64(a)/100(b/c)	11	[26]
	Mammary cancer	Hypermethylation	60(a)/90(b/c)	10	[26]
	Urothelial Cancer/UC	Hypermethylation	71(a)/57(b/c)	7	[26]
	Renal cell cancer	Hypermethylation	58(a)/100(b/c)	12	[26]
	Soft tissue Sarcomas	Hypermethylation	64(a)/45(b/c)	11	[26]
	Esophageal Squamous Cell Carcinoma/ESCC	Hypermethylation	67(a)/41(b/c)	54	[98]
miR-107	Head and Neck/Oral cancer/HNOC	Downregulation		4	[99]
	Acute Promyelocytic Leukemia/APL	Downregulation		26	[50]
	Pancreatic carcinoma cell lines	Hypermethylation		2	[100]
	Tongue squamous cell carcinoma/TSCC	Downregulation	n.d.	4	[99]
	Chronic Lymphocytic Leukemia/CLL	Hypermethylation	n.d.	50	[101]
	Pancreatic cancer	Upregulation		44 (+12 Controls)	[102]
miR-141	Colorectal cancer	Downregulation		10	[103]
	Mesenchymal breast cancer cell lines	Hypermethylation	100	4	[104]
	Epithelial breast cancer cell lines	Hypermethylation	0	4	[104]
	Lung cancer	Downregulation		10	[103]
	Bladder cancer	Hypermethylation		10 (+5 Controls)	[105]
miR-145	Prostate cancer	Hypermethylation, p53 Mutation	81	27	[106]

(continued)

Table 5.1 (continued)

miRNA-gene	Tumor type	Mechanism	Frequency [%]	n=	References
	Prostate cancer	Downregulation		63	[88]
miR-192	Colorectal cancer	p53 Mutation		34	[107]
	Colorectal cancer (MSI)	Downregulation		54 (+20 Controls)	[108]
	Multiple Myeloma/ MM	Hypermethylation		47 (+5 Controls)	[61]
miR-194	Colorectal Cancer with liver metastasis	p53 Mutation, SNP		30	[109]
	Multiple Myeloma/ MM	Hypermethylation		47 (+5 Controls)	[61]
miR-200a	Ovarian cancer	Downregulation		55	[110]
miR-200b	Colorectal cancer	Loss		30	[109]
	Ovarian cancer	Downregulation		55	[110]
	Lung cancer	Hypermethylation	25	24 (+ Controls)	[111]
	Bladder cancer	Hypermethylation		10 (+5 Controls)	[105]
miR-200c	Lung cancer	Hypermethylation	29	24 (+ Controls)	[111]
	Colorectal cancer	Hypermethylation, p53 mutation			[112]
	Mesenchymal breast cancer cell lines	Hypermethylation	100	4	[104]
	Epithelial breast cancer cell lines	Hypermethylation	0	4	[104]
	Lung cancer	Hypermethylation	25, -	24, 69	[111, 113]
	Bladder cancer	Hypermethylation		10 (+5 Controls)	[105]
miR-215	Colorectal cancer	Downregulation		34	[107]
miR-429	Colorectal cancer cell lines	Hypermethylation	50	2	[114]
	Breast cancer cell lines	Hypermethylation	50	2	[114]
	Lung cancer cell lines	Hypermethylation	33	3	[114]
	Ovarian cancer	Downregulation		55	[110]
Exportin 5	Breast cancer	Mutation		441 (+479 Controls)	[115]
	Hereditary nonpolypo-sis colon cancer	Downregulation	26	38	[116]
	Sporadic colon cancer (MSI+)	Downregulation	22	211	[116]
	Sporadic gastric cancer (MSI+)	Downregulation	28	58	[116]
	Sporadic endometrial tumors (MSI+)	Downregulation	13	30	[116]
Dicer/Drosha	Ovarian cancer	Downregulation	60/51	111	[117]
Dicer	Cystic nephroma, Wilm's tumor	Germline mutation	0	50	[118]
	Pulmonary pediatric cancer	Mutation	91	11 (+360 Controls)	[119]

(continued)

Table 5.1 (continued)

miRNA-gene	Tumor type	Mechanism	Frequency [%]	n=	References
Dicer	Lung cancer	Downregulation	67		[120]
	Colorectal cancer	Upregulation		237	[121]
	Acute Myeloid Leukemia/AML	Upregulation	86	71	[122]

Summary of the reported alterations in p53-regulated miRNAs in cancer. *Frequency* relates to the alteration indicated in the third column, *n* number of tumor samples/patients analyzed, *MSI* micro-satellite instable.
n.d. = not determind.

derived from different tumor types, as well as in primary prostate cancer and melanoma [25]. Also the expression of the miR-34 family members miR-34b and miR-34c, which are encoded by a common transcript, is down-regulated in many types of cancer [26]. A high frequency of silencing of the *miR-34b/c* promoter by CpG methylation has been found in colorectal cancer cell lines and colorectal tumor samples [94]. We also found CpG methylation of *miR-34b/c* in all 114 cases of primary colorectal cancers analyzed [26]. Interestingly, *miR-34b/c* methylation correlated with metastasis and poor survival for several types of cancer [124]. The reintroduction of *miR-34b/c* into cancer cell lines exhibiting *miR-34b/c* silencing inhibited their motility, reduced tumor growth, suppressed metastasis formation in a xenograft model and was associated with down-regulation of the respective target genes (e.g. c-MYC, E2F3, CDK6).

The *miR-34a* gene is located on chromosome 1p36, a region which is commonly deleted in human cancers, as for example in neuroblastoma [125], which often display loss of *miR-34a* expression [126].

5.3.3 Cancer-Specific Alterations of the *miR-200* Family

The miR-200 family encodes a highly conserved group of miRNAs, which control EMT by down-regulating the EMT-inducing transcription factors ZEB1 and ZEB2 [39]. The miR-200 family can be sub-divided into two clusters: miR-200c and miR-141 (located at chromosome 12p13), and miR-200a, miR-200b and miR-429 (located at chromosome 1p36). Expression of the miR-200c/141 cluster is frequently silenced by CpG methylation in breast cancer [104]. Interestingly, a correlation between methylation of the miR-200c promoter and invasiveness was determined in breast cancer cell lines. Down-regulation of the miR-200c/141 cluster was also described for breast cancer initiating cells [127] and EBV-associated gastric carcinomas [128]. As mentioned above, loss of 1p36 is a recurrent aberration especially in neuroblastoma, indicating that there may be two distinct mechanisms that down-regulate the expression of the miR-200 family.

5.3.4 Cancer-Specific Alterations of the *miR-192* Family

The p53-regulated miR-192 family is comprised of miR-192, miR-194-2, and miR-215, which induce p21 expression and cell cycle arrest in a p53-dependent manner [41]. The miR-192 family is down-regulated by an unknown mechanism in multiple myeloma (MM), which rarely shows mutation or deletion of p53 [61]. Reactivation of p53 in MM resulted in re-expression of miR-192, miR-194-2, and miR-215 and down-regulation of MDM2, which represents a target of these miRNAs [61]. Moreover, ectopic expression of miR-192 family members inhibited cell growth, migration and invasion of MM. Furthermore, the miR-192 family members are down-regulated in colon cancer, and induce apoptosis and senescence, although to a lesser extent than miR-34a [41]. The mechanism by which down-regulation of the miR-192 family occurs remained unclear in this study, but p53 inactivation [129] and a single nucleotide polymorphism (SNP) located within the miR-194-2 precursor [130] may contribute to this phenomenon.

5.3.5 Other p53-Induced miRNAs Inactivated in Cancer

Recently, the p53-inducible miR-145 was shown to be down-regulated by CpG methylation and p53 mutation in prostate cancer samples and cell lines [106].

5.3.6 Alterated Regulation of the miRNA Processing Machinery in Cancer

miR-107 was shown to directly target DICER1 mRNA, which encodes a central component of the miRNA processing machinery [131]. Ectopic expression of miR-107 enhances migration *in vitro* and allows metastatic dissemination of otherwise non-aggressive cells *in vivo*, whereas the loss of miR-107 opposes migration and metastasis of malignant cells. Moreover, it was shown that high levels of miR-107 are associated with metastasis and poor outcome in breast cancer. However, these observations are not compatible with mediation of p53-induced tumor suppression by miR-107. Nonetheless, these findings suggest that the deregulation of the miRNA processing machinery in cancer leads to metastasis and poor outcome, and predicts an anti-cancer activity of the majority of the miRNAs. In support of this conclusion, *DICER1* was characterized as an haplo-insufficient tumor suppressor gene in a tumor mouse model [132]. Furthermore, decreased expression of DICER1 correlates with poor prognosis in human lung cancer [120]. Interestingly, the p53 family member p63 transcriptionally controls DICER1 expression. Mutant p53 presumably interferes with this regulation, which leads to a reduction in DICER1 levels and reduces the levels of certain cancer-relevant miRNAs [64]. Mutant p53 may also interfere with the post-translational regulation of DROSHA by wild-type p53 and thereby affect the processing of selected, tumor suppressive miRNAs [48].

5.3.7 Mutations in the miRNA Processing Machinery in Cancer

Another possibility how the abundance of p53-regulated miRNAs could be altered in cancer is to constitutively change the processing of pri-miRNAs to miRNAs by genetic alterations in components of this pathway. For example, mutations of the nuclear export protein Exportin-5 resulted in the trapping of pre-miRNAs in the nucleus and reduced miRNA-processing [116]. As a result, numerous miRNAs were not fully processed and a diminished inhibition of the respective miRNA targets was detected. Notably, restoration of Exportin-5 function reversed the impaired export of pre-miRNA and had tumor-suppressive effects. Recently, several studies supported the notion that variations in the expression and mutations of miRNA processing components as Exportin-5 and DICER1 affect the outcome of breast [115], ovarian [117], cystic nephroma [118] and pediatric pulmonary cancer [119].

5.4 Approaches to Study p53-Regulated miRNAs and Their Targets

Although, numerous connections between p53 and miRNAs have been identified, the examples described above also illustrate that we have only begun to understand the role of miRNAs in tumor suppression mediated by p53. Therefore, additional efforts are necessary to obtain more details of the p53/miRNA network. A feasible strategy for a comprehensive, genome-wide identification of p53-regulated miRNAs and their associated target genes is the combination of the approaches depicted in Fig. 5.5. This strategy may in principle also apply to other transcription factors of interest besides p53. These analyses generate a large amount of bioinformatics data, which can be processed with the help of the algorithms indicated

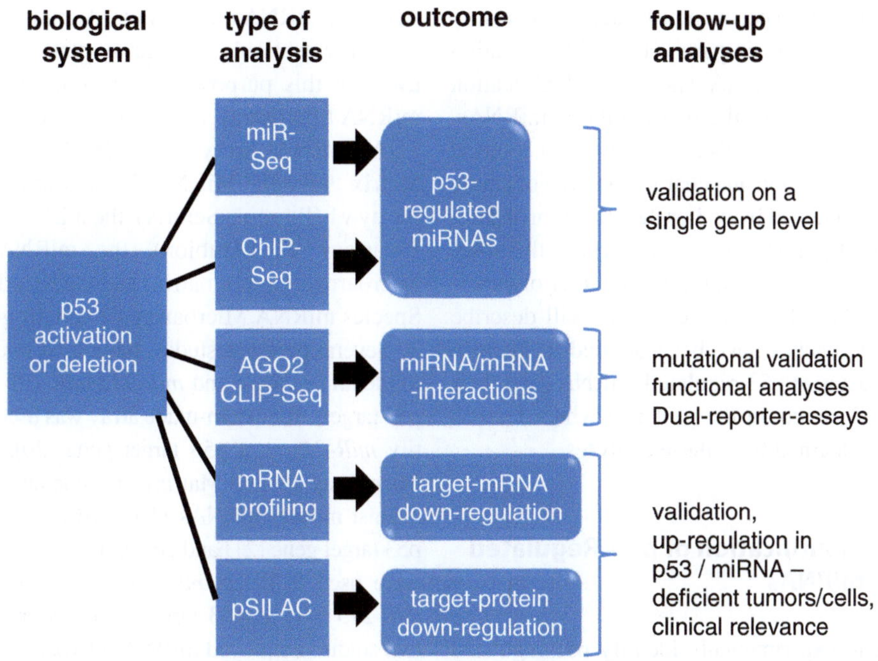

Fig. 5.5 **Analysis of p53-regulated miRNAs and their targets.** Summary of experimental approaches for the comprehensive identification and characterization of p53-regulated miRNAs. The approaches are described in detail in the text

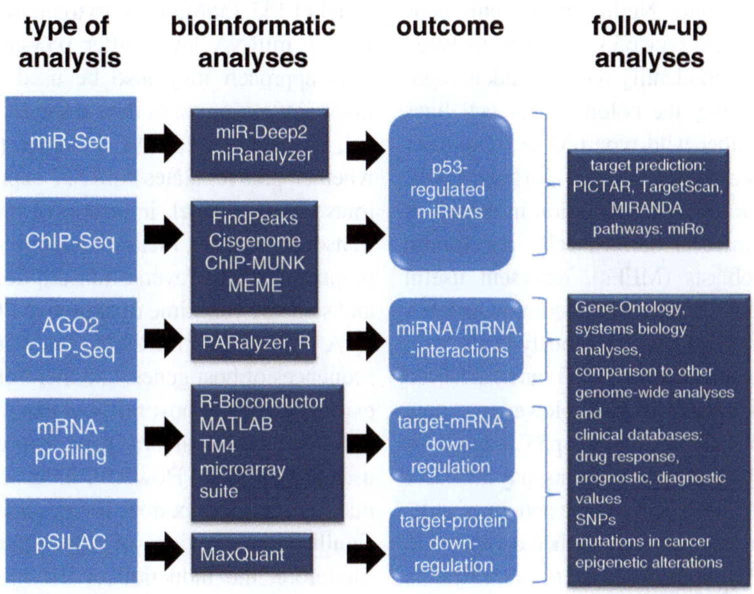

Fig. 5.6 **Bioinformatics characterization of p53-regulated miRNAs and their targets.** Summary of bioinformatics approaches for the comprehensive characterization of p53-regulated miRNAs. As indicated, the programs and websites Bioconductor [133], Cisgenome [134], ChIP-Munk [135], FindPeaks [136], MaxQuant [137], Meme [138], MirDeep2 [139], miRanalyzer [140], miRo [141], PARalyzer [142] and TM4 microarray suite [143] facilitate the analyses of data obtained by the experimental analyses described in the main text and in Fig. 5.5

in Fig. 5.6. The experimental strategy can be subdivided into two main parts: (1) the identification of p53-regulated miRNAs and (2) the identification of target mRNAs of the p53-regulated miRNAs. So far the studies in this area have rather focused on the identification and characterization of single miRNAs regulated by p53 or they have carried out one type of genome-wide approach, with subsequent confirmation of a limited number of candidates. In the following section we will describe which approaches have been applied to identify and characterize p53-regulated miRNAs and their associated targets in the past and which lessons have been learned from these analyses.

5.4.1 Identification of p53-Regulated miRNAs

In order to experimentally identify p53-regulated miRNAs cellular systems in which p53 activity can be turned on using conditional systems or pharmacological p53 activators should be employed. Endogenous p53 can either be activated by addition of DNA damaging substances or by p53-activators as the MDM2 inhibitor Nutlin-3a. Isogenic cells with and without wild-type p53 should be analyzed in parallel in order to identify p53-dependent regulations. For example, the colon cancer cell lines HCT-116 with either wild-type p53 expression or p53-knockout are useful for this purpose [144]. Alternatively, the miRNA expression in tissues of p53 knock-out mice or derived cells, e.g. mouse embryonic fibroblasts (MEFs), represent useful systems in order to identify p53-mediated miRNA regulations, as documented previously [18].

A highly specific activation of p53 can be achieved using ectopic expression of p53. However, certain post-translational modifications of p53 induced by treatment with DNA-damaging agents may not occur hereby. Therefore, differences in the pattern of miRNAs regulated by p53 may occur when compared to activation of p53 by stressors as oncogene activation and DNA damaging agents. In the past, we have used an episomal, doxycyclin-inducible expression system to re-express p53 in p53-deficient H1299 lung cancer cells [16].

Differential expression of miRNAs upon p53 activation can be monitored using specifically designed miRNA microarrays. A number of commercially available microarray platforms can be used for this purpose: for example the Human miRNA Microarray 1.0 (Agilent), the miRCURY LNA miRNA Array v9.2 (Exiqon), the Array Matrix 96-well MiRNA Expression Profiling Assay v1 (Illumina Sentrix), the mirVana miRNA Bioarrays v2 (Ambion), the miRNA 4X2K Microarray (Combimatrix) and the NCode Multi-Species miRNA Microarray v2 (Invitrogen).

Several previous studies have used microarrays to identify *miR-34* and *miR-215/miR-192* as direct p53 targets. A custom-made array was used to identify *miR-34*a as a p53 target gene [20], a 4X2K Microarray (CombiMatrix) that contained probes against mouse miRNAs identified *miR-34b/c* as a p53 target gene [21] and customized miRNA arrays were used to detect *miR-34a* [17] and *miR-192/miR-215* [41] as p53 target genes. More recently, two studies employed miRNA microarrays to identify members of the *miR-200* family as p53 targets [37, 38].

In addition, induction of mature miRNAs after p53 activation can be measured by stem-loop RT-qPCR assays. Hannon and colleagues used a panel of 145 TaqMan assays to monitor changes in mature miRNA levels after p53-activation [18]. This approach may also be used to verify the microarray expression data at the level of individual, processed miRNAs. In order to determine, whether p53 regulates miRNA expression at the transcriptional level, induction of the pri-miRNA transcript can be measured using total mRNA preparations after reverse transcription into cDNAs and standard real-time quantitative PCR (qPCR).

A subset of miRNAs lie within intronic sequences of host genes, and therefore differential expression of the host mRNAs can in principle be monitored by standard gene expression arrays used for mRNAs. However, induction of the primary host transcript does not necessarily lead to a significant induction of the mature miRNA. Therefore, the induction of the mature miRNA should be validated by stem-loop RT-qPCR assays. The above mentioned methods have in common that they only detect previously known miRNAs.

For the unbiased detection of all miRNA expressed in a certain state several Next Generation Sequencing (NGS) based approaches are currently being used. Small RNAs are isolated, ligated to adapters, reverse transcribed and amplified to generate libraries, which may be analysed using different NGS platforms, e.g. Solexa-sequencing (Illumina), 454-sequencing (Roche) or the SOLID system (Applied Biosystems). The adapters often contain distinct bar-codes, which allow multiplexing of several samples in one sequencing run generating up to several hundred million reads. The coverage which can be achieved by these analyses is presumably close to complete. In 2007 we reported a 454-sequencing approach to identify *miR-34a* as direct p53 target [16]. Although only ~200,000 sequencing reads per run were reached at that time, these were sufficient to identify many of the miRNAs displaying the most pronounced regulation by p53.

Since p53 may enhance the synthesis of miRNAs via directly influencing pre-miRNA processing the detection of differential expression of the mature miRNA is not sufficient to deduce a direct transcriptional regulation of the corresponding pri-miRNA by p53 [48]. Therefore, it is advantageous to obtain both miRNA and pri-miRNA profiles simultaneously in order to distinguish transcriptional from other modes of miRNA abundance regulation by p53.

5.4.2 Confirmation of Direct Regulation by p53 Using ChIP Approaches

The detection of p53 occupancy at the respective promoters of the genes encoding p53-regulated pri-miRNAs or other pre-cursor mRNAs can be achieved by chromatin-immunoprecipitation (ChIP) based techniques. These can either be performed on a gene-by-gene basis using qPCR-ChIP or on a genome-wide level by coupling ChIP with techniques as NGS, SAGE or hybridization to a promoter array. The disadvantage of the latter method is the limitation to previously characterized promoters.

The consensus sequence necessary for p53 binding consists of two copies of the RRRCWWGYYY motif separated by a small spacer of 0–21 nucleotides (R = pyrimidine; Y = purine; W = A/T; see also Fig. 5.1). However, among the validated p53 response elements identified in p53 target gene promoters, the majority displays slight deviations from the consensus sequence, indicating a certain flexibility in p53's binding requirements. Based on the consensus motif, potential p53 binding sites can be predicted using a variety of search algorithms. For example, the p53MH algorithm [145] and the MatInspector software (Genomatix) have been applied to identify p53-binding sites in the promoters of miRNA-encoding genes. The P53MH algorithm was used to identify a p53-binding site in the *miR-34b/c* promoter [21] and in the *miR-194-1/miR-215* cluster [41], whereas the two p53-binding sites in the *miR-145* promoter were identified using the MatInspector software [45].

Initially, binding of p53 to the predicted binding site was experimentally tested *in vitro* by gel shift assays. Furthermore, in order to test the requirement of the p53 response element, the promoter region encompassing the p53 binding site or its mutant version can be placed upstream of a *luciferase* ORF or an equivalent reporter gene. The responsiveness of these constructs to p53 can then be interrogated by co-transfection with p53-encoding plasmids into mammalian cells and a subsequent reporter assay. In order to test whether p53 binds to the predicted binding site in a native chromatin environment *in vivo*, chromatin immunoprecipitation (ChIP) assays have to be performed. This can either be done on a single gene basis by ChIP followed by semi-quantitative PCR or qPCR. Alternatively, p53 binding sites can also be identified on a genome-wide scale. In the initial genome-wide binding studies, immunoprecipitated DNA from the ChIP experiment was hybridized to high-density oligonucleotide tiling arrays (ChIP-on-Chip). For example, a ChIP-on-Chip approach was used to map p53 binding sites on human chromosomes 21 and 22 and identified 48 high confidence sites [146]. These results suggested the existence of ~1,600 putative p53 sites in the human genome. Indeed, when the same approach was applied to the complete genome 1,546 p53-binding sites were identified in actinomycin D treated U2OS cells [147].

The ChIP-PET method is an extension of the ChIP-on-Chip approach and is related to SAGE [148]. Short tags derived from immunoprecipitated

DNA fragments are converted into a DNA library. After further ligations the paired end ditags form concatemeres, which are subjected to capillary sequencing. The obtained tag-sequences are subsequently mapped to the genome and quantified. The ChIP-PET method was used to monitor p53 binding across the whole genome and identified more than 500 high-confidence p53 binding sites [149]. This resource was used by other laboratories to identify p53 binding sites in the *miR-34a* and *miR-34b/c* promoters [19, 20].

The methods mentioned above are currently replaced by a combination of ChIP and NGS (ChIP-Seq). Since the new sequencing devices achieve several hundred millions reads in one run it is possible to multiplex several time-points and experimental replicas in one single sequencing run. The identification of occupied p53-binding sites in the genome may be combined with detection of histone modifications indicating active transcription units and enhancers. This allows the assignment of orphan miRNAs derived from pri-miRNA transcribed from active promoters present in the vicinity, which have not been characterized before. Furthermore, the results obtained using the expression studies described above have to be compared to the DNA binding patterns of p53 in a genome-wide manner using bioinformatics approaches (see also Fig. 5.6).

5.4.3 Identification of miRNA Targets

After obtaining a set of p53-regulated miRNAs, the next step is to identify the physiologically relevant target mRNAs of these miRNAs. We suggest the systematic identification of miRNA-regulated target genes following p53 induction by an integrated approach that involves

(A) Identification and mapping of miRNA binding sites using biochemical techniques involving RISC isolation.
(B) Testing the functionality of these binding sites in the regulation of their respective target mRNAs using either microarrays or NGS as well as dual reporter assays.
(C) Proteomic approaches to measure changes in target abundance on the protein level indicating translational regulation in cases without decrease in the corresponding mRNA.

Similarly to the identification of p53-induced miRNAs described above, these approaches ideally should be performed in parallel as they complement each other. The identification and mapping of miRNA binding sites on mRNAs provides information as to whether a miRNA directly binds to its cognate target mRNA, but does not provide information about the regulation of the bound mRNA. Conversely, microarray and proteomic approaches provide information on the regulation of a given mRNA or protein, but do not *per se* distinguish between direct and indirect targets. Therefore, a combined approach that maps binding sites of p53-regulated miRNAs on mRNAs and validates the functionality of these binding sites regarding target regulation may comprehensively uncover the network of protein expression that is regulated by p53-induced miRNAs.

MiRNAs typically regulate their targets via association of a ~7 nucleotide stretch, the so-called seed-sequence, located in their 5′-portion with a complementary sequence in the 3′-UTR of the target mRNA. Additional base pairing may occur via nucleotides in the middle and 3′-portion of the miRNA. Since miRNAs only pair imperfectly with their respective target mRNAs, the number of theoretically possible targets is typically large and presumably most of the predicted targets are not significantly regulated by the respective miRNA. Several bioinformatics algorithms have been developed to predict miRNA targets with the intention to reduce the rate of false positive predictions by incorporating features as conservation between species. However, even these algorithms often predict hundreds of target mRNAs for a particular miRNA, most of which are presumably false positive hits.

Due to differences in the parameters used to weigh individual features involved in miRNA/mRNA interaction, different target prediction algorithms often result in only partially overlapping sets of predicted target genes. The algorithms TargetScan and Pictar [150, 151] place more weight on perfect, evolutionarily conserved seed matches, whereas PITA and RNA22 [152, 153] prioritize the ΔG of the miRNA/mRNA duplex and the accessibility of the site within the mRNA. Although algorithms like TargetScan and Pictar have been shown to have high predictive power when tested on experimentally obtained proteomic data [154–156], they may be less useful in the prediction of miRNA target sites that lack a perfect seed-sequence, are not evolutionarily conserved, or lie outside the 3′-UTR of the target gene. Therefore, the combined use of several different algorithms may be helpful to identify target mRNAs of a given miRNA.

The sets of predicted target mRNAs generated by different algorithms are typically being used to filter sets of differentially regulated genes that were identified by experimental perturbation of miRNA function, followed by unbiased genome- or proteome-wide measurements of changes in mRNA or protein abundance. As outlined in Fig. 5.5, miRNA binding sites can be mapped by isolation of miRNA target mRNAs via the association of RISC/miRNA/mRNA-complexes. This is typically accomplished by immunoprecipitation of RISC components such as Ago2, which can either be done via endogenous proteins or ectopically expressed epitope-tagged versions of the respective proteins [157–159]. The RISC/mRNA/miRNA complexes are precipitated and the associated mRNAs are identified either by hybridization to microarrays or by NGS technologies. However, this method does not directly lead to the identification of the actual miRNA binding site, since all RISC-bound mRNAs containing different miRNAs and their targets are immunoprecipitated and sequenced.

An improved version of these initial approaches is high-throughput sequencing of RNAs isolated by crosslinking and immunoprecipitation (HITS-CLIP) [160]: miRNA-bound RNAs are cross-linked to RISC by UV irradiation. The RISC/miRNA/mRNA complex is then immunoprecipitated with antibodies against RISC components such as Ago2. A RNAse-digest eliminates all mRNA fragments not protected by the RISC/miRNA complex. All miRNA seed-matching regions occupied by miRNA/RISC complexes are determined by NGS. Thereby, information is obtained not only regarding the bound mRNA target but also concerning the miRNA matching sequence, which allows to deduce the putative identity of the miRNAs. In the case of p53-induced miRNAs these miRNAs should be among those which are detected at increased levels after p53 activation.

In another version of an AGO2-IP based approach, named photoactivatable-ribonucleoside-enhanced crosslinking and immunoprecipitation (PAR-CLIP), cells are cultured with photo-reactive 4-thiouridine before UV-cross-linking [161]. 4-thiouridine is incorporated into the cellular RNA during transcription and leads to improved protein/mRNA cross-linking efficiencies. Since 4-thiouridine results in C-to-T transitions in the regions previously protected by AGO2/RISC complexes during reverse transcription, it allows to map the position of miRNA/RISC binding on the mRNA.

However, none of these approaches have been specifically applied to identify mRNA targets of p53-induced miRNAs yet. Furthermore, as all these approaches essentially rely on the isolation of the RISC complex, all miRNAs and their bound mRNA targets associated with RISC will be identified. Therefore, identification of mRNA targets of a particular miRNA from the obtained NGS data largely depends on the subsequent extraction of sequence features associated with that particular miRNA, i.e. either the presence of a hexameric seed sequence or the presence of other sequence features predicted to be targeted by miRNAs by algorithms, such as PITA or RNA22. A more direct, alternative approach involves the use of biotinylated miRNAs, which can be purified together with RISC in a tandem affinity purification approach [162, 163]. However, this approach may have limitations as the high concentrations of biotinylated miRNAs reached after transfection may result in false positive results.

As explained above, information on the miRNA binding site does not automatically mean that this particular binding site is physiologically relevant for target regulation. Therefore, miRNA-induced changes in either mRNA or protein abundance have to be confirmed by perturbation of miRNA expression. Experimental studies to identify target mRNAs of p53-regulated miRNAs should involve ectopic expression of miRNAs either by transfection of synthetic pre-miRNA molecules or inducible expression of pri-miRNA transcripts [16, 19, 23]. Furthermore, synthetic miRNA inhibitors (antagomirs) can be used to block miRNA function. Alternatively, and more elegantly, knock-out cell lines for individual miRNAs can be used to address this question. In addition, HCT116 DICER^{ex5}, a human colorectal cancer cell line harboring a hypomorphic *DICER* allele [164], has been used to validate the regulation of targets of p53-regulated miRNAs [18, 42].

A number of studies applied microarrays to identify targets of p53-induced miRNAs. For example, in the case of miR-34 [17–19, 165] and miR-215/miR-192 [42] mRNA expression profiles were generated after ectopic expression of the respective miRNA. However, mRNA-profiling based approaches are limited as they cannot detect miRNA targets that are solely regulated at the level of translational repression. On the other hand, assuming that miRNAs in most cases only cause modest decreases in protein translation, the miRNA-mediated regulation of proteins with long half-lives may not be detected by measuring steady-state protein levels using standard proteomic quantification as SILAC (***s****table* ***i****sotope* ***l****abeling by* ***a****mino acids in* ***c****ell culture*) [166]. This problem was solved by the introduction of pSILAC (***p****ulsed SILAC*), which facilitated the quantification of differences in protein translation rates caused by miRNAs [156]. With this approach, induction of miRNA expression is followed by a pulse of isotope-labeled amino acids which are incorporated into newly synthesized proteins. Subsequent mass spectrometric analysis of the proteome therefore allows to detect changes in protein translation rates caused by miRNA expression. In a recent study we applied this approach to identify target genes of the miR-34a miRNA [23]. Notably, numerous of the identified miR-34a targets were confirmed in an miRNA capture approach using biotinylated miR-34a as a bait [24]. Other quantitative proteomic methods like isotope-coded affinity tag (ICAT)-labeling following transfection with miR-34a have been used to identify miRNA targets [167]. One major drawback of all proteomic methods is their still limited ability to cover the entire proteome of the cell, as well as their strong bias for highly expressed proteins.

All transcriptome- or proteome-wide approaches to identify miRNA targets require subsequent validations such as qPCR or Western blot analyses to verify that a given mRNA or protein is indeed regulated following miRNA induction.

Direct regulation by a miRNA is often determined in dual-reporter assays. For this the 3′-UTR of the putative target mRNA is placed downstream of a *firefly* luciferase reporter gene. This reporter-construct is co-transfected either with miRNA mimics or miRNA inhibitors, and a *Renilla* luciferase vector for standardization. In case of specific, direct regulation the 3′-UTR reporter is repressed by ~20–80 %. In order to map and validate the seed-matching sequences these should be mutated in the context of the 3′-UTR sequence. The resulting constructs should ideally show resistance towards the respective miRNAs.

5.4.4 Follow-Up Analysis

Once p53-mediated regulation of miRNAs and their targets have been confirmed numerous additional analyses are possible to interrogate the physiological and pathophysiological relevance of the identified regulations. Co-expression of the p53-induced miRNA and a miRNA-resistant target mRNA can be used in rescue-experiments to determine the relevance of the respective down-regulation for cell biological phenotypes, as cell cycle arrest and/or apoptosis. Furthermore, the relevance of the respective miRNAs for p53-mediated effects can be tested using antagomirs specific for the respective miRNA. Finally, the importance of miRNA/target regulations for p53-mediated tumor suppression can be tested in miRNA knock-out mice in combination with tumor mouse models. However, these studies may take years. A recently published collection of ES cell lines with deletion of 392 miRNAs was generated to facilitate the rapid generation of knock-out mice and may therefore accelerate this type of analysis [168]. Furthermore, the inactivation of the respective miRNA encoding genes by CpG methylation or mutations in different tumor types may be analyzed. The miRNA inactivation can be correlated with the putative up-regulation of miRNA targets in the affected tumor samples and pathological features of the affected tumors. Detection of CpG-methylation and miRNA/target expression may also have prognostic and diagnostic value for cancer patients in the future.

5.4.5 Outlook

In the future technological developments may result in an increased sensitivity of mass-spectral analyses which could facilitate similar coverage rates of proteomic quantifications as are now reached by DNA-sequencing/hybridization based approaches. Furthermore, the integration of different bioinformatics platforms into a common program for mRNA/miRNA/DNA binding and protein quantification would make integrated analyses less complicated and laborious. Another useful tool would be a comprehensive ontology-like database for miRNA functions and targets. The miRo website is an example of such a tool [141]. In the future, more publicly available datasets of miRNA expression in cancer patient cohorts which allow to determine correlations with mutations, epigenetic changes and clinical data will become available. Taken together, these possibilities will hopefully lead to the rapid translation of knowledge derived from analysis of the p53/miRNA network into diagnostic and therapeutic applications.

Acknowledgements We thank the members of the Hermeking Lab for discussions, and Ralf Zimmer and Florian Erhard for advice concerning Fig. 5.6. Work in Heiko Hermeking's lab is supported by the German Israel Science Foundation (GIF), the Rudolf-Bartling-Stiftung, the Deutsche Krebshilfe and the Deutsche Forschungsgemeinschaft (DFG).

References

1. Soussi T (2011) TP53 mutations in human cancer: database reassessment and prospects for the next decade. Adv Cancer Res 110:107–139
2. Cheok CF et al (2011) Translating p53 into the clinic. Nat Rev Clin Oncol 8(1):25–37
3. Green DR, Kroemer G (2009) Cytoplasmic functions of the tumour suppressor p53. Nature 458(7242):1127–1130
4. Kruse JP, Gu W (2009) Modes of p53 regulation. Cell 137(4):609–622
5. Derheimer FA, Kastan MB (2010) Multiple roles of ATM in monitoring and maintaining DNA integrity. FEBS Lett 584(17):3675–3681
6. Menendez D, Inga A, Resnick MA (2009) The expanding universe of p53 targets. Nat Rev Cancer 9(10):724–737
7. Leung AK, Sharp PA (2007) microRNAs: a safeguard against turmoil? Cell 130(4):581–585
8. Leung AK, Sharp PA (2010) MicroRNA functions in stress responses. Mol Cell 40(2):205–215
9. Vogelstein B, Lane D, Levine AJ (2000) Surfing the p53 network. Nature 408(6810):307–310
10. Hermeking H (2012) MicroRNAs in the p53 network: micromanagement of tumor suppression. Not Rev Cancer 12(9):613–626
11. Hermeking H (2003) The 14-3-3 cancer connection. Nat Rev Cancer 3(12):931–943
12. Hermeking H (2010) The miR-34 family in cancer and apoptosis. Cell Death Differ 17(2):193–199
13. Vousden KH, Ryan KM (2009) p53 and metabolism. Nat Rev Cancer 9(10):691–700
14. Vousden KH, Prives C (2009) Blinded by the light: the growing complexity of p53. Cell 137(3):413–431
15. Riley T et al (2008) Transcriptional control of human p53-regulated genes. Nat Rev Mol Cell Biol 9(5):402–412
16. Tarasov V et al (2007) Differential regulation of microRNAs by p53 revealed by massively parallel sequencing: miR-34a is a p53 target that induces apoptosis and G1-arrest. Cell Cycle 6(13):1586–1593
17. Chang TC et al (2007) Transactivation of miR-34a by p53 broadly influences gene expression and promotes apoptosis. Mol Cell 26(5):745–752
18. He L et al (2007) A microRNA component of the p53 tumour suppressor network. Nature 447(7148): 1130–1134
19. Bommer GT et al (2007) p53-mediated activation of miRNA34 candidate tumor-suppressor genes. Curr Biol 17(15):1298–1307
20. Raver-Shapira N et al (2007) Transcriptional activation of miR-34a contributes to p53-mediated apoptosis. Mol Cell 26(5):731–743
21. Corney DC et al (2007) MicroRNA-34b and MicroRNA-34c are targets of p53 and cooperate in control of cell proliferation and adhesion-independent growth. Cancer Res 67(18):8433–8438

22. He X, He L, Hannon GJ (2007) The guardian's little helper: microRNAs in the p53 tumor suppressor network. Cancer Res 67(23):11099–11101
23. Kaller M (2011) Genome-wide characterization of miR-34a induced changes in protein and mRNA expression by a combined pulsed SILAC and microarray analysis. Mol Cell Proteomics 10(8):M111 010462
24. Lal A et al (2011) Capture of microRNA-bound mRNAs identifies the tumor suppressor miR-34a as a regulator of growth factor signaling. PLoS Genet 7(11):e1002363
25. Lodygin D et al (2008) Inactivation of miR-34a by aberrant CpG methylation in multiple types of cancer. Cell Cycle 7(16):2591–2600
26. Vogt M et al (2011) Frequent concomitant inactivation of miR-34a and miR-34b/c by CpG methylation in colorectal, pancreatic, mammary, ovarian, urothelial, and renal cell carcinomas and soft tissue sarcomas. Virchows Arch 458(3):313–322
27. Hwang CI et al (2011) Wild-type p53 controls cell motility and invasion by dual regulation of MET expression. Proc Natl Acad Sci U S A 108(34): 14240–14245
28. Schubert J, Brabletz T (2011) p53 spreads out further: suppression of EMT and stemness by activating miR-200c expression. Cell Res 21(5):705–707
29. Siemens H (2011) miR-34 and SNAIL form a double-negative feedback loop to regulate epithelial-mesenchymal transitions. Cell Cycle 10(24):4256–4271
30. Kim NH et al (2011) A p53/miRNA-34 axis regulates Snail1-dependent cancer cell epithelial-mesenchymal transition. J Cell Biol 195(3):417–433
31. Brabletz T (2012) MiR-34 and SNAIL: another double-negative feedback loop controlling cellular plasticity/EMT governed by p53. Cell Cycle 11(2):215
32. Liu C et al (2011) The microRNA miR-34a inhibits prostate cancer stem cells and metastasis by directly repressing CD44. Nat Med 17(2):211–215
33. Choi YJ (2011) miR-34 miRNAs provide a barrier for somatic cell reprogramming. Nat Cell Biol 13(11):1353–1360
34. Kim NH (2011) p53 and microRNA-34 are suppressors of canonical Wnt signaling. Sci Signal 4(197):ra71
35. Lize M, Klimke A, Dobbelstein M (2011) MicroRNA-449 in cell fate determination. Cell Cycle 10(17):2874–2882
36. Chang CJ et al (2011) p53 regulates epithelial-mesenchymal transition and stem cell properties through modulating miRNAs. Nat Cell Biol 13(3):317–323
37. Kim T et al (2011) p53 regulates epithelial-mesenchymal transition through microRNAs targeting ZEB1 and ZEB2. J Exp Med 208(5):875–883
38. Gregory PA et al (2008) The miR-200 family and miR-205 regulate epithelial to mesenchymal transition by targeting ZEB1 and SIP1. Nat Cell Biol 10(5):593–601
39. Keck T, Brabletz T (2011) Under stress: p53 controls EMT and stemness in pancreatic epithelial cells. Cell Cycle 10(11):1715
40. Braun CJ et al (2008) p53-responsive microRNAs 192 and 215 are capable of inducing cell cycle arrest. Cancer Res 68(24):10094–10104
41. Georges SA et al (2008) Coordinated regulation of cell cycle transcripts by p53-inducible microRNAs, miR-192 and miR-215. Cancer Res 68(24):10105–10112
42. Yamakuchi M et al (2010) P53-induced microRNA-107 inhibits HIF-1 and tumor angiogenesis. Proc Natl Acad Sci U S A 107(14):6334–6339
43. Bohlig L, Friedrich M, Engeland K (2011) p53 activates the PANK1/miRNA-107 gene leading to downregulation of CDK6 and p130 cell cycle proteins. Nucleic Acids Res 39(2):440–453
44. Sachdeva M et al (2009) p53 represses c-Myc through induction of the tumor suppressor miR-145. Proc Natl Acad Sci U S A 106(9):3207–3212
45. Xu N et al (2009) MicroRNA-145 regulates OCT4, SOX2, and KLF4 and represses pluripotency in human embryonic stem cells. Cell 137(4):647–658
46. Krizhanovsky V, Lowe SW (2009) Stem cells: the promises and perils of p53. Nature 460(7259):1085–1086
47. Suzuki HI et al (2009) Modulation of microRNA processing by p53. Nature 460(7254):529–533
48. Fabbri M et al (2011) Association of a microRNA/TP53 feedback circuitry with pathogenesis and outcome of B-cell chronic lymphocytic leukemia. JAMA 305(1):59–67
49. Careccia S et al (2009) A restricted signature of miRNAs distinguishes APL blasts from normal promyelocytes. Oncogene 28(45):4034–4040
50. Nishida N et al (2011) MicroRNA miR-125b is a prognostic marker in human colorectal cancer. Int J Oncol 38(5):1437–1443
51. Hu W et al (2010) Negative regulation of tumor suppressor p53 by microRNA miR-504. Mol Cell 38(5):689–699
52. Herrera-Merchan A (2010) miR-33-mediated downregulation of p53 controls hematopoietic stem cell self-renewal. Cell Cycle 9(16):3277–3285
53. Swarbrick A et al (2010) miR-380-5p represses p53 to control cellular survival and is associated with poor outcome in MYCN-amplified neuroblastoma. Nat Med 16(10):1134–1140
54. Tian S et al (2010) MicroRNA-1285 inhibits the expression of p53 by directly targeting its 3′ untranslated region. Biochem Biophys Res Commun 396(2):435–439
55. Kumar M et al (2011) Negative regulation of the tumor suppressor p53 gene by microRNAs. Oncogene 30(7):843–853
56. Yamakuchi M, Ferlito M, Lowenstein CJ (2008) miR-34a repression of SIRT1 regulates apoptosis. Proc Natl Acad Sci U S A 105(36):13421–13426
57. Bou Kheir T et al (2011) miR-449 inhibits cell proliferation and is down-regulated in gastric cancer. Mol Cancer 10:29
58. Fornari F et al (2009) MiR-122/cyclin G1 interaction modulates p53 activity and affects doxorubicin sensitivity of human hepatocarcinoma cells. Cancer Res 69(14):5761–5767

59. Afanasyeva EA et al (2011) MicroRNA miR-885-5p targets CDK2 and MCM5, activates p53 and inhibits proliferation and survival. Cell Death Differ 18(6):974–984
60. Pichiorri F et al (2010) Downregulation of p53-inducible microRNAs 192, 194, and 215 impairs the p53/MDM2 autoregulatory loop in multiple myeloma development. Cancer Cell 18(4):367–381
61. Xiao J et al (2011) miR-605 joins p53 network to form a p53:miR-605: Mdm2 positive feedback loop in response to stress. EMBO J 30(3):524–532
62. Mudhasani R et al (2008) Loss of miRNA biogenesis induces p19Arf-p53 signaling and senescence in primary cells. J Cell Biol 181(7):1055–1063
63. Su X et al (2010) TAp63 suppresses metastasis through coordinate regulation of Dicer and miRNAs. Nature 467(7318):986–990
64. Ory B, Ellisen LW (2011) A microRNA-dependent circuit controlling p63/p73 homeostasis: p53 family cross-talk meets therapeutic opportunity. Oncotarget 2(3):259–264
65. Salmena L et al (2011) A ceRNA hypothesis: the Rosetta Stone of a hidden RNA language? Cell 146(3):353–358
66. Rubio-Somoza I et al (2011) ceRNAs: miRNA target mimic mimics. Cell 147(7):1431–1432
67. Tay Y et al (2011) Coding-independent regulation of the tumor suppressor PTEN by competing endogenous mRNAs. Cell 147(2):344–357
68. Diederichs S, Haber DA (2006) Sequence variations of microRNAs in human cancer: alterations in predicted secondary structure do not affect processing. Cancer Res 66(12):6097–6104
69. Kuchenbauer F et al (2008) In-depth characterization of the microRNA transcriptome in a leukemia progression model. Genome Res 18(11):1787–1797
70. Mayr C, Bartel DP (2009) Widespread shortening of 3′UTRs by alternative cleavage and polyadenylation activates oncogenes in cancer cells. Cell 138(4):673–684
71. Mayr C, Hemann MT, Bartel DP (2007) Disrupting the pairing between let-7 and Hmga2 enhances oncogenic transformation. Science 315(5818):1576–1579
72. Bonci D et al (2008) The miR-15a-miR-16-1 cluster controls prostate cancer by targeting multiple oncogenic activities. Nat Med 14(11):1271–1277
73. Calin GA et al (2005) A microRNA signature associated with prognosis and progression in chronic lymphocytic leukemia. N Engl J Med 353(17): 1793–1801
74. Calin GA et al (2002) Frequent deletions and down-regulation of micro- RNA genes miR15 and miR16 at 13q14 in chronic lymphocytic leukemia. Proc Natl Acad Sci U S A 99(24):15524–15529
75. Stilgenbauer S et al (1998) Expressed sequences as candidates for a novel tumor suppressor gene at band 13q14 in B-cell chronic lymphocytic leukemia and mantle cell lymphoma. Oncogene 16(14):1891–1897
76. Kohlhammer H et al (2004) Genomic DNA-chip hybridization in t(11;14)-positive mantle cell lymphomas shows a high frequency of aberrations and allows a refined characterization of consensus regions. Blood 104(3):795–801
77. Bandi N et al (2009) miR-15a and miR-16 are implicated in cell cycle regulation in a Rb-dependent manner and are frequently deleted or down-regulated in non-small cell lung cancer. Cancer Res 69(13): 5553–5559
78. Bandi N, Vassella E (2011) miR-34a and miR-15a/16 are co-regulated in non-small cell lung cancer and control cell cycle progression in a synergistic and Rb-dependent manner. Mol Cancer 10:55
79. Amaral FC et al (2009) MicroRNAs differentially expressed in ACTH-secreting pituitary tumors. J Clin Endocrinol Metab 94(1):320–323
80. Bhattacharya R et al (2009) MiR-15a and MiR-16 control Bmi-1 expression in ovarian cancer. Cancer Res 69(23):9090–9095
81. Wada M et al (1999) Frequent chromosome arm 13q deletion in aggressive non-Hodgkin's lymphoma. Leukemia 13(5):792–798
82. Harrison CJ et al (2003) Cytogenetics of multiple myeloma: interpretation of fluorescence in situ hybridization results. Br J Haematol 120(6):944–952
83. Bottoni A et al (2005) miR-15a and miR-16-1 down-regulation in pituitary adenomas. J Cell Physiol 204(1):280–285
84. Zhang XJ et al (2010) Dysregulation of miR-15a and miR-214 in human pancreatic cancer. J Hematol Oncol 3:46
85. Musumeci M et al (2011) Control of tumor and microenvironment cross-talk by miR-15a and miR-16 in prostate cancer. Oncogene 30(41):4231–4242
86. Porkka KP et al (2011) The miR-15a-miR-16-1 locus is homozygously deleted in a subset of prostate cancers. Genes Chromosomes Cancer 50(7):499–509
87. Leite KR et al (2011) MicroRNA expression profiles in the progression of prostate cancer-from high-grade prostate intraepithelial neoplasia to metastasis. Urol Oncol, DOI: 10.106/j.urolonc
88. Chim CS et al (2010) Epigenetic inactivation of the miR-34a in hematological malignancies. Carcinogenesis 31(4):745–750
89. Suzuki H et al (2010) Methylation-associated silencing of microRNA-34b/c in gastric cancer and its involvement in an epigenetic field defect. Carcinogenesis 31(12):2066–2073
90. Wang Z et al (2011) DNA hypermethylation of microRNA-34b/c has prognostic value for stage non-small cell lung cancer. Cancer Biol Ther 11(5):490–496
91. Migliore C et al (2008) MicroRNAs impair MET-mediated invasive growth. Cancer Res 68(24):10128–10136
92. Cai KM et al (2010) Hsa-miR-34c suppresses growth and invasion of human laryngeal carcinoma cells via targeting c-Met. Int J Mol Med 25(4):565–571
93. Toyota M et al (2008) Epigenetic silencing of microRNA-34b/c and B-cell translocation gene 4 is associated with CpG island methylation in colorectal cancer. Cancer Res 68(11):4123–4132

94. Leucci E et al (2008) MYC translocation-negative classical Burkitt lymphoma cases: an alternative pathogenetic mechanism involving miRNA deregulation. J Pathol 216(4):440–450
95. Kubo T et al (2011) Epigenetic silencing of microRNA-34b/c plays an important role in the pathogenesis of malignant pleural mesothelioma. Clin Cancer Res 17(15):4965–4974
96. Corney DC et al (2010) Frequent downregulation of miR-34 family in human ovarian cancers. Clin Cancer Res 16(4):1119–1128
97. Chen X et al (2012) CpG island methylation status of miRNAs in esophageal squamous cell carcinoma. Int J Cancer 130(7):1607–1613
98. Wong TS et al (2008) Mature miR-184 as potential oncogenic microRNA of squamous cell carcinoma of tongue. Clin Cancer Res 14(9):2588–2592
99. Lee KH et al (2009) Epigenetic silencing of MicroRNA miR-107 regulates cyclin-dependent kinase 6 expression in pancreatic cancer. Pancreatology 9(3):293–301
100. Pallasch CP et al (2009) miRNA deregulation by epigenetic silencing disrupts suppression of the oncogene PLAG1 in chronic lymphocytic leukemia. Blood 114(15):3255–3264
101. Roldo C et al (2006) MicroRNA expression abnormalities in pancreatic endocrine and acinar tumors are associated with distinctive pathologic features and clinical behavior. J Clin Oncol 24(29):4677–4684
102. Baffa R et al (2009) MicroRNA expression profiling of human metastatic cancers identifies cancer gene targets. J Pathol 219(2):214–221
103. Neves R et al (2010) Role of DNA methylation in miR-200c/141 cluster silencing in invasive breast cancer cells. BMC Res Notes 3:219
104. Wiklund ED et al (2011) Coordinated epigenetic repression of the miR-200 family and miR-205 in invasive bladder cancer. Int J Cancer 128(6):1327–1334
105. Suh SO et al (2011) MicroRNA-145 is regulated by DNA methylation and p53 gene mutation in prostate cancer. Carcinogenesis 32(5):772–778
106. Karaayvaz M et al (2011) Prognostic significance of miR-215 in colon cancer. Clin Colorectal Cancer 10(4):340–347
107. Earle JS et al (2010) Association of microRNA expression with microsatellite instability status in colorectal adenocarcinoma. J Mol Diagn 12(4):433–440
108. Kahlert C et al (2011) Invasion front-specific expression and prognostic significance of micro RNA in colorectal liver metastases. Cancer Sci 102(10):1799–1807
109. Hu X et al (2009) A miR-200 microRNA cluster as prognostic marker in advanced ovarian cancer. Gynecol Oncol 114(3):457–464
110. Tellez CS et al (2011) EMT and stem cell-like properties associated with miR-205 and miR-200 epigenetic silencing are early manifestations during carcinogen-induced transformation of human lung epithelial cells. Cancer Res 71(8):3087–3097
111. Xi Y et al (2006) Prognostic values of microRNAs in colorectal cancer. Biomark Insights 2:113–121
112. Ceppi P et al (2010) Loss of miR-200c expression induces an aggressive, invasive, and chemoresistant phenotype in non-small cell lung cancer. Mol Cancer Res 8(9):1207–1216
113. Davalos V et al (2011) Dynamic epigenetic regulation of the microRNA-200 family mediates epithelial and mesenchymal transitions in human tumorigenesis. Oncogene 31(16):2062–2074
114. Leaderer D et al (2011) Genetic and epigenetic association studies suggest a role of microRNA biogenesis gene exportin-5 (XPO5) in breast tumorigenesis. Int J Mol Epidemiol Genet 2(1):9–18
115. Melo SA et al (2010) A genetic defect in exportin-5 traps precursor microRNAs in the nucleus of cancer cells. Cancer Cell 18(4):303–315
116. Merritt WM et al (2008) Dicer, Drosha, and outcomes in patients with ovarian cancer. N Engl J Med 359(25):2641–2650
117. Bahubeshi A, Tischkowitz M, Foulkes WD (2011) miRNA processing and human cancer: DICER1 cuts the mustard. Sci Transl Med 3(111):111ps46
118. Hill DA et al (2009) DICER1 mutations in familial pleuropulmonary blastoma. Science 325(5943):965
119. Karube Y et al (2005) Reduced expression of Dicer associated with poor prognosis in lung cancer patients. Cancer Sci 96(2):111–115
120. Faber C et al (2011) Overexpression of Dicer predicts poor survival in colorectal cancer. Eur J Cancer 47(9):1414–1419
121. Martin MG, Payton JE, Link DC (2009) Dicer and outcomes in patients with acute myeloid leukemia (AML). Leuk Res 33(8):e127
122. Klein U et al (2010) The DLEU2/miR-15a/16-1 cluster controls B cell proliferation and its deletion leads to chronic lymphocytic leukemia. Cancer Cell 17(1):28–40
123. Lujambio A et al (2008) A microRNA DNA methylation signature for human cancer metastasis. Proc Natl Acad Sci U S A 105(36):13556–13561
124. Thorstensen L et al (2000) Evaluation of 1p losses in primary carcinomas, local recurrences and peripheral metastases from colorectal cancer patients. Neoplasia 2(6):514–522
125. Welch C, Chen Y, Stallings RL (2007) MicroRNA-34a functions as a potential tumor suppressor by inducing apoptosis in neuroblastoma cells. Oncogene 26(34):5017–5022
126. Shimono Y et al (2009) Downregulation of miRNA-200c links breast cancer stem cells with normal stem cells. Cell 138(3):592–603
127. Shinozaki A et al (2010) Downregulation of microRNA-200 in EBV-associated gastric carcinoma. Cancer Res 70(11):4719–4727
128. de Krijger I et al (2011) MicroRNAs in colorectal cancer metastasis. J Pathol 224(4):438–447
129. Duan R, Pak C, Jin P (2007) Single nucleotide polymorphism associated with mature miR-125a alters the processing of pri-miRNA. Hum Mol Genet 16(9):1124–1131

130. Martello G et al (2010) A microRNA targeting dicer for metastasis control. Cell 141(7):1195–1207
131. Kumar MS et al (2009) Dicer1 functions as a haploinsufficient tumor suppressor. Genes Dev 23(23):2700–2704
132. Reimers M, Carey VJ (2006) Bioconductor: an open source framework for bioinformatics and computational biology. Methods Enzymol 411:119–134
133. Ji H et al (2008) An integrated software system for analyzing ChIP-chip and ChIP-seq data. Nat Biotechnol 26(11):1293–1300
134. Kulakovskiy IV et al (2010) Deep and wide digging for binding motifs in ChIP-Seq data. Bioinformatics 26(20):2622–2623
135. Fejes AP et al (2008) FindPeaks 3.1: a tool for identifying areas of enrichment from massively parallel short-read sequencing technology. Bioinformatics 24(15):1729–1730
136. Cox J, Mann M (2008) MaxQuant enables high peptide identification rates, individualized p.p.b.-range mass accuracies and proteome-wide protein quantification. Nat Biotechnol 26(12):1367–1372
137. Bailey TL (2002) Discovering novel sequence motifs with MEME. Curr Protoc Bioinformatics Chapter 2:Unit 2 4
138. Mackowiak SD (2011) Identification of novel and known miRNAs in deep-sequencing data with miRDeep2. Curr Protoc Bioinformatics Chapter 12:Unit 12 10
139. Hackenberg M, Rodriguez-Ezpeleta N, Aransay AM (2011) miRanalyzer: an update on the detection and analysis of microRNAs in high-throughput sequencing experiments. Nucleic Acids Res 39(Web Server issue):W132–W138
140. Lagana A et al (2009) miRo: a miRNA knowledge base. Database (Oxford) 2009:bap008
141. Corcoran DL et al (2011) PARalyzer: definition of RNA binding sites from PAR-CLIP short-read sequence data. Genome Biol 12(8):R79
142. Saeed AI et al (2006) TM4 microarray software suite. Methods Enzymol 411:134–193
143. Bunz F et al (1998) Requirement for p53 and p21 to sustain G2 arrest after DNA damage. Science 282(5393):1497–1501
144. Hoh J et al (2002) The p53MH algorithm and its application in detecting p53-responsive genes. Proc Natl Acad Sci U S A 99(13):8467–8472
145. Cawley S et al (2004) Unbiased mapping of transcription factor binding sites along human chromosomes 21 and 22 points to widespread regulation of noncoding RNAs. Cell 116(4):499–509
146. Smeenk L et al (2008) Characterization of genome-wide p53-binding sites upon stress response. Nucleic Acids Res 36(11):3639–3654
147. Velculescu VE et al (1995) Serial analysis of gene expression. Science 270(5235):484–487
148. Wei CL et al (2006) A global map of p53 transcription-factor binding sites in the human genome. Cell 124(1):207–219
149. Friedman RC et al (2009) Most mammalian mRNAs are conserved targets of microRNAs. Genome Res 19(1):92–105
150. Krek A et al (2005) Combinatorial microRNA target predictions. Nat Genet 37(5):495–500
151. Kertesz M et al (2007) The role of site accessibility in microRNA target recognition. Nat Genet 39(10):1278–1284
152. Miranda KC et al (2006) A pattern-based method for the identification of microRNA binding sites and their corresponding heteroduplexes. Cell 126(6):1203–1217
153. Alexiou P et al (2009) Lost in translation: an assessment and perspective for computational microRNA target identification. Bioinformatics 25(23):3049–3055
154. Baek D et al (2008) The impact of microRNAs on protein output. Nature 455(7209):64–71
155. Selbach M et al (2008) Widespread changes in protein synthesis induced by microRNAs. Nature 455(7209):58–63
156. Beitzinger M et al (2007) Identification of human microRNA targets from isolated argonaute protein complexes. RNA Biol 4(2):76–84
157. Hendrickson DG et al (2008) Systematic identification of mRNAs recruited to argonaute 2 by specific microRNAs and corresponding changes in transcript abundance. PLoS One 3(5):e2126
158. Karginov FV et al (2007) A biochemical approach to identifying microRNA targets. Proc Natl Acad Sci U S A 104(49):19291–19296
159. Chi SW et al (2009) Argonaute HITS-CLIP decodes microRNA-mRNA interaction maps. Nature 460(7254):479–486
160. Hafner M et al (2010) Transcriptome-wide identification of RNA-binding protein and microRNA target sites by PAR-CLIP. Cell 141(1):129–141
161. Orom UA, Lund AH (2007) Isolation of microRNA targets using biotinylated synthetic microRNAs. Methods 43(2):162–165
162. Orom UA, Lund AH (2010) Experimental identification of microRNA targets. Gene 451(1–2):1–5
163. Cummins JM et al (2006) The colorectal microRNAome. Proc Natl Acad Sci U S A 103(10):3687–3692
164. Tazawa H et al (2007) Tumor-suppressive miR-34a induces senescence-like growth arrest through modulation of the E2F pathway in human colon cancer cells. Proc Natl Acad Sci U S A 104(39):15472–15477
165. Ong SE et al (2002) Stable isotope labeling by amino acids in cell culture, SILAC, as a simple and accurate approach to expression proteomics. Mol Cell Proteomics 1(5):376–386
166. Chen QR et al (2011) Systematic proteome analysis identifies transcription factor YY1 as a direct target of miR-34a. J Proteome Res 10(2):479–487
167. Prosser HM et al (2011) A resource of vectors and ES cells for targeted deletion of microRNAs in mice. Nat Biotechnol 29(9):840–845

MicroRNAs in Melanoma Biology 6

Manfred Kunz

Abstract

Malignant melanoma is a highly aggressive tumour with increasing incidence and poor prognosis in the metastatic stage. In recent years, a substantial number of reports on individual miRNAs or miRNA patterns have been published providing strong evidence that miRNAs might play an important role in malignant melanoma and might help to better understand the molecular mechanisms of melanoma development and progression. A major preliminary finding was that melanoma-associated miRNAs are often located in genomic regions with frequent gains and losses in tumours. Detailed studies of different groups thereafter identified miRNAs with differential expression in benign melanocytes compared with melanoma cell lines or in benign melanocytic lesions compared with melanomas. Among these were *let-7a* and *b*, miR-23a and b, miR-148, miR-155, miR-182, miR-200c, miR-211, miR214, and miR-221 and 222. Some of these miRNAs target well-known melanoma-associated genes like the *NRAS* oncogene, microphthalmia-associated transcription factor (*MITF*), receptor tyrosine kinase c-KIT or AP-2 transcription factors (*TFAP2*). Although we are still far from a complete understanding of the role of miRNA-target gene interactions in malignant melanoma, these findings further underscore the notion of a direct involvement of miRNAs in melanoma biology. Very recently, a prognostic signature of six miRNAs has been identified consisting of miRNAs miR-150, miR-342-3p, miR-455-3p, miR-145, miR-155, and miR-497. High expression of these miRNAs was shown to be associated with improved long-term survival of metastatic patients.

Keywords

Melanoma • Tumour development • Non-coding RNAs • Oncogenes • Protein degradation

M. Kunz, M.D. (✉)
Department of Dermatology, Venereology and Allergology, University of Leipzig,
Philipp-Rosenthal-Str. 23-25, 04103 Leipzig, Germany
e-mail: manfred.kunz@medizin.uni-leipzig.de

U. Schmitz et al. (eds.), *MicroRNA Cancer Regulation: Advanced Concepts, Bioinformatics and Systems Biology Tools*, Advances in Experimental Medicine and Biology 774,
DOI 10.1007/978-94-007-5590-1_6, © Springer Science+Business Media Dordrecht 2013

6.1 Introduction

MicroRNAs are small non-coding 21–23 nt RNAs that, when bound to the 3′ untranslated region (3′UTR) of mRNA, induce mRNA degradation or inhibition of protein translation [1]. This kind of interfering RNAs was first described in the early 90s of the last century in experiments with *Caenorhabditis elegans* [2]. This organism produced small non-translated RNAs with high complementarity to protein-coding mRNAs. Further investigations showed that inactivating mutations in genes encoding for miRNAs *lin-4* and *let-7* led to developmental defects at different stages of larval development, arguing for a significant functional role of these non-coding RNAs. It was further shown that miRNAs exerted their effects via a direct interference with coding RNAs. At present, the term miRNA is commonly used for small non-coding RNA molecules which negatively regulate gene expression on transcriptional and translational level in a variety of different organisms.

miRNAs are either leading to degradation of target mRNA or to inhibition of mRNA translation into protein, in many cases depending on the level of complementarity between miRNAs and mRNAs [3]. MiRNA-mediated gene regulation is also influenced by the cooperation between multiple miRNA-recognition sites in target genes, the spacing between these, their position within the 3′UTR and the mRNA secondary structure [4]. At present, more than 700 miRNAs have been described in humans. Computer algorithms have provided genome-wide predictions of target sequences for miRNAs, and since the so-called *seed* sequence, which stands for the specifically binding minimal miRNA sequence, consists of only seven bases, the number of *in silico* predicted targets may run up to 1,000 targets per miRNA [3]. Thus, detailed functional analyses are necessary to validate each individual miRNA target. Evidence has also been provided that the number of miRNA binding sites in the 3′UTRs of target genes may correlate with the biological relevance of an individual gene and its role in controlling cellular protein networks [5, 6]). However, these findings await further investigations.

Because of the wide distribution of miRNAs in different organisms and their functional potential, miRNA expression patterns and the role of individual miRNAs for tumour development and progression have been investigated in recent years. The analysis of miRNA expression in malignant melanoma has been part of these investigations from the beginning and finally resulted in a series of comprehensive reports more recently.

6.2 Key Findings of miRNAs in Malignant Tumours

In one major large-scale study on miRNA expression patterns in different cancers including breast, colon, and lung cancer, different forms of leukaemia and malignant melanoma, tumour tissues showed a preponderance of down-regulated miRNAs as compared with benign tissues, suggestive for a role of many miRNAs as tumour suppressors [7]. Interestingly, in this study, miRNA patterns allowed a better discrimination between individual tumour entities than mRNA expression patterns. In the following, an increasing body of reports showed deregulated miRNA expression in a large series of different tumours [8–10]. In a key functional study, development of lung human carcinomas was associated with down-regulation of miRNA *let-7*. In *in vitro* experiments, *let-7* targeted expression and function of the *RAS* oncogene, which is known to play an important role in lung carcinomas [11]. *Let-7a* also showed significantly reduced expression in colon carcinomas, paralleled by enhanced RAS expression [12]. Interestingly, *let-7a* levels in lung cancer patients correlated with postoperative survival. Many miRNAs have classical oncogenes or other tumour-promoting genes as predicted targets, such as *RAS*, *c-Myc*, *BCL2*, and cell cycle dependent kinases (*CDKs*). Down-regulation of individual miRNAs might be one factor that contributes to over-expression and enhanced activity of some of these oncogenes with subsequent deregulated cellular growth and tumour development. MiRNA down-regulation in tumour development is not completely understood, but involves genetic as well as epigenetic mechanisms [13]. Evidence is accumulating that inactivation of some proteins of the miRNA machinery, e.g., proteins encoded by *DICER1*, *AGO2* and *XPO5* may contribute to down-regulation of miRNAs in malignant tumours [14].

However, miRNAs may also act as oncogenes and over-expression may be related to tumour development and progression. B-cell lymphomas often show up-regulation of miRNAs of the *miR-17~92* locus, which encodes a cluster of miRNAs that may cooperate with the MYC oncogene [15, 16]. In a mouse B-cell lymphoma model, enforced expression of the *miR-17~92* cluster acted with c-Myc expression to promote tumour development [15]. Moreover, over-expression of *miR-17~92* suppressed apoptosis in Myc-driven B-cell lymphomas in mice [17]. In these experiments, miR-19a and miR-19b within the *miR-17~92* cluster reflected the oncogenic properties of the entire cluster. A further outstanding example for oncogenic miRNAs is miR-155, which seems to play a role in some B-cell lymphomas [18, 19]. The oncogenic properties of miR-155 were shown in miR-155 transgenic mice which develop acute lymphoblastic leukemia, a high grade lymphoma, after a block at the pre-B stage of B-cell differentiation at an early age [20].

As exemplified by the *miR-17~92* cluster and miR-155, miRNAs may act as tumour promoters. A further candidate as tumour promoter is miR-21. MiR-21 is over-expressed in breast carcinomas and glioblastomas and is able to inhibit tumour cell apoptosis by targeting pro-apoptotic genes like *PTEN* and *TPM1* [21].

The term 'oncogene addiction', which has been coined in recent years, indicates that despite the fact that tumour development is a complex process, cancers often depend on the activity of a single oncogene [22, 23]. Targeting of this oncogene, e.g., by small molecule inhibitors, can lead to significant treatment responses and sometimes cure of the patients. In a recent report, oncomiR addiction to miR-21 has been demonstrated [24]. In this study, transgenic mice were generated conditionally expressing miR-21. Over-expression of miR-21 led to a pre-B malignant lymphoid-like phenotype, demonstrating that mir-21 may act as an oncogene. Interestingly, when miR-21 was inactivated, tumours regressed completely within a few days. These results demonstrated that tumours may become addicted to individual miRNAs which may act as oncomiRs, a finding that may have significant implications for future therapeutic approaches targeting miRNAs.

Although there is a considerable body of evidence that miRNAs play a role in development of different cancers, it has not been until recently that their role in tumour metastasis has been addressed [25, 26]. In one report, miR-10b was shown to make non-metastatic human breast carcinoma cells invasive and metastatic [25]. A comparison of miRNA expression patterns of non-metastatic and metastatic human breast cancer lines revealed differential expression of 20 miRNAs [27]. In a clinical study in this report, low levels of miR-335 and miR-126 were associated with dramatically reduced metastasis-free survival in breast cancer patients, indicating that these miRNAs might serve as predictive markers.

Taken together, miRNAs play an important role in tumour development and progression (metastasis). The knowledge of their role in melanoma biology significantly increased in recent years due to a series of large-scale clinical and functional studies. In order to better understand the biological and clinical context of these miRNAs, current knowledge of melanoma pathogenesis and treatment modalities are briefly reviewed.

6.3 Malignant Melanoma Pathogenesis

Malignant melanoma is a tumour of rapidly increasing incidence and high metastatic potential [28]. Current incidence is 10–15 patients per 100,000 individuals per year [29]. Although the aetiology and pathogenesis are not yet completely understood, much progress has been made in recent years in the understanding of the underlying molecular mechanisms. Activating mutations in *BRAF* and *NRAS* oncogenes have been identified in the majority of melanomas, with both being mutually exclusive [30, 31]. The presence of one or the other seems to be influenced by the localization of the lesion, with non-chronic sun-exposed tumours displaying higher rates of *BRAF* mutations and chronic sun-exposed lesions displaying higher rates of *NRAS* mutations [31]. These genetic alterations not only seem to contribute to melanoma biology but have also opened new therapeutic avenues, as activated BRAF may be targeted by small molecule inhibitors [32]. However, activating mutations of *BRAF* are only found in half of all cases, and in more than 30% of all melanomas no genetic alterations with presumed pathogenic relevance have been found up to now [31, 33]. Activating mutations of the receptor tyrosine kinase and growth factor receptor *c-KIT*, which are found in 80% of gastrointestinal stroma tumours, are present in 11% of acral, and 21% of mucosal, but only 2–4% of skin melanomas [33, 34]. The functional relevance of c-KIT for malignant melanoma remains to be determined as the loss of c-KIT expression correlates with melanoma progression.

BRAF and *NRAS* oncogenes are mediating their activity via the mitogen-activated protein kinase (MAPK) pathways which regulate cellular proliferation, differentiation, cell cycle progression and apoptosis. These pathways involve cell cycle molecules such as cyclin D1, cyclin E, and tumour suppressor molecules such as $p21^{WAF1/KIP1}$ and p^{27CIP1} [35]. Since a tightly controlled cell cycle is a major prerequisite for the suppression of tumour development, molecules involved in these processes have been in the focus of interest in the past years. Cell cycle molecules such as p53, $p16^{INK4a}$, $p21^{WAF1/CIP1}$, and cyclin-dependent kinases such as CDK4 have been demonstrated to be inactivated by mutation in different cancers including malignant melanoma [36, 37]. Compromised cell cycle regulation due to inactivation of tumour suppressor $p16^{INK4a}$ and $p14^{ARF}$ may play a role in a subset of malignant melanomas [38]. These pathways control the G1 phase of the cell cycle and are inhibited via genetic alteration of the respective gene locus on chromosome 9p21, which encodes both $p16^{INK4a}$ and $p14^{ARF}$ via alternative splicing of the same gene [39]. Activation of $p16^{INK4a}$ leads to inhibition of CDK4 and CDK6, with the consequence of a blockage of inactivating phosphorylation of tumour suppressor pRb. This results in an activated pRb with E2F transcription factors bound to it, which inhibits their activity. E2F transcription factors induce gene expression of genes involved in S phase progression such as cyclin A. Germline mutations of *CDKN2A* (which encodes for $p16^{INK4a}$) were identified in half of familial melanoma patients. Further evidence for a significant role of p16 inactivation in malignant melanoma comes from a well-known melanoma mouse model. Transgenic mice with an activating *Ras* mutation in combination with mutated *p16* develop primary melanomas, and in a significant percentage, distant metastases [40]. Inactivating mutations in the *TP53* gene, as identified in many epithelial tumours, seem to be of no relevance in malignant melanoma, but evidence has been provided that a correlation between *TP53* mutations and poor prognosis might exist. p53 is involved in $p14^{ARF}$ and MDM2 (murine double minute 2) regulation. At present, the role of p53 in malignant melanoma is still unclear and awaits further investigations.

Phosphatase and tensin homolog (*PTEN*) is a lipid phosphatase that inhibits activation of Akt kinase pathway by negative interference with phosphoinositide 3-kinase (PI3K). Loss of heterozygosity and inactivating mutations of the *PTEN* locus have been described in up to 30% of malignant melanomas [38]. Interestingly, more recent data showed that BRAF and PTEN may interact for melanoma development [41]. In this latter report, a mouse model is described for human melanoma using mice with conditional melanocyte-specific BRaf(V600E) expression. Upon induction, these mice develop benign melanocytic hyperplasias, but no melanomas. Expression of BRaf(V600E) in combination with *Pten* tumour suppressor gene silencing leads to melanoma development and metastasis in lymph nodes and lungs. Moreover, melanomas are prevented by inhibitors of mTorc1 or MEK1/2, downstream targets of PI3K and BRAF, respectively. Akt kinase, which is also involved in PI3K signalling, is constitutively activated in more than 60% of all melanomas, but this seems to be dependent on the amplification of AKT3, rather than inactivation of *PTEN*. Taken together, currently available data support the role of *BRAF* and

NRAS oncogenes (and c-KIT) in melanoma pathogenesis with a significant role for inactivated tumour suppressors like p16^{INK4a} and PTEN.

Downstream transcription factors involved in melanoma pathogenesis include MITF, TFAP-2 family and FOXO3. MITF, the most acknowledged protein, is a bHLH-Zip transcription factor of the Myc family involved in the regulation of melanocyte-specific genes such as tyrosinase and tyrosinase-related proteins [42]. Its role in melanoma pathogenesis is somewhat controversial, since increased levels of MITF reduce melanoma cell proliferation and tumourigenicity, while low levels induce cell cycle arrest and apoptosis [43]. Intermediate levels appear to promote melanoma cell survival. However, the role of MITF for melanoma pathogenesis was underlined by a study showing that significant number of human melanomas shows a genetic amplification of the *MITF* locus [44].

6.4 Current Treatment of Malignant Melanoma

Classical treatment of cutaneous lesions of malignant melanoma at early stages of the disease including regional lymph node metastasis is surgical excision [45]. However, surgery of distant metastases, in particular of metastases of internal organs such as lung, liver and spleen, does not significantly impact on the overall survival of patients and is not commonly recommended, although individual cases may profit from metastasis surgery [46]. After distant metastasis, median overall survival of patients is extremely poor (6–8 months) [47]. Clinical response rates of classical chemotherapeutic agents, either as single agents or as combinations, have been in the range of 10–15% [45, 48]. Response rates to newly developed protein kinase inhibitors such as the multi-kinase inhibitor sorafenib or MEK1/2 kinase inhibitors have been limited so far, not going beyond that of classical chemotherapies [48]. Thus, treatment in the metastatic stage has remained unsatisfying until very recently [32, 49].

Currently, the B-RAF oncogene seems to be the most promising target at least in cases with activating *BRAF*(V600E) mutations. These patients are treated with the recently developed specific BRAF inhibitor vemurafenib, which has recently been approved for treatment of metastatic melanoma in the U.S. and is expected to be approved in Europe in the near future [32, 49]. However, many of the patients treated with this agents experience recurrences, which seems to be due to molecular escape mechanisms of tumour cells switching from B-RAF to N-RAS or platelet-derived growth factor (PDGF) signalling. In an experimental set-up of melanoma cell cultures, it was shown that acquired resistance to vemurafenib develops due to mutually exclusive *PDGFRβ* upregulation or *NRAS* mutations, but not through secondary mutations in *BRAF* [50]. In these analyses, the mentioned BRAF-specific inhibitor was used to derive BRAF-inhibitor resistant sub-lines from B-RAF(V600E)-positive melanoma cell lines. Findings from these cells were verified in BRAF-inhibitor resistant tumours and tumour-matched, short-term cultures from clinical trial patients.

Interestingly, *PDGFRβ*-up-regulated tumour cells have low levels of activated RAS and, when treated with BRAF inhibitor, do not reactivate the MAPK pathway. In another cell subset, high levels of activated N-RAS resulting from mutations led to significant MAPK pathway activation upon BRAF inhibitor treatment. Based on these findings and a series of experiments with transfected cells, melanomas escape B-RAF(V600E) targeting not through secondary B-RAF(V600E) mutations but via receptor tyrosine kinase (RTK)-mediated activation (e.g., PDGFR) or activated RAS-mediated reactivation of the MAPK pathway. Interestingly, patients with wild-type BRAF did not respond to BRAF inhibitor treatment in the mentioned studies [32, 49]. Subsequent studies showed that wild-type *BRAF* cells activate CRAF signalling after BRAF inhibitor treatment [51].

Based on the fact that malignant melanoma is a highly immunogenic tumour, the immunological surveillance by T cells has been used for treatment of metastatic disease. This has been achieved by using specific antibodies directed against the common T lymphocyte antigen 4 (anti-CTLA4 antibodies) [52]. The CTLA4 antigen attenuates T-cell

Table 6.1 Differentially expressed miRNAs in malignant melanoma

No.	miRNAs	Targets[a]	References
1.	*Let-7a,b,d,e,g*	H-RAS, LIN28, HMGA2, ITGB3, CCND1, INK4A, PBX3, MYCN, COL1A2, c-MYC	[55, 58, 61]
2.	miR-137	JARID1B, CtBP1, Mib1, Ezh2, MITF, CDK6	[71, 72]
3.	miR-148	MITF, DNMT3b, CAMK2A, BCL2L11	[71]
4.	miR-155	MITF, CDKN1B, MEF2A, SATB1, SKI	[73, 88]
5.	miR-182	MITF, FOXO3, RGS17, CDKN1A, FOXO1	[58, 72, 77]
6.	miR-200c	BMI1, MARCKS, ZEB1, JAG2, ZEB2	[73, 75]
7.	miR-211	IGF2R, TGFBR2, NFAT5, KCNMA1, RAB22A	[76, 80]
8.	miR-214	NRAS, TFAP2C, EZH2, PLXNB1, PTEN	[86]
9.	miR-221, miR-222	CDKN1B, DDIT4, CDKN1C, PTEN, c-KIT	[73, 84]
	MiRNA patterns	Targets[a]	References
10.	mir-150	MYB, EGR2, NOTCH3	[88]
	mir-342-3p	DNMT1, AGPAT4*, ID4*	
	mir-455-3p	NFIB*, ZNF238*, FZD10*	
	mir-145	c-MYC, FSCN1, STAT1	
	mir-155	MITF, CDKN1B, MEF2A	
	mir-497	BCL2, CCND2, CCNE1*	
11.	miR-155	MITF, CDKN1B, SKI	[73]
	miR-200c	BMI1, MARCKS, ZEB1	
	miR-23a,b	FANCG, RUNX2, POX, PLAU, c-MET	

[a]Targets without asterisks indicate validated targets. Targets with asterisks indicate predicted targets by Targetscan (http://www.targetscan.org/), which are not yet validated

responses but after of blockage may lead to a dramatically enhanced T-cell response against melanoma cells. First clinical studies were indeed promising and the anti-CTLA4 antibody ipilimumab has recently been approved for melanoma treatment in the U.S. and Germany. Treatment with this compound improves mean overall survival of patients by 4 months [52]. Further clinical trials, especially combination therapies are ongoing and may further improve overall survival of melanoma patients. The combination of ipilimumab and dacarbacin, a common chemotherapeutic agent for malignant melanoma, has so far not improved overall survival of metastatic patients [53].

Very recently, a new avenue of tumour treatment has been opened by use of short interfering (si)RNAs, which were administered systemically to a melanoma patient via nanomolecules [54]. Local deposition of nanomolecules in tumour tissues of melanoma metastases could be verified and administration of these compounds was well tolerated. Studies on treatment efficacy remain to be performed. Overall, RNA interference might by an interesting therapeutic perspective for malignant melanoma in the near future as this technology is rapidly improving.

6.5 miRNAs in Malignant Melanoma

6.5.1 Introduction

A substantial number of reports addressing the role of miRNAs in melanoma pathogenesis have been published in past years, some of which have been summarized in recent reviews [55–57]. Table 6.1 and Fig. 6.1, respectively, show an overview of key miRNAs and intracellular pathways targeted by these in malignant melanoma. However, because of the complexity of molecular mechanisms influenced by miRNA activity and the heterogeneous experimental approaches, a complete picture of the role of miRNAs in malignant melanoma is still missing. Initial studies provided evidence for a possible role of miRNAs in malignant

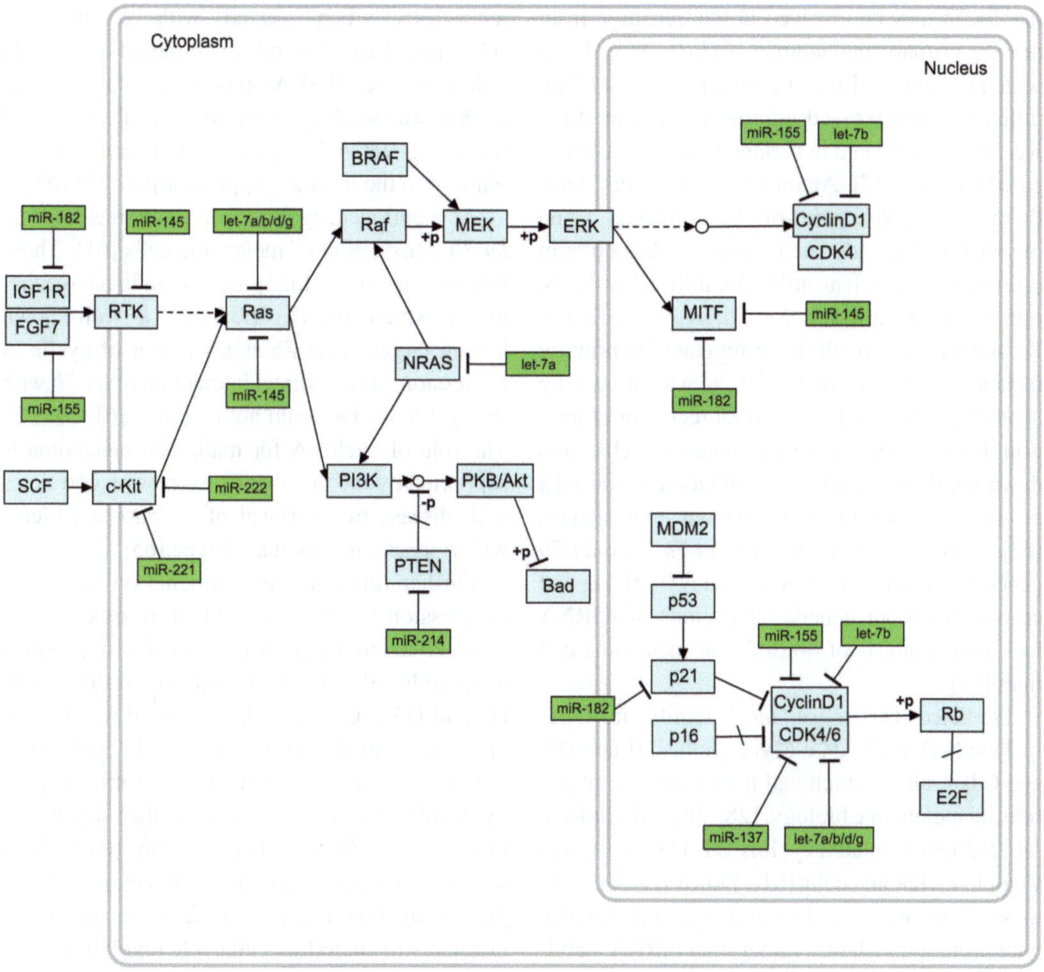

Fig. 6.1 Melanoma pathways and targeting miRNAs. Pathways active in malignant melanoma are shown together with miRNAs targeting individual components in these pathways. miRNAs are derived from studies mentioned in Table 6.1

melanoma by identifying amplifications and deletions in genomic areas which encode for miRNAs in melanoma cell lines [58]. In a subsequent report, miRNA patterns of ten melanoma cell lines of the U.S. National Cancer Institute (NCI) 60 cell line panel were compared with those of cell lines of other tumour entities [59]. In this study, miRNAs showed tissue-specific patterns, which allowed a clear differentiation between cell lines. In this latter study, a series of different benign tissues were used as controls, however, benign melanocytic nevi (or melanocytes) as precursors for melanoma development were not included in this study. In the following, a series of detailed investigations on individual miRNAs or miRNA patterns in malignant melanoma will be reviewed.

6.5.2 Expression and Function of miRNAs in Malignant Melanoma

6.5.2.1 *Let-7* Family

miRNAs of the *let-7* family seem to play an important role in different cancers such as lung and colon carcinoma [12, 60]. In an analysis of miRNA expression patterns in benign melanocytic nevi and primary melanomas using laser-microdissected tissue material, we found strong evidence that *let-7b* may be involved in the transition from nevi to primary melanomas [61]. In accordance with the report by Lu and co-workers, most of the differentially expressed microRNAs in our study were down-regulated in melanoma tissues as compared with nevi [7]. Among these were *let-7* family members and a series of other miRNAs, some of which had already been shown to play a role in other tumours, such as miR-15a, miR-17, miR-28, miR-106a, and miR-144. *Let-7a*, *let-7b*, *let-7d*, *let-7e*, and *let-7g* were all down-regulated in primary melanomas. Interestingly, *let-7a* was shown by another group to be an important regulator of integrin beta3 expression in melanoma cells, and down-regulation of *let-7a* by antagomirs induced a pro-invasive behaviour in benign melanocytes [62]. Reduced expression of *let-7a* and *let-7b* genes was also reported by the study of Zhang and co-workers about genetic aberrations of miRNA loci in an analysis of 40 different melanoma cell lines [58].

Predicted targets of *let-7* family miRNAs include N-RAS, RAF, c-Myc, cyclins D1 and D3, and Cdk4, all of which had been shown to play a role in melanoma biology [28, 38, 63]. Indeed, *let-7b* down-regulated cyclins D1, D3 and A, and Cdk4 in melanoma cells [61]. Down-regulation of these molecules was likely due to interference with protein translation rather than mRNA stability. mRNA stability had not been tested in our study, since *let-7b*-transfected HepG2 liver cancer cells did not show differences in mRNA expression of these molecules, indicating that mRNA stability is not directly affected by *let-7b* [60].

In support of these findings, cyclins D1 and D3 have been shown to be up-regulated in primary melanomas as compared with benign melanocytic nevi in a series of immunohistochemical studies [64–66]. Moreover, enhanced gene expression and copy number changes in the cyclin D1 region were found by comparative genomic hybridization of melanoma tissues, suggesting that over-expression of cyclins may be a central pathogenic mechanism in malignant melanoma [67]. The possible contribution of Cdk4 to melanoma pathogenesis was underlined by the fact that mice with an activating *cdk4* gene mutation develop primary melanomas and lymph-node metastases when treated with carcinogenic substances that only induce epithelial tumours in wild-type mice [68]. Moreover, a mutation in the *CDK4* gene leading to amino-acid substitution at residue 24 (R24C) negatively interferes with its binding to the tumour suppressor $p16^{INK4A}$ [63].

As mentioned, cyclin A was down-regulated by *let-7b* transfection of melanoma cells [61]. These findings are most likely due to indirect mechanisms which involve *HMGA2*, a well-known known target of *let-7b* and activator of cyclin A expression, since a direct interaction of *let-7b* with the cyclin 3′UTR could not be found [61, 69, 70]. The role of cyclin A for malignant melanoma is supported by the fact that its expression correlates with disease-free survival of melanoma patients with a special melanoma subtype [65].

Further functional experiments on cell cycle progression *in vitro* showed that over-expression of *let-7b* led to a significant reduction of proliferating cells [61]. It is well-understood that cyclin D1 and D3 expression play a central role in progression through G1 phase of cell cycle [37]. *Let-7b* also led to reduced colony forming capacity in soft agar assays. Taken together, down-regulation of *let-7b* and other *let-7* family members seems to impact on melanoma development and growth by interaction with different molecular mechanisms including cell cycle regulation.

6.5.2.2 MiR-148

Based on current knowledge, which attributes to MITF an important role in melanoma biology, a recent study tested a series of different miRNAs with conserved binding sites in the 3′UTR of the mouse *Mitf* gene [71]. Analyses were performed using a MITF 3′UTR reporter construct. miRNAs with highly conserved binding sites such as miR-27a, miR-25/32/92/363/367, miR-101/144, miR-124/506, miR-137, miR-148/152, and miR-124/506 (the latter with a less conserved binding site in the *Mitf* 3′UTR compared to all others) were included in this study. It was shown that miR-137 and miR-148 negatively affected *Mitf* expression in melanoma cells. In line with this, Bemis and co-workers showed that miR-137 may down-regulate MITF expression in melanoma cells [72]. None of the other miRNAs tested, although having highly conserved binding sites in the *Mitf* 3′UTR, affected reporter gene expression in HEK293 and 501mel melanoma cells, respectively. Thus, miR-137 and miR-148 seem to be regulators of *Mitf* expression at least in mice, which may be of relevance for the human situation, because of the high level of conservation of the *MITF* 3′UTR in different species.

6.5.2.3 MiR-155, miR-200c, miR-23a,b

Overall, there has been little overlap between candidate miRNAs and miRNA patterns in different melanoma studies. This might at least in part be due to the heterogeneity of the samples analysed in different studies and the limited number of samples in some of these studies, but might also due to different platforms used for miRNA expression analysis. In a recent report, authors tried to overcome these problems in a more comprehensive study by analysing expressing patterns in benign melanocytes, a series of melanoma cell lines and in tissue specimens of different stages of melanoma progression [73]. A normal human epidermal melanocyte culture (NHEM), nine melanoma cell lines and three pools of benign nevi and 17 primary and subcutaneous metastasis samples were analysed. First, genome-wide miRNA expression profiling was performed for NHEM cells and two melanoma cell lines, IGR39 and IGR37. IGR39 is a melanoma cell line derived from a primary melanoma and IGR37 is derived from a metastatic lesion of the same patient. Data from microarray analyses of cell lines were validated in the set of melanoma cell lines and NHEM by PCR technology. Tissue samples were analysed for the expression of 88 cancer-related miRNAs again using PCR array technology.

Several miRNAs were identified as potential candidates to distinguish between benign melanocytes and melanoma cell lines. MiR-155 and miR-146a showed decreased expression in all melanoma lines compared with NHEM. In contrast, miR-25 and miR-23a and miR-23b were consistently up-regulated. In the tissue sample analyses, miR-200c, miR-205 and miR23b were strongly down-regulated in melanoma samples as compared with benign nevi, in particular in metastatic lesions. In contrast, miR-363, miR-146a and miR-155 were up-regulated in all melanoma samples. The differences regarding miRNA expression in cell lines and tissues were interpreted as a cell culture phenomenon.

In an extension of these analyses, cDNA microarray analyses were performed for the mentioned two melanoma cell lines (IGR39 and IGR37) and melanocytes to identify putative target genes for transcriptional regulation of the newly identified miRNAs. An inverse correlation between miRNAs and target gene expression was detected for receptor tyrosine kinase c-KIT which was up-regulated in metastasis-derived IGR37 cells as compared with melanocytes and its targeting miRNA miR-221 which was down-regulated. Differentially expressed genes from cDNA array analyses belonged to 33 pathways. Out of these, authors chose "skin development and function" and found that MITF plays a central role in this gene network. Expression levels of MITF and targeting miRNAs like miR-23a and miR-23b as well as 4 MITF target genes (including miRNAs MiR-363 and MiR-146) were analysed in melanocytes and both cell lines IGR39 and IGR37. With the exception of miR-363, all target genes of MITF, namely *let-7i*, miR-146 and c-KIT correlated with its up-regulation in expression. These findings were suggestive that MITF and corresponding miRNAs might be key regulatory mechanism in malignant melanoma.

There was significant heterogeneity in miRNA expression profiles between cell lines and tissues. Thus, further experimental validation has to be performed. However some of the findings were in line with earlier reports supporting their biological relevance. MiR-200c was shown to be down-regulated in primary melanomas and metastatic melanomas as compared with nevi. Down-regulation of miR-200 family members during tumour progression has also been reported in other cancers [74]. Authors speculated that loss of miR-200c may contribute to epithelial-mesenchymal transition in malignant melanoma via release of repression of ZEB1, a transcription factor which represses E-cadherin transcription, a major prerequisite for epithelial-mesenchymal transition [75]. Interestingly, miR-211 was the most strongly up-regulated miRNA in the metastatic cell line IGR37 compared with the non-metastatic cell line IGR39 in array analyses. This again argues against a direct comparison of *in vitro* and *in vivo* findings, because miR-211 was shown to be down-regulated during metastasis in other experiments [76] (see below).

6.5.2.4 MiR-182

An important role of miR-182 for melanoma progression could be demonstrated in two recent studies [77, 78]. In search for melanoma-associated miRNAs, miRNA genes were analysed which are located in genomic regions with frequent gains and losses in melanoma tissues [58, 77]. Out of a series of differentially expressed miRNAs, miR-182 was significantly higher expressed in melanoma cell lines compared with benign melanocytes, and its expression correlated with tumour progression in tissue microarrays of benign melanocytic nevi, primary, and metastatic melanomas [77]. Functional experiments showed that miR-182 silencing suppressed the invasive capacity of different melanoma cell lines *in vitro* and induced apoptosis in melanoma cells. In contrast, miRNA-182 over-expressing melanoma cells showed increased anchorage-independent growth. Moreover, miR-182-transduced B16F10 melanoma cells showed increased lung metastasis, underscoring the relevance of miR-182 for distant metastasis.

FOXO3 (FKHRL1), FOXO1 (FKHR), MITF, CDKN2C (p18INK4C), CASP3, CASP2, and *FAS* are putative targets miR-182. Of these, only MITF had been shown before to be targeted by miR-182. By use of reporter gene assays with 3′UTRs of the respective gene, both *FOXO3* and *MITF* were shown to be direct targets of miR-182 [77]. It was suggested that FOXO3 and MITF might significantly contribute to the effects of up-regulated miR-182 in melanoma metastasis. Indeed, down-regulation of both genes by RNA interference technology enhanced the migratory and invasive potential of melanoma cell lines, similar to that of over-expressed miR-182. Moreover, the effects of over-expressed miR-182 on the migratory capacity could be reversed by over-expression of FOXO3 and MITF. Together, these experiments showed that miR 182 promotes cell viability of melanoma cells and induces enhanced invasive capacity. Down-regulation of both FOXO1 and MITF may contribute to these processes. However, since an enhanced oncogenic potential after miR-182 over-expression was also induced in melanoma cells with already low MITF expression, other mechanisms may also contribute to this process.

Using a mouse model of liver metastasis, immunocompromized mice received intra-splenic injections of A375 melanoma cells and subsequently intraperitoneal injections of chemically modified anti-miR-182 or negative control anti-miRNA [78]. Treatment of mice with anti-miR-182 resulted in significantly less liver metastasis compared with controls. Similar findings were observed when mice were pre-treated with three doses of ant-miR-182 oligonucleotides and followed by 3 weeks of miRNA-182 administration. Suppression of miR-182 in liver tissues was verified by *real-time* PCR and was accompanied by up-regulation of miR-182 targets ADCY6 and FOXO3. Moreover, mRNA expression profiles of anti-miR-182 treated tumours differed from those of controls, supporting the transcriptional impact of anti-miR-182 on gene expression profiles. Differentially expressed genes included genes involved in cell adhesion, migration and apoptosis. Up-regulated genes after anti-miR-182 treatment included *NFASC, CASP2, NCAM1* and *CLDN17*. CASP2, which had already been found to be a target of miR-182 in the above mentioned study of the same group, is a member of the caspase family of pro-apoptotic genes. Overall, the treatment of mice was well tolerated with no gross abnormalities besides slightly increased bilirubin and increased hepatic enzymes. Authors concluded that targeting of miR-182 might be a promising therapeutic strategy for metastatic melanoma in the future.

6.5.2.5 MiR-211

Melastatin, a member of the transient receptor potential (TRPM) cation channel family, has been shown to be down-regulated during melanoma progression and its expression inversely correlated with the prognosis and melanoma patients [79]. Intron 6 of the *TRPM1* gene harbours the miR-211 gene [59]. The inverse correlation of melastatin expression and melanoma progression was suggestive for a tumour suppressive role of melastatin. However, it remained to be determined whether melastatin or miR-211 exert the tumour-suppressive effects. A recent study showed that alterations in miR-211 levels, but not in melastatin levels, have a dramatic impact on the invasive potential of melanoma cells [76]. This study started with a miRNA library screen of the highly invasive melanoma cell line A375M by testing the impact of miRNAs on its migratory capacity. MiR-211 showed the strongest inhibition of melanoma cell migration of all tested miRNAs, followed by its paralog miR-204 [76]. In further experiments using melanoma short term cultures, two populations of melanomas were identified, one with greatly reduced miR-211 expression and high invasive potential, and another with only mildly reduced miR-211 expression and low invasive activity. Greatly reduced miR-211 levels were associated with more than 20-fold higher invasive activity. Interestingly, in melanoma cells with slightly or not reduced melastatin and miR-211 expression, inhibition of miR-211 but not that of melastatin increased melanoma invasiveness. Vice versa, in melanoma cells with a significant reduction of both genes, increasing miR-211 levels significantly reduced their invasive potential. Modulation of melastatin levels was without effect. Together, these findings provided strong evidence for a particular role of miR-211 in melanoma invasiveness and progression.

In search for gene targets that might explain these findings, a melanoma-specific metastasis gene network was analysed for overlapping patterns between the metastatic genes and miR-211 target genes. Six genes overlapped between miR-211 targets and the melanoma metastasis network (*IGF2R*, *NFAT5*, *TGFBR2*, *FBXW7*, *ANGPT1*, and *VHL*). Further functional experiments showed that knockdown of so-called central node genes of these targets had the same effect on melanoma cell invasiveness as up-regulated miR-211. Of these, *TGFB* has been shown to be related to melanoma progression via tissue and blood vessel invasion in earlier reports.

The role of miR-211 for malignant melanoma pathogenesis was further analysed in a study performing a miRNA expression profiling of the melanoma cell line WM1552C and the benign melanocytic cell line HEM-l [80]. MiRNA-211 was the most significant downregulated miRNA and showed the most consistent changes in expression levels between seven additional melanoma cell lines and melanocytes tested. MiR-211 expression levels in clinical melanoma samples (including primary, regional, lymph node and distal metastatic lesions), were reduced in 21 of 30 of these samples as compared with benign melanocytes. However, expression levels varied widely and high expression of miR-211 was also found in a number of primary and metastatic lesions. Predicted miRNA targets were then compared with differentially expressed genes in the WM1552C cell line and melanocyte line HEM-1. Overall, 26 putative target transcripts were identified whose expression levels were elevated relative to those in HEM-l. Over-expression of miR-211 in melanoma cell lines found 18 putative target transcripts for miR-211, which were down-regulated by expression of miR-211. Nine of these putative target transcripts overlapped between both analyses. These were: *ATP2B1*, *CDH2*, *GLIS3*, *KCNMA1*, *MEIS2*, *NCAM-1*, *NF-AT5*, *PRPF38B*, and *TCF12*. Some of these genes had been implicated in cancer biology. *CDH2, KCNMA1, NCAM-1 and NF-AT5* have been described to be involved in migration of metastatic cells or tissue invasion [81].

KCNMA1 was chosen for further analyses, because the 3′UTR of the KCNMA1 transcript contains strongly predicted target sites of miR-211. KCNMA1 is part of a K+ exporting ion channel, which is modulated by Ca++. KCNMA1 protein expression was significantly lower in normal melanocytes compared with all tested melanoma cell lines. The transcription factor MITF regulates the expression of the *TRPM1* gene and miR-211. In knockdown experiments, it was shown that the extent of reduction in MITF transcript levels correlated with the reduction of TRPM1 and miR-211 expression levels. Thus, the well-known tumour suppressive effect of MITF at high levels might at least in part be mediated via up-regulation of miR-211 and consecutive down-regulation of KCNMA1. MiR-211 over-expressing WM1552C cells showed reduced cell numbers compared with control cells and also a lower migratory capacity. Interestingly, over-expression of KCNMA1 partly restored the migratory capacity of miR-211 over-expressing melanoma cells. Taken together, the study of Mazar and co-workers linked miR-211 as a downstream target of MITF to tumour suppression via inhibition of potassium channel activity [80].

6.5.2.6 MiR-221 and -222

As mentioned above, c-KIT is often down-regulated during melanoma progression [82]. Since c-KIT is a target of miR-221 and miR-222 [83], a recent study addressed the question whether over-expression of both miRNAs might play a role during melanoma progression [84]. Endogenous levels of miR-221 and miR-222 were evaluated in normal human melanocytes and a panel of melanoma cell lines. MiR-221 and miR-222 expression was low in normal human melanocytes but increased according to the level of malignant transformation and aggressiveness of melanoma cells. *In vivo* validation showed that both microRNAs were expressed in primary melanomas and melanoma metastases, with metastases showing higher expression than primary tumours. Benign melanocytic nevi were negative.

Based on earlier findings on the role of tumour suppressor gene promyelocytic leukemia zinc finger (*PLZF*), authors hypothesized that PLZF, the expression of which is often lost during melanoma development, might be involved in miR-221 and miR-222 up-regulation during tumour progression [85]. Indeed, after PLZF over-expression in melanoma cell lines, both microRNAs were down-regulated. Further analysis of the regulatory regions of both miRNA genes identified two putative binding sites for the PLZF transcription factor in the 5′ region and in the intergenic region between both miRNAs. Electrophorectic mobility shift assays and chromatin immunoprecipitation showed that at least binding sites in the 5′ region were used by PLZF. Over-expression of miR-221 and miR-222 in the Me1402/R melanoma cell line which has low endogenous expression of both miRNAs increased proliferation rate and invasive capacity. Moreover, miRNA-transduced melanoma cells produced significantly greater tumours in mouse experiments than control cells. Tumour cell proliferation, invasive capacity and *in vivo* tumour formation could significantly be compromised by use of antagomirs directed against miR-221 and miR-222.

A subsequent analysis of the target gene c-KIT in different melanoma cell lines showed an inverse correlation between miR-221 and miR-222 and c-KIT expression depending on the level of aggressiveness of the cell with lines. Moreover, c-KIT expression was reduced by transfection of Me1402/R melanoma cells with miR-221 and miR-222. miR-221 and miR-222 expression also down-regulated MITF expression. Finally, the miR-221 and miR-222 target gene encoding for p27 was analysed in a miR-221 or 222 and PLZF transduced melanoma cell line. p27 was increased by PLZF expression but reduced by miRNA transduction. Functional experiments showed that the negative effects of miRNA antagomir treatment on cellular proliferation with consecutive up-regulation of p27 could be reversed by p27 silencing. Taken together, this study showed that transcription factor PLZF plays a significant role in miR-221 and miR-222 expression in melanoma cells. The induction of the malignant phenotype in melanoma cells in these experiments appeared to be mediated by miR-221- and miR-222-induced loss of c-KIT and p27 expression.

6.5.2.7 MiR-214

The particular role of miRNAs for melanoma metastasis was addressed in a more recent report [86]. By use of the poorly metastatic A375 parental cell line (A375P) and four highly metastatic variants, MA-1, MA-2, MC-1 and MC-2, authors proposed new mechanisms that may account for the metastatic capacity of melanoma cells involving the transcriptional repressor TFAP2C and enhanced cell migration and extravasation. A miRNA expression profiling was performed for the mentioned cell lines and miR-214 was found to be over-expressed in the metastatic cell lines. Interestingly, miR-214 showed strong enhancement of expression in tissue samples of mice directly derived from lung metastases, suggestive for a significant impact of the local microenvironment on miR-214 expression. *In vitro* assays showed that miR-214 overexpression in metastatic MA-2 cells further enhanced cell motility, migration and invasion as shown in wound healing or transwell migration assays. Moreover, a significantly higher number of macroscopic lung metastases was observed in immunocompromized mice for miR-214-overexpressing MA-2 cells compared with control cells. MiR-214 over-expression in A375P and MA-2 cells resulted in a 2–3-fold increase in transendothelial migration and increased extravasation in a lung metastasis model.

Based on a data bank search and functional testing of miR-214 targets combined with a proteomic approach focusing on cell surface proteins, 70 protein-coding genes were identified modulated by miR-214. These included *MMP2*, *CDH1*, *ITGA3*, *MET*, *PAK2*, *TFAP2A* and *TFAP2C*. Further experiments showed that TFAP2C silencing in MA-2 cells showed similar effects as miR-214 over-expression, while its forced expression in miR-214-transfected cells reversed enhanced migration and extravasation *in vivo*. Authors concluded that TFAP2C might be one of the main targets mediating miRNA-induced malignancy in melanoma. In line with this, earlier reports showed that melanoma progression was paralleled by loss of expression of TFAP2 proteins [87]. Among TFAP2-regulated genes were *CDH1*, *CTSD*, *EREG*, *TGFBI*, *CDH11*, *ICAM2*, *ITGAV*, *MMP2*, *TIMP1* and *TIMP2*, all of which are connected with cell movement and tumour biology, and were differentially expressed after miR-214 transfection [86]. Together, miR-214 seems to play an important role in melanoma metastasis via induction of enhanced migration and tissue invasion, which seems to involve TFAP2C and molecules regulated by this transcription factor.

6.5.3 Prognostic miRNA Patterns in Melanoma

A more recent report addressed the question, whether miRNAs or miRNA patterns in melanoma tissues might be predictive regarding long-time survival of melanoma patients [88]. In this study, 59 specimens of formalin-fixed paraffin-embedded metastatic melanoma lesions from different anatomic locations were analysed by miRNA expression profiling using 911 probes. A miRNA signature was identified in metastatic melanoma tissues which allowed the prediction of post-recurrence survival. High expression levels of a set of 18 miRNAs defined a longer survival group using a survival cut-off of 18 months for differentiation between both groups. The 18-miRNA signature of up-regulated miRNAs consisted of miR-150, miR-455-3p, miR-145, miR-342-3p, miR-497, miR-155, miR-342-5p, miR-143, miR-193a-3p, miR-146b-5p, miR-28-3p, miR-10b, miR-193b, miR-28-5p, miR-142-5p, miR-143*, miR-126, and miR-214. Further refinement led to a minimal number of a 6 miRNAs which allowed discrimination between both groups, consisting of miR-150, miR-342-3p, miR-455-3p, miR-145, miR-155, and miR-497. All of these miRNAs showed higher expression in metastatic lesions of the longer survival group.

In support of the biological relevance of these findings, it was further demonstrated that the 6 miRNA signature was even able to stratify stage III (lymph node metastasis) patients into a better and worse prognosis groups. This stratification was even better than the existing one based on the standard classification of stages IIIB and IIIC of the American Joint Committee on Cancer referring to patients with 2–3 lymph nodes affected (stage IIIB) and 4 or more lymph nodes or in transit metastases (stage IIIC). Interestingly, some of the miRNAs of the 6 miRNA signature were even predictive for primary tumours.

None of the signature miRNAs has been related to melanoma prognosis before, but some have been shown to be of relevance in other malignancies. MiR-155 has been shown to play an oncogenic role in hematopoietic malignancies and solid cancers [18, 20]. A recent study, however, showed that miR-155 has an anti-proliferative and pro-apoptotic effect in melanoma cell lines, suggestive for negative impact on melanoma cell aggressiveness [89]. Thus, miRNAs might exert different functions in different cancers.

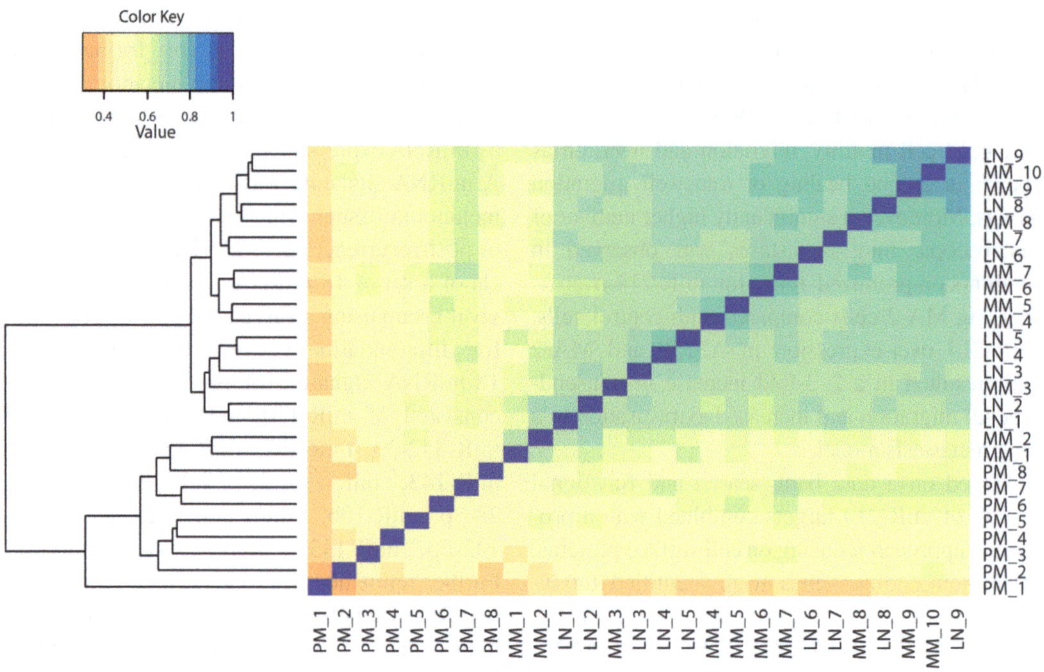

Fig. 6.2 MiRNA expression levels show high correlation in metastatic lesions. MiRNA expression patterns of primary melanomas, lymph node metastases and cutaneous metastases were generated by real-time PCR. Samples were compared based on normalized Ct values of miRNA signals in PCR reactions. Correlation coefficients of pairs of samples are shown (values are between 0 = no correlation and 1 = identical). Samples are hierarchically clustered depending on their correlation coefficient. Note that samples of primary melanomas (PM) do not correlate well, likely due to their biologic heterogeneity. Samples of lymph node (LN) and distant cutaneous metastases (MM) show higher levels of correlation and were clearly separated from primary tumours

MiR-145 was one of the most significantly downregulated regulated miRNAs in breast cancer versus normal control tissue.

In search for prognostic miRNAs or miRNA patterns, another report put a focus on lymph node metastases [90]. MiRNA expression profiling was performed in melanocyte cultures, melanoma cell lines and melanoma lymph node metastases. By applying the Prediction Analysis of Microarray (PAM) software, an expression signature of 10 miRNAs was identified that distinguished between melanocytes and lymph-node metastases. This signature consisted of under-expressed miR-192, *let-7i*, miR-194, miR-211, miR-602, miR-582, miR-454-3p, and miR-132, and over-expressed miR-126 and miR-801 in lymph node metastases. Nine miRNAs distinguished lymph node metastases and melanoma cell lines from melanocytes. Seven out of these (miR-192, miR-194, miR-132, miR-602, miR-211, let-7i, and miR-509) were under-expressed in lymph nodes and melanoma cell lines. In line with this, we were able to clearly distinguish between miRNA patterns in primary melanomas, lymph node and distant cutaneous metastases (unpublished; Fig. 6.2).

Interestingly, 6 miRNAs were identified that could predict disease outcome. Melanomas from patients of a short survival group showed under-expression of miR-191, whereas miR-193b, miR-365, miR-338, let-7i, and miR-193a were over-expressed. Kaplan-Meier survival curves regarding short survival were significant for low expression of miR-191 and high expression of miR-193b, miR-365 and *let-7i*. However, after validation by qRT-PCR, values of miR-365 and *let 7i* were not significant in Kaplan-Meier curves. The prognostic value of miR-191 and miR-193b regarding survival could be verified in an independent cohort of 16 lymph node metastases. Differential expression of miRNAs was also related to *BRAF* and *NRAS* mutation status. Low levels of miR-193a and a combination of both miR-193 and miR-338 were significantly associated with *BRAF* mutations. Taken together, the prognostic miRNAs found in this study may in future serve as a prognostic tool for patients with lymph node metastases.

References

1. Bartel DP (2004) MicroRNAs: genomics, biogenesis, mechanism, and function. Cell 116:281–297
2. Wightman B, Ha I, Ruvkun G (1993) Posttranscriptional regulation of the heterochronic gene lin-14 by lin-4 mediates temporal pattern formation in C. elegans. Cell 75:855–862
3. Bartel DP (2009) MicroRNAs: target recognition and regulatory functions. Cell 136:215–233
4. Grimson A, Farh KK-H, Johnston WK, Garrett-Engele P, Lim LP, Bartel DP (2007) MicroRNA targeting specificity in mammals: determinants beyond seed pairing. Mol Cell 27:91–105
5. Borneman AR, Leigh-Bell JA, Yu H, Bertone P, Gerstein M, Snyder M (2006) Target hub proteins serve as master regulators of development in yeast. Genes Dev 20:435–448
6. Liang H, Li W-H (2007) MicroRNA regulation of human protein protein interaction network. RNA 13:1402–1408
7. Lu J, Getz G, Miska EA, Alvarez-Saavedra E, Lamb J, Peck D, Sweet-Cordero A, Ebert BL, Mak RH, Ferrando AA, Downing JR, Jacks T, Horvitz HR, Golub TR (2005) MicroRNA expression profiles classify human cancers. Nature 435:834–838
8. Esquela-Kerscher A, Slack FJ (2006) Oncomirs – microRNAs with a role in cancer. Nat Rev Cancer 6:259–269
9. Volinia S, Calin GA, Liu C-G, Ambs S, Cimmino A, Petrocca F, Visone R, Iorio M, Roldo C, Ferracin M, Prueitt RL, Yanaihara N, Lanza G, Scarpa A, Vecchione A, Negrini M, Harris CC, Croce CM (2006) A microRNA expression signature of human solid tumors defines cancer gene targets. Proc Natl Acad Sci USA 103:2257–2261
10. Zhang W, Dahlberg JE, Tam W (2007) MicroRNAs in tumorigenesis. Am J Pathol 171:728–738
11. Johnson SM, Grosshans H, Shingara J, Byrom M, Jarvis R, Cheng A, Labourier E, Reinert KL, Brown D, Slack FJ (2005) RAS is regulated by the let-7 microRNA family. Cell 120:635–647
12. Akao Y, Nakagawa Y, Naoe T (2006) let-7 microRNA functions as a potential growth suppressor in human colon cancer cells. Biol Pharm Bull 29:903–906
13. Lopez-Serra P, Esteller M (2012) DNA methylation-associated silencing of tumor-suppressor microRNAs in cancer. Oncogene 31:1609–1622
14. van Kouwenhove M, Kedde M, Agami R (2011) MicroRNA regulation by RNA-binding proteins and its implications for cancer. Nat Rev Cancer 11:644–656
15. He L, Thomson JM, Hemann MT, Hernando-Monge E, Mu D, Goodson S, Powers S, Cordon-Cardo C, Lowe SW, Hannon GJ, Hammond SM (2005) A microRNA polycistron as a potential human oncogene. Nature 435:828–833
16. Mendell JT (2008) miRiad roles for the miR-17-92 cluster in development and disease. Cell 133:217–222
17. Mu P, Han Y-C, Betel D, Yao E, Squatrito M, Ogrodowski P, de Stanchina E, D'Andrea A, Sander C, Ventura A (2009) Genetic dissection of the miR-17 92 cluster of microRNAs in Myc-induced B-cell lymphomas. Genes Dev 23:2806–2811
18. Garzon R, Croce CM (2008) MicroRNAs in normal and malignant hematopoiesis. Curr Opin Hematol 15:352–358
19. Vasilatou D, Papageorgiou S, Pappa V, Papageorgiou E, Dervenoulas J (2010) The role of microRNAs in normal and malignant hematopoiesis. Eur J Haematol 84:1–16
20. Costinean S, Zanesi N, Pekarsky Y, Tili E, Volinia S, Heerema N, Croce CM (2006) Pre-B cell proliferation and lymphoblastic leukemia/high-grade lymphoma in E(mu)-miR155 transgenic mice. Proc Natl Acad Sci USA 103:7024–7029
21. Beitzinger M, Peters L, Zhu JY, Kremmer E, Meister G (2007) Identification of human microRNA targets from isolated argonaute protein complexes. RNA Biol 4:76–84
22. Weinstein IB, Joe A, Felsher D (2008) Oncogene addiction. Cancer Res 68:3077–3080
23. Hanahan D, Weinberg RA (2011) Hallmarks of cancer: the next generation. Cell 144:646–674
24. Medina PP, Nolde M, Slack FJ (2010) OncomiR addiction in an in vivo model of microRNA-21-induced pre-B-cell lymphoma. Nature 467:86–90
25. Ma L, Teruya-Feldstein J, Weinberg RA (2007) Tumour invasion and metastasis initiated by microRNA-10b in breast cancer. Nature 449:682–688
26. Ma L, Weinberg RA (2008) Micromanagers of malignancy: role of microRNAs in regulating metastasis. Trends Genet 24:448–456
27. Tavazoie SF, Alarcón C, Oskarsson T, Padua D, Wang Q, Bos PD, Gerald WL, Massagué J (2008) Endogenous human microRNAs that suppress breast cancer metastasis. Nature 451:147–152
28. Miller AJ, Mihm MC Jr (2006) Melanoma. N Engl J Med 355:51–65
29. Garbe C, Leiter U (2009) Melanoma epidemiology and trends. Clin Dermatol 27:3–9
30. Davies H, Bignell GR, Cox C, Stephens P, Edkins S, Clegg S, Teague J, Woffendin H, Garnett MJ, Bottomley W, Davis N, Dicks E, Ewing R, Floyd Y, Gray K, Hall S, Hawes R, Hughes J, Kosmidou V, Menzies A, Mould C, Parker A, Stevens C, Watt S, Hooper S, Wilson R,

Jayatilake H, Gusterson BA, Cooper C, Shipley J, Hargrave D, Pritchard-Jones K, Maitland N, Chenevix-Trench G, Riggins GJ, Bigner DD, Palmieri G, Cossu A, Flanagan A, Nicholson A, Ho JWC, Leung SY, Yuen ST, Weber BL, Seigler HF, Darrow TL, Paterson H, Marais R, Marshall CJ, Wooster R, Stratton MR, Futreal PA (2002) Mutations of the BRAF gene in human cancer. Nature 417:949–954
31. Curtin JA, Fridlyand J, Kageshita T, Patel HN, Busam KJ, Kutzner H, Cho K-H, Aiba S, Bröcker E-B, LeBoit PE, Pinkel D, Bastian BC (2005) Distinct sets of genetic alterations in melanoma. N Engl J Med 353:2135–2147
32. Chapman PB, Hauschild A, Robert C, Haanen JB, Ascierto P, Larkin J, Dummer R, Garbe C, Testori A, Maio M, Hogg D, Lorigan P, Lebbe C, Jouary T, Schadendorf D, Ribas A, O'Day SJ, Sosman JA, Kirkwood JM, Eggermont AMM, Dreno B, Nolop K, Li J, Nelson B, Hou J, Lee RJ, Flaherty KT, McArthur GA (2011) Improved survival with vemurafenib in melanoma with BRAF V600E mutation. N Engl J Med 364:2507–2516
33. Davies MA, Samuels Y (2010) Analysis of the genome to personalize therapy for melanoma. Oncogene 29:5545–5555
34. Carvajal RD, Antonescu CR, Wolchok JD, Chapman PB, Roman R-A, Teitcher J, Panageas KS, Busam KJ, Chmielowski B, Lutzky J, Pavlick AC, Fusco A, Cane L, Takebe N, Vemula S, Bouvier N, Bastian BC, Schwartz GK (2011) KIT as a therapeutic target in metastatic melanoma. JAMA 305:2327–2334
35. Coleman ML, Marshall CJ, Olson MF (2004) RAS and RHO GTPases in G1-phase cell-cycle regulation. Nat Rev Mol Cell Biol 5:355–366
36. Sherr C (2004) Principles of tumor suppression. Cell 116:235–246
37. Malumbres M, Barbacid M (2005) Mammalian cyclin-dependent kinases. Trends Biochem Sci 30:630–641
38. Chudnovsky Y, Khavari PA, Adams AE (2005) Melanoma genetics and the development of rational therapeutics. J Clin Invest 115:813–824
39. Sharpless NE, Kannan K, Xu J, Bosenberg MW, Chin L (2003) Both products of the mouse Ink4a/Arf locus suppress melanoma formation in vivo. Oncogene 22:5055–5059
40. Kabbarah O, Chin L (2006) Advances in malignant melanoma: genetic insights from mouse and man. Front Biosci 11:928–942
41. Dankort D, Curley DP, Cartlidge RA, Nelson B, Karnezis AN, Damsky WE Jr, You MJ, DePinho RA, McMahon M, Bosenberg M (2009) Braf(V600E) cooperates with Pten loss to induce metastatic melanoma. Nat Genet 41:544–552
42. Yasumoto K, Yokoyama K, Shibata K, Tomita Y, Shibahara S (1833) Microphthalmia-associated transcription factor as a regulator for melanocyte-specific transcription of the human tyrosinase gene. Mol Cell Biol 1995:15
43. Wellbrock C, Marais R (2005) Elevated expression of MITF counteracts B-RAF-stimulated melanocyte and melanoma cell proliferation. J Cell Biol 170:703–708
44. Garraway LA, Widlund HR, Rubin MA, Getz G, Berger AJ, Ramaswamy S, Beroukhim R, Milner DA, Granter SR, Du J, Lee C, Wagner SN, Li C, Golub TR, Rimm DL, Meyerson ML, Fisher DE, Sellers WR (2005) Integrative genomic analyses identify MITF as a lineage survival oncogene amplified in malignant melanoma. Nature 436:117–122
45. Thompson JF, Scolyer RA, Kefford RF (2005) Cutaneous melanoma. Lancet 365:687–701
46. Sondak VK, Han D, Deneve J, Kudchadkar R (2011) Current and planned multicenter trials for patients with primary or metastatic melanoma. J Surg Oncol 104:430–437
47. Rubin KM, Lawrence DP (2009) Your patient with melanoma: staging, prognosis, and treatment. Oncology (Williston Park, NY) 23:13–21
48. Hersey P, Bastholt L, Chiarion-Sileni V, Cinat G, Dummer R, Eggermont AMM, Espinosa E, Hauschild A, Quirt I, Robert C, Schadendorf D (2009) Small molecules and targeted therapies in distant metastatic disease. Ann Oncol 20(Suppl 6):vi35–40
49. Flaherty KT, Puzanov I, Kim KB, Ribas A, McArthur GA, Sosman JA, O'Dwyer PJ, Lee RJ, Grippo JF, Nolop K, Chapman PB (2010) Inhibition of mutated, activated BRAF in metastatic melanoma. N Engl J Med 363:809–819
50. Nazarian R, Shi H, Wang Q, Kong X, Koya RC, Lee H, Chen Z, Lee M-K, Attar N, Sazegar H, Chodon T, Nelson SF, McArthur G, Sosman JA, Ribas A, Lo RS (2010) Melanomas acquire resistance to B-RAF(V600E) inhibition by RTK or N-RAS upregulation. Nature 468:973–977
51. Heidorn SJ, Milagre C, Whittaker S, Nourry A, Niculescu-Duvas I, Dhomen N, Hussain J, Reis-Filho JS, Springer CJ, Pritchard C, Marais R (2010) Kinase-dead BRAF and oncogenic RAS cooperate to drive tumor progression through CRAF. Cell 140:209–221
52. Hodi FS, O'Day SJ, McDermott DF, Weber RW, Sosman JA, Haanen JB, Gonzalez R, Robert C, Schadendorf D, Hassel JC, Akerley W, van den Eertwegh AJM, Lutzky J, Lorigan P, Vaubel JM, Linette GP, Hogg D, Ottensmeier CH, Lebbé C, Peschel C, Quirt I, Clark JI, Wolchok JD, Weber JS, Tian J, Yellin MJ, Nichol GM, Hoos A, Urba WJ (2010) Improved survival with ipilimumab in patients with metastatic melanoma. N Engl J Med 363:711–723
53. Robert C, Thomas L, Bondarenko I, O'Day S, JW MD, Garbe C, Lebbe C, Baurain J-F, Testori A, Grob J-J, Davidson N, Richards J, Maio M, Hauschild A, Miller WH Jr, Gascon P, Lotem M, Harmankaya K, Ibrahim R, Francis S, Chen T-T, Humphrey R, Hoos A, Wolchok JD (2011) Ipilimumab plus dacarbazine for previously untreated metastatic melanoma. N Engl J Med 364:2517–2526
54. Davis ME, Zuckerman JE, Choi CHJ, Seligson D, Tolcher A, Alabi CA, Yen Y, Heidel JD, Ribas A (2010) Evidence of RNAi in humans from systemically administered siRNA via targeted nanoparticles. Nature 464:1067–1070

55. Mueller DW, Bosserhoff AK (2009) Role of miRNAs in the progression of malignant melanoma. Br J Cancer 101:551–556
56. Mueller DW, Bosserhoff A-K (2010) The evolving concept of "melano-miRs"-microRNAs in melanomagenesis. Pigment Cell Melanoma Res 23:620–626
57. Bar-Eli M (2011) Searching for the "melano-miRs": miR-214 drives melanoma metastasis. EMBO J 30:1880–1881
58. Zhang L, Huang J, Yang N, Greshock J, Megraw MS, Giannakakis A, Liang S, Naylor TL, Barchetti A, Ward MR, Yao G, Medina A, O'brien-Jenkins A, Katsaros D, Hatzigeorgiou A, Gimotty PA, Weber BL, Coukos G (2006) MicroRNAs exhibit high frequency genomic alterations in human cancer. Proc Natl Acad Sci USA 103:9136–9141
59. Gaur A, Jewell DA, Liang Y, Ridzon D, Moore JH, Chen C, Ambros VR, Israel MA (2007) Characterization of microRNA expression levels and their biological correlates in human cancer cell lines. Cancer Res 67:2456–2468
60. Johnson CD, Esquela-Kerscher A, Stefani G, Byrom M, Kelnar K, Ovcharenko D, Wilson M, Wang X, Shelton J, Shingara J, Chin L, Brown D, Slack FJ (2007) The let-7 microRNA represses cell proliferation pathways in human cells. Cancer Res 67:7713–7722
61. Schultz J, Lorenz P, Gross G, Ibrahim S, Kunz M (2008) MicroRNA let-7b targets important cell cycle molecules in malignant melanoma cells and interferes with anchorage-independent growth. Cell Res 18:549–557
62. Müller DW, Bosserhoff A-K (2008) Integrin beta 3 expression is regulated by let-7a miRNA in malignant melanoma. Oncogene 27:6698–6706
63. Wölfel T, Hauer M, Schneider J, Serrano M, Wölfel C, Klehmann-Hieb E, De Plaen E, Hankeln T, Meyer zum Büschenfelde KH, Beach D (1995) A p16INK4a-insensitive CDK4 mutant targeted by cytolytic T lymphocytes in a human melanoma. Science 269:1281–1284
64. Flørenes VA, Faye RS, Maelandsmo GM, Nesland JM, Holm R (2000) Levels of cyclin D1 and D3 in malignant melanoma: deregulated cyclin D3 expression is associated with poor clinical outcome in superficial melanoma. Clin Cancer Res 6:3614–3620
65. Flørenes VA, Maelandsmo GM, Faye R, Nesland JM, Holm R (2001) Cyclin A expression in superficial spreading malignant melanomas correlates with clinical outcome. J Pathol 195:530–536
66. Alonso SR, Ortiz P, Pollán M, Pérez-Gómez B, Sánchez L, Acuña MJ, Pajares R, Martínez-Tello FJ, Hortelano CM, Piris MA, Rodríguez-Peralto JL (2004) Progression in cutaneous malignant melanoma is associated with distinct expression profiles: a tissue microarray-based study. Am J Pathol 164:193–203
67. Sauter ER, Yeo U-C, von Stemm A, Zhu W, Litwin S, Tichansky DS, Pistritto G, Nesbit M, Pinkel D, Herlyn M, Bastian BC (2002) Cyclin D1 is a candidate oncogene in cutaneous melanoma. Cancer Res 62:3200–3206
68. Sotillo R, García JF, Ortega S, Martin J, Dubus P, Barbacid M, Malumbres M (2001) Invasive melanoma in Cdk4-targeted mice. Proc Natl Acad Sci USA 98:13312–13317
69. Tessari MA, Gostissa M, Altamura S, Sgarra R, Rustighi A, Salvagno C, Caretti G, Imbriano C, Mantovani R, Del Sal G, Giancotti V, Manfioletti G (2003) Transcriptional activation of the cyclin A gene by the architectural transcription factor HMGA2. Mol Cell Biol 23:9104–9116
70. Lee YS, Dutta A (2007) The tumor suppressor microRNA let-7 represses the HMGA2 oncogene. Genes Dev 21:1025–1030
71. Haflidadóttir BS, Bergsteinsdóttir K, Praetorius C, Steingrímsson E (2010) miR-148 regulates Mitf in melanoma cells. PLoS One 5:e11574
72. Bemis LT, Chen R, Amato CM, Classen EH, Robinson SE, Coffey DG, Erickson PF, Shellman YG, Robinson WA (2008) MicroRNA-137 targets microphthalmia-associated transcription factor in melanoma cell lines. Cancer Res 68:1362–1368
73. Philippidou D, Schmitt M, Moser D, Margue C, Nazarov PV, Muller A, Vallar L, Nashan D, Behrmann I, Kreis S (2010) Signatures of microRNAs and selected microRNA target genes in human melanoma. Cancer Res 70:4163–4173
74. Peter ME (2009) let-7 and miR-200 microRNAs: guardians against pluripotency and cancer progression. Cell Cycle 8:843–852
75. Hurteau GJ, Carlson JA, Spivack SD, Brock GJ (2007) Overexpression of the microRNA hsa-miR-200c leads to reduced expression of transcription factor 8 and increased expression of E-cadherin. Cancer Res 67:7972–7976
76. Levy C, Khaled M, Iliopoulos D, Janas MM, Schubert S, Pinner S, Chen P-H, Li S, Fletcher AL, Yokoyama S, Scott KL, Garraway LA, Song JS, Granter SR, Turley SJ, Fisher DE, Novina CD (2010) Intronic miR-211 assumes the tumor suppressive function of its host gene in melanoma. Mol Cell 40:841–849
77. Segura MF, Hanniford D, Menendez S, Reavie L, Zou X, Alvarez-Diaz S, Zakrzewski J, Blochin E, Rose A, Bogunovic D, Polsky D, Wei J, Lee P, Belitskaya-Levy I, Bhardwaj N, Osman I, Hernando E (2009) Aberrant miR-182 expression promotes melanoma metastasis by repressing FOXO3 and microphthalmia-associated transcription factor. Proc Natl Acad Sci USA 106:1814–1819
78. Huynh C, Segura MF, Gaziel-Sovran A, Menendez S, Darvishian F, Chiriboga L, Levin B, Meruelo D, Osman I, Zavadil J, Marcusson EG, Hernando E (2011) Efficient in vivo microRNA targeting of liver metastasis. Oncogene 30:1481–1488
79. Duncan LM, Deeds J, Cronin FE, Donovan M, Sober AJ, Kauffman M, McCarthy JJ (2001) Melastatin expression and prognosis in cutaneous malignant melanoma. J Clin Oncol 19:568–576
80. Mazar J, DeYoung K, Khaitan D, Meister E, Almodovar A, Goydos J, Ray A, Perera RJ (2010) The regulation of miRNA-211 expression and its role in melanoma cell invasiveness. PLoS One 5:e13779

81. Weaver AK, Bomben VC, Sontheimer H (2006) Expression and function of calcium-activated potassium channels in human glioma cells. Glia 54:223–233
82. Nazarian RM, Prieto VG, Elder DE, Duncan LM (2010) Melanoma biomarker expression in melanocytic tumor progression: a tissue microarray study. J Cutan Pathol 37(Suppl 1):41–47
83. Felli N, Fontana L, Pelosi E, Botta R, Bonci D, Facchiano F, Liuzzi F, Lulli V, Morsilli O, Santoro S, Valtieri M, Calin GA, Liu C-G, Sorrentino A, Croce CM, Peschle C (2005) MicroRNAs 221 and 222 inhibit normal erythropoiesis and erythroleukemic cell growth via kit receptor down-modulation. Proc Natl Acad Sci USA 102:18081–18086
84. Felicetti F, Errico MC, Bottero L, Segnalini P, Stoppacciaro A, Biffoni M, Felli N, Mattia G, Petrini M, Colombo MP, Peschle C, Carè A (2008) The promyelocytic leukemia zinc finger-microRNA-221/-222 pathway controls melanoma progression through multiple oncogenic mechanisms. Cancer Res 68:2745–2754
85. Felicetti F, Bottero L, Felli N, Mattia G, Labbaye C, Alvino E, Peschle C, Colombo MP, Carè A (2004) Role of PLZF in melanoma progression. Oncogene 23:4567–4576
86. Penna E, Orso F, Cimino D, Tenaglia E, Lembo A, Quaglino E, Poliseno L, Haimovic A, Osella-Abate S, De Pittà C, Pinatel E, Stadler MB, Provero P, Bernengo MG, Osman I (2011) Taverna D: microRNA-214 contributes to melanoma tumour progression through suppression of TFAP2C. EMBO J 30:1990–2007
87. Bar-Eli M (2001) Gene regulation in melanoma progression by the AP-2 transcription factor. Pigment Cell Res 14:78–85
88. Segura MF, Belitskaya-Lévy I, Rose AE, Zakrzewski J, Gaziel A, Hanniford D, Darvishian F, Berman RS, Shapiro RL, Pavlick AC, Osman I, Hernando E (2010) Melanoma microRNA signature predicts post-recurrence survival. Clin Cancer Res 16:1577–1586
89. Levati L, Alvino E, Pagani E, Arcelli D, Caporaso P, Bondanza S, Di Leva G, Ferracin M, Volinia S, Bonmassar E, Croce CM, D'Atri S (2009) Altered expression of selected microRNAs in melanoma: antiproliferative and proapoptotic activity of miRNA-155. Int J Oncol 35:393–400
90. Caramuta S, Egyházi S, Rodolfo M, Witten D, Hansson J, Larsson C, Lui W-O (2010) MicroRNA expression profiles associated with mutational status and survival in malignant melanoma. J Invest Dermatol 130:2062–2070

MicroRNAs in the Lung

7

Alexandra Sittka and Bernd Schmeck

Abstract

The lung constitutes one of the most delicate tissue structures in mammalian organisms to accomplish the vital function of gas exchange. On the other hand, its immense surface area, necessary in this context, exhibits the first line of defense against a variety of pro-inflammatory stimuli.

MicroRNAs (miRNAs) are a class of post-transcriptional regulators that revolutionized our view of gene expression regulation. By now, it is well established that miRNAs impair all known cellular and developmental processes. Extensive research over the last years revealed not only a fundamental role for miRNAs in lung development and homeostasis, but also in the process of lung inflammation. Lung inflammation occurs in response to stimuli very different in nature (e.g., physical, radioactive, infective, pro-allergenic, or toxic), and in some cases becomes manifest in chronic diseases (e.g., chronic bronchitis/chronic obstructive pulmonary disease (COPD), asthma and allergic airway diseases) or even lung cancer.

This review chapter will briefly describe the current knowledge concerning miRNA expression and their exerted target regulation in the course of lung inflammation and lung cancer.

Keywords

miRNA • Lung • Inflammation • Cancer • Homeostasis • COPD • Cystic fibrosis • Asthma

A. Sittka • B. Schmeck (✉)
Department of Molecular Pulmonology, German Center for Lung Research, Universities of Giessen & Marburg Lung Center, Philipps-University Marburg, Hans-Meerwein-Str. 2, 35043 Marburg, Germany
e-mail: bernd.schmeck@uni-marburg.de; alexandra.sittka@uni-marburg.de

7.1 Introduction

Among the ten most important causes of death worldwide, four involve the common pathophysiological aspect of inflammation in the

U. Schmitz et al. (eds.), *MicroRNA Cancer Regulation: Advanced Concepts, Bioinformatics and Systems Biology Tools*, Advances in Experimental Medicine and Biology 774,
DOI 10.1007/978-94-007-5590-1_7,

lung: pneumonia, chronic bronchitis/chronic obstructive pulmonary disease (COPD), lung cancer, and tuberculosis. Although thought to be a straightforward series of reactions, inflammation is a delicate matter of homeostasis and is controlled very tightly. Insufficient inflammation might result in overwhelming replication of infectious pathogens or defective/malignant host cells. Excessive inflammation can lead, acutely, to organ dysfunction (e.g., impaired gas exchange in the lung due to leukocyte influx) or multi-organ failure in sepsis or, chronically, to tissue remodelling or destruction (e.g., lung emphysema, requiring lung transplantation) or even the occurrence of malignancies.

This tight regulation of pulmonary inflammation is not trivial as the respiratory tracts, including the alveoli whose surface area is roughly equal to a tennis court, constitute a delicate tissue structure. This immense surface structure is required for every second of the vital function of gas exchange, but, on the other hand, also provides the primary line of defense against the environment, since it is constantly exposed to an airflow containing pro-inflammatory stimuli of complex nature, e.g., physical (low/high temperature and humidity), radioactive (radon as natural gas, e.g. in cellars), infective (bacteria, viruses, fungi, or their components, e.g., lipopolysaccharide in straw), pro-allergenic (pollen, house dust mite excrements), or toxic (cigarette smoke, air pollution, biomass combustion). From a (patho-) physiological point of view, gene expression in the lung should be expected to constitute a highly fine-tuned network including complex feedback mechanisms.

In general, gene expression is subjected to a vast variety of regulatory mechanisms. While regulation on the transcriptional level has been a focus of research for a long time, recently, the discovery of gene regulation by small non-coding RNAs has revolutionized the picture of gene expression. Ever since the first discovery of so-called microRNAs (miRNAs) and their role in post-transcriptional gene regulation in eukaryotes in the early 90ies (*lin-4* in *Caenorhabditis elegans*, [1, 2]), the number of regulatory RNAs has continuously expanded. The research of the last two decades has resolved many aspects of the biogenesis of non-coding RNAs. Moreover, their mode of function to exert target regulation has been the issue of extensive studies. However, target identification for specific miRNAs is still a challenging venture. It has been shown that gene regulation by miRNAs is an interweaved network: not only single miRNAs have been shown to target different mRNAs based on complementarities to shared binding motives in their 3′ untranslated region (3′UTR) [3], but also the regulation of one mRNA by multiple miRNAs is well documented. By now it is well established that miRNAs play key roles in almost every developmental and cellular process.

7.2 MicroRNAs in Lung Homeostasis

For the longest time no information has been available on the impact of miRNAs on lung development or under physiological conditions. One role of miRNAs in lung development appeared likely when *in situ* hybridization experiments revealed a significant expression of *Dicer1* and *Ago* family members in branching regions of the developing mouse lung [4]. The assumption was consolidated by findings of Harris and colleagues, who observed defective lung development in mice carrying a conditional *Dicer* knock-out in the lung epithelium [5]. Here, a significant disruption in epithelial morphogenesis and a strong increase in epithelial cell death were observed in knock-out mice. A comprehensive miRNA expression analysis, comparing neonatal and adult lung tissues of mice, revealed very distinct miRNA expression patterns in the lung [6]. Some miRNAs showing high expression in the adult mice lung were also highly expressed in human adult lung tissue, suggesting an evolutionarily conserved role in lung homeostasis (e.g., miR-26, -29a/b, -142-3p, and -187 are highly expressed in adult lung tissue in mice and humans, respectively [6]). In addition, neonatal mice lungs showed partial overlap in their miRNA expression profile when compared to human fetal lung tissues (e.g., miR-134, -296, -337, to name only a few). Therefore, a developmental role of microRNAs in the vertebrate lung seems undeniable. In a follow-up study, the group of Marc Lindsay analyzed the miRNA expression profile of aging mouse lungs (6 month vs. 18 month of age [7]). In the analysis, miR-26a appeared as the most highly expressed miRNA in lung tissue independent of age. Also, in their previous study, miR-26a was identified as the most highly expressed miRNA in adult mouse lung tissue as well as in human lung tissue [6], strengthening the assumption of miR-26a playing a central role in lung homeostasis. Nevertheless, no significant changes were observed in the miRNA expression profiles in mouse lungs during aging. Albeit highly expressed, for most of the miRNAs identified by Williams and co-workers no functional role in lung development or homeostasis could be assigned.

A miRNA cluster (miR-17~92) seems to play a central role in the development of the mouse lung. While the cluster is highly expressed in the developing lung at embryonic day 11.5 (E11.5), it declines gradually up to E17.5, and is barely expressed in the adult mouse lung [8]. Over-expression of the entire cluster in the lung epithelium of transgenic mouse embryos leads to the absence of terminal air sacs and increased cell numbers with respect to epithelial cells, as determined at E18.5. The data obtained suggest enhanced proliferation of distal progenitor cells as well as delayed differentiation of proximal cells [8]. Ventura and co-workers analyzed in detail the effect of miR-17~92 deletion on developing mice. They reported a strong phenotype of hypoplastic lung and a ventricular septal defect, leading to postnatal lethality [9]. A study from 2009 investigating the role of miR-17 and its homologues, miR-20a and miR-106a, in embryonic lung explants from mice revealed a strong epithelial structure phenotype and, therefore, a role of miR-17 family members during branching morphogenesis [10]. Another screen for miRNAs involved in lung development (from gestational day E16 up to 2 months old adults), performed in rats, resulted in four groups of miRNAs (each containing 4–8 miRNAs) either up- (group 1), down-regulated (group 2 and 3) or peaking at some point in development (group 4) during the course of lung development [11]. Here, the miR-17~92 cluster of miRNAs has also been part of the identified miRNAs (miR-17-5p, miR-18, and miR-20a were comprised in group 3). Further investigation on the role of miR-127 established a role in normal branching and terminal bud formation.

Aside from specific effects on lung development, several miRNAs have been identified as reducing proliferation of the lung cancer cell line A549. Over-expression of a variety of let-7 family members led to a significant decrease in the *in vitro* proliferation of A549 cells, comparable to the effect observed in c-Myc knockdown cells [12]. Ectopic expression of miR-29 family members had a similar effect on the A549 cells *in vitro*. Furthermore, transfection of members of the miR-29 family inhibited the growth of A549 engrafted tumors in mice [13]. Altogether, these results suggest that certain miRNAs act as tumor suppressors in lung cells. Furthermore, prevention of uncontrolled lung inflammation seems to be dependent, rather strictly, on proper miRNA expression. Mice deficient for miR-223 showed massive neutrophilia with highly increased inflammatory infiltration in the interstitium [14]. Another study revealed a role for miR-155 in lung homeostasis. Bic/miR-155 knockout mice showed increased lung remodelling, based on substantial collagen deposition and increased cell mass of myofibroblasts in bronchioles [15]. Further investigation revealed a defect in B- and T-cell responses as well as increased leukocyte numbers in fluids of bronchoalveolar lavages (BAL).

Taken together, there is a vast variety of microRNAs involved in the proper development and homeostasis of the lung and many more await identification.

7.3 MicroRNAs in Lung Inflammation

For more than a decade, the involvement of miRNAs in hematopoiesis has been a focus of research. The group led by David Bartel was the first to publish on the role of miRNAs in the differentiation of hematopoietic progenitor cells [16]. Ever since, the role, not only of miRNAs in cell differentiation in the hematopoietic system, but also in immune cell function, has been the subject of intensive research (reviewed in [17–19]).

The innate immune response provides the first line of defense against pathogens. The extremely large surface structure of the lung bears an enormous risk because of its contact with environmental particles and allergens as well as the invasion of pathogens. Pathogen-associated molecular patterns (PAMPs), like conserved microbial cell wall components or bacterial or viral nucleic acids, are recognized via binding to pattern recognition receptors (PRRs) which are located on the surface (e.g., Toll-like receptors, TLRs) as well as in the cytosol (e.g., NOD-like receptors, NLRs or RIG-like receptors, RLRs) of cells. Following binding of PAMPs to PRRs in concert with accessory proteins, an intracellular signalling cascade is initiated, leading to the activation of specific transcription factors, which in turn result in transcriptional activation of inflammation-related cyto- and chemokines to provide intercellular communication. The mounting of an immune response needs to be regulated very tightly since, aside from a preventive function with respect to diseases, an overwhelming immune response represents a serious risk and might cause severe tissue damage.

David Baltimore's group was the first to exploit the expression changes of various miRNAs in response to PAMPs in order to investigate the potential role of miRNAs in innate immunity. In an initial microarray analysis, they observed an increased expression of microRNA-146, -155, and -132, in response to stimulation of THP-1 cells (a human acute monocytic leukemia cell line) using LPS from *Escherichia coli* [20]. Time-course experiments revealed miR-146a as an immediate early-response gene in the inflammatory response. Further examination of the regulation of miR-146 displayed an induction upon stimulation with various microbial components and pro-inflammatory mediators (e.g., LPS, flagellin, and IL-1ß). Transcriptional activation of the miR-146a gene was shown to be strictly dependent on identified NFκB binding sites in the promoter region. MiR-146a seems to play a pivotal role in containing the mounting immune response, since IL-1 receptor-associated kinase 1 (IRAK1) as well as TNF receptor-associated factor 6 (TRAF6), both of which are involved in TLR-activated signalling cascades to activate NFκB, were identified as mRNA targets that are down-regulated upon binding of miR-146a [20]. A follow-up study performed by the same group analyzed the miRNA profile by means of microarray analysis of macrophages derived of murine bone marrow and stimulated with the synthetic analogue of viral dsRNA (poly(I:C)) or IFN-β, a cytokine highly expressed by the host upon viral infection. Both treatments revealed a significant increase in miR-155 expression upon 6 h of stimulation [21]. Aside from IFN- β, IFN-γ also induced miR-155 expression, even though both cytokines showed a delayed induction as compared to poly(I:C). Further studies revealed the requirement of TNF-α autocrine/paracrine signalling for IFN-mediated miR-155 induction, while poly(I:C) acted independently through signalling via the JNK pathway [21]. Upon LPS stimulation, an up-regulation of miR-155 as well as a slight down-regulation of miR-125b could also be observed in the RAW 264.7 macrophage cell line. Intraperitoneal inoculation of mice using LPS from *Salmonella enteritidis*, with subsequent miRNA analysis of isolated splenocytes confirmed these results *in vivo* [22].

In 2007 Mark Lindsay and colleagues were the first to verify the *in vitro* results in an *in vivo* situation using a mouse model of lung inflammation. Mice treated with aerosolized LPS, when compared to aerosolized saline-treated control animals, showed a significant increase of 104 miRNAs over the time course of 1, 3, and 6 h, respectively, with an average peak of expression at 3 h following treatment in RNA samples purified from whole mouse lungs. Cytokine and chemokine levels measured in BAL fluids peaked at 1 h before going back to baseline levels [23]. Altogether, they observed a LPS-induced expression of 46 different miRNAs at 3 h. Of these, 11 miRNAs were significantly increased at two time points (miR-21, -25, -27b, -100, -140, -142-3p, -181c, -187, -194, -214, and miR-224) while miRNA 223 showed a significant increase of expression at all three time points analyzed. *In situ* hybridization of miR-223 at 3 h following LPS-exposure showed a distinct increase of miR-223 expression in alveolar and bronchial epithelial cells, as well as in cells migrating into the bronchioles. A differential analysis of cell types in BAL fluids revealed that a substantial proportion of these cells represent neutrophils.

The observed chronological order of an increase in cytokine expression measured in BAL fluids followed by an increase in miRNA expression leads to the conclusion that miRNAs might be involved in the regulation rather than in the induction of an innate immune response [23]. Further *in vitro* studies verified miR-146 to be involved in a negative feedback loop regulation of IL-1ß-induced inflammatory response in the human lung epithelial cell line, A549. MiR-146a and -146b are both induced upon IL-1ß stimulation in A549 cells (even though induction of miR-146b occurs to a lesser extent) [24]. These changes are most likely to be attributed to the activating role of the transcription factor NFκB, since IL-1ß induced up-regulation of both miRNAs was abolished in the presence of dexamethasone, a corticosteroid known to attenuate the function of NFκB. Since other transcription factors were affected by the dexamethasone treatment, the attenuation of miRNA up-regulation might well be based on the synergy of multiple transcription factors. While induction of miR-146a was observed in different cell types, miR-146b induction could only be observed in A549 cells. Ectopic over-expression of miR-146a led to a significant decrease in IL-1ß induced IL-8 and RANTES secretion in A549 cells and, reciprocally, an increase of IL-8 secretion could be observed upon transfection of miRNA inhibitors, even though this was observed exclusively at high IL-1ß concentrations. This suggests regulation by miRNAs during the course of severe inflammation [24]. Differences in expression between the two isoforms of miR-146 upon IL-1ß stimulation could be attributed, in part, to different signalling pathways. While miR-146a seems to be regulated via NFκB as well as the c-jun N-terminal kinase (JNK)-1/2, miR-146b expression was observed to be up-regulated via MEK-1/2 and JNK-1/2 [25].

Besides these general considerations, inflammatory disease states differ in causative agents (bacteria, viruses, allergens, irritants, toxic substances, and unknown factors), pathomechanisms (hyperproliferation, tissue destruction etc.) and natural history (acute, chronic). Therefore, involvement of miRNAs can be expected to be diverse. Figure 7.1 provides an overview of the pathophysiological principles of different airway diseases and different miRNAs involved herein.

7.3.1 COPD

Chronic obstructive pulmonary disease (COPD) is a common cause of disability and mortality worldwide, and the number of patients is still increasing. Main causes are long term exposure to primary or secondary tobacco smoke, or biomass combustion (e.g., by cooking in developing countries). Recently, it has been discussed that COPD is not a solitary disease but rather a syndrome comprising (a) chronic, cortisone-insensitive, airway inflammation which causes mucus hypersecretion, and fixed bronchus constriction and remodelling, as well as (b) irreversible tissue destruction in terms of bronchiectasis and emphysema [26]. Both aspects progressively pave the way for respiratory and ventilatory failure, supported by a vicious cycle of impaired innate immune mechanisms and recurrent infectious disease exacerbations [27]. Treatment of COPD by long-acting airway dilatators is mainly symptomatic and life-long. In highly industrialized countries, increasing numbers of end-stage COPD patients receive long-term oxygen therapy, non-invasive mechanical ventilation, or even lung transplantation, resulting in a significant socio-economic burden.

Initial studies in rats revealed a statistically significant down-regulation of 24 miRNAs (miR-30c, -124a, -125a, -191, and members of the let-7 family, to name only a few) following exposure to environmental cigarette smoke (ECS) in the

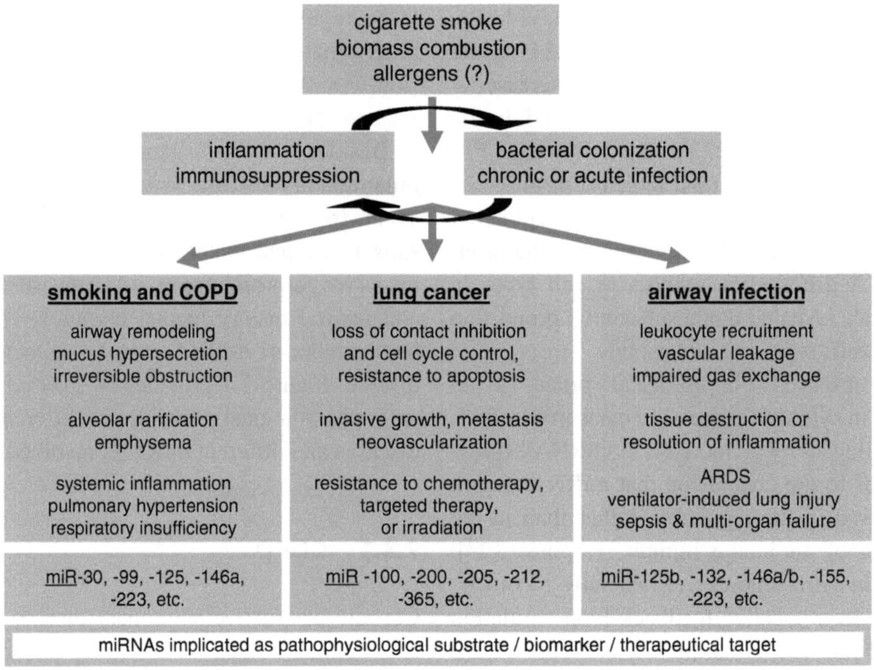

Fig. 7.1 miRNAs in conjunction with pathophysiological principles of inflammatory airway diseases

whole lung tissue while only one miRNA (miR-294) was increased following treatment [28]. Target analysis based on data available in the literature and the use of different target prediction programs showed that targets of the aforementioned 24 miRNAs that appeared down-regulated following ECS exposure were found to be involved in functions like stress response and cell proliferation, including oncogenes as well as tumor suppressor genes. A follow-up study by Izotti and colleagues conducted in mice also revealed predominant down-regulation of miRNAs in mice exposed to ECS [29]. Several miRNAs down-regulated in mice and rats (miR-30, miR-99, and miR-125) as well as two families of miRNAs that were down-regulated in rats only (miR-146 and miR-223) were also found to be down-regulated in humans in a study comparing miRNA profiles of human bronchial epithelial cells between current smokers and persons who had never smoked (Fig. 7.1). In line with the rodent studies, a large number of miRNAs appeared down-regulated in the population of current smokers (with miR-218, -15a, -199b, -125a/b being strongly down-regulated) [30]. Davidson and co-workers confirmed the down-regulation of miRNA-218 in association with smoking in a study investigating miRNA expression in human lung squamous cell carcinoma [31].

The role of miR-146a in COPD was investigated in more detail by Sato and colleagues. While, in general, down-regulation of miR-146a could be related to exposure to ECS in rats as well as to a smoking history in humans [28, 30], this effect is even more pronounced in a study comparing fibroblasts from a control population with fibroblasts from COPD patients. Both groups were closely matched for age and smoking status [32]. The inability of COPD patients to repair tissue could be partially attributed to an increase in prostaglandine (PG) E_2, an inflammatory mediator known to be highly expressed in COPD fibroblasts in culture [33]. Following treatment of control as well as COPD fibroblasts with pro-inflammatory cytokines (IL-1ß and TNF-α), a statistically significant increase in PGE_2 could be observed in both fibroblast populations, while an increase in PGE_2 expression was more pronounced in fibroblasts of COPD patients. This could be attributed to an increased expression of the prostaglandin-endoperoxide synthase 2 (a.k.a. cyclooxygenase-2, COX-2). Further analysis revealed *COX-2* as a direct target of miR-146a, which is induced to a lesser extent by pro-inflammatory cytokines in fibroblasts of COPD patients when compared to fibroblasts of the control group [32]. A study analyzing miRNA expression in the sputum of currently smoking COPD patients to control groups of current smokers and persons who had never smoked observed down-regulation of 27 miRNAs in the group of current smokers when compared to persons who had never smoked. Nevertheless, the comparison of COPD patients to current smokers found no statistically significant differences in miRNA profiles [34]. Taking into account the observations concerning miR-146a induction in COPD patients made by Sato and colleagues [32], it seems likely that even though reduced expression of miRNAs can be observed predominantly in smokers even without airflow obstruction, development of COPD might amplify downstream effects on miRNAs, leading to reduced miRNA induction [33].

7.3.2 Asthma and Allergic Airway Diseases

Asthma and allergic airway diseases, although not primarily mortal, constitute a huge loss in quality of life, and a socio-economic burden. Asthma is characterized by chronic airway inflammation, but paroxysmal and reversible airway obstruction. It can be provoked by drugs, exercise, or "intrinsic reasons", but mainly depends on inhalative allergen exposure. Sensitisation or predisposition can be facilitated by environmental factors even before birth, possibly involving epigenetic mechanisms, or repressed by early life endotoxin exposure. Therapy is based on allergen avoidance, symptomatic pharmacological bronchus dilatation, and suppression of inflammation by corticosteroids.

In 2009 Mattes and co-workers presented, in an initial study, the involvement of miRNAs in the development of allergic airway disease. In a well characterized mouse model of house dust mite (HDM)-induced allergic asthma they investigated the miRNA expression in the airway wall of HDM-challenged mice and reported a significant increase in miRNA-16, -21, and -126 expression as early as 24 h post-treatment when compared to vehicle control [35]. These miRNAs could be further increased following re-challenge of the sensitized mice using HDM. Follow-up experiments revealed signaling through TLR4/Myd88 as the leading pathway to increase expression of the aforementioned miRNAs. Several specific features of allergic airway inflammation, like airway hyper-responsiveness (AHR) to methacholine observed in wild type mice, could be observed in neither *Tlr4*$^{-/-}$, nor in *Myd88*$^{-/-}$ mice. Additionally, a significantly reduced recruitment of eosinophils to the airways, as well as reduction of mucus-producing cells and IL-5 and IL-4 levels, was observed. Treatment of wild type mice using anti-miR-126 led to comparable results, showing strong suppression of effector cytokines IL-4, -5, and -13, normally secreted from T_H2 cells during the course of allergic airway inflammation. Target analysis revealed several genes encoding for immunoglobulin chains, as well as down-regulation of Oct binding factor 1 (OBF.1, a.k.a. B-cell Oct binding protein 1, BOB.1) through miR-126, which also functions as a regulator for transcription factor PU.1, involved in down-regulation of TLR4 and T_H2 responses by suppression of GATA3 [35]. A recent report confirmed up-regulation of miR-21 in the airway walls of HDM-challenged mice in conjunction with up-regulation of let-7b and miR-145 [36]. However, exclusively inhibition of miR-145, but not miR-21 or let-7b, suppressed mucus-hypersecretion and eosinophilic inflammation. Treatment of HDM-challenged mice using anti-miR-145 also dramatically reduced the secretion of IL-5 as well as IL-13 from T_H2 cells, while antagomirs of miR-21 or let-7b showed no effect. Up-regulation of miR-21 was also observed in miRNA-expression profiling performed by Lu and colleagues after IL-13 induction in *Il-13* lung transgenic mice [37]. Of 21 total miRNAs identified as being differentially expressed, miR-21 showed the strongest up-regulation while miR-1 appeared as the microRNA showing the strongest down-regulation following doxycycline-induced IL-13 expression. Up-regulation of miR-21 as well as down-regulation of miR-1 could be further verified, by the same group, in two additional independent asthma models in mice. *In situ* hybridization showed that miR-21 up-regulation occurs mostly in monocytic cells. Target analysis revealed *Il-12p35* as the most prominent target being down-regulated via binding of miR-21 [37]. Surprisingly, a study by Mark Lindsay's lab comparing the miRNA expression profiles in airway biopsy samples from human patients with mild asthma versus healthy volunteers revealed no significant changes in microRNA expression. In addition, comparison of the miRNA expression profiles of patients with asthma did not significantly change after corticosteroid treatment, which significantly improved the mild asthmatic phenotype [38].

A genome-wide comprehensive study in mice based on next generation sequencing (NGS), compared short transcripts in naïve to allergen-challenged mice lungs and revealed a massive increase in mature miRNA transcripts in allergen-challenged lungs, with miRNAs of the let-7 family being most prominent in lung tissue, independent of the allergic state [39]. Aside from changes in miRNA expression patterns, substantial post-transcriptional modification of miRNAs could be observed in naïve as well as in allergen-challenged mice. Target prediction proposed *Il-13* as a target of the let-7 family miRNAs. Even though IL-13 levels are highly increased in allergen-exposed mouse lung, let-7 miRNA levels did not change. *In vitro* analysis of T_H2 cells isolated from mouse lung indeed revealed high levels of *Il-13* and accordingly low levels of let-7a (investigated as a representative for the let-7 family miRNAs). Further analysis using reporter constructs verified the direct interaction of mmu-let-7a with the 3′UTR of *Il-13*. The interaction could also be confirmed in the reporter system using the human *IL-13* 3′UTR in combination with hsa-let-7a. Nevertheless, application of anti-let-7-LNA to allergen-challenged mice markedly repressed the allergic phenotype, suggesting a proinflammatory role for members of the let-7 miRNA family in allergic lung disease in mice *in vitro* [39]. Kumar and colleagues confirmed regulation of IL-13 by let-7 family miRNAs in a later study [40]. Here, intranasal delivery of a let-7 mimic led to reduced IL-13 levels in mice with allergic airway inflammation and alleviated asthma features, suggesting that let-7 family members have an anti-inflammatory effect in allergic airway disease. However, how these results translate into humans needs further examination. Finally, we would like to mention a comprehensive study by Garbacki and co-workers analyzing miRNA- as well as mRNA regulation in a model of acute and chronic asthma. Here, different time-points (short-, intermediate-, and long-term exposure to the allergen) were taken into account, drawing a broad picture of miRNA-regulation during the process from early inflammation to chronically remodelled airways [41]. Throughout the entire study miR-146b appeared as the only miRNA being up-regulated at all time-points analyzed. MiRNAs being regulated at two time-points in the mouse model of asthma included miR-223, -690, -29c, -483, -574-5p, and -672. Analysis of predicted miRNA-mRNA pairs revealed different biological processes being regulated during the course of disease. These involved genes of transcriptional regulation, regulation of the cell cycle, protein metabolism, apoptosis, immunity, inflammation, and cell signalling. The work provided valuable comprehensive information on the molecular mechanisms in the development of asthma.

7.3.3 Cystic Fibrosis

Cystic fibrosis (CF) is the most prevalent genetic disease in Caucasian populations. Originally causing early death, medical progress enables more and more patients to grow up to adulthood. CF is inherited in an autosomal-recessive way. Causative mutations (e.g., ΔF508 as the most common mutation) alter the gene of the cystic fibrosis transmembrane conductance regulator

(CFTR), a chloride-channel expressed in the apical membrane glandular epithelium. This results in high viscosity of mucus in bronchi ("mucoid impactions") and pancreas ducts, enabling recurrent infections and causing chronic inflammation, tissue destruction, as well as impaired organ function.

In a comprehensive analysis of miRNA expression profiles of bronchial brushings from cystic fibrosis patients in comparison to non-CF brushings, Oglesby and co-workers identified 93 miRNAs as being significantly deregulated in 3 out of 5 CF-patients (56 miRNAs were down-regulated while 36 miRNAs were up-regulated) [42]. Further analysis was carried out on expression of miR-126 which appeared significantly decreased in four out of five CF-patients. Down-regulation of miR-126 could be confirmed in a CF tracheal airway epithelial cell line (CFTE) as well as in the CF bronchial epithelial (CFBE) cell line when compared to their non-CF counterpart (HTE and HBE). In addition, substantial expression of miR-126 was only detected in the HBE cell line when compared to six other human non-lung cell lines. Down-regulation of miR-126 could be induced in the normal bronchial epithelial cell line through induction of endoplasmic reticulum (ER) stress. *In silico* target predictions using different bioinformatic tools revealed TOM1 as a target of mir-126. Expression of TOM1 was shown to highly correlate with miR-126 expression *in vitro* as well as *in vivo* samples. Regulation of TOM1 by miR-126 was additionally confirmed using a reporter construct carrying the entire *TOM1* 3′ untranslated region (3′UTR) [42]. This is of particular interest since TOM1 has been shown in a two-hybrid-screen to interact with the adaptor protein Tollip [43]. Tollip in turn had been reported earlier to play a role in IL-1-dependent signalling [44]. A report from 2004 revealed that overexpression of TOM1 suppressed IL-1ß- as well as TNFα-induced activation of NFκB suggesting TOM1 acting as a common repressor [45]. However, this appears contradictory to the strong inflammatory response observed in Cystic Fibrosis. Therefore, additional studies need to be conducted to elucidate how down-regulation of miR-126, and therefore up-regulation of TOM1, may contribute to the pathology of Cystic Fibrosis.

It is well established, that the severe proinflammatory phenotype in CF is characterized by high levels of several proinflammatory cytokines (IL-8, IL-6 and TNFα) in CF airways [46–48]. In a recent study, Bhattacharyya and coworkers performed a miRNA profiling using the IB3-1 CF lung epithelial cell line in comparison to their wild type CFTR-repaired daughter cell line, IB3-1/S9 [49]. Out of 365 miRNAs analyzed, 22 showed significant differential expression (18 showing elevated levels, while four were reduced) in CF epithelial cells. The miRNAs miR-155 and let-7c showed the strongest (over fourfold) up-regulation in CF cells. Up-regulation of miR-155 could also be confirmed *ex vivo* in CF bronchial brushings (compared to normal bronchial epithelial cells) as well as in CF neutrophils compared to normal neutrophils (10.8-fold and 2.4-fold elevated, respectively). Further analysis revealed a direct effect of miR-155 on phosphatidylinositol-3,4,5-triphosphate 5-phosphatase 1 (SHIP 1), which in turn interferes with PI3K signalling to Akt. Therefore, up-regulation of miR-155 leads to down-regulation of *SHIP 1* mRNA. Reduction of SHIP 1 in turn leads to enhanced PI3K signaling to Akt, resulting in mRNA stabilization of *IL-8* mRNA and thereby to increased IL-8 protein levels [49].

Finally, one should mention that the *CFTR* gene itself is subjected to post-transcriptional regulation by microRNAs. Recently, two groups investigated the role of microRNAs in regulation of the *CFTR* gene [50, 51]. Both groups performed *in silico* analysis to identify miRNA responsive elements (MREs) in the *CFTR* 3′UTR. Even though the results are controversial with respect to regulation of *CFTR* 3′UTR by miR-101, both groups validated the predicted miR-494 as a negative regulator of CFTR. Despite showing partially inconsistent results for *CFTR* 3′UTR regulation using a reporter plasmid system in varying cell lines, Gillen and co-workers conclusively showed that three of their predicted miRNAs are highly expressed in primary human airway epithelial cells (miR-145, miR-331-3p, and miR-494) [50], including miR-494, which had also been identified in the screen by Megiorni and colleagues [51].

7.4 MicroRNA in Lung Cancer

Lung cancer is the predominant cause of neoplasia-related death in men, and is affecting more and more women. Due to a lack of efficient strategies for screening or early diagnosis, the prognosis is generally poor and cure rates are low. Cancer cellularity is diverse, comprising so-called small cell lung cancer (SCLC, of neuroendocrine origin), and non-small cell lung cancer (NSCLC, e.g., squamous cell cancer, adenocarcinoma). The causative agent is, for the most part, primary or secondary tobacco smoke; other causes include biomass combustion (e.g., cooking in developing countries), naturally occurring radon etc.

Changes in miRNA expression in lung cancer (Fig. 7.1) may be an epiphenomena of malignancies, or part of the causative chain of carcinogenesis. In that way, microRNA encoding DNA-regions may be spots of tumorigenic mutations themselves. MiRNAs involved in lung cancer could have diverse pathophysiological functions, e.g., in (1) down-regulating the expression of a tumor suppressor gene, (2) failing to suppress oncogenes, (3) promoting metastasis development, or (4) mediating resistance to an anti-cancer treatment.

MiRNA involvement in lung cancer pathogenesis was suggested since reduced Dicer expression correlates with poor survival in NSCLC [52]. In this line, Bishop et al. reported that NSCLC can be correctly classified by miR-205 in lung squamous cell carcinomas or lung adenocarcinomas, which is of relevance for therapy [53]. One example of cancer-correlated miRNA is miR-365, which down-regulates thyroid transcription factor 1 (TTF-1, associated with lung cancer) and high-mobility group AT-hook 2 (HMGA2, which promotes epithelial mesenchymal transition). Its DNA copy number is down-regulated in lung cancer samples [54]. Another mechanism of down-regulation of tumor-suppressive miRNAs lies in chromatin modifications. As an example, down-regulation of miR-212, which sensitizes lung cancer cells for apoptosis, seems to be mediated not by DNA hypermethylation but by histone modifications (H3K27me3 and H3K9me2) associated with transcriptionally inactive chromatin [55].

A great challenge of anti-tumor therapy is the chemo-resistance of some lung cancers. MiR-100 and miR-200b have been found to be down-regulated in cancer cells insensitive to the classical cancer drug docetaxel [56, 57]. Ectopic over-expression of miR-200b reversed the chemo-resistance *in vitro* and in a mouse model, similar to siRNA-mediated down-regulation of the miR-200b target E2F3. Accordingly, decreased miR-200b levels are associated with the poor prognosis of patients with pulmonary adenocarcinoma. So-called "targeted therapies" of lung cancer, addressing distinct molecular targets (e.g., inhibitors of receptor tyrosine kinases (TKI) like the epidermal growth factor receptor (EGFR)), have provoked great hope. Unfortunately, some lung cancers also display primary or acquired resistance to these approaches. Garofalo et al. could demonstrate an involvement of miRNAs in this process [58]: EGFR silencing down-regulates miR-30c, miR-221, and miR-222. In sensitive tumor cells, the synthetic EGFR-TKI gefitinib down-regulates these miRNAs, leading to increased proapoptotic factors BCL-2-like protein 11 (BIM) and apoptotic peptidase activating factor 1 (APAF1). Gefitinib-resistant tumor cells do not down-regulate miR-30c, miR-221 and miR-222 after treatment, and artificial over-expression of these miRNAs renders cells insensitive to treatment.

Another hallmark of cancer progression is metastasis. In a murine system, there seems to be an antagonistic interdependence of the miR-200 family and the transcription factor GATA3 [59]. GATA factors inhibit miR-200 expression and promote epithelial-to-mesenchymal transition (EMT) as well as metastasis. On the other hand, miR-200 targets *Gata3*, thereby blocking EMT and metastasis. Metastatic endothelial recruitment and neovascularization can be suppressed by miR-126, which in turn is down-regulated in many human cancers [60]. First results suggest that miR-378 might be a marker for brain metastasis in NSCLC, a common and severe complication in lung cancer [61].

Recently, a potential therapeutic application of miRNA in lung cancer has been demonstrated by Babar and colleagues: the miR-155 level is elevated in (hypoxic) lung cancer cells and renders these cells insensitive to irradiation, a common therapeutic procedure in this disease [62]. Ectopic over-expression of miR-155 in turn radio-sensitizes lung cancer cells.

Besides potential therapeutic use of miRNAs in lung cancer, e.g., in connection with drugs acting on epigenetic phenomena, the usage of miRNAs as biomarkers seems to be quite imminent. Analysis of miRNAs might facilitate the diagnosis of cancer as well as the determination of the organ origin, especially in adenocarcinoma (lung, stomach, gut, thyroid gland etc.), the underlying mutations, the grade of malignity, the probability and location of metastasis, and the sensitivity to different treatment regimes. However, it is necessary to distinguish potential lung cancer related miRNA changes from alterations which are provoked by cigarette smoke alone, other smoking-associated diseases like COPD, aging, or mere chronic inflammation, e.g., in atherosclerosis [63]. Different types of biological samples may be of use for biomarker characterization:

- Patient serum has been used in several studies now, e.g., by microarray analysis [64]. By deep-sequencing technology, 8 miRNAs have been detected as diagnostic markers [65].
- In exosomes, miRNA is detectable and seems to have a high stability [66].
- Even in whole blood samples, miRNAs have been measured, although this means an integration of RNA from serum, leukocytes, a huge majority of erythrocytes, and other circulating cells (stem cells, tumor cells etc.). Using a next generation sequencing approach, Keller et al. could detect 76 previously unknown miRNAs in peripheral blood of which 7 were altered in patients with lung cancer [67].

Taken together, miRNAs are quite likely to have a clinical usage as biomarkers in lung cancer patients in the near future. However, the details of the most advantageous patient material, the most efficient technology, and the right portfolio of miRNAs, have to be clarified in further studies.

Based on (1) the current knowledge of miRNA involvement in many lung diseases, and (2) the tremendous complexity and challenges on the way to their usage in diagnosis and treatment, we strongly urge special efforts in the line of Medical Systems Biology/Systems Medicine in this field. Interestingly, the German Federal Government has just established such a platform within the German Centre for Lung Research (Deutsches Zentrum für Lungenforschung).

Acknowledgements We thank many collaborators for fruitful discussion, especially Annalisa Marsico, Julio Vera Gonzales, Martin Vingron, and Xin Lai. Part of this work has been funded by BMBF (Forsys Lung - FKZ 0315256) and DFG (SFB/TR-84, IRTG 1673) to B.S. We would like to apologize to all colleagues whose excellent contributions to the field could not be included in this text due to space constraints.

References

1. Lee RC, Feinbaum RL, Ambros V (1993) The C. elegans heterochronic gene lin-4 encodes small RNAs with antisense complementarity to lin-14. Cell 75(5):843–854
2. Wightman B, Ha I, Ruvkun G (1993) Posttranscriptional regulation of the heterochronic gene lin-14 by lin-4 mediates temporal pattern formation in C. elegans. Cell 75(5):855–862
3. Moss EG, Lee RC, Ambros V (1997) The cold shock domain protein LIN-28 controls developmental timing in C. elegans and is regulated by the lin-4 RNA. Cell 88(5):637–646
4. Lu J, Qian J, Chen F, Tang X, Li C, Cardoso WV (2005) Differential expression of components of the microRNA machinery during mouse organogenesis. Biochem Biophys Res Commun 334(2):319–323
5. Harris KS, Zhang Z, McManus MT, Harfe BD, Sun X (2006) Dicer function is essential for lung epithelium morphogenesis. Proc Natl Acad Sci USA 103(7):2208–2213
6. Williams AE, Moschos SA, Perry MM, Barnes PJ, Lindsay MA (2007) Maternally imprinted microRNAs are differentially expressed during mouse and human lung development. Dev Dyn 236(2):572–580
7. Williams AE, Perry MM, Moschos SA, Lindsay MA (2007) MicroRNA expression in the aging mouse lung. BMC Genomics 8:172
8. Lu Y, Thomson JM, Wong HY, Hammond SM, Hogan BL (2007) Transgenic over-expression of the

microRNA miR-17-92 cluster promotes proliferation and inhibits differentiation of lung epithelial progenitor cells. Dev Biol 310(2):442–453
9. Ventura A, Young AG, Winslow MM, Lintault L, Meissner A, Erkeland SJ, Newman J, Bronson RT, Crowley D, Stone JR et al (2008) Targeted deletion reveals essential and overlapping functions of the miR-17 through 92 family of miRNA clusters. Cell 132(5):875–886
10. Carraro G, El-Hashash A, Guidolin D, Tiozzo C, Turcatel G, Young BM, De Langhe SP, Bellusci S, Shi W, Parnigotto PP et al (2009) MiR-17 family of microRNAs controls FGF10-mediated embryonic lung epithelial branching morphogenesis through MAPK14 and STAT3 regulation of E-Cadherin distribution. Dev Biol 333(2):238–250
11. Bhaskaran M, Wang Y, Zhang H, Weng T, Baviskar P, Guo Y, Gou D, Liu L (2009) MicroRNA-127 modulates fetal lung development. Physiol Genomics 37(3):268–278
12. Johnson CD, Esquela-Kerscher A, Stefani G, Byrom M, Kelnar K, Ovcharenko D, Wilson M, Wang X, Shelton J, Shingara J et al (2007) The let-7 microRNA represses cell proliferation pathways in human cells. Cancer Res 67(16):7713–7722
13. Fabbri M, Garzon R, Cimmino A, Liu Z, Zanesi N, Callegari E, Liu S, Alder H, Costinean S, Fernandez-Cymering C et al (2007) MicroRNA-29 family reverts aberrant methylation in lung cancer by targeting DNA methyltransferases 3A and 3B. Proc Natl Acad Sci USA 104(40):15805–15810
14. Johnnidis JB, Harris MH, Wheeler RT, Stehling-Sun S, Lam MH, Kirak O, Brummelkamp TR, Fleming MD, Camargo FD (2008) Regulation of progenitor cell proliferation and granulocyte function by microRNA-223. Nature 451(7182):1125–1129
15. Rodriguez A, Vigorito E, Clare S, Warren MV, Couttet P, Soond DR, van Dongen S, Grocock RJ, Das PP, Miska EA et al (2007) Requirement of bic/microRNA-155 for normal immune function. Science 316(5824):608–611
16. Chen CZ, Li L, Lodish HF, Bartel DP (2004) MicroRNAs modulate hematopoietic lineage differentiation. Science 303(5654):83–86
17. Havelange V, Garzon R (2010) MicroRNAs: emerging key regulators of hematopoiesis. Am J Hematol 85(12):935–942
18. Malumbres R, Lossos IS (2010) Expression of miRNAs in lymphocytes: a review. Methods Mol Biol 667:129–143
19. Navarro F, Lieberman J (2010) Small RNAs guide hematopoietic cell differentiation and function. J Immunol 184(11):5939–5947
20. Taganov KD, Boldin MP, Chang KJ, Baltimore D (2006) NF-kappaB-dependent induction of microRNA miR-146, an inhibitor targeted to signaling proteins of innate immune responses. Proc Natl Acad Sci USA 103(33):12481–12486
21. O'Connell RM, Taganov KD, Boldin MP, Cheng G, Baltimore D (2007) MicroRNA-155 is induced during the macrophage inflammatory response. Proc Natl Acad Sci USA 104(5):1604–1609
22. Tili E, Michaille JJ, Cimino A, Costinean S, Dumitru CD, Adair B, Fabbri M, Alder H, Liu CG, Calin GA et al (2007) Modulation of miR-155 and miR-125b levels following lipopolysaccharide/TNF-alpha stimulation and their possible roles in regulating the response to endotoxin shock. J Immunol 179(8):5082–5089
23. Moschos SA, Williams AE, Perry MM, Birrell MA, Belvisi MG, Lindsay MA (2007) Expression profiling in vivo demonstrates rapid changes in lung micro RNA levels following lipopolysaccharide-induced inflammation but not in the anti-inflammatory action of glucocorticoids. BMC Genomics 8:240
24. Perry MM, Moschos SA, Williams AE, Shepherd NJ, Larner-Svensson HM, Lindsay MA (2008) Rapid changes in microRNA-146a expression negatively regulate the IL-1beta-induced inflammatory response in human lung alveolar epithelial cells. J Immunol 180(8):5689–5698
25. Perry MM, Williams AE, Tsitsiou E, Larner-Svensson HM, Lindsay MA (2009) Divergent intracellular pathways regulate interleukin-1beta-induced miR-146a and miR-146b expression and chemokine release in human alveolar epithelial cells. FEBS Lett 583(20):3349–3355
26. McDonough JE, Yuan R, Suzuki M, Seyednejad N, Elliott WM, Sanchez PG, Wright AC, Gefter WB, Litzky L, Coxson HO et al (2011) Small-airway obstruction and emphysema in chronic obstructive pulmonary disease. N Engl J Med 365(17):1567–1575
27. Sethi S, Murphy TF (2008) Infection in the pathogenesis and course of chronic obstructive pulmonary disease. N Engl J Med 359(22):2355–2365
28. Izzotti A, Calin GA, Arrigo P, Steele VE, Croce CM, De Flora S (2009) Downregulation of microRNA expression in the lungs of rats exposed to cigarette smoke. FASEB J 23(3):806–812
29. Izzotti A, Calin GA, Steele VE, Croce CM, De Flora S (2009) Relationships of microRNA expression in mouse lung with age and exposure to cigarette smoke and light. FASEB J 23(9):3243–3250
30. Schembri F, Sridhar S, Perdomo C, Gustafson AM, Zhang X, Ergun A, Lu J, Liu G, Bowers J, Vaziri C et al (2009) MicroRNAs as modulators of smoking-induced gene expression changes in human airway epithelium. Proc Natl Acad Sci USA 106(7):2319–2324
31. Davidson MR, Larsen JE, Yang IA, Hayward NK, Clarke BE, Duhig EE, Passmore LH, Bowman RV, Fong KM (2010) MicroRNA-218 is deleted and downregulated in lung squamous cell carcinoma. PLoS One 5(9):e12560
32. Sato T, Liu X, Nelson A, Nakanishi M, Kanaji N, Wang X, Kim M, Li Y, Sun J, Michalski J et al (2010) Reduced miR-146a increases prostaglandin Ein chronic obstructive pulmonary disease fibroblasts. Am J Respir Crit Care Med 182(8):1020–1029
33. Togo S, Holz O, Liu X, Sugiura H, Kamio K, Wang X, Kawasaki S, Ahn Y, Fredriksson K, Skold CM et al

(2008) Lung fibroblast repair functions in patients with chronic obstructive pulmonary disease are altered by multiple mechanisms. Am J Respir Crit Care Med 178(3):248–260

34. Pottelberge GR, Mestdagh P, Bracke KR, Thas O, Durme YM, Joos GF, Vandesompele J, Brusselle GG (2011) MicroRNA expression in induced sputum of smokers and patients with chronic obstructive pulmonary disease. Am J Respir Crit Care Med 183(7):898–906
35. Mattes J, Collison A, Plank M, Phipps S, Foster PS (2009) Antagonism of microRNA-126 suppresses the effector function of TH2 cells and the development of allergic airways disease. Proc Natl Acad Sci USA 106(44):18704–18709
36. Collison A, Mattes J, Plank M, Foster PS (2011) Inhibition of house dust mite-induced allergic airways disease by antagonism of microRNA-145 is comparable to glucocorticoid treatment. J Allergy Clin Immunol 128(1):160–167, e164
37. Lu TX, Munitz A, Rothenberg ME (2009) MicroRNA-21 is up-regulated in allergic airway inflammation and regulates IL-12p35 expression. J Immunol 182(8):4994–5002
38. Williams AE, Larner-Svensson H, Perry MM, Campbell GA, Herrick SE, Adcock IM, Erjefalt JS, Chung KF, Lindsay MA (2009) MicroRNA expression profiling in mild asthmatic human airways and effect of corticosteroid therapy. PLoS One 4(6):e5889
39. Polikepahad S, Knight JM, Naghavi AO, Oplt T, Creighton CJ, Shaw C, Benham AL, Kim J, Soibam B, Harris RA et al (2010) Proinflammatory role for let-7 microRNAS in experimental asthma. J Biol Chem 285(39):30139–30149
40. Kumar M, Ahmad T, Sharma A, Mabalirajan U, Kulshreshtha A, Agrawal A, Ghosh B (2011) Let-7 microRNA-mediated regulation of IL-13 and allergic airway inflammation. J Allergy Clin Immunol 128(5):1077–1085, e1071-1010
41. Garbacki N, Di Valentin E, Huynh-Thu VA, Geurts P, Irrthum A, Crahay C, Arnould T, Deroanne C, Piette J, Cataldo D et al (2011) MicroRNAs profiling in murine models of acute and chronic asthma: a relationship with mRNAs targets. PLoS One 6(1):e16509
42. Oglesby IK, Bray IM, Chotirmall SH, Stallings RL, O'Neill SJ, McElvaney NG, Greene CM (2010) MiR-126 is downregulated in cystic fibrosis airway epithelial cells and regulates TOM1 expression. J Immunol 184(4):1702–1709
43. Yamakami M, Yoshimori T, Yokosawa H (2003) Toml, a VHS domain-containing protein, interacts with tollip, ubiquitin, and clathrin. J Biol Chem 278(52):52865–52872
44. Burns K, Clatworthy J, Martin L, Martinon F, Plumpton C, Maschera B, Lewis A, Ray K, Tschopp J, Volpe F (2000) Tollip, a new component of the IL-1RI pathway, links IRAK to the IL-1 receptor. Nat Cell Biol 2(6):346–351
45. Yamakami M, Yokosawa H (2004) Toml (target of Myb 1) is a novel negative regulator of interleukin-1- and tumor necrosis factor-induced signaling pathways. Biol Pharm Bull 27(4):564–566
46. Dean TP, Dai Y, Shute JK, Church MK, Warner JO (1993) Interleukin-8 concentrations are elevated in bronchoalveolar lavage, sputum, and sera of children with cystic fibrosis. Pediatr Res 34(2):159–161
47. Richman-Eisenstat JB, Jorens PG, Hebert CA, Ueki I, Nadel JA (1993) Interleukin-8: an important chemoattractant in sputum of patients with chronic inflammatory airway diseases. Am J Physiol 264(4 Pt 1):L413–418
48. Bonfield TL, Panuska JR, Konstan MW, Hilliard KA, Hilliard JB, Ghnaim H, Berger M (1995) Inflammatory cytokines in cystic fibrosis lungs. Am J Respir Crit Care Med 152(6 Pt 1):2111–2118
49. Bhattacharyya S, Balakathiresan NS, Dalgard C, Gutti U, Armistead D, Jozwik C, Srivastava M, Pollard HB, Biswas R (2011) Elevated miR-155 promotes inflammation in cystic fibrosis by driving hyperexpression of interleukin-8. J Biol Chem 286(13):11604–11615
50. Gillen AE, Gosalia N, Leir SH, Harris A (2011) MicroRNA regulation of expression of the cystic fibrosis transmembrane conductance regulator gene. Biochem J 438(1):25–32
51. Megiorni F, Cialfi S, Dominici C, Quattrucci S, Pizzuti A (2011) Synergistic post-transcriptional regulation of the cystic fibrosis transmembrane conductance regulator (CFTR) by miR-101 and miR-494 specific binding. PLoS One 6(10):e26601
52. Karube Y, Tanaka H, Osada H, Tomida S, Tatematsu Y, Yanagisawa K, Yatabe Y, Takamizawa J, Miyoshi S, Mitsudomi T et al (2005) Reduced expression of Dicer associated with poor prognosis in lung cancer patients. Cancer Sci 96(2):111–115
53. Bishop JA, Benjamin H, Cholakh H, Chajut A, Clark DP, Westra WH (2010) Accurate classification of non-small cell lung carcinoma using a novel microRNA-based approach. Clin Cancer Res 16(2):610–619
54. Qi J, Rice SJ, Salzberg AC, Runkle EA, Liao J, Zander DS, Mu D (2012) MiR-365 regulates lung cancer and developmental gene thyroid transcription factor 1. Cell Cycle 11(1):177–186
55. Incoronato M, Urso L, Portela A, Laukkanen MO, Soini Y, Quintavalle C, Keller S, Esteller M, Condorelli G (2011) Epigenetic regulation of miR-212 expression in lung cancer. PLoS One 6(11):e27722
56. Feng B, Wang R, Chen LB (2012, epub was 2011) MiR-100 resensitizes docetaxel-resistant human lung adenocarcinoma cells (SPC-A1) to docetaxel by targeting Plk1. Cancer Lett 317:184–191
57. Feng B, Wang R, Song HZ, Chen LB (2012) MicroRNA-200b reverses chemoresistance of docetaxel-resistant human lung adenocarcinoma cells by targeting E2F3. Cancer 118:3365–3376
58. Garofalo M, Romano G, Di Leva G, Nuovo G, Jeon YJ, Ngankeu A, Sun J, Lovat F, Alder H, Condorelli G et al (2011) EGFR and MET receptor tyrosine kinase-altered microRNA expression induces tumorigenesis and gefitinib resistance in lung cancers. Nat Med 18:74–82

59. Yang Y, Ahn YH, Gibbons DL, Zang Y, Lin W, Thilaganathan N, Alvarez CA, Moreira DC, Creighton CJ, Gregory PA et al (2011) The Notch ligand Jagged2 promotes lung adenocarcinoma metastasis through a miR-200-dependent pathway in mice. J Clin Invest 121(4):1373–1385
60. Png KJ, Halberg N, Yoshida M, Tavazoie SF (2011) A microRNA regulon that mediates endothelial recruitment and metastasis by cancer cells. Nature 481:190–194
61. Chen LT, Xu SD, Xu H, Zhang JF, Ning JF, Wang SF (2012) MicroRNA-378 is associated with non-small cell lung cancer brain metastasis by promoting cell migration, invasion and tumor angiogenesis. Med Oncol 29:1673–1680
62. Babar IA, Czochor J, Steinmetz A, Weidhaas JB, Glazer PM, Slack FJ (2011) Inhibition of hypoxia-induced miR-155 radiosensitizes hypoxic lung cancer cells. Cancer Biol Ther 12(10):908–914
63. Russ R, Slack FJ (2012) Cigarette-smoke-induced dysregulation of MicroRNA expression and its role in lung carcinogenesis. Pulm Med 2012:791234
64. Lodes MJ, Caraballo M, Suciu D, Munro S, Kumar A, Anderson B (2009) Detection of cancer with serum miRNAs on an oligonucleotide microarray. PLoS One 4(7):e6229
65. Chen X, Ba Y, Ma L, Cai X, Yin Y, Wang K, Guo J, Zhang Y, Chen J, Guo X et al (2008) Characterization of microRNAs in serum: a novel class of biomarkers for diagnosis of cancer and other diseases. Cell Res 18(10):997–1006
66. Rabinowits G, Gercel-Taylor C, Day JM, Taylor DD, Kloecker GH (2009) Exosomal microRNA: a diagnostic marker for lung cancer. Clin Lung Cancer 10(1):42–46
67. Keller A, Leidinger P, Borries A, Wendschlag A, Wucherpfennig F, Scheffler M, Huwer H, Lenhof HP, Meese E (2009) MiRNAs in lung cancer – studying complex fingerprints in patient's blood cells by microarray experiments. BMC Cancer 9:353

The E2F1-miRNA Cancer Progression Network

8

Susanne Knoll, Stephan Emmrich, and Brigitte M. Pützer

Abstract

The transcription factor E2F1 exhibits dual properties, acting as a tumor suppressor and oncogene. Cellular stress such as DNA damage or mitogenic signaling leads to the activation of E2F1 as a mediator of apoptosis in the context of a conserved cellular anti-tumorigenic safeguard mechanism. However in highly aggressive chemoresistant tumors like malignant melanoma and prostate/bladder cancer it switches off this role and acts as promoter of cancer progression. Possible reasons for E2F1 mediated aggressiveness are defects in cell death pathways caused by epigenetic inactivation of important tumor suppressor genes, which often occur in late stage cancer and contribute to chemoresistance. Nevertheless exact mechanisms underlying E2Fs role in invasiveness and metastasis are largely unknown. Different reports hint towards the existence of feedback loops between E2F1 and microRNAs (miRNAs or miRs). MiRs are activated by E2F1 and either the transcription factor itself or cellular genes necessary for the growth regulating function of E2F1 are inhibited by different miRNAs. This mutual regulation possibly influences the balance between E2F1s proapoptotic versus prosurvival function. In the following we will summarize some miRNA-E2F1-interactions contributing to a complex regulatory network.

Keywords

E2F1 transcription factor • microRNA • Apoptosis • Cancer

8.1 The E2F History

The E2F family of transcription factors plays a critical role in the control of cell cycle progression by regulating the timely expression of

S. Knoll • S. Emmrich • B.M. Pützer (✉)
Department of Vectorology and Experimental Gene Therapy, Biomedical Research Center, Rostock University Medical Center, Schillingallee 69, 18055 Rostock, Germany
e-mail: brigitte.puetzer@med.uni-rostock.de

U. Schmitz et al. (eds.), *MicroRNA Cancer Regulation: Advanced Concepts, Bioinformatics and Systems Biology Tools*, Advances in Experimental Medicine and Biology 774,
DOI 10.1007/978-94-007-5590-1_8,

genes required for DNA synthesis at the G1/S phase boundary [1]. E2F activity itself is controlled through association with hypophosphorylated retinoblastoma protein RB and the pocket proteins p107 and p130. During cell cycle progression, D-type cyclin associated kinases initiate phosphorylation of RB family proteins, which results in the release of E2F and the transactivation of E2F regulated genes [2]. The E2F family consists of nine members including both "activator E2Fs" (*E2F1-3a*) that are potent transcriptional activators driving G0 cells to cycle, and "repressor E2Fs" (*E2F3b-8*) with weak activation potential that appear to be involved primarily in gene silencing of quiescent or differentiated cells. These family members perform distinct, perhaps overlapping, functions in the control of cell cycle progression and have unique roles during development, tissue homeostasis, and apoptosis [3]. During the course of tumor development cells sustain mutations that disrupt their normal growth control mechanisms. Notably, the p16INK4/RB/E2F pathway is defective in the vast majority of human tumors. This results in deregulated and hyperactive E2F in transformed cells. However, the role of E2F proteins in determining cell fate is not restricted to their effects on cell cycle progression. Compelling evidence indicates that particularly E2F1 can also efficiently induce apoptosis, and depending on other oncogenic mutations that are present, either promotes or inhibits tumorigenesis [3, 4].

In agreement with its oncogenic activity increased expression of E2F1 causes neoplastic transformation of rodent cells and tumor development in tissues from transgenic mice [5, 6]. Moreover, deregulation of E2F1 through overexpression or RB inactivation has been demonstrated to provoke DNA damage and thus could contribute to cancer by inducing mutations [7]. In contrast, a large body of evidence for its function as a fail-safe mechanism that engages cell death pathways to protect organisms from oncogenic transformation and tumor development comes from knockout mice. The apoptotic response to deregulated E2F is best shown by the observation that RB deficient mouse embryos have increased apoptosis, which is suppressed by the loss of E2F1 [8]. Notably, loss of E2F1 impairs the development of pituitary and thyroid tumors in RB-heterozygous mice but promotes tumor incidence in other tissues, implicating that the role of E2F1 in tumorigenesis might be context dependent and tissue specific [9]. Apoptosis related to E2F1 is mediated in a p53-dependent manner and independent of p53. In most cases, induction of cell death by E2F1 occurs via direct transcriptional activation of genes encoding proapoptotic proteins such as p14ARF, a positive regulator of p53, p73, Apaf-1, BH3-only proteins, and caspases or through inhibition of survival and antiapoptotic signaling mediated by NF-κB, Bcl-2, and GRP78 [10–18]. According to its tumor suppressor function, E2F1, in analogy to p53, determines the cellular response to genotoxic stress. Treatment of cells with DNA damaging agents can induce endogenous E2F1 through posttranslational stabilization of the protein, and aberrant expression of this transcription factor has been shown to increase the sensitivity of certain neoplastic cell types to apoptosis when treated combined with genotoxic drugs [19, 20]. Thus, the final decision of whether deregulated E2F1 activity leads to cell survival or death most likely depends on the genetic status or molecular background of a cell. In addition, integration of external signals plays an important role in determining the sensitivity to E2F1 induced apoptosis. Supported by earlier studies, both the PI3K/AKT pathway and the EGFR/Ras/Raf pathway can inhibit E2F1-induced apoptosis [21, 22]. Referring to this, our recent findings indicating that knockdown of E2F1 in highly metastatic melanoma cells inhibits cell migration, invasion and pulmonary metastasis are intriguing [23]. E2F1 induced malignant progression of tumor cells occurs independent from its proliferative activity and involves direct transcriptional activation of the epidermal growth factor receptor EGFR. Together, these data encourage the hypothesis that E2F1 strictly functions as a promoter of survival when pathways that mediate E2F1s apoptotic activity are disabled.

8.2 MiRs in Human Cancer

MicroRNAs are ~20–24 nucleotide (nt) RNAs that negatively regulate eukaryotic gene expression at the post-transcriptional level. They were first discovered in *C. elegans* and were shown to regulate expression of partially complementary mRNAs [24]. MiRs use base-pairing to guide RNA-induced silencing complexes (RISCs) to specific messages with fully or partly complementary sequences. The repression of targeted messages is a common outcome of RISC recruitment and occurs through translational inhibition, accelerated exonucleolytic mRNA decay or site-specific endonucleolytic cleavage in miR–mRNA pairs [25]. For miRNA biogenesis, primary (pri)-miRNA transcripts with stem-loop regions are usually produced by RNA polymerase II, and occasionally by RNA polymerase III [26, 27]. The stem-loop precursor (pre)-miRNA is released by a cleavage event, which is catalyzed by the nuclear microprocessor complex containing the RNase III Drosha and in mammals the cofactor DGCR8. Pre-miRs are bound by the nuclear export factor exportin5, which mediates their transport to the cytoplasm. A distinct RNase III, Dicer, subsequently produces a ~22 base-pair duplex RNA that is composed of the eventual mature miRNA, base-paired to the so-called miRNA* strand [28]. In miRNA duplexes, the strand with the weakest 5′-end base pairing is selected as the mature miRNA and loaded onto an Argonaute (Ago) protein [29]. Agos are guided by the incorporated miRNA to the target mRNA, where the miRISC mediates the repressive effect. Watson-Crick base-pairing to the 5′-end of miRs, especially to the so-called 'seed' that comprises nucleotides 2–7, is crucial for targeting. Imperfect miR–mRNA hybrids with central bulges (nucleotides 9–12) enable translational inhibition or exonucleolytic mRNA decay, although the factors that govern the prevalence of one specific mechanism remain unknown [30]. Owing to their relaxed base-pairing requirements each miR might have hundreds of target transcripts involved in numerous biological categories.

It has become clear that besides their role as regulators in a variety of developmental and physiological processes, miRs are implicated in the development of cancer [31]. About 50% of the annotated human miRs map within fragile regions of chromosomes, which are areas of the genome that are associated with various human cancers [32]. Recent evidence indicates that components of the miRNA machinery and miRs themselves are involved in many cellular processes that are altered in cancer such as differentiation, proliferation and apoptosis [33]. Some miRs exhibit differential expression levels in tumors and have shown capability to affect cellular transformation, tumorigenesis, and metastasis acting either as oncogenes or tumor suppressors depending on their target genes. For example, *miR-15a* and *miR-16-1* have been associated with chronic lymphocytic leukemia [34]. *MiR-21* and *miR-17-5p* are upregulated, while *miR-143* and *miR-145* are downregulated in colorectal cancer [35–37]. *MiR-21* has been reported for its anti-apoptotic effect in glioma and breast cancer, and the let-7 family of miRs is described as a prognostic factor in lung cancer [38–40]. The discovery of these so called oncomirs is currently one of the major goals in cancer research.

8.3 The miR-E2F Interactome: MicroRNAs Regulating E2F1

MicroRNAs target hundreds of mRNAs leading to a complex network, which regulates various cellular processes. Among diverse miR regulated transcription factors, E2F1 represents a particularly interesting target as multiple links between miR activity and cancer development have been described, and E2F1 has dual function related to cancer [41, 42]. On one hand, E2F1 is a mediator of apoptosis, as shown in clinical trials where its high expression leads to increased survival of patients with adjuvant chemoradiation therapy [43]. Contrariwise, in highly aggressive chemoresistant tumors like malignant melanoma, prostate, and bladder cancer it exhibits oncogenic properties [23, 44, 45]. Possible reasons for this are defects in cell death pathways caused by epigenetic inactivation of important tumor suppressor genes [46, 47]. Nevertheless, exact mechanisms concerning E2F1 induced malignant progression are largely unknown. MiR-E2F1-interactions might contribute either to the tumor suppressive or to the oncogenic function of E2F1 [48]. In the following section we will summarize some miRs regulating E2F1, which are divided into oncogenic ones that inhibit apoptosis through E2F1 in early stages of cancer progression and tumor suppressive miRs shown to block the oncogenic outcome of E2F1 in late stage tumors.

8.4 Oncogenic miRs Inhibiting E2F1-Induced Apoptosis

8.4.1 MicroRNA-17-Family Cluster

One of the first studies which revealed that E2F1 is targeted by miRs was by O'Donnell et al. [49]. It was shown that E2F1 is negatively regulated by two miRs of the *miR-17~92* cluster, *miR-17-5p* and *miR-20a*, that arise from the polycistronic transcript *c13orf25* placed in 13q31.3. This chromosomal region is often amplified in B-cell lymphomas and other malignancies, leading to the upregulation of miRs from this cluster [50, 51]. Consistent with the idea that the *miR-17~92* cluster plays an important role in proliferation and survival, its forced expression promotes the high proliferation and undifferentiated phenotype of normal lung cells [52]. In RB inactivated small-cell lung cancer (SCLC), *miR-17~92* counterbalances DNA damage, thereby favouring genomic instability and the acquisition of further oncogenic features [53]. In agreement, knockdown of *miR-17-5p* and *miR-20a* resulted in a significant induction of apoptosis [54]. With respect to the interaction between *miR-17~92* and E2F1, ectopic expression of *miR-17-5p* and *miR-20a* in tumor cells resulted in a strong decrease of E2F1 protein levels without affecting E2F1 mRNA abundance [49]. Accordingly, the levels of *miR-17-5p* were found to inversely correlate with the E2F1 levels in tumor samples from colon cancer patients, suggesting that these miRs promote malignancy in many tissues by rendering cells insensitive to the apoptotic abilities of E2F1 [55].

Translation of E2F1 in malignant HepG2 and HeLa cells is also limited through *miR-106b* and *miR-93* mapping to the *miR-106b~25* cluster [56]. Previous studies indicated that a knockdown of the complete *miR-106b~25* cluster results in reduced proliferation and suppressed anchorage-independent growth in liver cancer, while proapototic phenotypes were induced by antisense-mediated inhibition of these miRs in gastric cancers [56, 57]. Therefore, it can be concluded that *miR-106b~25* like *miR-17~92* prevents excessively high E2F1 expression that may otherwise cause tumor cell apoptosis. In addition, miRNAs from the *miR-106b~25* and the *miR-17~92* cluster are emerging as key modulators of TGFβ tumor suppressor signaling in gastrointestinal and other tumors, interfering with cell cycle arrest and apoptosis when overexpressed [57, 58].

8.5 Tumor Suppressive miRs Blocking the Oncogenic Outcome of E2F1

8.5.1 MicroRNA-106a

In contrast to the above mentioned miRNAs with a generally growth promoting effect mediated by downregulation of E2F1 in its proapoptotic function, *miR-106a* acts as a tumor suppressor by negatively regulating E2F1 expression in glioma cells. Here, E2F1 seems to provide its tumorigenic function in the context of *mir-106a* regulation as low expression of *mir-106a* significantly correlated with high levels of E2F1 protein in high-grade gliomas and vice versa. In this type of cancer *mir-106a* negatively regulates E2F1 levels via translational suppression, thereby inhibiting proliferation and inducing apoptosis [59].

8.5.2 MircoRNA-330-3p

MiR-330-3p acts as a tumor suppressor through negative regulation of E2F1 expression in prostate cancer cells [60]. *MiR-330* belongs to a poorly conserved miRNA family that maps to chromosome 19q13, a locus strongly linked to the aggressiveness of prostate cancer. E2F1 is significantly overexpressed in metastatic prostate cancers and tumors with high levels of E2F1 were shown to have low *miR-330-3p* levels. In this case, *miR-330-3p* exhibits its tumor suppressive activity through E2F1 mediated repression of Akt activation, which has been shown to communicate the major effect of E2F1 in conferring survival advantages [61]. By inhibiting E2F1 the miRNA decreases phosphorylation of Akt kinases, thereby blocking Akt mediated survival signaling and activating apoptosis.

8.5.3 MicroRNA-34 Family

The miR-34 family is composed of three evolutionary conserved members *miR-34a*, *miR-34b*, and *miR-34c*. A first link to the tumor suppressor activity of these miRNAs arose from the observation that their reintroduction in neuroblastoma cell lines inhibits cell proliferation through the induction of caspase-dependent apoptosis. *MiR-34a* was subsequently shown to be directly induced by p53 and to contribute to p53 mediated cell death on promoting apoptosis [62–66]. At the same time, global gene expression microarray analyses have implicated that *miR-34a* functions as a potent suppressor of cell proliferation by modulation of E2F signaling [66]. In colon cancer cells, tumor suppressive *miR-34a* caused senescence-like growth arrest by downregulating E2F1, which is associated with the accumulation of p53 and its downstream target p21. A significant increase of *miR-34a* was also found upon irradiation of hematopoietic lymphoid tissues, paralleled by a decrease in the expression of its prosurvival target genes *NOTCH1*, *MYC*, *CCND1*, and *E2F3* [67]. The nongenotoxic activator of p53 pathway, Nutlin-3, significantly reduced transcription of both B-Myb and E2F1 in p53wt leukemic cells, thus leading to suppression of cell proliferation by p53 [68]. This effect is mediated by *miR-34a* that is upregulated upon Nutilin-3 treatment. Beside *miR-34a* both other family members *miR-34b/c* that are epigenetically silenced by DNA methylation in cancer cell lines of lymph node metastases originated from melanoma, colon, and head and neck were shown to abolish cell motility and tumor growth, and to inhibit metastasis formation in xenograft models after reintroduction [69]. The observed effect was associated with the downregulation of their oncogenic target genes *MYC*, *CDK6*, and *E2F3*. These studies commonly suggest a tumor suppression mechanism by this miRNA family, which is encompassed by modulation of E2F activity. Current findings indicate that E2F1 and p53 share the property of being activated in response to DNA damage followed by the induction of an overlapping set of proapoptotic target genes [70]. Although E2F1 further enhances p53 activity through p14ARF and p73, E2F activity is attenuated by p53 induced expression of the cyclin-dependent kinase inhibitor p21 [71]. Hence, p53 synergizes only with the proapoptotic activity of E2F1, while antagonizing cell cycle progression through this transcription factor. As such, inhibition of E2F1s prosurvival function is recapitulated by the activity of *miR-34* microRNA, thus constituting a dual safety mechanism to counter E2F1 hyperproliferative signaling by supporting p53 apoptotic activity. Since *miR-34a* functions in p53 knockout cells and cell lines with mutant p53 as well, this mechanism may be an essential barrier to cancer progression also in cells that have lost p53 function [66].

8.5.4 MicroRNA-223

MiR-223 is another interesting tumor suppressive miRNA with a pivotal role in granulopoiesis. Pulikkan and colleagues reported first that this miR acts as key effector of the tumor suppressor CCAAT enhancer binding protein α (C/EBPα) in acute myeloid leukemia (AML) [72]. Here, C/EBPα induces *miR-223*, which in turn inhibits tumorigenic E2F1, thereby blocking myeloid cell cycle progression. However, on the other hand C/EBPα has been shown to be deregulated by different oncoproteins or mutated in approximately 10% of AML patients [73–75], leading to weak *miR-223* expression and high E2F1 protein levels. This kind of deregulation may result in the shift from differentiation to proliferation and finally AML.

8.5.5 MicroRNA-15 Family

MiRNA-195 belongs to the *miR-15* family, which members are downregulated in various cancers including hepatocellular carcinoma (HCC). This family harbours the potent tumor suppressors *miR-15* and *miR-16*, both promoting apoptosis by downregulating antiapoptotic B cell lymphoma 2 (Bcl-2) protein and through repression of cell proliferation by targeting cell cycle genes like *CCND1*, *CCND3*, *CCNE*, and *CDK6* [76, 77]. Moreover, the sibling *miR-195* has been shown to dramatically suppress the ability of HCC and colorectal carcinoma cells in colony formation and tumor development in nude mice by blocking G1 to S transition [78]. Their investigation characterized multiple G1/S transition-related molecules, including *CCND1*, *CDK6*, and *E2F3* as direct targets of *miR-195*. In addition, this miR represses RB phosphorylation and the transactivation of downstream E2F target genes, suggesting that *miR-195*, in concordance with its anti-proliferative effect, mediates tumor suppression via inhibition of RB/E2F-signaling.

8.5.6 MicroRNA-205

MiR-205 represents an additional tumor suppressor candidate. Overexpression of *miR-205* reduced E2F1 protein levels in malignant melanoma, thereby blocking their proliferative capacity mediated by E2F regulated Akt phosphorylation. Furthermore, enhanced expression of *miR-205* induced apoptosis, repressed melanoma cell proliferation, colony formation and tumor cell growth, and triggered senescence [79]. Since *miR-205* is located on 1q32, a locus that is often lost in skin cancer, its significant downregulation in metastatic melanoma compared to primary tumors or nevi is obvious. This may be responsible for E2F1 activation frequently observed in metastatic melanoma.

8.6 The Loops: E2F1 Regulates microRNAs

Since miRs themselves are transcribed by RNA polymerase II, they can be regulated by transcription factors. This fact predicts that interesting regulatory loops can be established between genes coding for classic transcription factors and genes coding for miRNAs. According to this, a new level of regulation has been identified, indicating that some miRs are not only important modulators of E2F1 mRNA translation but also themselves are regulated by the E2F1 transcription factor in an autoregulatory feedback loop.

8.6.1 The E2F1–miR-17~92/miR-106b~25–Myc Circuit

E2F1 directly binds to the promoter of the *miR-17~92* cluster activating its transcription, while as described above, miRs encoded by this polycistronic cluster in turn negatively modulate translation of E2F1 mRNA via binding sites in its 3′-untranslated region [80, 81]. This suggests that E2F1, which stimulates its own transcription in a positive autoregulatory loop, prevents its abnormal accumulation in a negative feedback loop by activation of *miR-17~92*. In general, such a fail-safe mechanism against high E2F activity might be important, since deregulated E2F1 can lead to cell death or malignant transformation depending on the cellular context. A similar interaction has been proposed for E2F1 and the *miR-17~92* paralog clusters *miR-106b~25* and *miR-106a~363* [58, 82]. The reciprocal regulation between E2F1 and the miRNA polycistrons results in largely context-dependent net expression levels of the feedback loop components in cancers. Typically miRs can confer a balancing effect on signaling systems that is regulating the relative levels of pathway players to achieve optimal activity. Furthermore miRs may impart a buffering effect, where they protect molecular communication from fluctuations. Since the process of aberrant degenerateness requires several distinct and heritable gene regulatory alterations with phenotypic impact on signaling pathways to yield cancer cells, the combinatorial outcome of these changes is likely to overcome the balancing and buffering effect of the E2F/*miR-17* loop and to produce abnormal expression levels in response to different cellular scenarios.

In this negative feedback loop another layer of complexity was added due to the fact that c-Myc

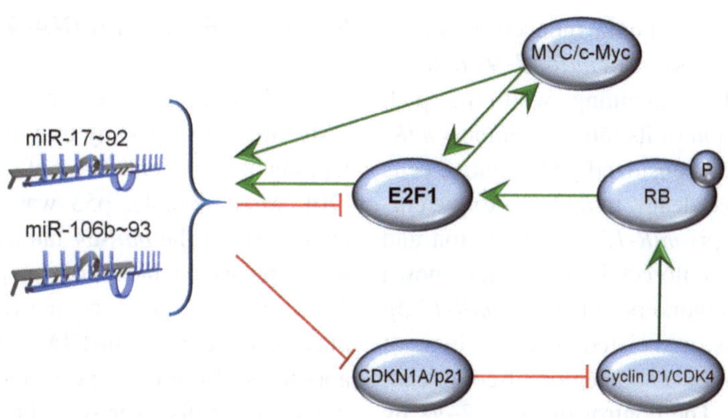

Fig. 8.1 Schematic diagram of the E2F1–miR-17~92/miR-106b~25–Myc loop. This regulatory feedback loop represents the complex interaction of E2F1 with the oncogenic miRNA clusters *miR-17~92* and *miR-106b~25*. OncomiRs of these clusters are induced by E2F1 and c-Myc, and vice versa inhibit E2F1s proapoptotic function. In addition, miR-17~92 can increase Myc enhanced proliferation by targeting *CDKN1A*/p21, resulting in the activation of cyclin D1/CDK4-dependent RB phosphorylation and subsequent release of E2F1. *Green*, activation; *red*, inhibition

activates expression of members from all three paralogous microRNA clusters *miR-17~92*, *miR-106a~363*, and *miR-106b~25* [49, 83]. Thus, the positive regulatory feedback loop of Myc and E2F1 is additionally regulated by Myc induced miRNA clusters [84] (Fig. 8.1). The primary function of oncogenic *miR-17~92* lies in the regulation of cell cycle activities [85]. During mitosis precise timing of E2F1 expression dictates entry into S-phase and accurate timing of E2F1 accumulation requires converging signals from the RB/E2F pathway and the c-Myc regulated *miR-17* and *miR-20a* miRNAs to circumvent a G1 checkpoint arising from the untimely accumulation of E2F1 [86, 87]. In addition, *miR-17~92* can increase Myc enhanced proliferation by targeting *CDKN1A* and consequently activating the cyclin D1/CDK4 complex to release RBs inhibition on E2F [88]. Consistent with the idea that there are multiple levels of regulation during cell cycle transitions, microRNAs in the *miR-106b* family promote G1 exit and cell cycle progression by direct downregulation of *CDKN1A* [89]. Genomic amplification and elevated expression of *miR-17~92* occurs in several human B-cell lymphomas and its enforced expression in mice cooperates with c-Myc to promote formation of B-cell lymphomas [90, 91]. Specifically, *miR-19* was identified as important oncogenic component of this cluster, both necessary and sufficient for promoting c-Myc induced lymphomagenesis by repressing apoptosis [90, 92]. Since E2F1 is known to activate transcription of the *MYC* proto-oncogene, and vice versa, transcriptional activation of the *miR-17~92* and *miR-106b~25* clusters by both transcription factors may represent a way to maintain the level of miRNAs proportional to E2F1 activity [93–95]. The consequences of the coupling between the E2F1/Myc positive feedback loops and the E2F1/Myc/*miR-17~92* negative feedback loop have been analyzed using a mathematical model [96]. This model predicts that *miR-17~92* is critical in regulating the position of the off-on switch in E2F1/Myc protein levels and in determining the on levels of these proteins. Due to the negative feedback loop in the network, large-amplitude protein oscillations were shown to coexist with the off steady state levels, allowing the system to respond through apoptosis to dangerously large perturbations.

8.6.2 p53 Acts as Regulator of E2F1 Controlled miRNAs

Several lines of evidence suggest that the *c13orf25* locus is not only controlled by potentially oncogenic transcription factors. In this context, it was reported that the tumor suppressor protein p53 binds to the *miR-17~92* promoter to repress its transcription during hypoxia induced apoptosis [97]. Enhanced expression of *miR-17~92* reduced apoptosis in cells containing wild-type p53, whereas suppression of its family members *miR-17-5p* and *miR-20a* sensitized p53-negative cells to hypoxia induced death. In addition to the correlation between pri-*miR-17~92* expression and the p53 status in colorectal carcinomas shown here, Diaz and co-workers found that *miR-17-5p* overexpression is associated with a loss of heterozygosity in the *TP53* region of colon cancer patients [55]. The control of *miR-17~92* by p53 further understates the role of this cluster as crucial regulator of cell cycle progression and its function during malignancy. As p53 and E2F1 are both involved in the induction of apoptosis, mutually regulate each other, and both modulate expression of the *miR-17~92* cluster in an opposite manner, this establishes another fascinating network of interactions. Specifically, the Rotter group has shown that p53 repressed miRNAs are involved with E2F in a feed-forward loop promoting proliferation [82]. They demonstrated that the *miR-106b/93/25* polycistron and its paralogs are coordinately activated by E2F1, and importantly, established E2F1 as the mediator of the p53-dependent repression of *miR-106b/93/25*, suggesting that this mechanism underlies the repression of the two additional paralogous polycistrons. Indicative of the significance of this finding, expression of these miRNAs was downregulated in senescent cells and in breast cancers harboring wild-type p53. Together, this regulatory network results in a feed-forward loop similar to that observed with c-Myc, with the difference that p53 represses the miRNA clusters instead of activating them as in case of c-Myc. Thus, it appears conceivable that transcriptional repression of the proliferative *miR-106b/93/25* cluster and their paralogs mediates part of the antiproliferative effects of p53, while Myc exerts its oncogenic function through E2F1 mediated transcriptional activation of these miRs. The involvement of p53 and c-Myc and their opposing modes of target control perfectly reflect their roles in cell proliferation and tumorigenesis with p53 being the classical tumor suppressor and c-Myc the strong oncogene.

8.6.3 E2F1 and miRNA-449 Cluster

Both E2F1 and p53 are strong inducers of apoptosis, at least in part by transactivating an overlapping but not identical set of proapoptotic genes [70]. Most notably, p53 was found to induce expression of the *miR-34* family that contributes to apoptosis on induction by p53 [62–66, 98]. These findings raised the question whether E2F1 may also activate miRNAs that contribute to apoptosis. In this respect, the miRNAs *miR-449a/b* were discovered as direct transcriptional targets of E2F1 [99, 100]. The *miR-449* cluster encoding the highly conserved *miR-449a/b* and the much later described *miR-449c*, structurally resembles the p53 inducible *miR-34* family, thus being classified as one family of microRNAs [101]. In agreement with a putative tumor suppressive function, *miR-449a* and its isomer reduced proliferation and promoted apoptosis by at least partially p53-independent mechanisms [99]. Similar to *miR-34a*, the *miR-449a* was shown capable of negatively regulating the Sirt1 deacetylase, thus augmenting the highly active acetylated p53 and RB levels [99, 102]. In addition, *miR-449a/b* inhibits E2F activity through feedback loops by targeting *CDK2* and *CDC25A*, two oncoproteins that positively regulate phosphorylation of the RB pocket protein, which binds to and inactivates E2F in its hypophosphorylated form [103].

Since *miR-34* and *miR-449* activate p53 while attenuating the activity of E2F1, this strongly argues that they fortify both the proapoptotic activities of E2F1 and p53, but also the mutual regulation of these transcription factors [100]. This asymmetric regulation is already characterized by the fact that E2F1 stabilizes p53 and accessorily enhances the transactivation of p53-responsive genes, whereas p53 leads to E2F1 inactivation through increased transcription of the CDK inhibitor p21. This effect is enforced by *miR-449*. In detail, E2F1 transactivates in response to DNA damage *miR-449,* which activates the p53 pathway, thereby inducing the expression of *miR-34*. In turn *miR-449* and *miR-34* inhibit the E2F pathway in a negative feedback loop and promote growth arrest [101].

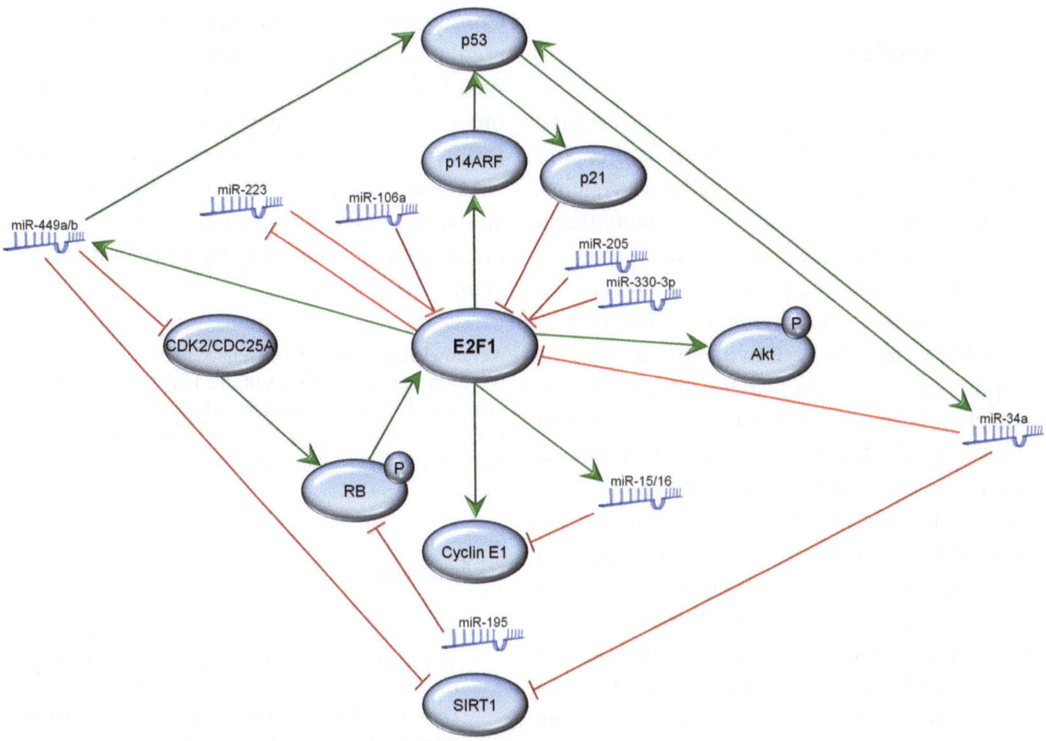

Fig. 8.2 Regulatory network of tumor suppressive miRs and E2F1. MiRs that inhibit the oncogenic function of E2F1 either by directly targeting the transcription factor or indirectly via reactivation of proapoptotic molecules such as p53. *Green*, activation; *red*, inhibition

8.6.4 E2F1 and miR-15 Family

Recently, Ofir et al. demonstrated that expression of *miR-15* and *miR-16* is regulated by E2F1 [77]. Specifically, E2F1 overexpression and induction of endogenous E2Fs resulted in elevated levels of *miR-15a*, *miR-16-1* and *miR-15b*, *miR-16-2*. ChIP analyses revealed a direct binding of E2F1 to their host gene promoters *SMC4* (structural maintenance of chromosomes 4) and *DLEU2*, regulating the expression of these miRNAs. As aforementioned *miR-15* and *miR-16* are known to inhibit cyclin E, a pivotal E2F target gene that plays a critical role in G1/S transition [104]. These data support the existence of a regulatory loop where E2F1 transcriptionally regulates expression of cyclin E, *miRNAs-15*, and *miRNAs-16*, and these miRNAs then repress cyclin E expression. As E2F1 activates both cyclin E and repressors of cyclin E, this regulatory loop is similar to the one described above, where E2F1 regulates the levels of *miR-449a/b* which restrict E2F activity.

8.6.5 C/EBPα–miRNA-223–E2F1-Loop

As already mentioned E2F1 is negatively regulated by *miR-223* in granulopoiesis and deregulation of *miR-223* results in AML. Once this deregulation occurs, E2F1 additionally represses *miR-223* in a negative feedback loop by direct binding to the *miR-223* promoter leading to myeloid cell-cycle progression and transformation [72]. Hence, *miR-223* overexpression might be used as a therapeutic tool to inhibit E2F1 protein levels in AML. The complex interactions between E2F1 and tumor suppressive miRs are summarized in Fig. 8.2.

8.7 Conclusions

E2F1 represents an important transcription factor, which is well known to exert its proapoptotic functions in different cellular systems, thereby enhancing the efficiency of chemotherapeutic treatment [43]. However, once it switches off this role it acts as an oncogene. The regulatory mechanisms underlying this discrepancy are widely unknown. Obviously, defects in apoptotic signaling pathways contribute to the malignant behavior of E2F1 [23, 44, 45], but for a detailed molecular background further investigation is needed. MiRs are a new class of regulatory molecules, which are involved in a variety of cellular processes. As negative regulators of numerous molecules, miRs contribute to cell cycle regulation, development, and cancer [31]. This article outlined relevant interactions between miRs and E2F1, possibly influencing its complex cell context-dependent function. E2F1 regulating miRNAs are divided into oncogenic and tumor suppressive ones, inhibiting either E2F1s proapoptotic or tumorigenic activity. Since tumor suppressive miRs are often downregulated in late stage tumors, they indirectly contribute to the oncogenic function of E2F1. Interestingly, this regulation is often mutual. Different miRs and E2F1 build upregulatory feedback loops leading to the fine-tuning of E2F1 activities. For example E2F1 induces the miR-17~92 cluster, which in turn inhibits translation of the transcription factor, thus preventing its abnormal accumulation in the cell [49, 80, 81]. In addition to direct miRNA-E2F1 interactions, indirect interactions between different miRs and E2F1 regulating proteins also contribute to its cell context-dependent behavior. Important regulators of E2F1 are p53 as a typical tumor suppressor and c-Myc as a strong oncogene [71, 84]. These proteins are either regulated by or regulate themselves E2F1-dependent and -independent miRNAs, thus being involved in the complex network of miRNA-E2F1-mediated signaling [49, 82, 83]. In sum, miRNAs represent a novel level of E2F1 regulation, which is already described by different existing regulatory feedback loops. Consequently miRs give new insights into E2F1 regulation and the molecular background for its final functional outcome, therefore representing interesting tools for new therapeutic strategies. Enforced expression of tumor suppressive miRNAs like, for example miR-205, silenced in apoptosis resistant metastatic tumors could increase chemosensitivity by inhibition of oncogenic E2F1 pathways.

Acknowledgements Work related to this review was supported by grants from the Deutsche Krebshilfe and FORUN program of Rostock University Medical Faculty. S. Knoll receives a fellowship of the Landesgraduiertenförderung des Landes Mecklenburg-Vorpommern.

References

1. DeGregori J (2002) The genetics of the E2F family of transcription factors: shared functions and unique roles. Biochim Biophys Acta 1602:131–150
2. Dyson N (1998) The regulation of E2F by pRB-family proteins. Genes Dev 12:2245–2262
3. Stanelle J, Putzer BM (2006) E2F1-induced apoptosis: turning killers into therapeutics. Trends Mol Med 12:177–185
4. Johnson DG, Degregori J (2006) Putting the oncogenic and tumor suppressive activities of E2F into context. Curr Mol Med 6:731–738
5. Pierce AM, Schneider-Broussard R, Gimenez-Conti IB et al (1999) E2F1 has both oncogenic and tumor-suppressive properties in a transgenic model. Mol Cell Biol 19:6408–6414
6. Xu G, Livingston DM, Krek W (1995) Multiple members of the E2F transcription factor family are the products of oncogenes. Proc Natl Acad Sci USA 92:1357–1361
7. Pickering MT, Kowalik TF (2006) Rb inactivation leads to E2F1-mediated DNA double-strand break accumulation. Oncogene 25:746–755
8. Tsai KY, Hu Y, Macleod KF et al (1998) Mutation of E2f-1 suppresses apoptosis and inappropriate S phase entry and extends survival of Rb-deficient mouse embryos. Mol Cell 2:293–304
9. Yamasaki L, Bronson R, Williams BO et al (1998) Loss of E2F-1 reduces tumorigenesis and extends the lifespan of Rb1(+/−)mice. Nat Genet 18:360–364
10. Bates S, Phillips AC, Clark PA et al (1998) p14ARF links the tumour suppressors RB and p53. Nature 395:124–125
11. Eischen CM, Packham G, Nip J et al (2001) Bcl-2 is an apoptotic target suppressed by both c-Myc and E2F-1. Oncogene 20:6983–6993
12. Furukawa Y, Nishimura N, Furukawa Y et al (2002) Apaf-1 is a mediator of E2F-1-induced apoptosis. J Biol Chem 277:39760–39768

13. Hershko T, Ginsberg D (2004) Up-regulation of Bcl-2 homology 3 (BH3)-only proteins by E2F1 mediates apoptosis. J Biol Chem 279:8627–8634
14. Irwin M, Marin MC, Phillips AC et al (2000) Role for the p53 homologue p73 in E2F-1-induced apoptosis. Nature 407:645–648
15. Nahle Z, Polakoff J, Davuluri RV et al (2002) Direct coupling of the cell cycle and cell death machinery by E2F. Nat Cell Biol 4:859–864
16. Phillips AC, Ernst MK, Bates S et al (1999) E2F-1 potentiates cell death by blocking antiapoptotic signaling pathways. Mol Cell 4:771–781
17. Racek T, Buhlmann S, Rust F et al (2008) Transcriptional repression of the prosurvival endoplasmic reticulum chaperone GRP78/BIP by E2F1. J Biol Chem 283:34305–34314
18. Stiewe T, Putzer BM (2000) Role of the p53-homologue p73 in E2F1-induced apoptosis. Nat Genet 26:464–469
19. Engelmann D, Knoll S, Ewerth D et al (2010) Functional interplay between E2F1 and chemotherapeutic drugs defines immediate E2F1 target genes crucial for cancer cell death. Cell Mol Life Sci 67:931–948
20. Lin WC, Lin FT, Nevins JR (2001) Selective induction of E2F1 in response to DNA damage, mediated by ATM-dependent phosphorylation. Genes Dev 15:1833–1844
21. Hallstrom TC, Mori S, Nevins JR (2008) An E2F1-dependent gene expression program that determines the balance between proliferation and cell death. Cancer Cell 13:11–22
22. Hallstrom TC, Nevins JR (2003) Specificity in the activation and control of transcription factor E2F-dependent apoptosis. Proc Natl Acad Sci USA 100:10848–10853
23. Alla V, Engelmann D, Niemetz A et al (2010) E2F1 in melanoma progression and metastasis. J Natl Cancer Inst 102:127–133
24. Lee RC, Feinbaum RL, Ambros V (1993) The C. elegans heterochronic gene lin-4 encodes small RNAs with antisense complementarity to lin-14. Cell 75:843–854
25. Carthew RW, Sontheimer EJ (2009) Origins and mechanisms of miRNAs and siRNAs. Cell 136:642–655
26. Borchert GM, Lanier W, Davidson BL (2006) RNA polymerase III transcribes human microRNAs. Nat Struct Mol Biol 13:1097–1101
27. Lee Y, Kim M, Han J et al (2004) MicroRNA genes are transcribed by RNA polymerase II. EMBO J 23:4051–4060
28. Kim VN, Han J, Siomi MC (2009) Biogenesis of small RNAs in animals. Nat Rev Mol Cell Biol 10:126–139
29. Tomari Y, Matranga C, Haley B et al (2004) A protein sensor for siRNA asymmetry. Science 306:1377–1380
30. Filipowicz W, Bhattacharyya SN, Sonenberg N (2008) Mechanisms of post-transcriptional regulation by microRNAs: are the answers in sight? Nat Rev Genet 9:102–114
31. Esquela-Kerscher A, Slack FJ (2006) Oncomirs – microRNAs with a role in cancer. Nat Rev Cancer 6:259–269
32. Calin GA, Sevignani C, Dumitru CD et al (2004) Human microRNA genes are frequently located at fragile sites and genomic regions involved in cancers. Proc Natl Acad Sci USA 101:2999–3004
33. Medina PP, Slack FJ (2008) microRNAs and cancer: an overview. Cell Cycle 7:2485–2492
34. Calin GA, Ferracin M, Cimmino A et al (2005) A MicroRNA signature associated with prognosis and progression in chronic lymphocytic leukemia. N Engl J Med 353:1793–1801
35. Bandres E, Cubedo E, Agirre X et al (2006) Identification by real-time PCR of 13 mature microRNAs differentially expressed in colorectal cancer and non-tumoral tissues. Mol Cancer 5:29
36. Lanza G, Ferracin M, Gafa R et al (2007) mRNA/microRNA gene expression profile in microsatellite unstable colorectal cancer. Mol Cancer 6:54
37. Volinia S, Calin GA, Liu CG et al (2006) A microRNA expression signature of human solid tumors defines cancer gene targets. Proc Natl Acad Sci USA 103:2257–2261
38. Chan JA, Krichevsky AM, Kosik KS (2005) MicroRNA-21 is an antiapoptotic factor in human glioblastoma cells. Cancer Res 65:6029–6033
39. Eder M, Scherr M (2005) MicroRNA and lung cancer. N Engl J Med 352:2446–2448
40. Iorio MV, Ferracin M, Liu CG et al (2005) MicroRNA gene expression deregulation in human breast cancer. Cancer Res 65:7065–7070
41. Chen HZ, Tsai SY, Leone G (2009) Emerging roles of E2Fs in cancer: an exit from cell cycle control. Nat Rev Cancer 9:785–797
42. Engelmann D, Putzer BM (2010) Translating DNA damage into cancer cell death – a roadmap for E2F1 apoptotic signalling and opportunities for new drug combinations to overcome chemoresistance. Drug Resist Updat 13:119–131
43. Lee J, Park CK, Park JO et al (2008) Impact of E2F-1 expression on clinical outcome of gastric adenocarcinoma patients with adjuvant chemoradiation therapy. Clin Cancer Res 14:82–88
44. Lee JS, Leem SH, Lee SY et al (2010) Expression signature of E2F1 and its associated genes predict superficial to invasive progression of bladder tumors. J Clin Oncol 28:2660–2667
45. Tuve S, Wagner SN, Schittek B et al (2004) Alterations of DeltaTA-p 73 splice transcripts during melanoma development and progression. Int J Cancer 108:162–166
46. Chin L, Pomerantz J, Polsky D et al (1997) Cooperative effects of INK4a and ras in melanoma susceptibility in vivo. Genes Dev 11:2822–2834
47. Soengas MS, Capodieci P, Polsky D et al (2001) Inactivation of the apoptosis effector Apaf-1 in malignant melanoma. Nature 409:207–211
48. Emmrich S, Putzer BM (2010) Checks and balances: E2F-microRNA crosstalk in cancer control. Cell Cycle 9:2555–2567

49. O'Donnell KA, Wentzel EA, Zeller KI et al (2005) c-Myc-regulated microRNAs modulate E2F1 expression. Nature 435:839–843
50. He L, Thomson JM, Hemann MT et al (2005) A microRNA polycistron as a potential human oncogene. Nature 435:828–833
51. Ota A, Tagawa H, Karnan S et al (2004) Identification and characterization of a novel gene, C13orf25, as a target for 13q31-q32 amplification in malignant lymphoma. Cancer Res 64:3087–3095
52. Lu Y, Thomson JM, Wong HY et al (2007) Transgenic over-expression of the microRNA miR-17-92 cluster promotes proliferation and inhibits differentiation of lung epithelial progenitor cells. Dev Biol 310:442–453
53. Ebi H, Sato T, Sugito N et al (2009) Counterbalance between RB inactivation and miR-17-92 overexpression in reactive oxygen species and DNA damage induction in lung cancers. Oncogene 28:3371–3379
54. Matsubara H, Takeuchi T, Nishikawa E et al (2007) Apoptosis induction by antisense oligonucleotides against miR-17-5p and miR-20a in lung cancers overexpressing miR-17-92. Oncogene 26:6099–6105
55. Diaz R, Silva J, Garcia JM et al (2008) Deregulated expression of miR-106a predicts survival in human colon cancer patients. Genes Chromosomes Cancer 47:794–802
56. Li Y, Tan W, Neo TW et al (2009) Role of the miR-106b-25 microRNA cluster in hepatocellular carcinoma. Cancer Sci 100:1234–1242
57. Petrocca F, Visone R, Onelli MR et al (2008) E2F1-regulated microRNAs impair TGFbeta-dependent cell-cycle arrest and apoptosis in gastric cancer. Cancer Cell 13:272–286
58. Petrocca F, Vecchione A, Croce CM (2008) Emerging role of miR-106b-25/miR-17-92 clusters in the control of transforming growth factor beta signaling. Cancer Res 68:8191–8194
59. Yang G, Zhang R, Chen X et al (2011) MiR-106a inhibits glioma cell growth by targeting E2F1 independent of p53 status. J Mol Med (Berl) 89:1037–1050
60. Lee KH, Chen YL, Yeh SD et al (2009) MicroRNA-330 acts as tumor suppressor and induces apoptosis of prostate cancer cells through E2F1-mediated suppression of Akt phosphorylation. Oncogene 28:3360–3370
61. Chaussepied M, Ginsberg D (2004) Transcriptional regulation of AKT activation by E2F. Mol Cell 16:831–837
62. Chang TC, Wentzel EA, Kent OA et al (2007) Transactivation of miR-34a by p53 broadly influences gene expression and promotes apoptosis. Mol Cell 26:745–752
63. He L, He X, Lim LP et al (2007) A microRNA component of the p53 tumour suppressor network. Nature 447:1130–1134
64. Hermeking H (2010) The miR-34 family in cancer and apoptosis. Cell Death Differ 17:193–199
65. Raver-Shapira N, Marciano E, Meiri E et al (2007) Transcriptional activation of miR-34a contributes to p53-mediated apoptosis. Mol Cell 26:731–743
66. Tazawa H, Tsuchiya N, Izumiya M et al (2007) Tumor-suppressive miR-34a induces senescence-like growth arrest through modulation of the E2F pathway in human colon cancer cells. Proc Natl Acad Sci USA 104:15472–15477
67. Ilnytskyy Y, Zemp FJ, Koturbash I et al (2008) Altered microRNA expression patterns in irradiated hematopoietic tissues suggest a sex-specific protective mechanism. Biochem Biophys Res Commun 377:41–45
68. Zauli G, Voltan R, di Iasio MG et al (2011) miR-34a induces the downregulation of both E2F1 and B-Myb oncogenes in leukemic cells. Clin Cancer Res 17:2712–2724
69. Lujambio A, Calin GA, Villanueva A et al (2008) A microRNA DNA methylation signature for human cancer metastasis. Proc Natl Acad Sci USA 105:13556–13561
70. Polager S, Ginsberg D (2009) p53 and E2f: partners in life and death. Nat Rev Cancer 9:738–748
71. Timmers C, Sharma N, Opavsky R et al (2007) E2f1, E2f2, and E2f3 control E2F target expression and cellular proliferation via a p53-dependent negative feedback loop. Mol Cell Biol 27:65–78
72. Pulikkan JA, Dengler V, Peramangalam PS et al (2010) Cell-cycle regulator E2F1 and microRNA-223 comprise an autoregulatory negative feedback loop in acute myeloid leukemia. Blood 115:1768–1778
73. Nerlov C (2004) C/EBPalpha mutations in acute myeloid leukaemias. Nat Rev Cancer 4:394–400
74. Pabst T, Mueller BU, Zhang P et al (2001) Dominant-negative mutations of CEBPA, encoding CCAAT/enhancer binding protein-alpha (C/EBPalpha), in acute myeloid leukemia. Nat Genet 27:263–270
75. Schuster MB, Porse BT (2006) C/EBPalpha: a tumour suppressor in multiple tissues? Biochim Biophys Acta 1766:88–103
76. Cimmino A, Calin GA, Fabbri M et al (2005) miR-15 and miR-16 induce apoptosis by targeting BCL2. Proc Natl Acad Sci USA 102:13944–13949
77. Ofir M, Hacohen D, Ginsberg D (2011) MiR-15 and miR-16 are direct transcriptional targets of E2F1 that limit E2F-induced proliferation by targeting cyclin E. Mol Cancer Res 9:440–447
78. Xu T, Zhu Y, Xiong Y et al (2009) MicroRNA-195 suppresses tumorigenicity and regulates G1/S transition of human hepatocellular carcinoma cells. Hepatology 50:113–121
79. Dar AA, Majid S, de Semir D et al (2011) miRNA-205 suppresses melanoma cell proliferation and induces senescence via regulation of E2F1 protein. J Biol Chem 286:16606–16614
80. Sylvestre Y, De Guire V, Querido E et al (2007) An E2F/miR-20a autoregulatory feedback loop. J Biol Chem 282:2135–2143
81. Woods K, Thomson JM, Hammond SM (2007) Direct regulation of an oncogenic micro-RNA clus-

ter by E2F transcription factors. J Biol Chem 282:2130–2134

82. Brosh R, Shalgi R, Liran A et al (2008) p53-Repressed miRNAs are involved with E2F in a feed-forward loop promoting proliferation. Mol Syst Biol 4:229
83. Fernandez PC, Frank SR, Wang L et al (2003) Genomic targets of the human c-Myc protein. Genes Dev 17:1115–1129
84. Matsumura I, Tanaka H, Kanakura Y (2003) E2F1 and c-Myc in cell growth and death. Cell Cycle 2:333–338
85. Monzo M, Navarro A, Bandres E et al (2008) Overlapping expression of microRNAs in human embryonic colon and colorectal cancer. Cell Res 18:823–833
86. Osada H, Takahashi T (2011) let-7 and miR-17-92: small-sized major players in lung cancer development. Cancer Sci 102:9–17
87. Pickering MT, Stadler BM, Kowalik TF (2009) miR-17 and miR-20a temper an E2F1-induced G1 checkpoint to regulate cell cycle progression. Oncogene 28:140–145
88. Fontana L, Fiori ME, Albini S et al (2008) Antagomir-17-5p abolishes the growth of therapy-resistant neuroblastoma through p21 and BIM. PLoS One 3:e2236
89. Ivanovska I, Ball AS, Diaz RL et al (2008) MicroRNAs in the miR-106b family regulate p21/CDKN1A and promote cell cycle progression. Mol Cell Biol 28:2167–2174
90. Olive V, Bennett MJ, Walker JC et al (2009) miR-19 is a key oncogenic component of mir-17-92. Genes Dev 23:2839–2849
91. Tagawa H, Karube K, Tsuzuki S et al (2007) Synergistic action of the microRNA-17 polycistron and Myc in aggressive cancer development. Cancer Sci 98:1482–1490
92. Mu P, Han YC, Betel D et al (2009) Genetic dissection of the miR-17 92 cluster of microRNAs in Myc-induced B-cell lymphomas. Genes Dev 23:2806–2811
93. Adams MR, Sears R, Nuckolls F et al (2000) Complex transcriptional regulatory mechanisms control expression of the E2F3 locus. Mol Cell Biol 20:3633–3639
94. Leone G, DeGregori J, Sears R et al (1997) Myc and Ras collaborate in inducing accumulation of active cyclin E/Cdk2 and E2F. Nature 387:422–426
95. Thalmeier K, Synovzik H, Mertz R et al (1989) Nuclear factor E2F mediates basic transcription and trans-activation by E1a of the human MYC promoter. Genes Dev 3:527–536
96. Aguda BD, Kim Y, Piper-Hunter MG et al (2008) MicroRNA regulation of a cancer network: consequences of the feedback loops involving miR-17-92, E2F, and Myc. Proc Natl Acad Sci USA 105:19678–19683
97. Yan HL, Xue G, Mei Q et al (2009) Repression of the miR-17-92 cluster by p53 has an important function in hypoxia-induced apoptosis. EMBO J 28:2719–2732
98. Corney DC, Flesken-Nikitin A, Godwin AK et al (2007) MicroRNA-34b and MicroRNA-34c are targets of p53 and cooperate in control of cell proliferation and adhesion-independent growth. Cancer Res 67:8433–8438
99. Lize M, Pilarski S, Dobbelstein M (2010) E2F1-inducible microRNA 449a/b suppresses cell proliferation and promotes apoptosis. Cell Death Differ 17:452–458
100. Yang X, Feng M, Jiang X et al (2009) miR-449a and miR-449b are direct transcriptional targets of E2F1 and negatively regulate pRb-E2F1 activity through a feedback loop by targeting CDK6 and CDC25A. Genes Dev 23:2388–2393
101. Lize M, Klimke A, Dobbelstein M (2011) MicroRNA-449 in cell fate determination. Cell Cycle 10:2874–2882
102. Yamakuchi M, Ferlito M, Lowenstein CJ (2008) miR-34a repression of SIRT1 regulates apoptosis. Proc Natl Acad Sci USA 105:13421–13426
103. Feng M, Yu Q (2010) miR-449 regulates CDK-Rb-E2F1 through an auto-regulatory feedback circuit. Cell Cycle 9:213–214
104. Ohtani K, DeGregori J, Nevins JR (1995) Regulation of the cyclin E gene by transcription factor E2F1. Proc Natl Acad Sci USA 92:12146–12150

Modeling microRNA-Transcription Factor Networks in Cancer

9

Baltazar D. Aguda

Abstract

An increasing number of transcription factors (TFs) and microRNAs (miRNAs) is known to form feedback loops (FBLs) of interactions where a TF positively or negatively regulates the expression of a miRNA, and the miRNA suppresses the translation of the TF messenger RNA. FBLs are potential sources of instability in a gene regulatory network. Positive FBLs can give rise to switching behaviors while negative FBLs can generate periodic oscillations. This chapter presents documented examples of FBLs and their relevance to stem cell renewal and differentiation in gliomas. Feed-forward loops (FFLs) are only discussed briefly because they do not affect network stability unless they are members of cycles. A primer on qualitative network stability analysis is given and then used to demonstrate the network destabilizing role of FBLs. Steps in model formulation and computer simulations are illustrated using the miR-17-92/Myc/E2F network as an example. This example possesses both negative and positive FBLs.

Keywords

Mathematical modeling • Feedback loops • Feedforward loops • miR-17-92 • E2F • Myc • p53 • Cancer zone • Qualitative network

Abbreviations

CZ	cancer zone
dTF	differentiation transcription factor module
FFL	feed-forward loop
FBL	feedback loop
P_A	apoptosis factors
P_C	cell cycle factors

B.D. Aguda (✉)
Neuro-Oncology Branch, Center for Cancer Research,
National Cancer Institute, National Institutes of Health,
Bethesda 20892, MD, USA
e-mail: agudabd@mail.nih.gov; bdaguda@gmail.com

U. Schmitz et al. (eds.), *MicroRNA Cancer Regulation: Advanced Concepts, Bioinformatics and Systems Biology Tools*, Advances in Experimental Medicine and Biology 774,
DOI 10.1007/978-94-007-5590-1_9, © Springer Science+Business Media Dordrecht 2013

qNET qualitative network
sTF stem cell transcription factor module
TF transcription factor

9.1 Introduction

Decoding genes and their expression into proteins (that mostly carry out the business of living) is run by a molecular machinery in which RNA polymerases and ribosome organelles are core components. This machinery is regulated and intricately orchestrated by a web of molecular interactions and processes, including epigenetic modifications (*e.g.*, DNA methylation, histone acetylation and phosphorylation), binding of transcription factors with DNA, and processing of various RNA transcripts. A lot of attention has been focused recently on non-coding microRNAs (miRNAs) for several reasons: they are endogenous, many are conserved across animal species, they target about a third of the genes in the human genome, and many have been implicated in various human cancers [1–3]. It is generally observed that miRNAs only have minor influence on the protein levels of their targets; however, miRNAs can have profound influence on cell-fate determination [4]. Perhaps this is because miRNAs exert their influence where it matters most – that is, miRNAs may be members of gene regulatory network modules that control switches between cellular states. As illustrated in this chapter, these switches can arise from network instabilities. A primer on network stability analysis will be given and then applied to network modules involving transcription factors (TFs) and miRNAs. The modules are small (mostly involving direct interactions) and can form feed-forward loops (FFLs) or feedback loops (FBLs). FFLs will be discussed briefly, but the focus of this chapter is on FBLs because they can generate instabilities. Examples of FBLs associated with gliomas are discussed.

This chapter illustrates how one can create a phenomenological kinetic model from a qualitative network that lacks information on mechanisms and model parameter values. A detailed mathematical modeling of the miR-17-92/E2F/Myc network is provided to illustrate the various steps in model building and analysis. This network is essential in the control of the G1-S transition of the mammalian cell cycle and is often compromised in human cancers, including gliomas.

9.2 Loops in miRNA-TF Interactions

Interactions among miRNAs, Transcription Factors (TFs) and Targets can organize into feed-forward loops (FFLs) or feedback loops (FBLs). 'Targets' refer to proteins whose expressions are positively or negatively regulated by the TFs. As will be explained in Sect. 9.4, FFLs are not expected to affect the stability of networks so only a brief discussion will be given of their subtle effects on the maintenance (homeostasis), fine-tuning, reinforcement or precision of the steady states of Target proteins. The instabilities and switching behaviors associated with FBLs will be discussed in more detail in the remaining sections of this chapter.

FFLs. All the possible FFLs involving miRNA, TF and Target, are shown in Fig. 9.1. Tsang et al. [5] presented evidence that these FFLs are *network motifs* in the sense that they occur more frequently in gene regulatory networks than would be expected by chance alone. Figure 9.1a shows the case of a miRNA and the Target mRNA being co-regulated by a common TF. This co-regulation would increase the chances of co-occurrence of a miRNA and its Target mRNA, which is an obvious requirement for the miRNA to find its Target. Edge 1 represents the transcription of the miRNA positively or negatively regulated by a TF, edge 2 represents the inhibition of the translation of Target by the miRNA, and edge 3 represents the composite of transcription giving the Target mRNA and its translation to the Target protein. The solid circles at the end of edges 1 and 3 signify either an activation or inhibition (these possibilities are enumerated in Fig. 9.1c, d).

Figure 9.1b shows the case where the miRNA regulates both the TF and its Target mRNA. Edge 1 in Fig. 9.1b depicts the inhibition of the translation of the mRNA that is translated as TF protein. Edges 2 and 3 are defined as in Fig. 9.1a.

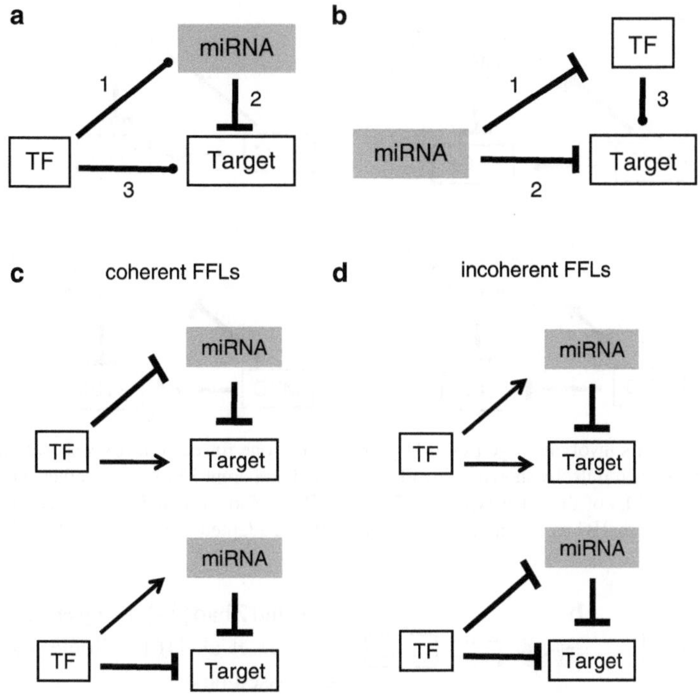

Fig. 9.1 Possible ways that a miRNA interacts with a TF (transcription factor) to Target pathway. A miRNA always suppresses the translation of the Target, as shown by the hammerhead. (**a**) The feed-forward loop (FFL) configuration. The *dots* on edges 1 and 3 could either be *arrows* (activate) or *hammerheads* (inhibit). (**b**) Case with the miRNA is the origin of the FFL. The miRNA suppresses the translation of both TF and its Target. (**c**) The two coherent FFLs. (**d**) The two incoherent FFLs

The two FFLs in Fig. 9.1c are referred to as *coherent* FFLs because the pathway to Target via the miRNA has the same net effect as the direct TF-to-Target pathway. The miRNA-mediated pathway in coherent FFLs serves to reinforce the TF-mediated regulation of the Target expression, as could be intuitively gleaned from the structure of the FFL. For example, the bottom coherent FFL in Fig. 9.1c can be used to suppress 'leaks' in transcription of Target (see [5]).

The two FFLs in Fig. 9.1d are referred to as *incoherent* FFLs because the miRNA-mediated pathway to Target has an effect opposite to that of the TF-to-Target pathway. These FFLs may act as 'noise buffers' to maintain or fine-tune Target steady states and keep uniformity of expression within a cell population [5–7].

Examples of coherent and incoherent FFLs are shown in Fig. 9.2, with the oncogene c-Myc as the TF. These examples are taken from experimentally verified regulatory interactions assembled in the database of El Baroudi et al. [8]. Myc and its targets shown in this figure are genes associated with gliomas. VEGF (Fig. 9.2a) is a pro-angiogenic factor that has increased expression in malignant human glioma consistent with this highly vascularized tumor [9]. The retinoblastoma protein (product of the RB1 gene; Fig. 9.2b) is a tumor suppressor involved in arresting the cell cycle – one mechanism of which is binding and subsequent inhibition of the E2F1 (Fig. 9.2c) transcription factor. The Rb/E2F pathway is deregulated in many human cancers, including gliomas [10]. PTEN (Fig. 9.2d) is another tumor suppressor commonly deleted or mutated in human cancers, including glioblastomas (reviewed in [11]). Note that all the FFLs in Fig. 9.2 support the oncogenic property of Myc: (a) promotes angiogenesis, (b) and (d) are inhibitions of tumor suppressors, and (c) promotes the expression of the proliferation transcription factor E2F1.

The case of Fig. 9.1b with edge 3 being an arrow (*i.e.*, TF inducing expression of Target) has been invoked to explain the phenomenon of

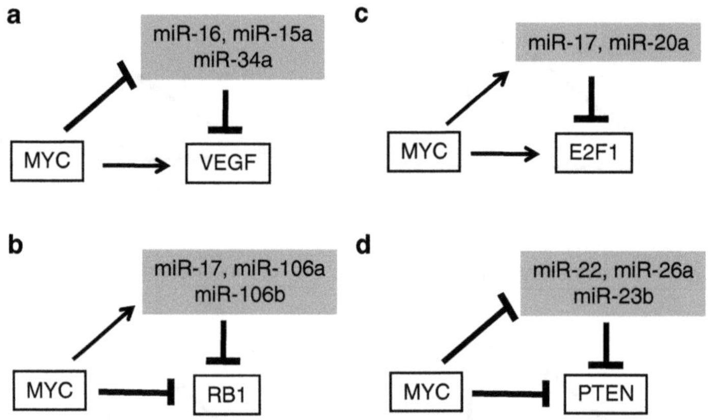

Fig. 9.2 Examples of FFLs involving Myc as the TF. (**a**) Coherent FFLs (one for each miRNA shown) with VEGF as Target. (**b**) Coherent FFLs of the other type (one FFL for each miRNA shown) with RB1 as Target. (**c**) Incoherent FFLs (one for each miRNA shown) with E2F1 as Target. (**d**) Incoherent FFLs of the other type (one FFL for each miRNA shown) with PTEN as Target (Figures (**a**), (**b**), and (**d**) are adapted from El Baroudi et al. [8])

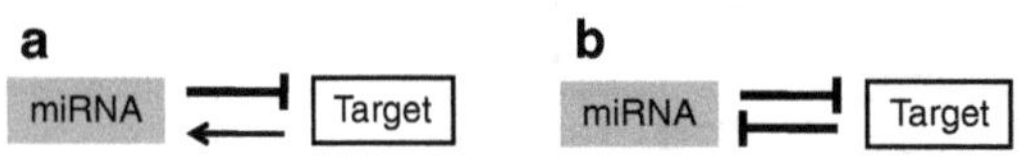

Fig. 9.3 Two types of feedback loops (FBLs) between a miRNA and its Target. The hammerhead means suppression of the translation of the Target mRNA. The *arrow* means transcription of the miRNA is induced by the Target protein acting as a TF. (**a**) A negative FBL. (**b**) A positive FBL

'miRNA-target spatiotemporal avoidance' [4]. This phenomenon has been observed in Drosophila where certain miRNAs are essential in defining tissue boundaries in space and time [4]. Also, the top coherent FFL in Fig. 9.1c may provide a mechanism for 'spatial avoidance'. An example of a mechanism for 'temporal avoidance' is the top incoherent FFL in Fig. 9.1d when there is a delay in the transcription of the miRNA by the TF – a delay compared to the expression of the Target, and there is a temporal shut-down of the Target once the miRNA goes up [4].

FBLs. The two possible feedback loops (FBLs) are shown in Fig. 9.3. FBLs will be discussed in more detail in the next section.

Recently, various databases of validated and predicted FFLs and FBLs involving miRNAs and TFs have been created and made available on the internet [8, 12, 13]. FFLs and FBLs associated with glioblastoma have been reviewed by Gong, Sun and Zhao [14], Dong et al. [15] and Gonzalez-Gomez et al. [16]. Glioma-associated FBLs are discussed next.

9.3 FBLs Associated with Gliomas

Glioblastoma multiforme is an aggressive primary brain tumor with mean survival time of just over a year after diagnosis. Recent gene expression profiling studies are strongly suggesting that miRNAs play a significant role in gliomagenesis, particularly in the regulation of neural stem cell-related pathways [16, 17]. In this section, glioma-relevant examples are given of feedback loops (FBLs) between a miRNA and its target.

9.3.1 Negative FBLs

miR-17-92 and E2F1. The negative FBL between the transcription factor E2F1 and the miR-17-92 cluster is shown in Fig. 9.4a. In this FBL, E2F1 induces the expression of miR-17-92 and, in return, some members of the miR-17-92 cluster target and inhibit the translation of E2F1 mRNA. The miR-17-92 cluster is significantly overexpressed in human glioblastomas [18–20]. Gliomas have also been described as 'addicted' to E2F1 because of their increasing dependence on this TF

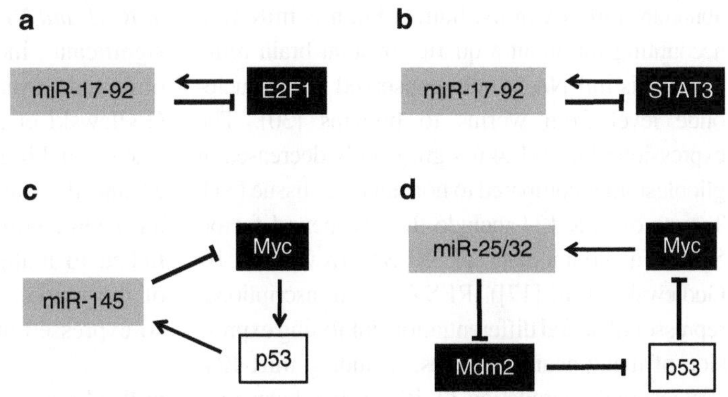

Fig. 9.4 Examples of negative FBLs associated with glioma. Gray boxes are miRNAs, black boxes are oncogenes, and white boxes are tumor suppressor genes

for growth and survival [10]. Modeling the dynamics of this FBL has been carried out [21]. The potential of this negative FBL to generate sustained periodic oscillations and switching behavior (in conjunction with a positive FBL involving E2F1) will be discussed in detail in Sect. 9.5.

miR-17-92 and STAT3. A negative FBL between miR-17-92 and STAT3 is shown in Fig. 9.4b. STAT3 is a transcription factor and a known embryonic stem cell regulator that can function either as a tumor suppressor or as an oncogene in gliomas, depending on the tumor's genetic profile [22–24]. The expression of miR-17-92 is positively induced by STAT3 [25]. On the other hand, STAT3 mRNA translation is inhibited by miR-17-5p, a member of the miR-17-92 cluster [26].

miR-145, p53 and Myc. A negative FBL involving miR-145, Myc, and the tumor suppressor gene p53 is shown in Fig. 9.4c. The expression of miR-145 is induced by p53, and miR-145 inhibits Myc expression [27]. Myc upregulates the activity of p53 in various ways. In return, p53 inhibits Myc via several pathways, including miR-145 [28]. These negative FBLs between Myc and p53 may be essential in coordinating cell differentiation and proliferation in neural stem cells [28]. This topic will be discussed further in Sect. 9.3.3.

miR-25/32, p53 and Myc. A negative FBL exists between Myc and miR-25/32 as recently shown by Suh et al. [29]. These authors reported that Myc (as well as E2F1) transcriptionally activates miR-25 and miR-32, and in return these miRs

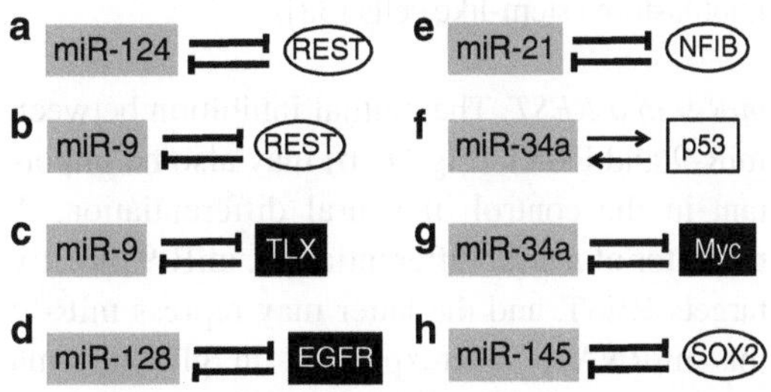

Fig. 9.5 Examples of positive FBLs associated with glioma. Gray boxes are miRNAs, black boxes are oncogenes, and white boxes are tumor suppressor genes

effectively induce p53 accumulation by downregulating Mdm2 (see Fig. 9.4d). Note that Fig. 9.4d can also be interpreted as a negative FBL between p53 and miR-25/32. Suh et al. [29] further demonstrated that overexpression of miR-25/32 in glioblastoma cells inhibited growth of these cells in mouse brain *in vivo*.

9.3.2 Positive FBLs

Examples of glioma-associated positive FBLs between miRNAs and their targets are shown in Fig. 9.5. Most of these examples are of the mutual inhibition type, except the example in Fig. 9.5f of the mutual activation type (the arrow from miR-34a to p53 is an indirect pathway). Not all the miRNA targets shown are transcription factors.

miR-124 and REST. The mutual inhibition between miR-124 and REST (Fig. 9.5a) may be essential in the control of neural differentiation. The most abundant miRNA in the human brain is miR-124, accounting for about a quarter of total brain miRNAs. This miRNA is fully conserved at the nucleotide level from worms to humans [30]. The expression of miR-124 is significantly decreased in glioblastomas compared to normal brain tissue [31]. Targets of miR-124 include the anti-neural factor, SCP1, a component of REST (reviewed by Godlewski et al. [17]). REST is a transcriptional repressor of neural differentiation, inhibiting expression of many neuronal genes, including miR-124. Proteosomal degradation of REST has been proposed as a mechanism for specifically targeting glioblastoma stem-like cells [32].

miR-9 and REST. The mutual inhibition between miR-9 and REST (Fig. 9.5b) may also be important in the control of neural differentiation. A mediator of neural differentiation, miR-9 directly targets REST, and the latter may repress miR-9; this miRNA is overexpressed in glioblastoma (reviewed by Godlewski et al. [17]).

miR-9 and TLX. Also called NR2E1, TLX is a nuclear receptor that binds DNA as a monomer at hormone response elements. TLX and miR-9 mutually antagonize each other (reviewed by Qu and Shi [33], Park et al. [34]). As has been mentioned, miR-9 induces neural differentiation. On the other hand, TLX promotes self-renewal of stem cells and has been reported to play a role in glioma initiation and progression (reviewed by Park et al. [34]). Thus, the mutual inhibition between miR-9 and TLX (Fig. 9.5c) as well as between miR-9 and REST (Fig. 9.5b) may prove to be critical in a stem cell decision whether to differentiate or self-renew.

miR-128 and EGFR. The epidermal growth factor receptor (EGFR) gene amplification and overexpression is a striking feature of glioblastomas, occurring in about 40 % of cases (reviewed recently in [35]). Significant decrease in expression of miR-128 has been observed in aggressive human gliomas, and it has recently been reported that miR-128 represses gliomagenesis [36]. The mutual inhibitory relationship between miR-128 and EGFR (Fig. 9.5d) was identified by Papagiannakopoulos et al. [36].

miR-21 and NFIB. First reported in glioblastoma, significant increases in miR-21 have been observed in many types of cancer (reviewed by Godlewski et al. [17], Moore and Zhang [37]). The mutual inhibitory relationship between miR-21 and the transcription factor NFIB (Fig. 9.5e) has been confirmed by Fujita et al. [38]. NFIB is linked to malignant gliomas and the regulation of the markers, GFAP and B-FABP, which are co-expressed in these tumors [39].

miR-34a and p53. The interaction between miR-34a and p53 is of the mutual activation type. miR-34a is a confirmed transcriptional target of p53 [40, 41]. This miRNA inhibits the translation of SIRT1 mRNA, and SIRT1 protein downregulates p53 activity (this sequence of two negative interactions is the reason why the pathway from miR-34a to p53 is depicted as an arrow in Fig. 9.5f). A positive feedback loop thus exists between miR-34a and p53 [42]. The tumor suppressor p53 is the most commonly mutated gene in human cancers, and is also intimately involved in gliomagenesis [43] (also reviewed by Aguda et al. [28]). It has been shown that miR-34a inhibits glioblastoma growth [41, 44].

miR-34a and Myc. A mutual inhibition exists between Myc and miR-34a (Fig. 9.5g). Myc represses the expression of miR-34a [45, 46], and miR-34a inhibits Myc expression (reviewed by Sotillo et al. [46]). Myc is intimately involved in maintaining the stem-like properties of glioma-propagating cells [47] (reviewed by Aguda et al. [28]).

miR-145 and SOX2. Fang et al. [48] recently presented evidence of a positive (double negative) FBL between SOX2 and miR-145, as shown in Fig. 9.5h. This FBL potentially generates a bistable switch in gliomas. SOX2 is one of the core pluripotent stem cell transcription factors (the others are OCT4 and NANOG) maintaining the stemness of embryonic and adult stem cells (reviewed recently by Heng et al. [49]). Overexpression or gene amplification of SOX2 have been observed in glioma tumors and cell lines [50].

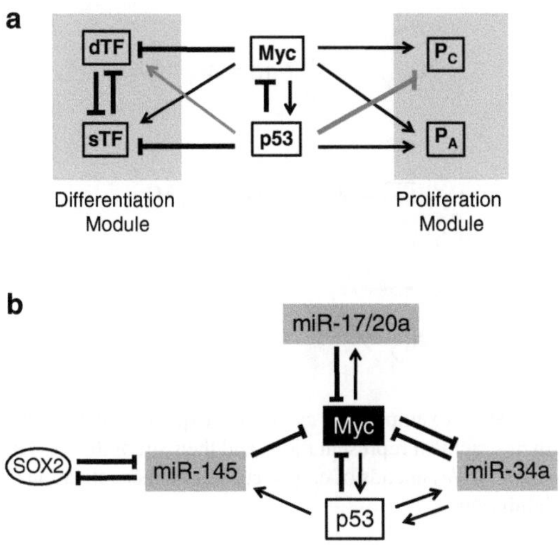

Fig. 9.6 The Myc-p53 control system of cell proliferation and differentiation, and some miRNAs affecting it. (**a**) Representation of the control system using a modular qualitative network. sTF = module of stem cell transcription factors; dTF = module of differentiation transcription factors; P_C = cell cycle factors; P_A = apoptosis factors. (**b**) miRNAs interacting with Myc and p53, and with the pluripotent stem cell transcription factor Sox2

9.3.3 miRNAs in the Context of the Myc-p53 Control System of Cell Proliferation and Differentiation

A network model of the interactions between the transcription factors Myc and p53 has been proposed as a central control mechanism for coordinating cell proliferation and differentiation in mammalian cells [28]. The modular qualitative network (qNET) is shown in Fig. 9.6a. The modules include a proliferation module (composed of cell cycle factors, P_C, and apoptosis factors, P_A) and a differentiation module (composed of sub-modules of stem cell transcription factors, sTF, and differentiation transcription factors, dTF). Three of the glioma-associated miRNAs discussed in the previous sections are shown in Fig. 9.6b, regulating Myc, p53 and SOX2 (a member of the sub-module sTF). A pro-differentiation function (equivalently, anti-stemness function) of p53 is depicted by the inhibition (hammerhead) of sTF by p53 in Fig. 9.6a. Examples of this inhibition would be the path via miR-145 [48] and p53's suppression of another pluripotency factor, NANOG [51–54]. Because of the proposed key role of the Myc-p53 negative FBL in controlling cell proliferation and differentiation [28], the influence of miR-145, miR-34a and miR-17/20a may prove critical in regulating cell-fate decisions, especially when the cellular conditions or parameters are such that the qNET is in the vicinity of switching points between cellular states or phenotypes. An explicit example of this will be illustrated in Sect. 9.5 using the FBL between Myc and miR-17/20a.

9.4 Primer on Qualitative Network Analysis

A network composed of nodes with directed interactions depicted as arrows or hammerheads is referred to as a *qualitative network* (qNET). Besides visually aiding one's intuition on the behavior of the network, the qNET structure can be analyzed to predict potential instabilities. This analysis requires a precise definition of what arrows and hammerheads are in order to employ the standard mathematical method of linear stability analysis summarized here.

A network is *unstable* if a perturbation of a steady state of the network gets amplified in time. Let **x** be a vector of the *n* state variables $x_1, x_2, \ldots, x_n$ (*e.g.*, concentrations of *n* molecules) that interact with each other. Assuming deterministic conditions, the dynamics of the interaction network is usually modeled as a system of nonlinear ordinary differential equations:

$$\frac{d\mathbf{x}}{dt} = \mathbf{f}(\mathbf{x}) \tag{9.1}$$

where $\mathbf{f}(\mathbf{x})$ is a vector with components $f_1(\mathbf{x}), f_2(\mathbf{x}), \ldots, f_n(\mathbf{x})$ which are generally nonlinear functions. Let $\mathbf{x}_s$ be a steady state defined as the state where all these functions vanish, *i.e.*, $\mathbf{f}(\mathbf{x}_s) = \mathbf{0}$. Let $\boldsymbol{\xi} = \mathbf{x} - \mathbf{x}_s$ be the deviation (perturbation) from the steady state. The local dynamics near $\mathbf{x}_s$ is described by the following set of linear equations:

$$\frac{d\xi}{dt} = \mathbf{M}\xi \tag{9.2}$$

where the element M_{ij} of the Jacobian matrix $\mathbf{M}$ is equal to $\partial f_i/\partial x_j$ evaluated at the steady state $\mathbf{x}_s$. The stability of the steady state $\mathbf{x}_s$ depends on the eigenvalues, λ, of $\mathbf{M}$. If at least one eigenvalue has a positive real part, then $\mathbf{x}_s$ is *unstable* (*i.e.*, the perturbation ξ grows); otherwise, if all eigenvalues have negative real parts then $\mathbf{x}_s$ is asymptotically *stable* (*i.e.*, the perturbation ξ decreases with time and the system returns to the steady state). The eigenvalues λ are the roots of the characteristic polynomial $P(\lambda)$:

$$P(\lambda) = \det(\lambda \mathbf{I} - \mathbf{M}) = \lambda^n + \alpha_1 \lambda^{n-1} + \alpha_2 \lambda^{n-2} + \cdots + \alpha_n = 0 \quad (9.3)$$

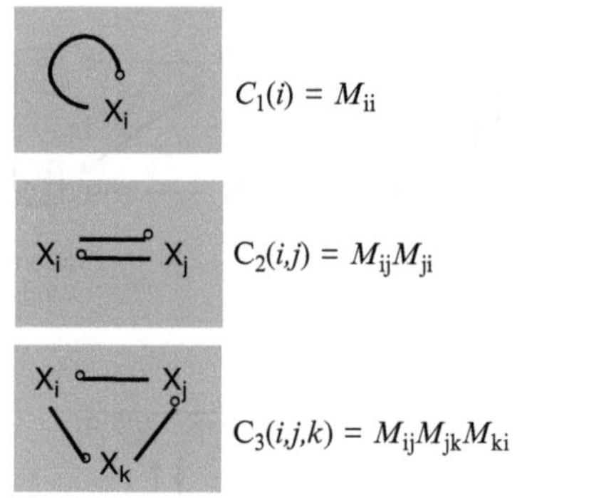

Fig. 9.7 Examples of cycles in a qualitative network, their graphical representations and their strength C_i's. The *small circles* at ends of edges represent either activation or inhibition

To see how the eigenvalues are related to the structure of the qNET, note that the coefficients of the α_i's in $P(\lambda)$ can be written as follows:

$$\alpha_1 = \sum_{i=1}^{n} [-C_1(i)] \quad (9.4)$$

$$\alpha_2 = \sum_{i \neq j} [-C_1(i)][-C_1(j)] + \sum_{i \neq j} [-C_2(i,j)] \quad (9.5)$$

$$\alpha_3 = \sum_{i\neq j, j\neq k, k\neq i} [-C_1(i)][-C_1(j)][-C_1(k)] + \sum_{i\neq j, j\neq k, k\neq i} [-C_1(i)][-C_2(j,k)] + \sum_{i\neq j, j\neq k, k\neq i} [-C_3(i,j,k)] \quad (9.6)$$

where

$$C_1(i) = M_{ii}, \quad C_2(i,j) = M_{ij}M_{ji}, \quad C_3(i,j,k) = M_{ij}M_{jk}M_{ki}, \ldots \quad (9.7)$$

In graphical terms, $C_1(i)$ is referred to as the *strength of the 1-cycle* involving species i (species i affects itself); $C_2(i,j)$ is the strength of the 2-cycle between species i and j; $C_3(i,j,k)$ is the strength of the 3-cycle among species i, j and k; and so on (see Fig. 9.7).

Note that a positive (negative) M_{ij} means that the *rate of change of species* x_i (*i.e.*, dx_i/dt) increases (decreases) with increasing x_j. A positive M_{ij} is depicted as an arrow from species X_j to X_i in a qNET diagram to show the activation of X_i by X_j. A negative M_{ij} means that X_j inhibits X_i, and is depicted as a hammerhead from X_j to X_i. Also, the magnitude of M_{ij} is referred to as the *strength of the interaction*, and $C_i(\cdot)$ as the *strength of the i-cycle*.

The stability of a qNET can now be expressed in terms of the strengths of its component cycles. It is seen from Eq. 9.3 that the eigenvalues are functions of the α_i's which are themselves functions of the strengths of the cycles in the qNET. Clearly, *the cycles determine the stability of the steady state of a qNET*. Interactions in qNETs that are not members of cycles do not affect the stability of the network. This is the reason why FFLs are not sources of instabilities in qNETs.

Next, the eigenvalues can now be expressed in terms of the cycles. A *destabilizing cycle* is one that makes the real part of an eigenvalue more positive if the cycle strength is increased. Cycles that keep all the real part of the eigenvalues negative (or some of them more negative) are *stabilizing cycles*.

A convenient theorem to use for determining the stability of a qNET is the *Routh-Hurwitz Theorem* [55, 56]. It states that the number of eigenvalues with positive real part is equal to the total of the number of sign changes in two sequences, $\{1, \Delta_1, \Delta_3, \ldots\}$ and $\{1, \Delta_2, \Delta_4, \ldots\}$, where the Δ_i's are the Hurwitz determinants. The ith Hurwitz determinant is the determinant of the ith principal minor of the following square matrix $\mathbf{H}$ associated with the characteristic polynomial $P(\lambda)$ in Eq. 9.3.

$$\mathbf{H} = \begin{bmatrix} \alpha_1 & \alpha_3 & \alpha_5 & .. & .. & 0 & .. & 0 \\ 1 & \alpha_2 & \alpha_4 & & & .. & & .. \\ 0 & \alpha_1 & \alpha_3 & & & 0 & & .. \\ .. & 1 & & .. & & \alpha_n & .. & .. \\ .. & 0 & & & .. & \alpha_{n-1} & 0 & .. \\ .. & .. & 0 & & & \alpha_{n-2} & \alpha_n & .. \\ .. & .. & .. & & & \alpha_{n-3} & \alpha_{n-1} & 0 \\ 0 & 0 & 0 & .. & .. & \alpha_{n-4} & \alpha_{n-2} & \alpha_n \end{bmatrix} \quad (9.8)$$

For example, for $n = 3$, the Hurwitz determinants are:

$$\Delta_1 = \alpha_1, \quad \Delta_2 = \alpha_1\alpha_2 - \alpha_3, \quad \Delta_3 = \alpha_3\Delta_2 \quad (9.9)$$

To illustrate the use of the Routh-Hurwitz Theorem, consider the following simple example involving the pluripotent stem cell transcription factor SOX2 and miR-145, as shown in Fig. 9.8. Note that negative 1-cycles are included to account for the degradation of these molecules (these cycles can also be referred to as self-regulatory loops). It will be shown that an instability arises when a certain relationship between the strengths of the 1-cycles and the 2-cycle between SOX2 and miR-145 occurs.

Let X_1 = SOX2 and X_2 = miR-145. The strengths of the 1-cycles are $C_1(1) = M_{11}$ and $C_1(2) = M_{22}$. The strength of the 2-cycle is $C_2(1,2) = M_{12}M_{21}$. The Hurwitz determinants are $\Delta_1 = \alpha_1 = -C_1(1) - C_1(2)$ and $\Delta_2 = \alpha_1\alpha_2$ where $\alpha_2 = C_1(1)C_1(2) - C_2(1,2)$. Since $C_1(1) < 0$ and $C_1(2) < 0$, Δ_1 cannot become negative, and therefore the sequence $\{1, \Delta_1\}$ cannot change sign. The sequence $\{1, \Delta_2\}$ can change sign once (giving one eigenvalue with positive real part) if Δ_2 becomes negative which happens when

$$C_2(1,2) > C_1(1)C_1(2) \quad (9.10)$$

Thus, when the strength of the positive 2-cycle is greater than the product of the two negative 1-cycles, an instability is predicted to arise. This statement is also true for all the positive FBLs in Fig. 9.5, for both mutual antagonisms and mutual activations. Modeling the mutual antagonism between SOX2 and miR-145 has been performed recently by Fang et al. [48]; these authors demonstrated that such a positive FBL exhibits switching behavior between two locally stable states.

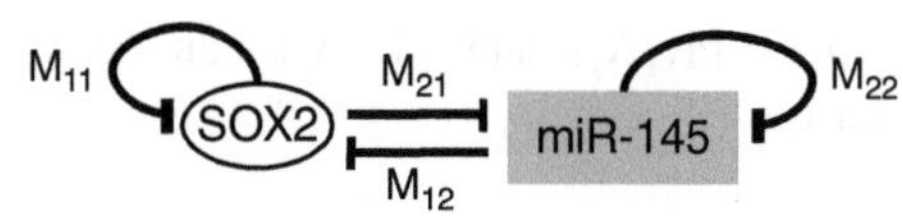

Fig. 9.8 A positive FBL between miR-145 and SOX2, including their self-regulatory 1-cycles. See text for more details

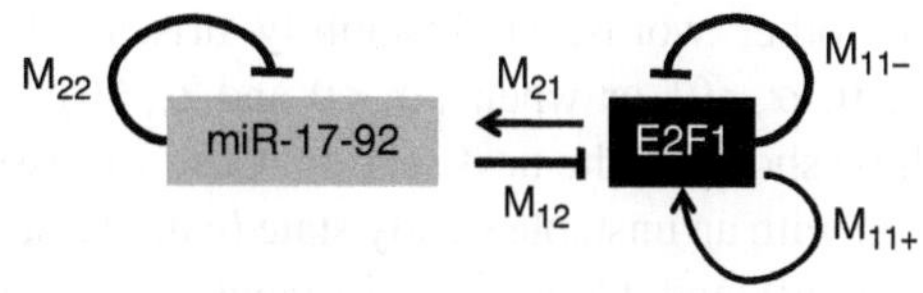

Fig. 9.9 A negative FBL between miR-17-92 and E2F1, including their self-regulatory 1-cycles. See text for more details

Another example, shown in Fig. 9.9, illustrates the important fact that a qNET, despite lacking in mechanistic details or parameter values, may already predict the potential for instability. The next section will show details of the instability predicted by the qNET in Fig. 9.9. Note that there is an autocatalytic (positive) loop involving E2F1 with strength M_{11+}, representing the fact that the E2F1 protein can induce the expression of its own gene. The first-order decay of the levels of miR-17-92 and E2F1 are also shown, with strengths M_{22} and M_{11-}, respectively.

Let X_1 = E2F1 and X_2 = miR-17-92. The strengths of the 1-cycles are $C_1(1) = M_{11} = M_{11+} + M_{11-}$ (with $M_{11+} > 0$ and $M_{11-} < 0$) and $C_1(2) = M_{22} < 0$. The strength of the 2-cycle is $C_2(1,2) = M_{12}M_{21} < 0$. The Hurwitz determinants are $\Delta_1 = \alpha_1 = -C_1(1) - C_1(2)$ and $\Delta_2 = \alpha_1\alpha_2$, where $\alpha_2 = C_1(1)C_1(2) - C_2(1,2)$. The determinant Δ_1 is negative if the sum of the strengths of the negative 1-cycles is less than that of the positive 1-cycle, *i.e.*, $(-M_{22}) + (-M_{11-}) < M_{11+}$. The second determinant Δ_2 can become negative in two ways: (1) if $\Delta_1 = \alpha_1 > 0$ and $\alpha_2 < 0$; (2) if $\alpha_2 > 0$ and $\Delta_1 = \alpha_1 < 0$. Thus, sign changes for the sequences $\{1, \Delta_1\}$ and $\{1, \Delta_2\}$ can occur in two cases:

Case I. $\Delta_2 < 0$ (with $\alpha_2 < 0$) and $\Delta_1 > 0$. This case gives one eigenvalue with positive real part.
Case II. $\Delta_1 < 0$ and $\Delta_2 < 0$ (with $\alpha_2 > 0$). This case gives two eigenvalues with positive real part.

In other words, an instability arises when $\{\alpha_1 > 0, \alpha_2 < 0\}$ or when $\{\alpha_1 < 0$ and $\alpha_2 > 0\}$. As will be shown in the next section, Case I is associated with an unstable steady state (called a *saddle point*) and Case II is associated with an unstable steady state that spirals out to periodic oscillatory states.

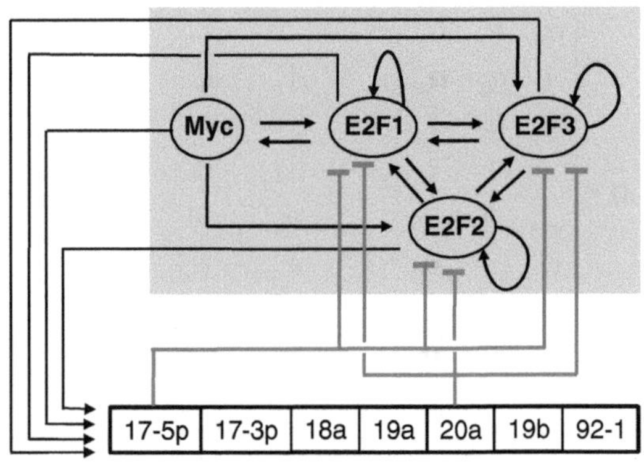

Fig. 9.10 The transcription factors Myc, E2F1, E2F2, and E2F3 induce the expression of miR-17-92 (shown as a cluster of seven miRNAs). Translation of E2F1, E2F2, and E2F3 is suppressed by miR-17-5p and miR-20a as depicted by the hammerheads (Figure from Fig. 9.2 in [21])

9.5 Modeling the miR-17-92/E2F/Myc Network

This section illustrates how to formulate a dynamical model of a negative FBL between miR-17-92 and the transcription factors E2Fs and Myc. Despite the lack of details on the mechanism and kinetic parameters of the network, a set of phenomenological equations can be written that are consistent with the available qualitative information. Mathematical techniques such as non-dimensionalization of the differential equations, use of time-delay equations, steady state stability and bifurcation analysis are discussed. The significance of studying the Myc/E2F/miR-17-92 network is highlighted by its role in regulating the G1-S transition of the mammalian cell cycle, particularly as a potential switching mechanism for a G1 checkpoint (also called the *Restriction Point*) that is often compromised in cancers [21, 57, 58]. The discussion below is mostly based on the paper of Aguda et al. [21].

9.5.1 Network Model Formulation

Upon growth-factor stimulation, cascades of intracellular signal transduction pathways are activated in quiescent cells, resulting in the expression of early-response genes such as *c-myc*. The corresponding protein, Myc, is a TF that induces the expression of other TF genes promoting the G1-S transition in the cell cycle, including E2F1, E2F2 and E2F3. Interestingly, each of these E2Fs induces its own expression, as well as the other two E2F family members (*e.g.*, the E2F1 protein is a TF that binds the *e2f1* gene promoter inducing expression of the gene). Furthermore, E2F1 induces the expression of Myc. Thus, the set {Myc, E2F1, E2F2, E2F3} forms an autocatalytic system of TFs (the set is shown in the gray box of Fig. 9.10). This autocatalytic system of TFs has been proposed to be necessary for a cell's commitment to S-phase [57]. In addition, there are positive feedback loops in the so-called Rb/E2F pathway that may contribute to the sharpness of the switching behavior in CDK (cyclin-dependent kinase) activities (particularly, CDK4, CDK6, and CDK2) associated with the *Restriction Point* in the mammalian cell cycle [57, 59, 60].

As shown in Fig. 9.10, Myc, E2F1, E2F2 and E2F3 promote the transcription of the miR-17-92 cluster (single transcript shown at the bottom of Fig. 9.10) that is subsequently processed to give seven mature miRNAs, namely, miR-17-5p, miR-17-3p, miR-18a, miR-19a, miR-20a, miR-19b, and miR-92-1. Two of these mature miRNAs, miR-17-5p and miR-20a, have been shown to suppress the translation of E2F1, E2F2, E2F3 as shown by the hammerheads in Fig. 9.10. Thus, negative FBLs are formed between the

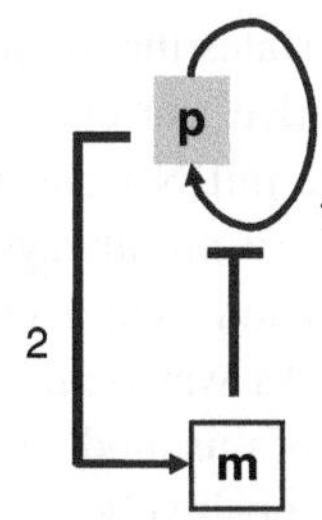

Fig. 9.11 A two-variable model of the network in Fig. 9.10. The autocatalytic protein module (shown in gray box in Fig. 9.10) is represented by p. The miR-17-92 cluster is symbolized by m (Figure from Fig. 9.2 in [21])

autocatalytic TFs and miR-17-5p/miR-20a. Simplifying the mathematical model, but keeping the essential structure of the qNET, an abstraction of the miRNA-TF network in Fig. 9.10 can be carried out to give the two-variable model shown in Fig. 9.11 [21].

9.5.2 Dimensionless Equations and Time-Delays

The two variables in Fig. 9.11 correspond to an autocatalytic protein module, **p** (corresponding to the autocatalytic Myc/E2Fs), and to a miRNA, **m** (corresponding to miR-17-92). Step 1 represents the protein acting as a TF for its own gene's expression. It is to be noted that the rate of the protein's expression cannot be a function of the instantaneous concentration of the protein because of a time delay associated with transcription and translation of the gene; in other words, when these intermediate processes are 'hidden' in the phenomenological rate equations, the rate of change in the concentration of **p** must be a function of the concentration of **p** at some previous time (t-Δ) where t is the current time and Δ is the *time delay*. In the simple model suggested earlier [21], the rate of step 1 is a phenomenological function that combines the autocatalytic property of **p** and the inhibition of the expression of **p** by **m**; hence, for the rate function for step 1, the concentration of **m** to be considered is also that at (t–Δ).

Not explicitly showing time delays in the rate equations for now, a set of phenomenological dynamical equations for the model is the following:

$$\frac{dp}{dt} = \alpha + \left(\frac{k_1 p^2}{\Gamma_1 + p^2 + \Gamma_2 m} \right) - \delta p$$
$$\frac{dm}{dt} = \beta + k_2 p - \gamma m \qquad (9.11)$$

where p is the concentration of the protein, m is the concentration of the miRNA, α is a constant constitutive rate of expression of the protein, β is a constant constitutive rate of input of the miRNA, δ is a coefficient of the rate of degradation of the protein, γ is a coefficient of the rate of degradation of the miRNA. The second term on the right-hand side of dm/dt corresponds to the rate of synthesis of **m** due to **p**, which is simply modeled as the linear term k_2p. The second term on the right-hand side of dp/dt represents two important features of the network: (a) the autocatalytic character of **p** according to a Hill-type function with exponent of 2, and (b) the inhibition of the expression of **p** by **m** represented by putting m in the denominator (*i.e.*, if m increases, the rate dp/dt decreases).

Although this model seems to be quite simple, it already has eight parameters. One way to reduce this number is by non-dimensionalizing the variables and parameters. Indeed, the ratios or other combinations of parameters are what essentially determine the qualitative behavior of the system. A dimensionless version of Eq. 9.11 is given in Eq. 9.12 below:

$$\varepsilon \frac{d\phi}{d\tau} = \alpha' + \left(\frac{\kappa \phi^2}{\Gamma'_1 + \phi^2 + \Gamma'_2 \mu} \right) - \phi$$
$$\frac{d\mu}{d\tau} = 1 + \phi - \mu \qquad (9.12)$$

where the dimensionless variables and parameters are

$$\phi = \left(\frac{k_2}{\beta} \right) p, \quad \mu = \left(\frac{\gamma}{\beta} \right) m, \quad \tau = \gamma t$$
$$\varepsilon = \frac{\gamma}{\delta}, \quad \alpha' = \left(\frac{k_2}{\delta \beta} \right) \alpha, \quad \kappa = \frac{k_1 k_2}{\delta \beta}$$
$$\Gamma'_1 = \left(\frac{k_2^2}{\beta^2} \right) \Gamma_1, \quad \Gamma'_2 = \left(\frac{k_2^2}{\beta \gamma} \right) \Gamma_2 \qquad (9.13)$$

Note that there are now only five dimensionless parameters. Furthermore, there are only four parameters that determine the steady states (compared to the eight parameters in Eq. 9.11).

If time delays are now considered, the ϕ and μ on the right-hand side of $d\phi/d\tau$ in Eq. 9.12 must be values at $(t{-}\Delta)$ where Δ is the time delay discussed above. Also, note that the ϕ and μ on the right-hand side of $d\mu/d\tau$ are not subject to this time delay because it is assumed that the expression of **p** (including its inhibition by **m**) occurs on a much slower time scale. Explicitly, the $d\phi/d\tau$ in Eq. 9.12 is written as follows:

$$\varepsilon\frac{d}{d\tau} = \alpha' + \frac{\kappa[\phi(\kappa-\Delta)]^2}{\Gamma'_1 + [\phi(\tau-\Delta)]^2 + \Gamma'_2\mu(\tau-\Delta)} - \phi(\tau) \quad (9.14)$$

where Δ scales as τ.

9.5.3 Steady State Bifurcation Diagrams and Role of miR-17-92 in Switching Behavior

In the analysis of a dynamical model, a first step is to determine the steady states of the system. A steady state that is stable (in the sense that all perturbations from it eventually die out) represents a long-term behavior of the system. The steady states of Eq. 9.12 are defined as the values of ϕ and μ that make both $d\phi/d\tau$ and $d\mu/d\tau$ vanish. From the second expression in Eq. 9.12, the steady state μ_s is given by

$$\mu_s = 1 + \phi_s \quad (9.15)$$

Substituting Eq. 9.15 into the first expression of Eq. 9.12 gives the following cubic polynomial whose non-negative roots are the steady states ϕ_s:

$$\phi_s^3 + c_2\phi_s^2 + c_1\phi_s + c_0 = 0 \quad (9.16)$$

where

$$\begin{aligned} c_2 &= \Gamma'_2 - (\alpha' + \alpha) \\ c_1 &= \Gamma'_1 + \Gamma'_2(1-\alpha') \\ c_0 &- -\alpha'(\Gamma'_1 + \Gamma'_2) \end{aligned} \quad (9.17)$$

Equation 9.15 states that there is a direct proportionality (positive correlation) between the steady state of the miRNA and that of its target protein. So one must *not* always expect that the levels of a miRNA and its target protein are negatively correlated. As was demonstrated in computer simulations of this model [21], $\phi(t)$ and $\mu(t)$ can also be either positively or negatively correlated in non-steady state conditions.

The cubic polynomial in ϕ_s suggests the possibility of having three coexisting positive steady states. The necessary (but not sufficient) conditions for having three positive roots are $c_2<0$, $c_1>0$, and $c_0<0$. These conditions on the c_i's can be summarized as follows

$$(\Gamma'_2 - \kappa) < \alpha' < \left(1 + \frac{\Gamma'_1}{\Gamma'_2}\right) \quad (9.18)$$

Thus, it is possible that there is a range of α' where multiple steady states coexist for the same set of parameter values.

Interpreting α' as a parameter associated with growth-factor signaling, it is chosen as the bifurcation parameter in further analysis below. This parameter can be conveniently controlled in laboratory experiments (see for example, [60]). The steady states as functions of α', and for various Γ_2', are shown in Fig. 9.12. The parameter Γ_2' is linked with the inhibition of the translation of the target protein by the miRNA, and is referred to as the *coefficient of inhibition* by the miRNA.

The cases of $\Gamma_2' = 1.8$ and 2.0 in Fig. 9.12 give S-shaped steady-state curves having two 'knees'; in between these knees are ranges of α' where there are three steady states of ϕ (and μ). Linear stability analysis (discussed in the previous section) shows that the middle branches of these S-shaped curves are composed of unstable steady states, whereas the upper and lower branches are stable (when time delay is zero); in other words, the system is *bistable* between the 'knees'. (The unstable steady states are of the *saddle point* type, *Case I*, discussed in the last example of Section 4). As shown in Fig. 9.12, as Γ_2' is decreased, the S-shaped curve shifts to the left leading to the disappearance of the left 'knee' (*e.g.*, for $\Gamma_2' = 1.5$,

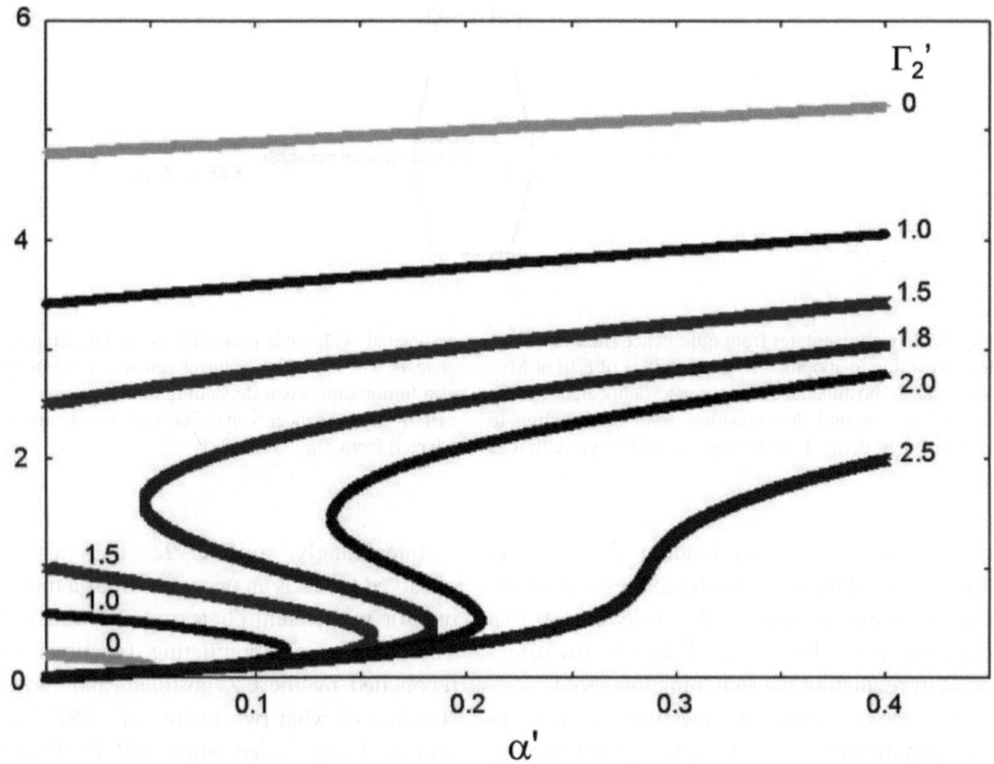

Fig. 9.12 Graphs of the steady state ϕ_s versus α' determined from Eq. 9.12 by setting the *right-hand sides* of the equations to zero. Each curve is for a different value of the parameter Γ_2' as shown. Other parameter values: $\kappa=5$, $\Gamma_1'=1$. (Figure from Fig. 9.4 in [21])

1, and 0). Also note that as Γ_2' increases, the bistable range for α' narrows until both 'knees' vanish and bistability is lost (*e.g.*, when $\Gamma_2'=2.5$).

The role of miR-17-92 in regulating the switching behavior of the system is further illustrated in Fig. 9.12. The value of α' corresponding to the right 'knee' of an S-shaped curve is viewed as a 'switching threshold' for growth factor-induced protein expression beyond which the system jumps to the upper branch of steady states. (In the paper of Yao et al. [60] where the authors propose a bistable model due to positive feedback loops involving E2F1, this switch is associated with the G1-S transition in the cell cycle). Let this threshold be called α_{ON}'. As shown in Fig. 9.12, α_{ON}' decreases as Γ_2' decreases. (In cancer parlance, decreasing Γ_2' leads to decreased growth-factor requirement for cell proliferation.) Let the steady state value on the upper branch corresponding to α_{ON}' be called ϕ_s^{ON}. As Γ_2' decreases, ϕ_s^{ON} increases. When Γ_2' is set to zero (corresponding, for example, to a knockout of the miRNA), α_{ON}' approaches zero. Very small threshold values of α_{ON}' may mean that the switch to the upper branch of steady states could be driven by random noise which is highly undesirable. Furthermore, the value of ϕ_s^{ON} increases as Γ_2' decreases which is also a bad situation when normal downstream events depend on controlled values of ϕ_s^{ON}. Thus, this model is a good illustration of the important role of the miRNA in fine-tuning both α_{ON}' and ϕ_s^{ON}.

As demonstrated above, the effect of miR-17-92 on the predicted bistable switching behavior of the system can be manipulated by varying the coefficient of inhibition Γ_2'. Experimentally, Γ_2 can be varied by mutating the sequence of the

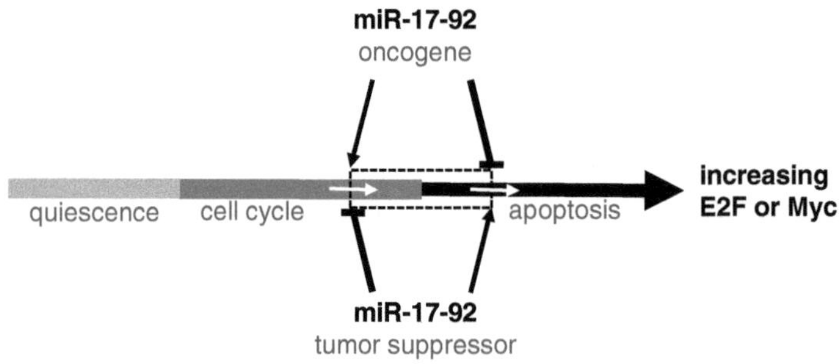

Fig. 9.13 Cellular states from quiescence (non-dividing) to cell cycling to apoptosis as the activities of E2F or Myc increase. A postulated *cancer zone* (delineated by the *dashed box* around the transition from cell cycling to apoptosis) is defined as a range of E2F/Myc activities associated with high probabilities of initiating carcinogenesis. The miR-17-92 cluster can act as an oncogene or as a tumor suppressor depending how it affects entry and exit from the cancer zone. See text for details (Figure adapted from Fig. 9.3 in [21])

miRNA that affects its binding to its target mRNA. In addition, from the expression of the dimensionless parameter Γ_2' (see Eq. 9.13), changing the values of k_2, β and γ also offers ways of regulating the switching threshold.

The next important question is how to interpret the meaning of the changes in the magnitudes of ϕ_s^{ON} as Γ_2' is varied. More specifically, what are the cell physiological consequences of increasing activities of Myc or the E2Fs when the inhibition coefficients of mature members of the miR-17-92 cluster (against Myc and the E2Fs) are decreased? Next, a discussion of the *cancer zone* postulate is given to answer this question [21].

9.5.4 The Cancer Zone Postulate

Myc and E2F1 promote cell cycle progression and, if overexpressed, apoptosis – hence, these TFs have been referred to as oncogenes and as tumor suppressor genes (see [58] and references therein). According to many reported observations, as Myc/E2F1 activities increase, a cell can be made to transit from quiescence, to cell cycling, and eventually to apoptosis (Fig. 9.13). It has been proposed [21] that between cell cycling and apoptosis is a *range* of Myc and/or E2F activities associated with significant probabilities of initiating carcinogenesis; this range is referred to as the *cancer zone* (CZ).

Interestingly, miR-17-92 has also been observed to act as an oncogene or as a tumor suppressor in different contexts (reviewed in [21]). These seemingly conflicting functions can be reconciled by the CZ postulate and the understanding of what two factors of miRNA inhibition are being varied when miR-17-92 is being demonstrated as an oncogene or as a tumor suppressor. Figure 9.13 illustrates how these two factors are involved in the control of entry into or exit from the CZ (delineated by the dashed region).

In Fig. 9.13, miR-17-92 is viewed as an oncogene if it induces entry into the CZ or prevents exit from the CZ (arrow and hammerhead above the CZ). On the other hand, miR-17-92 is a tumor suppressor if it does exactly the opposite (hammerhead and arrow below the CZ). The mechanisms for entering or exiting the CZ are explained in Fig. 9.14. In this figure, the gray horizontal band (between ϕ_s ~2.8 and ~3.75) is arbitrarily assigned the range of ϕ corresponding to the CZ. Note that the different cell states are drawn on the right side of this figure (these ranges are arbitrary and are for illustration purposes only).

miR-17-92 as oncogene: First, the arrow showing that this miRNA induces entry into CZ (top of Fig. 9.13) corresponds to the increase in the miRNA and protein steady states as α' is increased (see Eq. 9.15 which states that the graph of μ_s vs α' is similar to that of Fig. 9.14). In other words,

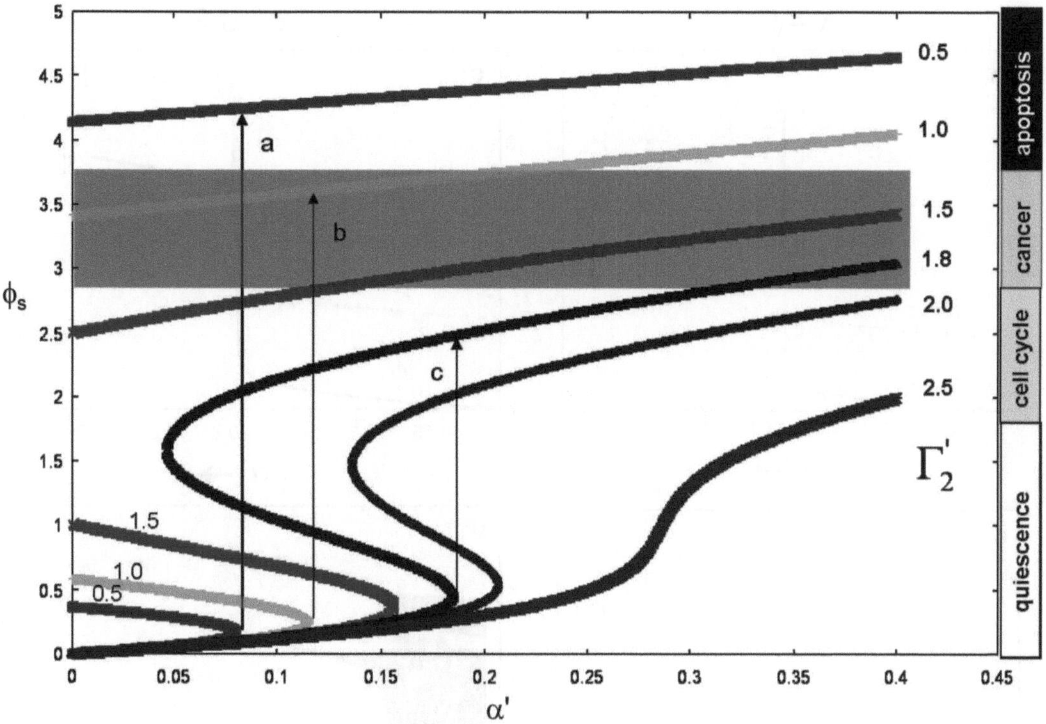

Fig. 9.14 This figure is the same as Fig. 9.12 (with identical parameter values used) except the vertical labels on the *right-hand side* showing ranges of cellular states (as in Fig. 9.13). The cancer zone is labeled *cancer* with arbitrarily chosen range indicated by the gray horizontal bar. See text for details (Figure adapted from Fig. 9.6 in [21])

starting with ϕ_s below CZ, as α' increases, the ϕ_s^{ON} increases and could enter the CZ from below (given the right values of Γ_2'). Second, if the system is near the upper boundary of the CZ, its exit from the CZ (into apoptosis) is prevented by an increase in Γ_2'. For example, looking at the upper branch of the curve for $\Gamma_2'=1.0$ in Fig. 9.14, the system will exit from the CZ if α' is increasing; this exit is prevented if Γ_2' is increased (say, to $\Gamma_2'=1.5$ as shown in the figure). In summary, the oncogenic property of this miRNA is manifested when there is an increase in miRNA level (when below the CZ lower boundary) as α' is increased, or an increase in the miRNA inhibition coefficient (when below the CZ upper boundary) as α' is increased.

miR-17-92 as tumor suppressor gene: The discussion here is similar to the above, although everything is reversed. The tumor suppressor function of this miRNA is manifested when there is an increase in the miRNA inhibition coefficient (when below the CZ lower boundary) as α' is increased, or an increase in the miRNA level (when near the CZ upper boundary) as α' is increased.

9.5.5 Periodic Oscillations

As predicted in the stability analysis of the last example in Sect. 9.4, the two-variable model can exhibit sustained periodic oscillations for some parameter values. The conditions for the existence of the instability that gives rise to these oscillations are those of Case II (two eigenvalues with positive real part). These oscillations are shown in the lower right panel of Fig. 9.15 when there is time delay and for decreasing α' (in fact, the same oscillations can be obtained for increasing α' for some initial

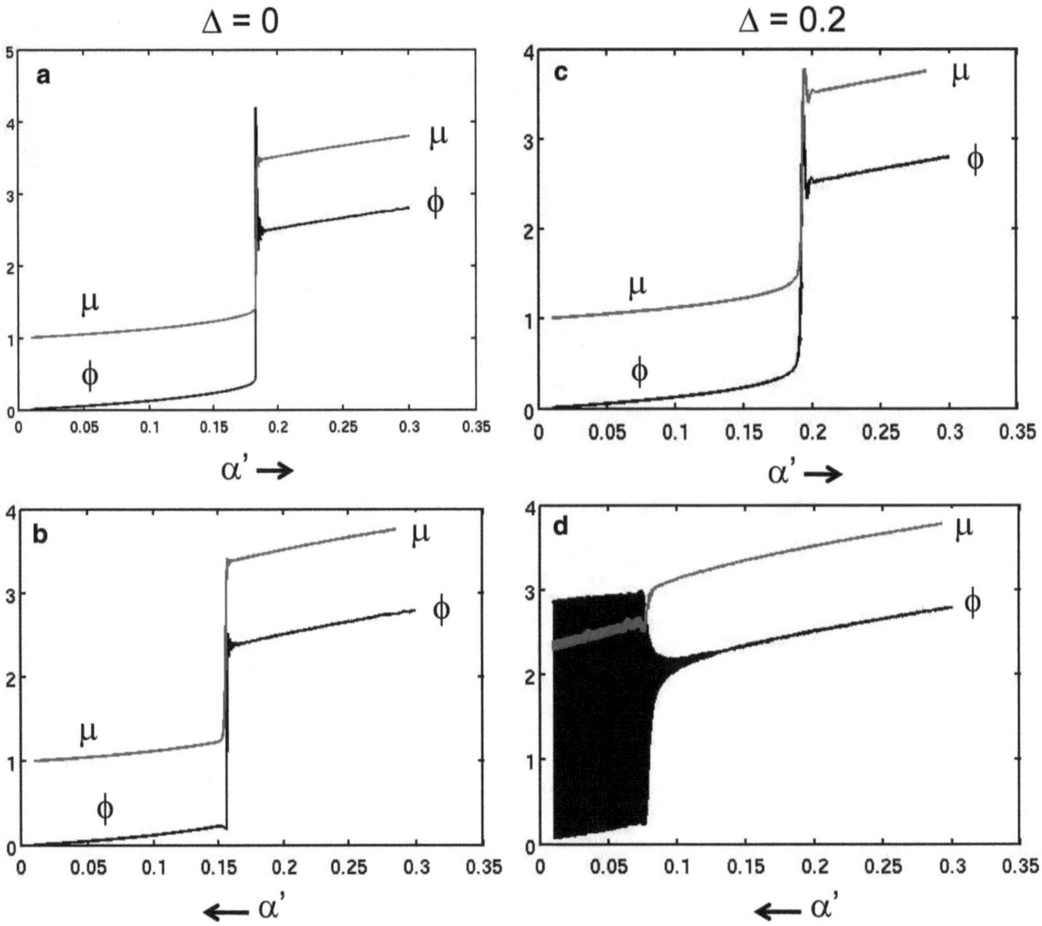

Fig. 9.15 Equations 9.12 and 9.14 are integrated numerically for slowly changing values of α' in the increasing (**a** and **c**) and decreasing (**b** and **d**) directions, without time delay ($\Delta=0$, *left column*) and with time delay ($\Delta=0.2$, *right column*). Sustained periodic oscillations are obtained in (**d**). Other parameter values: $\varepsilon=0.02$, $\kappa=5$, $\Gamma_1'=1$, $\Gamma_2'=1.8$ (Figure from Fig. 9.7 in [21])

conditions other than those used in the upper right panel). The maximum and minimum of the oscillations for a given α' are shown (for example, fixing the value $\alpha'=0.05$ will give oscillations of ϕ between ~0.2 and ~2.9).

9.5.6 Role of miR-17-92 in the G1 Checkpoint

The S-shaped curves in Fig. 9.12, exhibiting bistability, are primarily generated by the positive FBL in Fig. 9.11. As mentioned earlier, there are several positive FBLs that may contribute to driving the G1-S transition in the mammalian cell cycle – including FBLs that involve the E2Fs, the tumor suppressor Rb (Retinoblastoma protein), and various cyclin-dependent kinases (CDK4, CDK6, CDK2). Modeling studies of the Rb/E2F/CDK network have been carried out previously [57, 59] predicting and demonstrating its sharp switching behavior. Indeed, the bistable behavior of the network has been validated experimentally by Yao et al. [60].

One can claim that the single positive feedback loop in the two-variable model in Fig. 9.11 may be a qualitative representation of the several positive loops in the Rb/E2F/CDK network.

Thus, the results of the analysis of the simple two-variable model presented above may carry over to a more complex model that includes details of the Rb/E2F/CDK network.

9.6 Summary

This chapter has given an overview of miRNA-TF interactions that contain loops. Feed-forward loops (FFLs) are discussed only briefly because they are shown not to affect network stability (if they are not members of cycles in the qNET); however, FFLs may play important roles in buffering transcriptional noise or fine-tuning target protein levels. FFLs involve the co-regulation by a common TF of the expression of a miRNA and its target mRNA. Examples of coherent and incoherent FFLs are given. The focus of this chapter is on feedback loops (FBLs) between a miRNA and its Target – the Target mRNA being translated to a TF protein which positively or negatively regulates the transcription of the miRNA. FBLs are shown to affect network stability. Examples of positive and negative FBLs associated with gliomas are discussed.

The method of qualitative network (qNET) analysis is summarized. This method's most important lesson is that, although the interactions are qualitative and no mechanisms or rate parameters are specified, one can still say something about the stability or instability of a qNET based on its structure. It was shown that only cycles in a qNET diagram are the ones that determine the stability of the network. This is the main reason for focusing on FBLs in miRNA-TF interactions. Positive FBLs have the potential to be toggle or bistable switches, while negative FBLs can generate sustained periodic oscillations.

A detailed analysis of a mathematical model of the miR-17-92/E2F/Myc network is given. This network is associated with the G1-S transition in the mammalian cell cycle, and is part of the Rb/E2F pathway that is often compromised in cancer. The steps in creating a simple network model of only two variables and its phenomenological dynamical equations are illustrated. Computer simulations of the kinetics of this model exhibited bistability and switching behavior, as well as oscillations when time delays are included in the negative FBL. With the introduction of the *cancer zone* postulate, this model is able to explain why miR-17-92 can act as an oncogene or as a tumor suppressor depending on the conditions.

References

1. Esquela-Kerscher A, Slack FJ (2006) Oncomirs—microRNAs with a role in cancer. Nat Rev Cancer 6:259–269
2. Cho WC (2007) OncomiRs: the discovery and progress of microRNAs in cancers. Mol Cancer 6:60
3. Krutovskikh VA, Herceg Z (2010) Oncogenic microRNAs (OncomiRs) as a new class of cancer biomarkers. Bioessays 32:894–904
4. Shalgi R, Brosh R, Oren M, Pilpel Y, Rotter V (2009) Coupling transcriptional and post-transcriptional miRNA regulation in the control of cell fate. Aging (Albany NY) 1:762–770
5. Tsang J, Zhu J, van Oudenaarden A (2007) MicroRNA-mediated feedback and feedforward loops are recurrent network motifs in mammals. Mol Cell 26:753–767
6. Osella M, Bosia C, Corá D, Caselle M (2011) The role of incoherent microRNA-mediated feedforward loops in noise buffering. PLoS Comput Biol 7:e1001101
7. Herranz H, Cohen SM (2010) MicroRNAs and gene regulatory networks: managing the impact of noise in biological systems. Genes Dev 24:1339–1344
8. El Baroudi M, Corà D, Bosia C, Osella M, Caselle M (2011) A curated database of miRNA mediated feed-forward loops involving MYC as master regulator. PLoS One 6:e14742
9. Takano S, Yamashita T, Ohneda O (2010) Molecular therapeutic targets for glioma angiogenesis. J Oncol 2010:351908
10. Alonso MM, Alemany R, Fueyo J, Gomez-Manzano C (2008) E2F1 in gliomas: a paradigm of oncogene addiction. Cancer Lett 263:157–163
11. Koul D (2008) PTEN signaling pathways in glioblastoma. Cancer Biol Ther 7:1321–1325
12. Le Bechec A, Portales-Casamar E, Vetter G, Moes M, Zindy P-J, Saumet A, Arenillas D, Theillet C, Wasserman W, Lecellier C-H, Friederich E (2011) MIR@NT@N: a framework integrating transcription factors, microRNAs and their targets to identify sub-network motifs in meta-regulation network model. BMC Bioinform 12:67
13. Friard O, Re A, Taverna D, Bortoli MD, Cora' D (2010) CircuitsDB: a database of mixed microRNA/transcription factor feed-forward regulatory circuits in human and mouse. BMC Bioinform 11:435

14. Gong X, Sun J, Zhao Z (2011) Gene regulation in glioblastoma: a combinatorial analysis of microRNAs and transcription factors. Int J Comput Biol Drug Des 4:111–126
15. Dong H, Luo L, Hong S, Siu H, Xiao Y, Jin L, Chen R, Xiong M (2010) Integrated analysis of mutations, miRNA and mRNA expression in glioblastoma. BMC Syst Biol 4:163
16. González-Gómez P, Sánchez P, Mira H (2011) MicroRNAs as regulators of neural stem cell-related pathways in glioblastoma multiforme. Mol Neurobiol 44:235–249
17. Godlewski J, Newton HB, Chiocca EA, Lawler SE (2010) MicroRNAs and glioblastoma; the stem cell connection. Cell Death Differ 17:221–228
18. Malzkorn B, Wolter M, Liesenberg F, Grzendowski M, Stühler K, Meyer HE, Reifenberger G (2010) Identification and functional characterization of microRNAs involved in the malignant progression of gliomas. Brain Pathol 20:539–550
19. Ernst A, Campos B, Meier J, Devens F, Liesenberg F, Wolter M, Reifenberger G, Herold-Mende C, Lichter P, Radlwimmer B (2010) De-repression of CTGF via the miR-17-92 cluster upon differentiation of human glioblastoma spheroid cultures. Oncogene 29:3411–3422
20. Lages E, Guttin A, El Atifi M, Ramus C, Ipas H, Dupré I, Rolland D, Salon C, Godfraind C, de Fraipont F, Dhobb M, Pelletier L, Wion D, Gay E, Berger F, Issartel J-P (2011) MicroRNA and target protein patterns reveal physiopathological features of glioma subtypes. PLoS One 6:e20600
21. Aguda BD, Kim Y, Piper-Hunter MG, Friedman A, Marsh CB (2008) MicroRNA regulation of a cancer network: consequences of the feedback loops involving miR-17-92, E2F, and Myc. Proc Natl Acad Sci USA 105:19678–19683
22. de la Iglesia N, Puram SV, Bonni A (2009) STAT3 regulation of glioblastoma pathogenesis. Curr Mol Med 9:580–590
23. Liu Y, Li C, Lin J (2010) STAT3 as a therapeutic target for glioblastoma. Anticancer Agents Med Chem 10:512–519
24. Brantley EC, Benveniste EN (2008) Signal transducer and activator of transcription-3: a molecular hub for signaling pathways in gliomas. Mol Cancer Res 6:675–684
25. Brock M, Trenkmann M, Gay RE, Michel BA, Gay S, Fischler M, Ulrich S, Speich R, Huber LC (2009) Interleukin-6 modulates the expression of the bone morphogenic protein receptor type II through a novel STAT3-microRNA cluster 17/92 pathway. Circ Res 104:1184–1191
26. Foshay KM, Gallicano GI (2009) MiR-17 family miRNAs are expressed during early mammalian development and regulate stem cell differentiation. Dev Biol 326:431–443
27. Sachdeva M, Zhu S, Wu F, Wu H, Walia V, Kumar S, Elble R, Watabe K, Mo Y-Y (2009) p53 represses c-Myc through induction of the tumor suppressor miR-145. Proc Natl Acad Sci USA 106:3207–3212
28. Aguda BD, Kim Y, Kim HS, Friedman A, Fine HA (2011) Qualitative network modeling of the Myc-p53 control system of cell proliferation and differentiation. Biophys J 101:2082–2091
29. Suh S-S, Yoo JY, Nuovo GJ, Jeon Y-J, Kim S, Lee TJ, Kim T, Bakàcs A, Alder H, Kaur B, Aqeilan RI, Pichiorri F, Croce CM (2012) MicroRNAs/TP53 feedback circuitry in glioblastoma multiforme. Proc Natl Acad Sci USA 109:5316–5321
30. Gao F-B (2008) Posttranscriptional control of neuronal development by microRNA networks. Trends Neurosci 31:20–26
31. Silber J, Lim DA, Petritsch C, Persson AI, Maunakea AK, Yu M, Vandenberg SR, Ginzinger DG, James CD, Costello JF, Bergers G, Weiss WA, Alvarez-Buylla A, Hodgson JG (2008) MiR-124 and miR-137 inhibit proliferation of glioblastoma multiforme cells and induce differentiation of brain tumor stem cells. BMC Med 6:14
32. Zhang P, Lathia JD, Flavahan WA, Rich JN, Mattson MP (2009) Squelching glioblastoma stem cells by targeting REST for proteasomal degradation. Trends Neurosci 32:559–565
33. Qu Q, Shi Y (2009) Neural stem cells in the developing and adult brains. J Cell Physiol 221:5–9
34. Park H-J, Kim J-K, Jeon H-M, Oh S-Y, Kim S-H, Nam D-H, Kim H (2010) The neural stem cell fate determinant TLX promotes tumorigenesis and genesis of cells resembling glioma stem cells. Mol Cells 30:403–408
35. Hatanpaa KJ, Burma S, Zhao D, Habib AA (2010) Epidermal growth factor receptor in glioma: signal transduction, neuropathology, imaging, and radioresistance. Neoplasia 12:675–684
36. Papagiannakopoulos T, Friedmann-Morvinski D, Neveu P, Dugas JC, Gill RM, Huillard E, Liu C, Zong H, Rowitch DH, Barres BA, Verma IM, Kosik KS (2012) Pro-neural miR-128 is a glioma tumor suppressor that targets mitogenic kinases. Oncogene 31:1884–1895
37. Moore LM, Zhang W (2010) Targeting miR-21 in glioma: a small RNA with big potential. Expert Opin Ther Targets 14:1247–1257
38. Fujita S, Ito T, Mizutani T, Minoguchi S, Yamamichi N, Sakurai K, Iba H (2008) miR-21 Gene expression triggered by AP-1 is sustained through a double-negative feedback mechanism. J Mol Biol 378:492–504
39. Brun M, Coles JE, Monckton EA, Glubrecht DD, Bisgrove D, Godbout R (2009) Nuclear factor I regulates brain fatty acid-binding protein and glial fibrillary acidic protein gene expression in malignant glioma cell lines. J Mol Biol 391:282–300
40. Chang T-C, Wentzel EA, Kent OA, Ramachandran K, Mullendore M, Lee KH, Feldmann G, Yamakuchi M, Ferlito M, Lowenstein CJ, Arking DE, Beer MA, Maitra A, Mendell JT (2007) Transactivation of miR-34a by p53 broadly influences gene expression and promotes apoptosis. Mol Cell 26:745–752

41. Li Y, Guessous F, Zhang Y, Dipierro C, Kefas B, Johnson E, Marcinkiewicz L, Jiang J, Yang Y, Schmittgen TD, Lopes B, Schiff D, Purow B, Abounader R (2009) MicroRNA-34a inhibits glioblastoma growth by targeting multiple oncogenes. Cancer Res 69:7569–7576
42. Yamakuchi M, Lowenstein CJ (2009) MiR-34, SIRT1 and p53: the feedback loop. Cell Cycle 8:712–715
43. Zheng H, Ying H, Yan H, Kimmelman AC, Hiller DJ, Chen A-J, Perry SR, Tonon G, Chu GC, Ding Z, Stommel JM, Dunn KL, Wiedemeyer R, You MJ, Brennan C, Wang YA, Ligon KL, Wong WH, Chin L, DePinho RA (2008) p53 and Pten control neural and glioma stem/progenitor cell renewal and differentiation. Nature 455:1129–1133
44. Guessous F, Zhang Y, Kofman A, Catania A, Li Y, Schiff D, Purow B, Abounader R (2010) microRNA-34a is tumor suppressive in brain tumors and glioma stem cells. Cell Cycle 9:1031–1036
45. Chang T-C, Yu D, Lee Y-S, Wentzel EA, Arking DE, West KM, Dang CV, Thomas-Tikhonenko A, Mendell JT (2008) Widespread microRNA repression by Myc contributes to tumorigenesis. Nat Genet 40:43–50
46. Sotillo E, Laver T, Mellert H, Schelter JM, Cleary MA, McMahon S, Thomas-Tikhonenko A (2011) Myc overexpression brings out unexpected antiapoptotic effects of miR-34a. Oncogene 30:2587–2594
47. Zheng H, Ying H, Yan H, Kimmelman AC, Hiller DJ, Chen A-J, Perry SR, Tonon G, Chu GC, Ding Z, Stommel JM, Dunn KL, Wiedemeyer R, You MJ, Brennan C, Wang YA, Ligon KL, Wong WH, Chin L, dePinho RA (2008) Pten and p53 converge on c-Myc to control differentiation, self-renewal, and transformation of normal and neoplastic stem cells in glioblastoma. Cold Spring Harb Symp Quant Biol 73:427–437
48. Fang X, Yoon J-G, Li L, Yu W, Shao J, Hua D, Zheng S, Hood L, Goodlett DR, Foltz G, Lin B (2011) The SOX2 response program in glioblastoma multiforme: an integrated ChIP-seq, expression microarray, and microRNA analysis. BMC Genomics 12:11
49. Heng J-CD, Orlov YL, Ng H-H (2010) Transcription factors for the modulation of pluripotency and reprogramming. Cold Spring Harb Symp Quant Biol 75:237–244
50. Annovazzi L, Mellai M, Caldera V, Valente G, Schiffer D (2011) SOX2 expression and amplification in gliomas and glioma cell lines. Cancer Genomics Proteomics 8:139–147
51. Lin T, Chao C, Saito S, Mazur SJ, Murphy ME, Appella E, Xu Y (2005) p53 induces differentiation of mouse embryonic stem cells by suppressing Nanog expression. Nat Cell Biol 7:165–171
52. Kuijk EW, van Mil A, Brinkhof B, Penning LC, Colenbrander B, Roelen BAJ (2010) PTEN and TRP53 independently suppress Nanog expression in spermatogonial stem cells. Stem Cells Dev 19:979–988
53. Brandner S (2010) Nanog, Gli, and p53: a new network of stemness in development and cancer. EMBO J 29:2475–2476
54. Moon J-H, Kwon S, Jun EK, Kim A, Whang KY, Kim H, Oh S, Yoon BS, You S (2011) Nanog-induced dedifferentiation of p53-deficient mouse astrocytes into brain cancer stem-like cells. Biochem Biophys Res Commun 412:175–181
55. Gantmacher F (1959) Applications of the theory of matrices. Interscience, New York
56. Clarke BL (1980) Stability of complex reaction networks. Adv Chem Phys 43:1–213
57. Aguda BD, Tang Y (1999) The kinetic origins of the restriction point in the mammalian cell cycle. Cell Prolif 32:321–335
58. Aguda BD, Algar CK (2003) A structural analysis of the qualitative networks regulating the cell cycle and apoptosis. Cell Cycle 2:538–544
59. Aguda BD, Goryachev AB (2007) From pathways databases to network models of switching behavior. PLoS Comput Biol 3:1674–1678
60. Yao G, Lee TJ, Mori S, Nevins JR, You L (2008) A bistable Rb-E2F switch underlies the restriction point. Nat Cell Biol 10:476–482

Coordinated Networks of microRNAs and Transcription Factors with Evolutionary Perspectives

10

Hisakazu Iwama

Abstract

MicroRNAs (miRNAs) and transcription factors (TFs) are two major classes of *trans*-regulators in gene regulatory networks. Coordination between miRNAs and TFs has been demonstrated by individual studies on developmental processes and the pathogenesis of various cancers. Systematic computational approaches have an advantage in elucidating global network features of the miRNA-TF coordinated regulation. miRNAs and TFs have distinct molecular and evolutionary properties. In particular, miRNA genes have a rapid turnover of birth-and-death processes during evolution, and their effects are widespread but modest. Therefore, miRNAs and TFs are considered to have different contributions to their coordination. The miRNA-TF coordinated feedforward circuits are considered to cause significant increases in redundancy but drastically reduce the target gene repertoire, which poses the question, to what extent is miRNA-TF coordination beneficial? Evolutionary analyses provide wide perspectives on the features of miRNA-TF coordinated regulatory networks at a systems level.

Keywords

MicroRNA • Transcription factor • Coordinated regulation, regulatory network • Redundancy • Natural selection

Abbreviations

miRNA	microRNA
TF	transcription factor
TFBS	transcription factor binding site
PSSM	position-specific scoring matrix
PWM	position-weight matrix
pri-miRNA	primary miRNA
pre-miRNA	precursor miRNA

H. Iwama, M.D., D. Med. Sci. (✉)
Life Science Research Center, Kagawa University, Ikenobe 1750-1, Miki-cho, Kita-gun, Kagawa 761-0793, Japan

Faculty of Medicine, Kagawa University, Ikenobe 1750-1, Miki-cho, Kita-gun, Kagawa 761-0793, Japan
e-mail: iwama@med.kagawa-u.ac.jp

U. Schmitz et al. (eds.), *MicroRNA Cancer Regulation: Advanced Concepts, Bioinformatics and Systems Biology Tools*, Advances in Experimental Medicine and Biology 774,
DOI 10.1007/978-94-007-5590-1_10, © Springer Science+Business Media Dordrecht 2013

RISC	RNA-induced silencing complex
UTR	untranslated region
HITS-CLIP	high-throughput sequencing of RNAs isolated by crosslinking immunoprecipitation
PAR-CLIP	photoactivatable-ribonucleoside-enhanced crosslinking and immunoprecipitation
TE	transposable element
MITE	miniature inverted-repeat transposable element
Myr	million years
GO	Gene Ontology
FFL	feedforward loop
FFC	feedforward circuit
Y1H	yeast one-hybrid

10.1 Introduction

This chapter focuses on coordinated gene regulatory networks comprising microRNAs (miRNAs) and transcription factors (TFs) with special reference to evolutionary viewpoints. To facilitate understanding of these evolutionary viewpoints, an overview of the origin and turnover of miRNAs, and nonconserved miRNA targeting is provided. The species under study are limited to animals (Kingdom *Animalia*), unless specified elsewhere. Readers interested in evolutionary viewpoints, should go to Sects. 10.3, 10.4, and 10.6. Some might hold a fixed notion that miRNAs are conserved among various species and bearing critical functions in cells. This is almost certainly correct, but only represents a tip of the miRNA iceberg, thus revealing further facets as we explore more about miRNAs.

10.2 Two Major Layers of Gene Regulation

10.2.1 Two Classes of *Trans*-Regulators

miRNAs and TFs are now considered as the two major classes of the *trans*-regulators. Since the discovery of miRNAs as *trans*-regulators in *Caenorhabditis elegans* in 1993 [1, 2], there has been a rapid accumulation of knowledge concerning miRNAs. The number of human miRNAs registered in miRBase (http://www.mirbase.org) [3–5] is 1,424 (Release 17, as of August 2011). This number is comparable to the number of curated TF genes (1,391), and the upper bound of the number of TFs is estimated to be 1,700–1,900 in humans [6]. Furthermore, more than 30% of human protein coding genes have been estimated to be under miRNAs' regulatory influence [7]. From these, we realize that miRNAs and TFs constitute two major layers of gene regulatory networks; therefore, this chapter regards the *trans*-regulators, *i.e.* miRNAs and TFs, as the mainstays of gene regulation.

10.2.2 Molecular Mechanisms of TFs vs. miRNAs in Brief

TFs are proteins that regulate gene transcription by binding to genomic *cis*-regulatory DNA sequences that usually reside in the upstream region of genes. These sequences are called TF binding sites (TFBSs), and are short DNA stretches of between ~5 and 15 bps. The bindings of TFs to TFBSs are sequence-specific; however, a considerable degree of degeneracy is allowed, leading to correspondingly varied affinities. TFBSs and their degeneracy patterns are often represented by position-specific scoring matrices (PSSM) or position-weight matrices (PWM), which are available from various databases, for example, JASPAR [8, 9] (http://jaspar.genereg.net/.) or TRANSFAC [10, 11] (http://www.gene-regulation.com/pub/databases.html). A TF can function either as an activator, a repressor, or both, depending on the type of the TF or the context of its binding to its target gene. On the other hand, a miRNA acts as a repressor agent. For animals, miRNAs (in the mature form, *i.e.* mature miRNAs) are single-stranded RNAs of about 22–25 nucleotides long that are typically transcribed by RNA polymerase II [12] as a longer primary transcript, primary miRNA (pri-miRNA). In the nucleus, the catalytic RNase III domain of an enzyme, Drosha, cleaves the pri-miRNA to yield hairpin-like precursor miRNAs (pre-miRNAs) which are about 70-nt long [13]. Subsequently, the pre-miRNAs are transported to the cytoplasm, where an enzyme, Dicer, excises away the loop to produce 22–25 nt RNA duplex (stem) stretches [14]. In many cases, one strand of the duplex RNA functions as a mature miRNA by being incorporated into the RNA-induced silencing complex (RISC) [15]. Deep sequencing technologies have revealed a considerable number of cases in which both strands of the duplexes are functional [3, 16]. The important features of mature miRNAs are: (I) miRNAs posttranscriptionally repress target transcripts or repress the transcripts' translation; (II) this repression is caused by (nearly) exact base-pairing between the seed sequences of mature miRNAs and the target regions in the 3′ UTR (untranslated region) of transcripts; and (III) the seed sequence is 6–8 nt long located in the 5′-end region of the mature miRNA [17, 18].

10.2.3 Identifying *Trans*-Regulators: TFs vs. miRNAs

For identifying TFs (as *trans*-regulators), we can utilize comprehensive information on protein domains, such as those in the InterPro database [19, 20] (http://www.ebi.ac.uk/interpro/). The use of protein domain information and protein three-dimensional structures provide us with a rich source of information for predicting TFs. In addition, the relatively high degree of similarity at the amino acid sequence level among members of a TF family that share the same DNA-biding domain often make the identification of other TFs by similarity-based searches relatively easy [21, 22], although manual curation is necessary to accurately annotate true TFs. On the other hand, miRNAs, as *trans*-regulators, are hard to identify automatically because: (I) primary miRNAs have simple structure; (II) are short in length; and (III) a vast number of genomic DNA stretches can yield similar hairpin-loop transcripts [16, 23]. High throughput sequencing methods are expected to provide sufficient power to determine the miRNA repertoire for each species. In fact, they are now major sources that provide significant amounts of novel miRNA candidates; however, they are often hindered by the elusiveness of low copy number miRNA candidates and/or by their expression in narrowly limited tissues or developmental stages [3]. We also take 'mirtrons' into consideration as an alternative source of miRNAs. Mirtrons are spliced directly from introns through the function of a lariat-debranching enzyme. Accordingly, they bypass Drosha cleavage and enter the canonical miRNA pathway after being transported to the cytoplasm [24, 25].

10.2.4 Identifying miRNA Target Sites and TFBSs

Computationally predicting miRNA target sites and TFBSs, *i.e. cis*-elements to miRNAs and TFs, respectively, presents a challenge. The main reasons are (I) the short *cis*-element length, ~5–15 nt long for TFBSs and ~6–8 nt long for miRNA target sites (mainly seed sequence complementary sites [23, 26]), (II) TFBSs allow degenerate sequence motifs and miRNA target sites allow a mismatched base (mostly one mismatch). These circumstances cause computational prediction to yield a huge amount of false positive results if only one genome sequence is subjected to prediction (i.e. *de novo* prediction). Another complication is the accessibility of *trans*-regulators to *cis*-elements. For TFs, it is widely recognized that histone modifications alter chromatin structure and changes the accessibility of TFs to the genomic DNAs (known as the 'histone code') [27, 28]. Therefore, the mere existence of TFBSs is a weak signal for functioning *cis*-elements. For miRNAs, the accessibility of mature miRNAs to the 3′ UTRs of transcripts is also an important factor [29]. The location of the miRNA target site within the 3′ UTR and the other sequence contextual features are also important predictors for miRNA target site prediction [30]. Simple counting of seed sequence matches to 3′ UTR sequences obviously yields a huge number of hits [31] that include possible less-functional sites, although matching of the seed sequence to a region in the 3′ UTR of a transcript is a strong and main predictor of the miRNA target site [23, 26].

10.3 Evolutionary Features of miRNA

10.3.1 Evolutionary Information Is Useful for Prediction

Evolutionary aspects are important, as exemplified by their utility in overcoming the difficulties in *de novo* prediction of the miRNA repertoire, miRNA target sites, and TFBSs (see Sects. 10.2.3 and 10.2.4). If a region of the genome is conserved among various species, the region is generally considered to bear certain function(s) that are important for the species. This is because mutations that occurred in functional regions are harmful in most cases, therefore such mutations are considered to be rejected or selected out by purifying selection over an evolutionary time span. Thus, searching such evolutionarily conserved genomic sequences for functional elements significantly increases the sensitivity and specificity. This type of search strategy is termed phylogenetic shadowing or phylogenetic footprinting. Many prediction applications make use of information on evolutionary conservation in various schemes, which apparently boosts their detection power [23]. However, factors other than the conserved seed matches to 3′ UTRs may need to be taken into consideration. Some lines of experimental evidence have identified cases in which miRNAs target to regions located outside of the 3′ UTR, [32, 33] and a 3′-part of the mature miRNA sequence, in addition to the seed sequence, may participate in targeting transcripts [34]. Furthermore, the genome annotation of the 3′ UTR, on which many prediction applications depend, is incomplete in accuracy and in comprehensiveness. In this respect, newly devised high-throughput methods to identify the 3′ ends of transcripts, such as polyA capture followed by sequencing [35] and poly(A)-position profiling by sequencing (3P-seq) [36], which were used for *C. elegans*, would remarkably improve the 3′ UTR annotations of genomes.

10.3.2 Evolutionary Conservation Sacrifices Species-Specific Information

Integrating evolutionary conservation information greatly increases predictive power; however, it sacrifices the functional properties that occur in one species only or in a limited group of species. In fact, ten times as many such species-specific (or nonconserved) miRNA target sites are estimated to exist compared to conserved miRNA target sites for mouse [37]. Thus, if we only analyze the predicted evolutionarily conserved miRNA target sites, we will miss the entire picture of the miRNA regulation. High-throughput sequencing-based approaches, such as argonaute HITS-CLIP (high-throughput sequencing of RNAs isolated by crosslinking immunoprecipitation) [38] or PAR-CLIP (photoactivatable-ribonucleoside-enhanced crosslinking and immunoprecipitation) [39] can now be used to comprehensively identify species-specific miRNA target sites. They are expected to complement the drawbacks identified in *de novo* computational predictions (see Sects. 10.2.3 and 10.2.4). Although the high throughput sequencing methods are valuable in providing empirical evidence-based information on miRNA target sites, including species-specific ones, they have their own drawbacks. They are not comprehensive in terms of different cell types, developmental stages, or cellular conditions, because each data set only captures one slice of the various cellular states corresponding to a snapshot of the vast space of gene expression statuses. Accumulation of these slices will reveal the whole picture of gene expression.

10.3.3 Evolutionary Origins of miRNA

For animals, some sources of new miRNA genes have been proposed. A mechanism involving gene duplications at miRNA loci is one source of new miRNAs [40]. For example, in primates, a cluster of miRNAs on the X chromosome has undergone tandem duplications followed by possible adaptive evolution [41, 42]. (Note that, for plants, new miRNA genes often originate by a mechanism of inverted duplication of a protein-coding gene. Such pairs of inverted portions provide the hair pin-like structure, and some of them evolve to form miRNA genes and their corresponding targets [43, 44].) Some studies proposed that transposable elements (TEs) are the sources of new miRNAs [45–47]. In primates, hsa-mir-548 is proposed to be derived from a type of TE, a short miniature inverted-repeat transposable element (MITE) [47]. A possible large source of new miRNA genes is thought to reside in the non-coding genomic DNAs that can yield random hairpin structures of transcripts, which in turn shape into authentic miRNAs. In the genomes of higher organisms, 100s of 1,000s of regions are estimated to potentially encode random hairpin structures of RNAs [16, 23]. A large proportion of the non-coding genomic regions are known to be transcribed [48, 49]; therefore, it is feasible that significant numbers of those random hairpins give rise to nascent miRNAs.

10.3.4 Rapid Turnover of miRNA Genes

A particular characteristic of miRNAs as *trans*-regulators is their fast turnover of the birth-and-death processes during evolution. The term 'turnover' here stands for the processes over an evolutionary time span and not for cellular molecular processes of miRNAs. Lu et al. reported a high turnover rate of newly emerged, species-specific miRNAs based on gross small RNA sequencing of three *Drosophila* species [50]. Although it was subsequently suggested that this result included a possible overestimation of the birth and death rates [51, 52] (Lu et al.'s reply to [51]), the notion of the evolutionarily transient nature of miRNAs is particularly important. Precise estimates of turnover rates per Myr (million years) may fluctuate; however, the species-specific transient nature derived from abundant random hairpins could make a clear distinction from TF genes. (It should be noted that the term 'birth-and-death evolution' originally explained a form of multigene family evolution where gene duplication events create new genes, some of which are maintained for a long time, while the others are deleted or inactivated by mutations (see review [53]). Thus, the creation of new miRNA genes from random hairpins might be a conceptual extension of the original notion.) Furthermore, not only miRNA genes themselves, but also miRNA targets sites, are likely to be subject to a notably high turnover rate [54], because of the short length of the miRNA target sites and relatively low degree of conservation of the 3′ UTR sequences. Accordingly, we should note the transient nature of miRNAs in considering the influence of miRNAs on regulatory networks.

10.3.5 Weak Expression of Nascent miRNAs Fosters Their Survival

As newly emerged miRNAs join a regulatory network, it is likely that the network is perturbed. This interfering influence is considered to be alleviated by the lower level of expression of the nascent miRNAs [50, 55]. An estimated time span of a miRNA to become highly expressed is about 50 Myr, based on a dataset from *Drosophila* miRNA sequencing [52]. (This should be compared with the time span of about 100 Myr of the human lineage divergence from the common ancestor of human and mouse.) Many nascent miRNAs thus have the opportunity to survive for some time because their weak expressions elicit relatively small conflicts with the existing networks. In this course, most of those young miRNAs are drifting neutrally (their survivals or losses are by-chance phenomena in a population of the organism), or others may be evolving in an adaptive direction. A small fraction of them may then acquire certain functions, leading to their maintenance for a long time under the influence of natural selection. Others, lacking functions, will disappear from the population by chance or become extinct because of deleterious mutations. The longer surviving, functional miRNAs become expressed more strongly and their expressions may be further tuned in accordance with appropriate spatiotemporal functional contexts. For young miRNAs, some reports suggest that adaptive evolution (where nucleotide changes that are beneficial for the organism's survival are fixed in the population more rapidly and preferentially than by chance) play important roles [52, 56].

10.3.6 Nonconserved miRNA Target Sites

To date, functional investigations have mostly concentrated on conserved miRNAs and conserved miRNA target sites; therefore, miRNAs are prone to be presupposed as conserved among species. However, nonconserved or species-specific miRNAs constitute the larger part of miRNA families. In addition, a huge number of nonconserved miRNA target sites are also present. A mere 6- or 7-nt stretch of seed sequence complementarity is sufficient to interfere with transcripts, regardless of whether it is conserved or not [37]; thus, the nonconserved miRNA target sites amount to ~10 times the number of conserved ones [23, 37, 57]. It is obvious that a living cell does not refer to other species' genomes for conservation. Accordingly, a huge number of nonconserved miRNA target sites are under the influence of miRNAs. During a long evolutionary time course, those nonconserved target sites have been selectively avoided when the nonconserved sites reside in the 3′ UTRs of genes that are co-expressed with the corresponding miRNAs spatiotemporally. This phenomenon is called 'selective avoidance' [37]. The cases in which ubiquitously expressed housekeeping genes tend to have significantly shorter 3′ UTRs (thus less space for miRNA target sites) show an apparent extreme manifestation of selective avoidance [58]. The widely disseminating regulatory effects of nonconserved miRNAs have influenced the evolution of 3′ UTR sequences. It is important to recognize that the existence and turnover of nonconserved miRNA target sites is one of the driving forces of evolution.

10.4 Basic Concepts of miRNA-TF Coordination

10.4.1 miRNA Networks Have a Widespread, but Modest, Effect

miRNAs exhibit a paradoxical nature as regulators. High conservation across species for many miRNAs seems to indicate their functional importance. Furthermore, many studies have shown critical roles of miRNAs. On the other hand, miRNAs have also been shown to be relatively dispensable. In *C. elegans*, Miska et al. reported that the majority of miRNAs into which deletion mutations were introduced did not result in grossly abnormal phenotypes [59]. They attributed this resilience to redundancy of miRNA regulatory connections and/or redundancy of the pathways regulated by miRNAs. Concerning the output quantity of proteins, Nakahara et al. demonstrated that in *Drosophila*, oocytes lacking the dicer-1 gene (essential for miRNA biogenesis) had relatively small changes in protein outputs [60]. Studies involving direct systematic protein output measurements showed that a single miRNA can influence hundreds of proteins, which are almost all products of genes targeted by the miRNA; however, the extent of the effects in the protein output reduction were small [61, 62].

Thus, it could be said that the influences of miRNAs are widespread, but modest. This is why miRNAs are considered to function mainly as rheostats, to fine-tune or buffer gene expression rather than acting as decisive switches. This led to the hypothesis that miRNA regulatory networks are subsidiary, rather than stand alone, in many cases. Herein lies the reason we should study the coordination of miRNAs with the TF regulatory networks.

10.4.2 Coordinated Networks: By Chance or by Rule(s)?

From the point of view of evolution, there seems to be an abyss for researchers investigating these regulatory systems. One viewpoint is that even a highly complex regulatory system is only a phase of by-chance processes in which mutations occur, drift, and fix neutrally along with recombination in a population [63, 64]; thus, complex regulatory systems are 'byproducts' of floating phases [65]. The other view is that there should be some rule(s) or principle(s) according to which complex regulatory systems are built [66], and thus ordered systems could be reduced to adaptive components. Bearing these two opposing premises in mind, the following sections analyze pioneering studies on miRNA-TF coordinated regulation.

In these studies, there is a common pivotal concept: the concept of network motifs, or certain topologies of small regulatory circuits. A network motif is a frequently recurring circuit motif in networks. It is defined as a specific circuit (or a subgraph) that appears in a network significantly more frequently than it is expected to appear in random networks. Such motifs have been proposed as simple building blocks that constitute complex networks [66–71]. Therefore, this train of thought belongs to the latter of the two concepts outlined above, and stands on the grounds that there should be some design principle(s) or rule(s) for complex regulatory networks, whether the grounds are explicit or implicit.

10.5 Pioneering Studies on miRNA-TF Coordinated Regulation

This section focuses on three important, pioneering studies on miRNA-TF coordinated regulatory networks. Not only main findings, but also technically important points are explained.

10.5.1 miRNA-TF Coordination from Microarray Data

Using gene expression microarray data of human and mouse, John Tsang et al. [72] showed widespread miRNA-TF coordinated regulation in human and mouse tissues or in specific cells. This was a pioneering systematic analysis conducted in the search of miRNA-TF coordinated regulations based on large-scale experimental data. This study adopted three key cautious features in conducting their research. Firstly, they focused on intronic miRNAs. miRNAs are divided according to their genomic context into two main categories. One category includes the miRNAs that reside in introns (intronic miRNAs), and the other consists of miRNAs that locate between genes (intergenic miRNAs). Tsang et al. used expression data of the protein-coding genes in which miRNAs were embedded as a proxy to plumb the level of the miRNA expression for intronic miRNAs to overcome the paucity of miRNA expression data at that time. With few exceptions (*e.g.* miR-7 during *Drosophila* embryogenesis [73]), the expression of intronic miRNAs is well correlated with that of the host gene [73–76]. Secondly, they extended their investigation using individual homogeneous cells to avoid the mixture of heterogeneous cell types that is often used tissue-level resolution analyses. Thirdly, they circumvented inaccurate, noisy computational predictions of miRNA target sites for their main investigations. However, information on the number of seed matches to the target genes and the proportion of conserved matches, was used for the scoring scheme, termed conservation enrichment (CE).

They concluded that the expression of a miRNA tended to correlate with that of its target gene, although the direction of the correlation could be either positive or negative. They categorized the miRNA-TF regulatory circuits into type I and II (Fig. 10.1). In type I circuits (Fig. 10.1a–c), the transcription rates of the miRNA and the TF are modulated in the same way (either positive or negative) by the upstream regulator. In contrast, in type II circuits, the miRNA and the TF are modulated in opposite directions in terms of the transcription rate. Based on the gene expression datasets of human and mouse, they revealed that both type I and II circuits showed expressions significantly biased to be correlated between miRNAs and target genes either positively or negatively. Interestingly, they found that a significant preferential enrichment of type I circuits was revealed only in brain tissues, particularly in adult

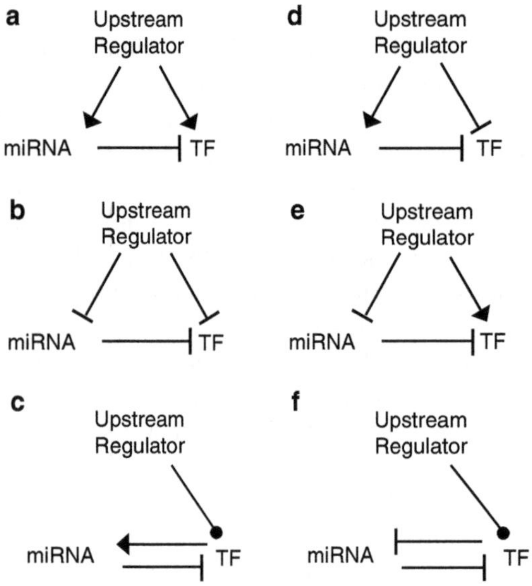

Fig. 10.1 Two types of miRNA-TF regulatory networks indicated by Tsang et al. [72]. An upstream regulator (or input) regulates the transcription rate of a miRNA and a TF. In type I circuits (panels **a–c**), the transcription rates of the miRNA and the TF are modulated in the same direction (either positive or negative) by the upstream regulator. In contrast, in type II circuits (panels **d–f**), the miRNA and the TF are modulated in opposite directions in terms of the transcription rate. An *arrow* stands for a positive regulation, a *line* terminated with a *bar* stands for a negative regulation, and a *line* terminated with a *closed circle* represents either positive or negative regulation

mature neurons (the dataset from [77]). They found that this type-I circuit preference disappeared in an expression dataset of developing neurons (the dataset from [78]). Based on this difference, they inferred that type-I circuits may play roles in homeostasis of mature neurons by maintaining protein steady state and regulating local translation. According to this scheme, they argued that type I circuits could have noise-buffering functions, while the type II circuits could act as a toggle switch that fixes the circuit state after a transient signal. They suggested that the "miRNA-containing" type I and II circuits are recurrent motifs. The use of expression data is certainly meaningful in avoiding the false positives of computational target predictions; however, the expression level of each gene usually reflects the summation of influences of multiple circuits that affect the particular target gene rather than a single circuit. Nonetheless, this work was a pioneering systematic presentation of miRNA-TF coordination, and was particularly valuable because it was based on experimental data.

10.5.2 Intensive Computations Depict miRNA-TF Coordination

At almost the same time as Tsang et al. (see Sect. 10.5.1), Shalgi et al. [79] made the full use of computational approaches to investigate the characteristics of miRNA networks and miRNA-TF networks. One of the basic ideas underlying this study was the concept that miRNAs' regulatory effects play pivotal roles in 'canalization of noise' that ensures a reduction in phenotypic variations during development [80]. They focused on conserved miRNAs and the conserved miRNA target sites, using TargetScan (http://www.targetscan.org/) [23], see also [30, 81] and PicTar (http://pictar.mdc-berlin.de/) [57], see also [82] datasets.

Concerning the global structure of miRNA networks, they proposed a set of miRNA 'target hub' genes that are targeted significantly more frequently by miRNAs. Significantly, these 'target hub' genes tend to encode transcription factors and transcription regulatory proteins, and are often involved in developmental processes. This miRNA network property parallels the feature of TF networks in which genes with longer conserved promoter regions (*i.e.* likely to have more TFBSs) showed a strong tendency to encode TF genes, particularly those related to developmental processes [83]. Lewis et al. [84] showed that a significantly higher proportion of miRNA target genes are involved in 'transcription' and 'regulation of transcription' by Gene Ontology (GO) analysis; thus, the concept of the miRNA target hub suggested by Shalgi et al. seems to be consistent with the finding of Lewis et al. In dealing with the number of miRNA target sites in the 3′ UTR, a problem lies in the wide variation in 3′ UTR length. They used a null model by adopting degree-preserving randomization for the statistical test, and successfully showed that the miRNA target hub is defined also by its high miRNA target density.

In this study, for example, the degree-preserving randomization was applied to the real network matrix that consisted of a set of connections 'miRNA → target genes'. In this case, the degree-preserving randomization means that for each gene, the number of genes targeted by a miRNA is kept the same as the real matrix, and also for each miRNA, the number of genes that the miRNA targets is kept the same as the real matrix. This randomization method preserves the indegree and outdegree of each miRNA and of each target gene as in the original matrix [66–71] and see review [69]. Therefore, this method can control for the possibility that any resultant significance is merely attributable to the skew in the degree distribution of the original network. They also suggested that the degree distribution in the miRNA coregulation network is subject to a power law, which indicates that the miRNA coregulation network is scale free. This suggestion, together with the notion of the miRNA target hub gene, does not seem to agree with the indication by Martinez et al. (see Sect. 10.5.3).

Concerning miRNA-TF coordinated regulations, Shalgi et al. found several important network motifs (see the last paragraph of Sect. 10.4.2) of gene regulatory circuits that comprise miRNAs and TFs. They searched for a series of various types of feedforward circuits comprising a miRNA and a TF (or two TFs), as shown in Fig. 10.2. They first identified significant miRNA-TF pairs. Each of these pairs was defined so that the miRNA target site and the TFBS should co-occur significantly more often in the target gene compared to the background probabilities of a gene having a miRNA target site and having a TFBS. They then successfully identified significant network motifs, as shown in panels a though d in Fig. 10.2. Interestingly, reciprocal regulation between a TF and a miRNA constitutes a feedforward circuit (Fig. 10.2c), which they denoted as "FFL (feedforward loop, FFL is a synonymous to feedforward circuit (FFC) in this chapter) miR↔TF", significantly more often than the random expectation.

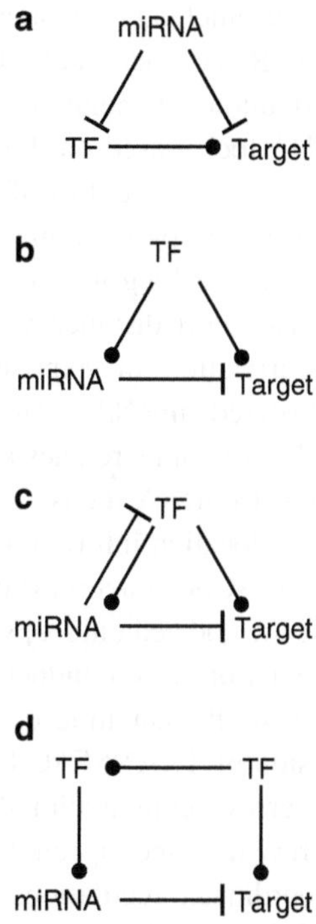

Fig. 10.2 Networks of miRNA-TF coordinated regulation indicated by Shalgi et al. [79]. Circuit graphs showing the overrepresented miRNA-TF coordinated regulatory networks based on the predicted conserved miRNA target sites and predicted conserved TFBSs. A *line* terminated with a *bar* stands for a negative regulation, and a *line* terminated with a *closed circle* represents either positive or negative regulation

They also examined correlations in the expression levels between miRNAs and TFs, using microarray data for miRNAs [85] and TFs [86]. They found both positive and negative strong correlations between miRNA and TF expressions for significant TF-miRNA pairs and also for TF-miRNA pairs that constitute FFCs. These findings are consistent with the correlation pattern found by Tsang et al. [72] (see Sect. 10.5.1).

Predicting TFBSs that regulate miRNAs is a challenging task, because the locations and spans of regulatory regions of miRNA loci are, to date, elusive. Another problem is that clustered miRNA loci are transcribed together as a polycistron and possibly have a similar expression pattern among them. In this case, assigning a regulatory region to each of the clustered miRNAs makes little sense. Accordingly, Shalgi et al. made a histogram of the genomic nucleotide distance between neighboring miRNA loci and uncovered a bimodal distribution of neighboring distance between miRNA loci. Based on this distribution pattern, they determined a cutoff of 10-kb as the distance within which two neighboring miRNA loci are expected to belong to one cluster. They showed that this cutoff distance yielded a relatively good correlation of expression pattern among the clustered miRNAs. They then used 10-kb and 5-kb upstream stretches as the regulatory regions for a miRNA locus. They obtained conserved TFBS location information from tfbsConsSites and tfbsCons Factors (http://genome.ucsc.edu/) (see the updated guide, [87]), in which the TFBS prediction was conducted using all kinds of PSSMs for human, mouse, and rat registered in a version of TRANSFAC [88], see also [89]. Their intensive computational approaches successfully revealed the specific features of miRNA-TF coordinated regulations.

10.5.3 Genome-Scale Map of TFs Targeting to miRNAs

With a genome-scale experimental data set of TFBSs that regulate miRNAs of *C. elegans*, Martinez et al. elucidated important properties of TF-miRNA coordinated regulatory networks [90]. They performed genome-scale yeast one-hybrid (Y1H) assays and identified, with high-confidence, 347 transcriptional connections of 116 TFs targeting 63 miRNA promoter regions that span 300 bp to 2 kb upstream sequences (the dataset is available from http://edgedb.umassmed.edu [91]). This dataset of miRNAs' proximal TFBSs were used to draw connections of TFs to miRNAs (TF→miRNA). It revealed that the distribution of both the number of TFs that target a miRNA (*i.e.* the indegree) and the number of target genes that a miRNA regulates (*i.e.* the outdegree) are similar to those of *C. elegans* protein-coding gene networks [92, 93]. From this similarity, they proposed that the structure of miRNA regulatory networks is similar to the structure of networks comprising protein-coding genes.

Martinez et al. merged computationally predicted conserved miRNA target sites residing in the 3′ UTRs. They adopted a stringent strategy in which they used only the predicted target sites supported by at least two computational prediction methods. These prediction methods included TargetScan [23, 30, 81], miRanda (http://www.microrna.org/) [94–97], RNAhybrid (http://bibiserv.techfak.uni-bielefeld.de/rnahybrid/) [98, 99], and PicTar [see [57] for vertebrates, [100] for *Drosophila* species, [101] for nematodes species, and [82] for human coexpression]. By this approach, they obtained 252 miRNA→TF connections. They suggested that TF genes do not constitute hubs that are targeted by miRNAs (this notion does not seem to agree with that proposed by Shalgi et al. (see Sect. 10.5.2)); however, among the genes (including TF genes) targeted by miRNAs, included a small number of miRNA target hubs that were enriched in TFs. The latter notion was consistent with the notion of Shalgi et al. [79].

Martinez et al. posed a question as to why feedback circuits, which are basic mechanisms for organisms, were less abundant in the pure transcriptional regulatory networks investigated [96, 102]. With regard to one possible answer, they identified 23 mutual feedback circuits comprising a TF and a miRNA (miRNA ↔ TF). They showed that the degree of overrepresentation of miRNA-TF mutual feedback circuits (miRNA ↔ TF) was statistically significant. From this, they suggested that miRNAs are a post-transcriptional missing links that form the feedback motifs. They confirmed that the miRNA ↔ TF circuit is a significant network motif in real *C. elegans* regulatory networks, even by the most stringent method of "edge switching" (the same method of randomization as the degree-preserving randomization (see the third paragraph of Sect. 10.5.2.)). Martinez et al. performed the randomization separately for the miRNA network and the TF network, in which they regarded the two networks as a bipartite graph, thus miRNA→miRNA and TF→TF connections were excluded. Afterwards, Yu et al. suggested that mutual feedback, TF↔TF, in which both of the TFs are targeted by a miRNA had the highest degree of overrepresentation among all the possible three-node topologies in which at least one miRNA and one TF are included [103]. In Yu et al.'s results, it is interesting that the transcriptional feedback is overrepresented with the coordination of a miRNA. Concerning the simplest form of transcriptional feedback, *i.e.* autoregulation, its presence or absence was shown to be a critical factor in the degree of representation of FFCs comprising TFs for human-mouse conserved networks [104]. Although these two studies are based on computational predictions, in higher multicellular organisms, the notion that transcriptional feedback is sparse may need further analysis. However, as Martinez et al. suggested, it is particularly important that miRNAs play key roles in feedback mechanisms in gene regulatory networks.

Another interesting finding was that the miRNAs and TFs that constitute the miRNA ↔ TF mutual feedback circuit have a higher indegree and outdegree than those not participating in the miRNA ↔ TF circuit. From this, they introduced a new network property of "flux capacity", which they defined as the product of the indegree and the outdegree of a node (*i.e.* a miRNA or a TF) and denoted it as Fc. They suggested that the concept 'flux capacity' is a good indicator for the participation of the miRNA-TF mutual feedback circuit in the miRNA-integrated higher order networks.

They verified the mutual regulation of mir-43 ↔ LIN-26, by showing that LIN-26 transcriptionally activated mir-43, and that mir-43 post-transcriptionally represses LIN-26, with rigorous experiments. They also confirmed the coexpression of both mir-43 and LIN-26, which supports, they suggested, a possible function for stable expression of the mutual feedback circuit with an activator TF incorporated. Importantly, they reported that a mir-43-deleted mutant showed no detectable phenotypic changes, and they suggested that this was because of redundancy contributed by other genes (and potentially miRNAs). Although this possible redundancy in miRNA regulatory networks was not the main focus of their research, it seems to be important in considering why and how miRNA regulatory networks are so intricate or complex. Martinez et al. further referred to Miska et al. [59] "Most single miRNA mutants do not confer a detectable phenotype." For TFs as well, they cautiously suggested, as the possible downside of Y1H assays, that members of a TF family function redundantly *in vivo* [105]. They also demonstrated that FLH-1 and FLH-2, members of the same TF family, redundantly regulate several miRNAs in the *C. elegans* embryo [106].

Martinez and Walhout argued [107] that the redundant and relatively dispensable nature of miRNAs in particular, together with the lack of miRNA hubs, "agree with the hypothesis that miRNAs do not function as master regulators, but rather predominantly function to fine-tune gene expression instead of establishing crucial developmental gene expression programs [80, 108]." Importantly, Martinez et al. experimentally identified TFBSs in miRNAs' promoters and elucidated significant, mutual feedback regulatory circuits between miRNAs and TFs. In the following section, the redundancy of regulatory circuits which Martinez et al. referred to will be considered in more detail, with perspectives from evolution.

10.6 miRNA Networks Alter Significantly in Coordination with TF Networks

This section focuses on the differences in the contributions between miRNAs and TFs to their coordinated regulation. To examine these differences, the redundancy and the gene repertoire targeted by the miRNA-TF coordinated networks are important properties. From an evolutionary viewpoint, how adaptive or beneficial the miRNAs' coordination with TFs could be is discussed.

10.6.1 miRNAs and TFs Have Different Contributions to the Coordinated Networks

miRNAs seem to present paradoxical features (see Sect. 10.3.1 in detail). Experiments have shown that miRNAs function in many crucial biological processes. Strong sequence conservation over diverse species indicates their functional importance. However, knockout experiments of miRNA genes cause only modest phenotypic alterations and have small effects on protein syntheses. Considered together with these points, the high turnover rate of miRNAs (see Sect. 10.2.4) prompts us to ask what are the differences between the contributions of miRNAs and TFs to regulatory networks as a whole.

In this respect, we (Iwama et al. [109]) investigated the differences in influence of selection on miRNA and TF networks, by focusing on FFCs comprising a TF, a miRNA, and a target gene (see Fig. 10.3). We exhaustively predicted 386,241 TFBSs and 35,850 miRNA target sites conserved between human and mouse, by phylogenetic footprinting (see Sect. 12.2.1) (Fig. 10.4) using ReAlignerV [110] and TRANSFAC PSSM for TFBSs, and ReAlignerVR [109] and PITA [29] for miRNA target sites. From these, we detected non-redundant connections: 124,736 for conserved TF→target connections and 34,298 miRNA→target (including TFs) connections. We then compiled the TF→target connections and miRNA→target connections into two distinct matrices, M and μ (Fig. 10.5), in which the presence or absence of each connection is represented as 1 or 0, respectively. These two matrices stand for the real human regulatory networks, although they are predicted ones. To differentially assess the influence of selection, we devised a partial randomization approach in which only one matrix, M or μ, was randomized,

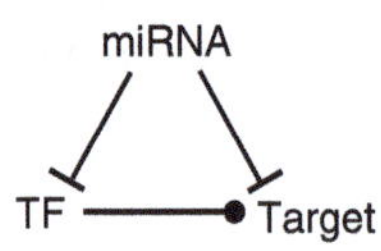

Fig. 10.3 Network graph of miRNA-TF coordinated regulations examined by Iwama et al. [109]. A *line* terminated with a *bar* stands for a negative regulation, and a *line* terminated with a *closed circle* represents either positive or negative regulation

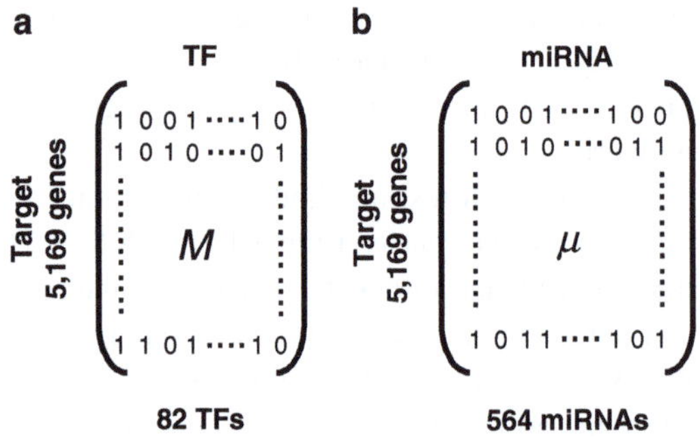

Fig. 10.5 Schemes of connection matrices M and μ used in [109]. (**a**) Matrix M stands for conserved transcriptional connections of the 83 TFs to the 5,169 genes. (**b**) Matrix μ stands for conserved posttranscriptional connections of the 564 miRNAs to the 5,169 genes. Presence or absence of a targeting connection is represented as 0 or 1, respectively, for each element of the matrices

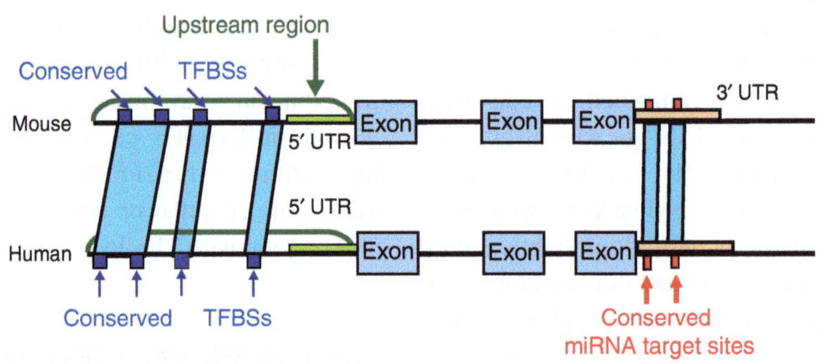

Fig. 10.4 Schema illustrating phylogenetic footprinting for miRNA target sites and TFBSs. Two horizontal *black lines* represent genome DNA sequences in the direction of 5′ (*left*) to 3′ (*right*) for human and mouse, which were used in the research in [109] (see Sect. 10.4). *Blue-shadowed parallelograms* stand for evolutionarily conserved regions in which conserved miRNA target sites and TFBSs were searched for. In this case, the upstream regions are defined as the regions immediately upstream of the first exon. Therefore the upstream region includes the 5′ UTR sequence. There are alternative cases in which the sequence upstream of the transcription start site is deemed as the upstream region

a

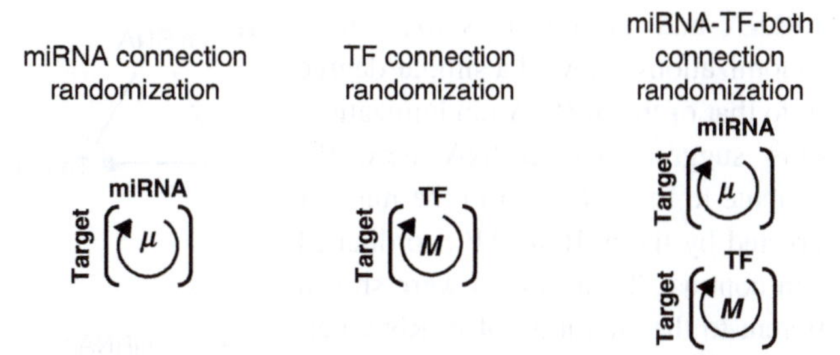

b

	miRNA Z	miRNA P	TF Z	TF P	Both Z	Both P
	Z	P	Z	P	Z	P
No. of cFFCs	0.28	0.390	-0.64	0.262	0.09	0.463
No. of cFFC-targeted genes	**-13.94**	1.8×10^{-44}	-1.13	0.128	**-13.44**	1.8×10^{-41}
cFFC redundancy	**4.47**	3.9×10^{-6}	0.30	0.383	**4.15**	1.5×10^{-5}

Fig. 10.6 Three schemes of randomizations and the degree of deviation of the real networks from the random expectation. (**a**) Schematic representations of three methods of connection matrix randomizations. (**b**) The degree of deviation of the real network from the random expectation is shown as *Z* and *P* values for each network properties. The *Z* and *P* values were computed based on each series of 1,000 randomized matrices generated in a degree-preserving manner. The *P* values were estimated assuming normal distribution over each series of 1,000 randomized matrices. The Z values are color-coded according to the *colored bar* beneath the table (This figure was modified from Fig. 2 in Iwama H. et al. [109] by permission of Oxford University Press/on behalf of The Society for Molecular Biology and Evolution)

whereas the other was kept unchanged (and *vice versa*). In this way, when we compared the real networks with the matrices in which only miRNA connections (μ) are randomized, allowing us to assess the influence caused by miRNA network alterations separately from TF network alterations (and *vice versa*). Randomizations were performed in a degree-preserving manner (see the third paragraph of Sect. 5.2) to generate 1,000 random matrices. When both matrices (M and μ) were randomized, the total influence caused by the whole regulatory networks under study was assessed. In particular, we focused on two network properties, (I) changes in the degree of redundancy that the FFCs caused, and (II) changes in the number of genes targeted by the FFCs. It should be noted that, because we examined evolutionarily conserved networks, the changes in the network properties that we could capture were brought about by selected-out connections, *i.e.* connections not maintained after the divergence of human and mouse from their common ancestor. This is often the case where studies are done based on computational predictions, because most of the prediction methods use evolutionary conservation information (see Sect. 2.1), whether it is mentioned explicitly or not.

10.6.2 Target Repertoire Shrinks and Redundancy Increases by TF-miRNA Coordination

The miRNA connection randomization demonstrated a significant reduction in the number of target genes in the real human networks, as shown in Fig. 10.6a, b. The number of target genes here represents the non-redundant count. It means that even if multiple FFCs target one gene, the number of target genes is 'one', thus, it is not the number of FFCs that target the gene. The *Z* scores represent how far the values observed in the real networks deviated from the random expectation, with the scale of the standard deviation over the 1,000 sufficiently randomized matrices. Randomizations of both miRNA and TF matrices *i.e.* whole network randomizations, showed a similar degree of deviation to that of the miRNA randomization. These results suggest that miRNA networks mainly contribute to the reduction in the number of genes targeted by the miRNA-TF coordinated regulation. In contrast, TF networks were shown to be indifferent to the alteration of miRNA networks, suggesting little contribution of TF network alteration to the reduction in target gene number. In the same way, we observed that the redundancy caused by miRNA-TF coordinated FFCs was significantly raised, primarily by miRNA network alterations, rather than those of TFs.

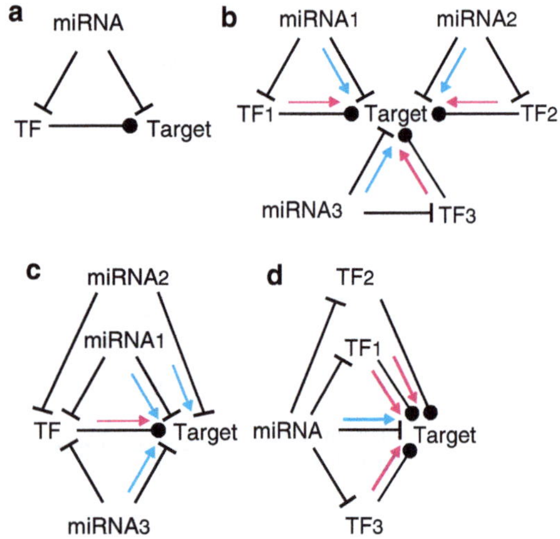

Fig. 10.7 Network graphs showing a unitary miRNA-TF coordinated FFC and three different types of FFC redundancy. (**a**) A unitary FFC under analysis. (**b**) Three unitary FFCs independently make a redundancy of three. (**c**) Redundancy of three is made through three miRNAs with one TF, whereas in (**d**), the same degree of redundancy is made through TFs. A *blue arrow* stands for a miRNA-derived targeting path and a *red arrow* for a TF-derived targeting path (This figure was modified from Fig. 3 in Iwama H. et al. [109] by permission of Oxford University Press/on behalf of The Society for Molecular Biology and Evolution)

10.6.3 Increased Redundancy Is Derived from miRNAs

To further confirm that the redundancy was mainly increased by miRNAs, we enumerated the miRNA-derived redundant connections separately from TF-derived redundant connections. As shown in Fig. 10.7, redundant FFC paths can be classified into three representative cases. In panel b, three unitary FFCs (each represented in panel a) independently have effects on one target. For this case, redundancy of three is made by three miRNAs and three TFs. However, in the case of panel c, redundancy of three is made by three miRNAs and one TF. In panel d, the same degree of redundancy of three is caused by three TFs and one miRNA. Using these differences, miRNA-derived FFC redundancy is calculable as (the number of FFCs)/(the number of TF→target connections). In the same way, the TF-derived FFC redundancy is given by (the number of FFCs)/(the number of miRNA→target connections). Based on these values, Fig. 10.8 shows that miRNA-TF FFC redundancy is mostly derived from miRNAs. Therefore, one of the major roles of miRNA connections is considered to be the addition of redundancy to the stable TF networks through the coordination of miRNA-TFs.

The redundancy provided by miRNAs is often explained in adaptive ways. By 'adaptive', we mean beneficial for an organism's survival or fitness. For example, miRNA redundant regulatory pathways would enable the fine-tuning of gene expression [17, 111–113], to shut off leaky expression by transcription factor regulation [17, 114–116], or to have roles in buffering the variation elicited by genetic and environmental perturbations [35, 80].

10.6.4 Coordination and Redundancy Are Not Necessarily Beneficial Consequences

Significant reductions in the number of target genes were shown for TF-miRNA coordinated FFCs (Fig. 10.6b). This poses a question. If the marked increase in miRNA-derived redundancy in FFCs is adaptive for the organism, why is the number of those circuits' target genes so drastically decreased? In this research, the decrease in target genes means that many of them have been selected-out during evolution. Thus, we may need to consider that most of the miRNA connections are substantially detrimental or at least not

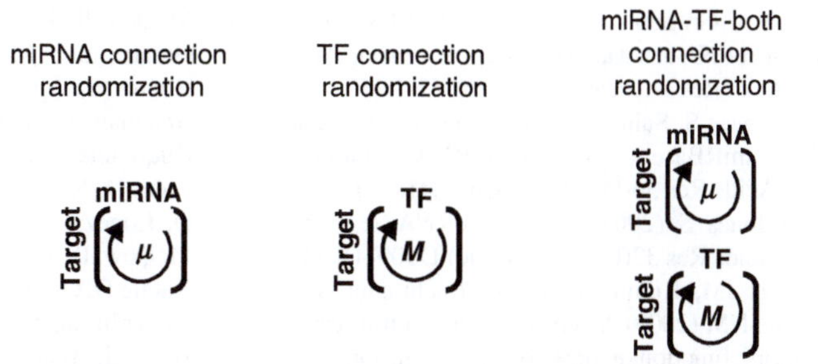

	Z	P	Z	P	Z	P
miRNA-derived cFFC redundancy	**35.10**	2.6×10^{-270}	0.08	0.468	**34.24**	3.5×10^{-257}
TF-derived cFFC redundancy	0.98	0.164	0.00	0.500	0.96	0.168

-16 -12 -8 -4 -1.64 0 1.64 4 8 12 16 20 24 28 32 36 40

Fig. 10.8 Degree of deviation of miRNA-derived and TF-derived FFC redundancies for three types of randomization schemes. The FFC redundancy is demonstrated to be mainly derived from miRNAs. Randomization schemes, the methods of estimation of *Z* and *P* values, and the color-coding scale are the same as those in Fig. 10.6 (This figure was modified from Fig. 2 in Iwama H. et al. [109] by permission of Oxford University Press/on behalf of The Society for Molecular Biology and Evolution)

beneficial. A possible key factor could lie in the rapid turnover of miRNA connections. In the research dealing with the conserved networks between human and mouse, miRNA connections that emerged after human–mouse divergence were out of range of the analyses. Therefore, the results could indicate that the miRNA connections that were tolerated at the time of human–mouse divergence have been further selected out after divergence, because the majority of them would be detrimental or at least not beneficial. However, if we consider the cases where those miRNA connections are part of redundant paths, such that they parallel beneficial connections, then the redundant paths become, at least, not detrimental. Therefore, for miRNAs, being part of redundant paths in FFCs can be considered as a way of escaping from being selected out. This may indicate that the redundancy-adding role of miRNAs provides a niche for many miRNAs' survival, avoiding conflicts with the stable TF networks. This scenario does not require adaptive roles of miRNA redundancy, but rather it is by chance that the redundancy has appeared. This indicates that redundancy and coordinated networks do not always accompany beneficial functions.

10.7 Concluding Remarks

Studies on miRNA-TF coordinated networks could include various scientific fields and tenets, which would provide us with wider perspectives on miRNAs. This theme might challenge the scientific ground upon which each of us stands; are there principles under which complex systems, including life, appear, or can we only understand the principles on the chance processes through which life has appeared?

Acknowledgment This work was supported by Grant-in-Aid for Scientific Research (MEXT) KAKENHI 23570273.

References

1. Lee RC, Feinbaum RL, Ambros V (1993) The *C. elegans* heterochronic gene lin-4 encodes small RNAs with antisense complementarity to lin-14. Cell 75(5):843–854
2. Wightman B, Ha I, Ruvkun G (1993) Posttranscriptional regulation of the heterochronic gene *lin-14* by *lin-4* mediates temporal pattern formation in *C. elegans*. Cell 75(5):855–862

3. Kozomara A, Griffiths-Jones S (2011) miRBase: integrating microRNA annotation and deep-sequencing data. Nucleic Acids Res 39(Database Issue):D152–D157
4. Griffiths-Jones S, Saini HK, van Dongen S, Enright AJ (2008) miRBase: tools for microRNA genomics. Nucleic Acids Res 36(Database Issue):D154–D158
5. Griffiths-Jones S (2004) The microRNA registry. Nucleic Acids Res 32(Database Issue):D109–D111
6. Vaquerizas JM, Kummerfeld SK, Teichmann SA, Luscombe NM (2009) A census of human transcription factors: function, expression and evolution. Nat Rev Genet 10(4):252–63
7. Lewis BP, Burge CB, Bartel DP (2005) Conserved seed pairing, often flanked by adenosines, indicates that thousands of human genes are microRNA targets. Cell 120(1):15–20
8. Sandelin A, Alkema W, Engstrom P, Wasserman WW, Lenhard B (2004) JASPAR: an open-access database for eukaryotic transcription factor binding profiles. Nucleic Acids Res 32(Database issue):D91–94
9. Bryne JC, Valen E, Tang MH, Marstrand T, Winther O, da Piedade I, Krogh A, Lenhard B, Sandelin A (2008) JASPAR, the open access database of transcription factor-binding profiles: new content and tools in the 2008 update. Nucleic Acids Res 36(Database issue):D102–106
10. Wingender E, Dietze P, Karas H, Knüppel R (1996) TRANSFAC: a database on transcription factors and their DNA binding sites. Nucleic Acids Res 24(1):238–241
11. Matys V et al (2006) Transfac and its module transcompel: transcriptional gene regulation in eukaryotes. Nucleic Acids Res 34(Database issue): 108–110
12. Lee Y, Kim M, Han J, Yeom KH, Lee S, Baek SH, Kim VN (2004) MicroRNA genes are transcribed by RNA polymerase II. EMBO J 23(20):4051–4060
13. Lee Y, Ahn C, Han J, Choi H, Kim J, Yim J, Lee J, Provost P, Radmark O, Kim S, Kim VN (2003) The nuclear RNase III Drosha initiates microRNA processing. Nature 425(6956):415–419
14. MacRae I, Zhou K, Li F, Repic A, Brooks A, Cande W, Adams P, Doudna J (2006) Structural basis for double-stranded RNA processing by Dicer. Science 311(5758):195–198
15. Gregory RI, Chendrimada TP, Cooch N, Shiekhattar R (2005) Human RISC couples microRNA biogenesis and posttranscriptional gene silencing. Cell 123(4):631–640
16. Stark A, Kheradpour P, Parts L, Brennecke J, Hodges E, Hannon GJ, Kellis M (2007) Systematic discovery and characterization of fly microRNAs using 12 *Drosophila* genomes. Genome Res 17(12):1865–1879
17. Bartel DP (2004) MicroRNAs: genomics, biogenesis, mechanism, and function. Cell 116(2):281–297
18. Mallory AC, Reinhart BJ, Jones-Rhoades MW, Tang G, Zamore PD, Barton MK, Bartel DP (2004) MicroRNA control of *PHABULOSA* in leaf development: importance of pairing to the microRNA 5′ region. EMBO J 23(16):3356–3364
19. McDowall J, Hunter S (2011) InterPro protein classification. Methods Mol Biol 694:37–47
20. Hunter S, Apweiler R, Attwood TK, Bairoch A, Bateman A, Binns D, Bork P, Das U, Daugherty L, Duquenne L, Finn RD, Gough J, Haft D, Hulo N, Kahn D, Kelly E, Laugraud A, Letunic I, Lonsdale D, Lopez R, Madera M, Maslen J, McAnulla C, McDowall J, Mistry J, Mitchell A, Mulder N, Natale D, Orengo C, Quinn AF, Selengut JD, Sigrist CJ, Thimma M, Thomas PD, Valentin F, Wilson D, Wu CH, Yeats C (2009) InterPro: the integrative protein signature database. Nucleic Acids Res 37(Database issue):D211–215
21. Kummerfeld SK, Teichmann SA (2006) DBD: a transcription factor prediction database. Nucleic Acids Res 34(Database issue):D74–D81
22. Reece-Hoyes JS, Deplancke B, Shingles J, Grove CA, Hope IA et al (2005) A compendium of C. elegans regulatory transcription factors: a resource for mapping transcription regulatory networks. Genome Biol 6:R110
23. Bentwich I et al (2005) Identification of hundreds of conserved and nonconserved human microRNAs. Nat Genet 37(7):766–770
24. Okamura K, Hagen JW, Duan H, Tyler DM, Lai EC (2007) The mirtron pathway generates microRNA-class regulatory RNAs in *Drosophila*. Cell 30(1):89–100
25. Berezikov E, Chung W, Willis J, Cuppen E, Lai E (2007) Mammalian mirtron genes. Mol Cell 28(2):328–336
26. Rajewsky N (2006) microRNA target predictions in animals. Nat Genet 38(Suppl):S8–S13
27. Strahl BD, Allis CD (2000) The language of covalent histone modifications. Nature 403(6765):41–5
28. Jenuwein T, Allis CD (2001) Translating the histone code. Science 293(5532):1074–1080
29. Kertesz M, Iovino N, Unnerstall U, Gaul U, Segal E (2007) The role of site accessibility in microRNA target recognition. Nat Genet 39(10): 1278–1284
30. Grimson A, Farh KK, Johnston WK, Garrett-Engele P, Lim LP, Bartel DP (2007) MicroRNA targeting specificity in mammals: determinants beyond seed pairing. Mol Cell 27(1):91–105
31. Iwama H, Masaki T, Kuriyama S (2007) Abundance of microRNA target motifs in the 3'-UTRs of 20527 human genes. FEBS Lett 581(9):1805–1810
32. Tay Y, Zhang J, Thomson AM, Lim B, Rigoutsos I (2008) MicroRNAs to Nanog, Oct4 and Sox2 coding regions modulate embryonic stem cell differentiation. Nature 455(7216):1124–1128
33. Lal A, Navarro F, Maher CA, Maliszewski LE, Yan N, O'Day E, Chowdhury D, Dykxhoorn DM, Tsai P, Hofmann O, Becker KG, Gorospe M, Hide W, Lieberman J (2009) miR-24 Inhibits cell proliferation by targeting E2F2, MYC, and other cell-cycle genes via binding to "seedless" 3'UTR microRNA recognition elements. Mol Cell 35(5):610–625

34. Thomas M, Lieberman J, Lal A (2010) Desperately seeking microRNA targets. Nat Struct Mol Biol 17(10):1169–1174
35. Mangone M, Manoharan AP, Thierry-Mieg D, Thierry-Mieg J, Han T, Mackowiak SD, Mis E, Zegar C, Gutwein MR, Khivansara V, Attie O, Chen K, Salehi-Ashtiani K, Vidal M, Harkins TT, Bouffard P, Suzuki Y, Sugano S, Kohara Y, Rajewsky N, Piano F, Gunsalus KC, Kim JK (2010) The landscape of *C. elegans* 3'UTRs. Science 329(5990):432–435
36. Jan CH, Friedman RC, Ruby JG, Bartel DP (2011) Formation, regulation and evolution of *Caenorhabditis elegans* 3'UTRs. Nature 469(7328):97–101
37. Farh KK, Grimson A, Jan C, Lewis BP, Johnston WK, Lim LP, Burge CB, Bartel DP (2005) The widespread impact of mammalian MicroRNAs on mRNA repression and evolution. Science 310(5755):1817–1821
38. Chi SW, Zang JB, Mele A, Darnell RB (2009) Argonaute HITS-CLIP decodes microRNA-mRNA interaction maps. Nature 460(7254):479–486
39. Hafner M, Landthaler M, Burger L, Khorshid M, Hausser J, Berninger P, Rothballer A, Ascano M Jr, Jungkamp AC, Munschauer M, Ulrich A, Wardle GS, Dewell S, Zavolan M, Tuschl T (2010) Transcriptome-wide identification of RNA-binding protein and microRNA target sites by PAR-CLIP. Cell 141(1): 129–141
40. Tanzer A, Stadler PF (2004) Molecular evolution of a microRNA cluster. J Mol Biol 339(2):327–335
41. Zhang R, Peng Y, Wang W, Su B (2007) Rapid evolution of an X-linked microRNA cluster in primates. Genome Res 17(5):612–617
42. Li J, Liu Y, Dong D, Zhang Z (2010) Evolution of an X-linked primate-specific micro RNA cluster. Mol Biol Evol 27(3):671–683
43. Fahlgren N, Howell MD, Kasschau KD, Chapman EJ, Sullivan CM, Cumbie JS, Givan SA, Law TF, Grant SR, Dangl JL, Carrington JC (2007) High-throughput sequencing of Arabidopsis microRNAs: evidence for frequent birth and death of miRNA genes. PLoS One 2(2):e219
44. Allen E, Xie Z, Gustafson AM, Sung GH, Spatafora JW, Carrington JC (2004) Evolution of microRNA genes by inverted duplication of target gene sequences in *Arabidopsis thaliana*. Nat Genet 36(12):1282–1290
45. Piriyapongsa J, Marino-Ramirez L, Jordan IK (2007) Origin and evolution of human microRNAs from transposable elements. Genetics 176(2):1323–1337
46. Piriyapongsa J, Jordan IK (2008) Dual coding of siRNAs and miRNAs by plant transposable elements. RNA 14(5):814–821
47. Piriyapongsa J, Jordan IK (2007) A family of human microRNA genes from miniature inverted-repeat transposable elements. PLoS One 2(2):e203
48. Bentwich I, Avniel A, Karov Y, Aharonov R, Gilad S, Barad O, Barzilai A, Einat P, Einav U, Meiri E, Sharon E, Spector Y, Bentwich Z (2005) The transcriptional landscape of the mammalian genome. Science 309(5740):1559–1563
49. Manak JR, Dike S, Sementchenko V, Kapranov P, Biemar F, Long J, Cheng J, Bell I, Ghosh S, Piccolboni A, Gingeras TR (2006) Biological function of unannotated transcription during the early development of *Drosophila melanogaster*. Nat Genet 38(10):1151–1158
50. Lu J, Shen Y, Wu Q, Kumar S, He B, Shi S, Carthew RW, Wang SM, Wu CI (2008) The birth and death of microRNA genes in *Drosophila*. Nat Genet 40(3):351–355
51. Berezikov E, Liu N, Flynt AS, Hodges E, Rooks M, Hannon GJ, Lai EC (2010) Evolutionary flux of canonical microRNAs and mirtrons in *Drosophila*. Nat Genet 42(1):6–9
52. Lu J, Shen Y, Carthew RW, San MW, Wu C-I (2010) Reply to "Evolutionary flux of canonical microRNAs and mirtrons in *Drosophila*". Nat Genet 42(1):9–10
53. Nei M, Rooney AP (2005) Concerted and birth-and-death evolution of multigene families. Annu Rev Genet 39:121–152
54. Wu CI, Shen Y, Tang T (2009) Evolution under canalization and the dual roles of microRNAs: a hypothesis. Genome Res 19(5):734–743
55. Liang H, Li WH (2009) Lowly expressed human microRNA genes evolve rapidly. Mol Biol Evol 26(6):1195–1198
56. Lu J, Fu Y, Kumar S, Shen Y, Zeng K, Xu A, Carthew R, Wu CI (2008) Adaptive evolution of newly emerged micro-RNA genes in *Drosophila*. Mol Biol Evol 25(5):929–938
57. Krek A, Grün D, Poy MN, Wolf R, Rosenberg L, Epstein EJ, MacMenamin P, da Piedade I, Gunsalus KC, Stoffel M, Rajewsky N (2005) Combinatorial microRNA target predictions. Nat Genet 37(5): 495–500
58. Eisenberg E, Levanon EY (2003) Human housekeeping genes are compact. Trends Genet 19(7):362–365
59. Miska EA, Alvarez-Saavedra E, Abbott AL, Lau NC, Hellman AB, McGonagle SM, Bartel DP, Ambros VR, Horvitz HR (2007) Most *Caenorhabditis elegans* microRNAs are individually not essential for development or viability. PLoS Genet 3(12):e215
60. Nakahara K, Kim K, Sciulli C, Dowd SR, Minden JS, Carthew RW (2005) Targets of microRNA regulation in the *Drosophila* oocyte proteome. Proc Natl Acad Sci USA 102(34):12023–12028
61. Selbach M, Schwanhäusser B, Thierfelder N, Fang Z, Khanin R, Rajewsky N (2008) Widespread changes in protein synthesis induced by microRNAs. Nature 455(7209):58–63
62. Baek D, Villen J, Shin C, Camargo FD, Gygi SP, Bartel DP (2008) The impact of microRNAs on protein output. Nature 455(7209):64–71
63. Lynch M (2007) The evolution of genetic networks by non-adaptive processes. Nat Rev Genet 8(10): 803–813
64. Lynch M (2007) The frailty of adaptive hypotheses for the origins of organismal complexity. Proc Natl Acad Sci USA 104(Suppl 1):8597–8604

65. Zuckerkandl E (1997) Neutral and nonneutral mutations: the creative mix–evolution of complexity in gene interaction systems. J Mol Evol 44(4):470
66. Alon U (2006) Introduction to systems biology: design principles of biological circuits. CRC Press, Boca Raton
67. Shen-Orr SS, Milo R, Mangan S, Alon U (2002) Network motifs in the transcriptional regulation network of *Escherichia coli*. Nat Genet 31:64–68
68. Milo R et al (2002) Network motifs: simple building blocks of complex networks. Science 298(5594):824–827
69. Alon U (2007) Network motifs: theory and experimental approaches. Nat Rev Genet 8(6):450–461
70. Newman ME, Strogatz SH, Watts DJ (2001) Random graphs with arbitrary degree distributions and their applications. Phys Rev E 64:026118
71. Milo R, Itzkovitz S, Kashtan N, Levitt R, Shen-Orr S, Ayzenshtat I, Sheffer M, Alon U (2004) Superfamilies of evolved and designed networks. Science 303(5663):1538–1542
72. Tsang J, Zhu J, van Oudenaarden A (2007) MicroRNA-mediated feedback and feedforward loops are recurrent network motifs in mammals. Mol Cell 26(5):753–767
73. Aboobaker AA, Tomancak P, Patel N, Rubin GM, Lai EC (2005) *Drosophila* microRNAs exhibit diverse spatial expression patterns during embryonic development. Proc Natl Acad Sci USA 102(50):18017–18022
74. Baskerville S, Bartel DP (2005) Microarray profiling of micro-RNAs reveals frequent coexpression with neighboring miRNAs and host genes. RNA 11(3):241–247
75. Li X, Carthew RW (2005) A microRNA mediates EGF receptor signaling and promotes photoreceptor differentiation in the *Drosophila* eye. Cell 123(7):1267–1277
76. Rodriguez A, Griffiths-Jones S, Ashurst JL, Bradley A (2004) Identification of mammalian microRNA host genes and transcription units. Genome Res 14(10):1902–1910
77. Sugino K, Hempel CM, Miller MN, Hattox AM, Shapiro P, Wu C, Huang ZJ, Nelson SB (2006) Molecular taxonomy of major neuronal classes in the adult mouse forebrain. Nat Neurosci 9(1):99–107
78. Arlotta P, Molyneaux BJ, Chen J, Inoue J, Kominami R, Macklis JD (2005) Neuronal subtype-specific genes that control corticospinal motor neuron development in vivo. Neuron 45(2):207–221
79. Shalgi R, Lieber D, Oren M, Pilpel Y (2007) Global and local architecture of the mammalian microRNA-transcription factor regulatory network. PLoS Comput Biol 3(7):e131
80. Hornstein E, Shomron N (2006) Canalization of development by microRNAs. Nat Genet 38(Suppl):S20–24
81. Friedman RC, Farh KK-H, Burge CB, Bartel DP (2009) Most mammalian mRNAs are conserved targets of MicroRNAs. Genome Res 19(1):92–105
82. Chen K, Rajewsky N (2006) Natural selection on human microRNA binding sites inferred from SNP data. Nat Genet 38(12):1452–1456
83. Iwama H, Gojobori T (2004) Highly conserved upstream sequences for transcription factor genes and implications for the regulatory network. Proc Natl Acad Sci USA 101(49):17156–17161
84. Lewis BP, Shih IH, Jones-Rhoades MW, Bartel DP, Burge CB (2003) Prediction of Mammalian MicroRNA Targets. Cell 115(7):787–798
85. Barad O, Meiri E, Avniel A, Aharonov R, Barzilai A et al (2004) MicroRNA expression detected by oligonucleotide microarrays: system establishment and expression profiling in human tissues. Genome Res 14(12):2486–2494
86. Su AI, Wiltshire T, Batalov S, Lapp H, Ching KA et al (2004) A gene atlas of the mouse and human protein-encoding transcriptomes. Proc Natl Acad Sci USA 101(16):6062–6067
87. Fujita PA, Rhead B, Zweig AS, Hinrichs AS, Karolchik D, Cline MS, Goldman M, Barber GP, Clawson H, Coelho A, Diekhans M, Dreszer TR, Giardine BM, Harte RA, Hillman-Jackson J, Hsu F, Kirkup V, Kuhn RM, Learned K, Li CH, Meyer LR, Pohl A, Raney BJ, Rosenbloom KR, Smith KE, Haussler D, Kent WJ (2011) The UCSC genome browser database: update 2011. Nucleic Acids Res 39(Database issue):D876–882
88. Matys V, Fricke E, Geffers R, Gossling E, Haubrock M et al (2003) TRANSFAC: transcriptional regulation, from patterns to profiles. Nucleic Acids Res 31(1):374–378
89. Wingender E (2008) The TRANSFAC project as an example of framework technology that supports the analysis of genomic regulation. Brief Bioinform 9(4):326–332
90. Martinez NJ, Ow MC, Barrasa MI, Hammell M, Sequerra R, Doucette-Stamm L, Roth FP, Ambros VR, Walhout AJ (2008) A *C. elegans* genome-scale microRNA network contains composite feedback motifs with high flux capacity. Genes Dev 22(18):2535–2549
91. Barrasa MI, Vaglio P, Cavasino F, Jacotot L, Walhout AJM (2007) EDGEdb: a transcription factor–DNA interaction database for the analysis of *C. elegans* differential gene expression. BMC Genomics 8:21
92. Deplancke B, Mukhopadhyay A, Ao W, Elewa AM, Grove CA, Martinez NJ, Sequerra R, Doucette-Stam L, Reece-Hoyes JS, Hope IA et al (2006) A gene-centered *C. elegans* protein–DNA interaction network. Cell 125(6):1193–1205
93. Vermeirssen V, Barrasa MI, Hidalgo C, Babon JAB, Sequerra R, Doucette-Stam L, Barabasi AL, Walhout AJM (2007) Transcription factor modularity in a gene-centered *C. elegans* core neuronal protein–DNA interaction network. Genome Res 17(7):1061–1071
94. Enright AJ, John B, Gaul U, Tuschl T, Sander C, Marks DS (2003) miRanda algorithm: MicroRNA targets in Drosophila. Genome Biol 5(1):R1

95. John B, Enright AJ, Aravin A, Tuschl T, Sander C, Marks DS (2004) Human MicroRNA targets. PLoS Biol 2(11):e363
96. Betel D, Wilson M, Gabow A, Marks DS, Sander C (2008) microRNA target predictions: the microRNA. org resource: targets and expression. Nucleic Acids Res 36(Database Issue):D149–153
97. Betel D, Koppal A, Agius P, Sander C, Leslie C (2010) mirSVR predicted target site scoring method: comprehensive modeling of microRNA targets predicts functional non-conserved and non-canonical sites. Genome Biol 11:R90
98. Rehmsmeier M, Steffen P, Hochsmann M, Giegerich R (2004) Fast and effective prediction of microRNA/target duplexes. RNA 10(10):1507–1517
99. Krüger J, Rehmsmeier M (2006) RNAhybrid: microRNA target prediction easy, fast and flexible. Nucleic Acids Res 34(Web Server issue):W451–454
100. Grün D, Wang YL, Langenberger D, Gunsalus KC, Rajewsky N (2005) microRNA target predictions in seven *Drosophila* species. PLoS Comp Biol 1:e13
101. Lall S, Grün D, Krek A, Chen K, Wang YL, Dewey CN, Sood P, Colombo T, Bray N, Macmenamin P, Kao HL, Gunsalus KC, Pachter L, Piano F, Rajewsky N (2006) A genome-wide map of conserved microRNA targets in *C. elegans*. Curr Biol 16(5):460–471
102. Yeger-Lotem E, Sattath S, Kashtan N, Itzkovitz S, Milo R, Pinter RY, Alon U, Margalit H (2004) Network motifs in integrated cellular networks of transcription-regulation and protein-protein interaction. Proc Natl Acad Sci 101(16):5934–5939
103. Yu X, Lin J, Zack DJ, Mendell JT, Qian J (2008) Analysis of regulatory network topology reveals functionally distinct classes of microRNAs. Nucleic Acids Res 36(20):6494–6503
104. Iwama H, Murao K, Imachi H, Ishida T (2011) Transcriptional double-autorepression feedforward circuits act for multicellularity and nervous system development. BMC Genomics 12:228
105. Hollenhorst PC, Shah AA, Hopkins C, Graves BJ (2007) Genome-wide analyses reveal properties of redundant and specific promoter occupancy within the ETS gene family. Genes Dev 21(15):1882–1894
106. Ow MC, Martinez NJ, Olsen PH, Silverman HS, Barrasa MI, Conradt B, Walhout AJ, Ambros V (2008) The FLYWCH transcription factors FLH-1, FLH-2, and FLH-3 repress embryonic expression of microRNA genes in *C. elegans*. Genes Dev 22(18):2520–2534
107. Martinez NJ, Walhout AJ (2009) The interplay between transcription factors and microRNAs in genome-scale regulatory networks. Bioessays 31(4):435–445
108. Bartel DP, Chen CZ (2004) Micromanagers of gene expression: the potentially widespread influence of metazoan microRNAs. Nat Rev Genet 5(5):396–400
109. Iwama H, Murao K, Imachi H, Ishida T (2011) MicroRNA networks alter to conform to transcription factor networks adding redundancy and reducing the repertoire of target genes for coordinated regulation. Mol Biol Evol 28(1):639–646
110. Iwama H, Hori Y, Matsumoto K, Murao K, Ishida T (2008) ReAlignerV: web-based genomic alignment tool with high specificity and robustness estimated by species-specific insertion sequences. BMC Bioinform 9:112
111. Poy MN, Eliasson L, Krutzfeldt J, Kuwajima S, Ma X, MacDonald PE, Pfeffer S, Tuschl T, Rajewsky N, Rorsman P et al (2004) A pancreatic islet-specific microRNA regulates insulin secretion. Nature 432(7014):226–230
112. Karres JS, Hilgers V, Carrers I, Treisman J, Cohen SM (2007) The conserved microRNA miR-8 tunes atrophin levels to prevent neurodegeneration in *Drosophila*. Cell 131(1):136–145
113. Bartel DP (2009) MicroRNAs: target recognition and regulatory functions. Cell 136(2):215–233
114. Reinhart BJ, Slack FJ, Basson M, Pasquinelli AE, Bettinger JC, Rougvie AE, Horvitz HR, Ruvkun G (2000) The 21-nucleotide let-7 RNA regulates developmental timing in *Caenorhabditis elegans*. Nature 403(6772):901–906
115. Stark A, Brennecke J, Bushati N, Russell RB, Cohen SM (2005) Animal microRNAs confer robustness to gene expression and have a significant impact on 3′UTR evolution. Cell 123(6):1133–1146
116. Hornstein E, Mansfield JH, Yekta S, Hu JK, Harfe BD, McManus MT, Baskerville S, Bartel DP, Tabin CJ (2005) The microRNA miR-196 acts upstream of Hoxb8 and Shh in limb development. Nature 438(7068):671–674

Mathematical Modeling of microRNA–Mediated Mechanisms of Translation Repression

11

Andrei Zinovyev, Nadya Morozova, Alexander N. Gorban, and Annick Harel-Belan

Abstract

MicroRNAs can affect the protein translation using nine mechanistically different mechanisms, including repression of initiation and degradation of the transcript. There is a hot debate in the current literature about which mechanism and in which situations has a dominant role in living cells. The worst, same experimental systems dealing with the same pairs of mRNA and miRNA can provide ambiguous evidences about which is the actual mechanism of translation repression observed in the experiment. We start with reviewing the current knowledge of various mechanisms of miRNA action and suggest that mathematical modeling can help resolving some of the controversial interpretations. We describe three simple mathematical models of miRNA translation that can be used as tools in interpreting the experimental data on the dynamics of protein synthesis. The most complex model developed by us includes all known mechanisms of miRNA action. It allowed us to study possible dynamical patterns corresponding to different miRNA-mediated mechanisms of translation repression and to suggest concrete recipes on determining the dominant mechanism of miRNA action in the form of kinetic signatures. Using computational experiments and systematizing existing evidences from the literature, we justify a hypothesis about co-existence of distinct miRNA-mediated mechanisms of translation repression. The actually observed mechanism will be that acting on or changing the sensitive parameters of the translation process. The limiting place can vary from one experimental setting to another. This model explains the majority of existing controversies reported.

A. Zinovyev (✉)
Institut Curie, 26 rue d'Ulm, Paris
F-75248, France

INSERM, U900, Paris F-75248, France

Mines ParisTech, Fontainebleau F-77300, France
e-mail: Andrei.Zinovyev@curie.fr

N. Morozova • A. Harel-Belan
Laboratoire Epigenetique et Cancer CEA Saclay,
CNRS FRE 3377, 91191, Gif-sur-Yvette, France

A.N. Gorban
Centre for Mathematical Modelling, University
of Leicester, Leicester, UK

U. Schmitz et al. (eds.), *MicroRNA Cancer Regulation: Advanced Concepts, Bioinformatics and Systems Biology Tools*, Advances in Experimental Medicine and Biology 774,
DOI 10.1007/978-94-007-5590-1_11,

Keywords
microRNA mechanism • Translation • Mathematical modeling • Chemical kinetics • Limiting step • Relaxation time • Dominant system

11.1 Introduction

MicroRNAs (miRNAs) are short (21–23 nt long) non coding RNAs that are currently considered as key regulators of a wide variety of biological pathways, including development, differentiation and tumorigenesis. Recently, remarkable progress has been made in understanding miRNA biogenesis, function and mode of action. Mature miRNAs are incorporated into the RISC complex, whose key component is an Argonaute protein, and consequently regulate gene expression by guiding the RISC complex toward specific target mRNAs (see Fig. 11.1). However, the exact mechanism of this regulation is still a matter of debate. In the past few years, several possible mechanisms have been documented [1–11]. The most documented mechanisms are negative post-transcriptional regulation of mRNA by mRNA translation inhibition and/or mRNA decay, however, some observations show that miRNAs may also act at the transcriptional level.

There is a big controversy in the current literature about which mechanism and in which situations has a dominant role in living cells. The worst, same experimental systems dealing with

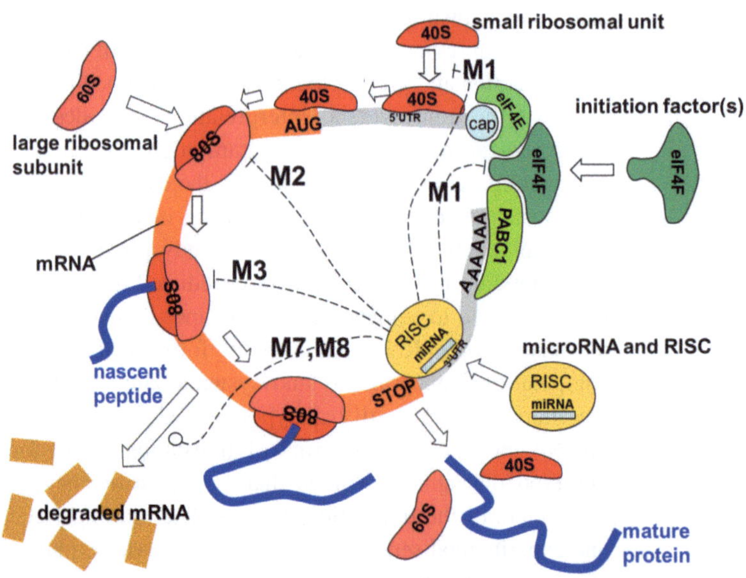

Fig. 11.1 Interaction of microRNA with protein translation process. Several (from nine documented) mechanisms of translation repression are shown: (M1) on the initiation process, preventing assembling of the initiation complex or recruiting the 40S ribosomal subunit; (M2) on the ribosome assembly; (M3) on the elongation process; (M7, M8) on the degradation of mRNA. There exist other mechanisms of microRNA action on protein translation (transcriptional, transport to P-bodies, ribosome drop-off, co-translational protein degradation and others) that are not visualized here. Here 40S and 60S are light and heavy components of the ribosome, 80S is the assembled ribosome bound to mRNA, eIF4F is an translation initiation factor, PABC1 is the Poly-A binding protein, "cap" is the mRNA cap structure needed for mRNA circularization (which can be the normal m7G-cap or artificial modified A-cap). The initiation of mRNA can proceed in a cap-independent manner, through recruiting 40S to IRES (Internal Ribosome Entry Site) located in 5′UTR region. The actual work of RNA silencing is performed by RISC (RNA-induced silencing complex) in which the main catalytic subunit is one of the Argonaute proteins (AGO), and miRNA serves as a template for recognizing specific mRNA sequences

the same pairs of mRNA and miRNA can provide controversial evidences about which is the actual mechanism of translation repression observed in the experiment. In this chapter we claim that using mathematical modeling can shed light on resolving contradicting experiment interpretations.

The structure of the chapter is the following:

First, we review the whole corpus of available experimental evidences suggesting existence of various mechanisms of miRNA action. Second, we give a detailed description of three mathematical models all describing the process of protein translation in the presence of miRNA. We start with the simplest linear model, suggested before by Nissan and Parker [12]. By analytical analysis of this simple model we demonstrate the importance of exploiting not only the stationary properties but also the dynamical properties in interpreting the experiments on miRNA-mediated silencing of translation. The second model of protein translation, also suggested first by Nissan and Parker and analyzed in [13] shows how recycling of initiation factors and ribosomal subuntis can be taken explicitly into account and to what limitation effects this can lead. We finalize the chapter by describing a mathematical model in which all nine known mechanisms of miRNA action are taken into account, developed by the authors of this chapter [14, 15]. Based on this model, we formulate practical recipes of distinguishing mechanisms of miRNA action by observing stationary and dynamical properties of three quantities: total amount of mRNA, amount of protein synthesized and the average number of ribosomes located on one transcript.

We analyze all three models following a common recipe. The purpose of the analysis is to obtain understanding of how the stationary states and the relaxation times of the model variables depend on model parameters. Though analysis of the stationary state is a well-known approach, analysis of relaxation time is a relatively poorly explored method in systems biology. By definition, the relaxation time is *the characteristic time needed for a dynamic variable to change from the initial condition to some close vicinity of the stationary state*. The relaxation time is a relatively easily observable quantity, and in some experimental methods it is an essential measurement (relaxometry, for example, see [16]). Most naturally the relaxation time is introduced in the case of a linear relaxation dynamics. For example, if a variable follows simple dynamics in the form $x(t) = A\left(1 - e^{-\lambda t}\right)$, where A is the steady-state value of x, then the relaxation time is $\tau = \frac{1}{\lambda}$ and it is the time needed for x to increase from the zero initial value to approximately $1/e \approx 63\%$ of the A value. Measuring the approximate relaxation time in practical applications consists in fitting the linear dynamics to the experimental time curves and estimating λ (for example, see [17]).

The most complete model allows us to simulate the scenario when several concurrent miRNA mechanisms act at the same time. We show that in this situation interpretation of a biological experiment might be ambiguous and dependent on the context of the experimental settings. This allows us to suggest a hypothesis that most of the controversies published in the literature can be attributed to the fact of co-occurrence of several miRNA mechanisms of action, when the observable mechanism acts on the limiting place of protein translation which can change from one experiment to another.

11.2 Review of Published Experimental Data Supporting Each of Proposed Mechanisms of microRNA Action

Protein translation is a multistep process which can be represented as sequence of stages (initiation, ribosome assembly, elongation, termination) involving circularization of mRNA, recruiting the mRNA cap structure and several protein initiation factors and ribosomal components. The process of normal translation can be regulated by small non-coding microRNAs through multiple mechanisms (Fig. 11.1).

Here we are reviewing available experimental data on all reported mechanisms of microRNA action, grouping them in a way which elucidates the main details supporting each of these proposed mechanisms.

11.2.1 M1: Cap-40S Initiation Inhibition

Inhibition of cap recruitment as a mechanism of mRNA repression was initially proposed by Pillai et al. [18], and since that time this mechanism was one of the most frequently identified [5, 6, 19–22].

The main evidence in favor of the cap-recognition and 40S assembly inhibition model was that IRES-driven or A-capped mRNA (see Fig. 11.1 for definitions of these terms) are refractory to microRNA inhibition, together with a shift toward the light fraction in the polysomal gradient. According to this, an initiation mechanism upstream of eIF4G recruitment by eIF4E was postulated and it was hypothesised that it suppresses the recognition of the cap by eIF4E. The very recent studies [22, 23] detailed GW182 involvement in the initiation suppression via cap-40S association, thus providing additional evidence for this mechanism.

11.2.2 M2: 60S Ribosomal Unit Joining Inhibition

It has also been proposed that microRNA could act in a later step of initiation, i.e., block the 60S subunit joining. This hypothesis, initially suggested by Chendrimada et al. [24], was next supported by in-vitro experiments showing a lower amount of 60S relative to 40S on inhibited mRNAs, while toe-printing experiments show that 40S is positioned at the *AUG* codon [25]. It is important to point out that, strictly speaking, there is no proof that miRNA affects the scanning for the *AUG* codon in this work, although some works interpret this data as an inhibition of scanning [12].

11.2.3 M3: Elongation Inhibition

Historically, the inhibition of translation elongation mechanism was the first proposed mechanism for microRNA action [26]. The major observation supporting this hypothesis was that the inhibited mRNA remained associated with the polysomal fraction, which was reproduced in different systems [27–30]. The idea of a post-initiation mechanism was further supported by the observation that some mRNAs can be repressed by microRNA even when their translation is cap-independent (see Fig. 11.1, mRNAs with an IRES or A-capped) [29–33].

Actually, in the work by Olsen and Ambros [26] there is no data supporting elongation inhibition rather than other post-initiation mechanisms (e.g. nascent polypeptide degradation), because the main conclusion is derived only by studying the polysomal profiles. But some evidences can be found in the work by Gu et al. [27], describing that on the same mRNA, when the ORF is prolongated downstream the binding site of miRNA (mutation in the stop codon), the inhibition by miRNA is lost. If a rare (slow) codon is introduced upstream the binding site, the inhibition is relieved, which shows that the presence of actively translating ribosomes on the binding site impairs the inhibition by miRNA. The presence of a normal polysomal distribution of the inhibited mRNA and sensitivity to EDTA (ethylenediaminetetraacetic acid) and puromycin indicating functional, translating polysomes, allowed the authors to suggest the "elongation" model. Also some data of Maroney et al. [29], could also imply that elongation is slowed down by microRNA (as the ribosome stays longer on the inhibited mRNA), but the authors discussed this point critically and were not able to reproduce it in vitro.

11.2.4 M4: Ribosome Drop-Off (Premature Termination)

First (and seems to be the only one till today) evidence of this mechanism was done by Petersen et al. [30], who observed no difference in polysomal profile in the presence of miRNA. Addition of puromycin, which necessitate peptidyl transferase activity to act, didn't change the polysomal profile in the presence or in the absence of the miRNA. The authors have concluded that polysomes are actively translating even in presence of miRNA. They were not be able to detect any peptide by radiolabelling and therefore postulated the ribosome drop-off mechanism.

However, Wang et al. [34] presented data also supporting premature termination: the read-through codon-stop and more rapid loss of polyribosome upon initiation block.

11.2.5 M5: Co-translational Nascent Protein Degradation

Initially, the idea of nascent protein degradation was proposed by Nottrott et al. [35], according to the presence of inhibited mRNA and AGO protein in polysomes, which suggests the action of miRNA on actively translated mRNA. However, no nascent peptide has ever been experimentally demonstrated [11, 30, 34, 35]; thus the nascent polypeptide degradation, if it exists, should occur extremely rapidly after the synthesis. Anyhow, being able to immunoprecipitate the nascent polypeptide together with the mRNA and the polysome in the case of normal translation, the authors failed to do so in the case of miRNA inhibition [35]. Pillai et al. in [11] showed that this degradation, if exists, should be proteasome-independent, and no specific protease or complex involved in this polypeptide degradation has ever been identified. Data supporting nascent polypeptide degradation are the following: (a) sedimentation of mRNA together with miRNA-RISC complexes in actively translating (puromycin-sensitive) polysomes; (b) blocking the initiation (in a cap-dependent manner), resulted in a shift in polysomal profile, suggesting that the repressed mRNA is actively transcribed. In Wang et al. [34] the authors also support nascent protein degradation showing polysomal distribution with puromycin sensitivity, but in the same paper they also present data supporting premature termination. Maroney et al. [29] and Gu et al. [27] presented experimental data which are very coherent with this line though not concluding that this mechanism is the dominating one: presence of miRNA/mRNA complex in polysomes, sensitivity to different conditions is an indication of translating ribosomes.

11.2.6 M6: Sequestration in P-Bodies

An effect of sequestration of mRNA targeted by AGO-microRNA complex in cytoplasmic structures called P-bodies was initially shown in [18, 36]. Next this result was confirmed in many studies characterising P-bodies as structures where the translational machinery is absent and the degradation machinery is functional [18, 36–42]. The main propositions about P-bodies' function was that they sequestrate targeted mRNA apart from translational machinery, or that P-bodies give a kinetics advantage for mRNA decay (local concentration of all needed enzymes). Two additional important points were elucidated in [43], showing that mRNA localised in the P-bodies, can be still associated with polysomes, and also that miRNA silencing is still possible when P-bodies are disrupted. This led to the conclusion that P-bodies are not required for but rather a consequence of microRNA-driven RNA degradation or translational inhibition. This conclusion is also supported by other studies [44] and is mostly accepted today. Moreover, only a small portion of miRNA, mRNA and RISC complex is localised inside macroscopic P-bodies [18, 44].

11.2.7 M7: mRNA Decay (Destabilisation)

Recently, starting from the first description by Lim et al. [45], a lot of data has revealed miRNA-mediated mRNA decay (destabilization) of targeted mRNA without direct cleavage at the binding site [45–53]. Also, most of the authors note that only a slight protein decrease can be obtained by translational inhibition only. When the protein level decreases by more than 33%, mRNA destabilization is the major component of microRNA-driven silencing [46]. Anyhow, all these data, concordant in the main point (mRNA decay mechanism), are different in details of its concrete mechanism (decay by mRNA deadenylation, decapping, or 5′–3′ degradation of the mRNA). In the review of Valencia-Sanchez et al. [54], it is concluded that the decapping followed by 5′ > 3′ degradation is the most plausible mechanism for the miRNA inhibition, while deadenylation could lead only to a decrease in the initiation efficiency by disrupting the loop between polyA and cap. Behn-Ansmant et al. [48], showed that GW182, an AGO partner in the microRNA pathway, triggers deadenilation and decapping of bound mRNA, which leads to mRNA decay. Filipowicz et al. [6] supports the idea of the degradation running mostly in 5′>3′ direction after deadenylation and decapping, in contradiction to [53], where it is claimed that deadenylation is the principal cause of the mRNA decay but degradation goes first in the 3′->5′ way. The latter work also indicated that the degradation mechanism is supposed to be only an addition to the translational inhibition and that translational inhibition has the same efficiency with or without degradation. Coller and Parker in 2004 proposed that as the poly(A) tail can enhance translation rates and inhibit mRNA decay, then the increase of deadenylation rates by miRNA/RISC could be counted as additional mechanism by which translation repression and mRNA decay could be stimulated [55]. Finally, Eulalio et al. showed that there could be two different cases in mRNA degradation by microRNA: in one the ongoing translation is required for the decay, and in the second the decay occurs in the absence of active translation, and assume that this depends on mRNAs undergoing the decay [49].

11.2.8 M8: mRNA Cleavage

mRNA cleavage (similar to what is observed with siRNA) can be observed when the sequence of microRNA is completely or almost completely complementary to its target binding site. The first proposition for this mechanism was made for plants [56, 57], and since that time, the miRNA-mediated mRNA cleavage was proved to be very common for plants, and much more rare in animals [58, 59].

Though the most of known mammalian microRNAs are only partially complementary to their targets, there is some data on miRNA-mediated mRNA cleavage, for example, for miR-196 [59]. A few other works (e.g., in mammals [54, 60] or in *C. elegans* [47]) also mentioned cleavage as a possible mechanism of microRNA repression in animals.

11.2.9 M9: Transcriptional Inhibition Through microRNA-Mediated Chromatin Reorganization Following by Gene Silencing

Although the first publication for siRNA-mediated transcriptional repression [61] was made in 2004, the first publication proving miRNA-mediated transcriptional repression in mammalian cells appeared only recently [62]. Around this time also appeared the first publication for miRNA-mediated transcriptional activation, showing that microRNA-373 induces expression of genes with complementary promoter sequences [63]. Since then very few evidences of miRNA-directed transcriptional gene silencing (TGS) in mammalian cells were obtained [62, 64].

11.2.10 Controversies Between the miRNA-Mediated Mechanisms of Translation Repression

It is important to note that it is extremely difficult to discriminate experimentally between different potential post-initiation mechanisms, such as elongation inhibition, premature ribosome dissociation ("ribosome drop-off") or normal elongation with nascent polypeptide degradation. Both elongation slowing down and nascent polypeptide degradation are supported by the fact that the mRNA-polysomal association is puromycin-sensitive, indicating polysomes' activity [29, 35] and by the observed requirement for microRNA binding in the unstranslated region [27]. Premature ribosome dissociation is supported by decreased read-through of inhibited mRNA [30]. Both ribosome drop-off and ribosomal "slowing down" are supported by the slight decrease in the number of associated ribosomes [29, 35]. But, eventually with premature drop-off, the polysomal profile will not be the same as in the case of nascent protein degradation, as one should have less ribosomes per mRNA.

Summarizing overview on the proposed mechanisms, we briefly emphasize the main controversial data.

1. First of all, even the question at which level (transcriptional, translational, etc.) the microRNA action takes place is still strongly debated. The most frequently reported, but also very contradictory in details, is the mechanism of gene repression by microRNAs occurring at the level of mRNA translation (this includes mechanisms of arrest at initiation and elongation steps, ribosome drop-off and nascent polypeptide degradation), but repression at the level of mRNA (before translation) have been also proposed as the principal one in many studies (this includes mechanisms of microRNA-mediated mRNA decay, sequestration of target mRNAs in P-bodies and rare in animals but frequent in plants mechanism of target mRNA cleavage). Moreover, it was proposed that some microRNAs mediate chromatin reorganization followed by transcriptional repression, which involves mechanisms strikingly different from the previous modes of repression. Finally, the transcriptional activation by microRNA [62, 63] and translational activation by microRNA have been also proposed [65, 66].
2. Currently, the action of microRNAs at the level of initiation of translation seems to be the most favourite one accordingly to many recent publications. Anyhow, the experimental data, supporting this mechanism, are also controversial in the result interpretations of different groups suggesting this mechanism. For example, it has been proposed that AGO2 protein could interact with the cap via the eIF4E-like domain and therefore compete with eIF4E for binding the cap [20]. However, this has been weakened by the recent finding that this domain could be involved in the binding with GW182, an important protein for miRNA action, and by crystallographic analysis showing that the folding will not allow such a interaction with the cap [23, 67].

   The main observation supporting the initiation mechanism is that mRNA with IRES or A-cap can't be inhibited by microRNA, but in the considerable number of works it was shown that some mRNAs can be repressed by a microRNA even when their translation is cap-independent [30–33].
3. For blocking the 60S subunit joining mechanism, it was shown that eIF6, an inhibitor of 60S joining, is required for microRNA action [24], but this was contradicted by other studies [23].
4. An interesting observation was reported in [68] about that the same mRNA targeted by the same microRNA can be regulated either at the initiation or the elongation step depending on the mRNA promoter. But next, using the same promoter described in [68], as leading to the initiation mechanism, the authors suggests the "elongation" model, according to the polysomal distribution on the inhibited mRNA [27].
5. Different results about mechanisms of microRNA action were obtained depending on the transfection method of the inhibited mRNA [33].
6. Karaa et al. describes the VEGF gene, which is endogenously regulated by a miRNA, miR16, acting on an IRES (see Fig. 11.1 legend) [32]. VEGF is translated from one of two IRES, and only one of these IRES allows inhibition by miR16. Therefore, inhibition by microRNA is possible even in IRES-driven translation, but not for all IRES-driven cases, even if those two IRES have been described as similar.
7. Kozak et al., reviewing different papers about miRNA-mediated inhibition, claimed a lot of experiments to be faulty [69] and reported that only few studies are based on reliable experiments could be considered, namely [19, 20, 24, 25, 52, 70]. The statement of the author that "other suggested mechanisms are not mentioned here because the speculations greatly exceed the facts" seems to concern [18, 30, 33, 35]. Together with this, the author is very critical about interpretations of the IRES experiments.
8. Olsen et al. has described inhibition of elongation step, based on the presence of polysomal distribution [26]. But, actually, there is no additional data supporting elongation inhibition rather than nascent polypeptide degradation,

because in both works the main (and different!) conclusion is driven only by studying the polysomal profiles.

9. In several studies it was shown that degradation and translational arrest can be coupled in many systems [18, 23, 49, 53, 71], but here the situation is also not completely understood: some mRNAs are repressed mostly at the translational level, others mostly at the stability level (with or without a requirement for concurrent translation inhibition), and some at both levels [60]. In some works it is suggested that microRNA-mediated mRNA decay is a consequence of translational repression, the other group of studies suggests that neither the destabilisation is a consequence of translational arrest, nor the translational repression is a consequence of degradation, but that the two mechanisms are concurrently occurring [5, 22, 23]. It has been concluded that the relative contributions of translational repression and decay differ depending on the presence or absence of the poly(A) tail [23]. However, in deciding whether the deadenylation is the cause or consequence of silencing, the authors again present controversial data interpretations [72].

Thus, the experimental data and summarizing conclusions about the mechanism by which microRNA repress mRNA expression are highly controversial, and though arise a question about interrelations between the different mechanisms and their possible concomitant action, do not consider it in the frame of one unique mechanism of microRNA action.

Using a series of mathematical models with increasing complexity, we show how mathematical modelling can help in interpreting the experimental results and even suggest some explanations of the ambiguous observations.

11.3 Modeling Notations and Assumptions

In this chapter we consider three mathematical models of miRNA action of increasing complexity:

1. *The simplest linear model of protein translation.* This model was first suggested in [12]. It allows distinguishing two types of miRNA-mediated mechanisms: those acting at the very early stage of translation initiation and those acting at a later stage.
2. *Non-linear model of protein translation taking into account recycling of ribosomes and initiation factors.* This model was first suggested in [12]. It allows distinguishing four types of miRNA-mediated mechanisms: acting at the very early stage of initiation, later stage of initiation, ribosome assembly step, elongation and termination (considered together as one step of translation).
3. *General model describing all known mechanisms of miRNA action.* This model was developed by the authors of this chapter [14, 15] and includes nine mechanisms of miRNA action. Using this model, we classify the existing mechanisms by their dynamical properties and suggest a tool to distinguish most of them based on experimental data.

Of course, any mathematical model is a significant simplification of biological reality. The first two models, for example, consider only a limited subset of all possible mechanisms of microRNA action on the translation process. All processes of synthesis and degradation of mRNA and microRNA are deliberately neglected in these models. Interaction of microRNA and mRNA is simplified: it is supposed that the concentration of microRNA is abundant with respect to mRNA. Interaction of only one type of microRNA and one type of mRNA is considered (not a mix of several microRNAs). The process of initiation is greatly simplified: all initiation factors are represented by only one molecule which is marked as eIF4F.

Finally, the classical chemical kinetics approach is applied, based on solutions of ordinary differential equations, which assumes sufficient and well-stirred amount of both microRNAs and mRNAs. Another assumption in the modeling is the mass action law assumed for the reaction kinetic rates.

It is important to underline the interpretation of certain chemical species considered in the system. The ribosomal subunits and the initiation factors in the model exist in free and bound forms. Moreover, the ribosomal subunits can be bound to several regions of mRNA (the initiation site, the start codon, the coding part). Importantly, several copies of fully assembled ribosomes can be bound to one mRNA. To model this situation, we have to introduce the following quantification rule for chemical species: amount of "ribosome bound to mRNA" means the total number of ribosomes translating proteins, which is not equal to the number of mRNAs with ribosome sitting on them, since one mRNA can hold several translating ribosomes (polyribosome). In this view, mRNAs act as *places* or *catalyzers*, where translation takes place, whereas mRNA itself formally is not consumed in the process of translation, but, of course, can be degraded or synthesized.

Let us introduce notations that will be used throughout the chapter for designation of chemical species:

1. *40S*, free small ribosomal subunit.
2. *60S*, free large ribosomal subunit.
3. *eIF4F*, free initiation factor.
4. *M,* free mRNA (models 1 and 2) and mRNA with free initiation site (model 3).
5. *P*, translated protein.
6. *B*, mRNA located in P-bodies.
7. *F*, state of mRNA when the small ribosomal subunit bound to the initiation site.
8. *A*, state of mRNA when the small ribosomal subunit bound to the start codon.
9. *R*, translating ribosome, located on mRNA.

Square brackets will denote the amounts of the corresponding species. For example, [*M*] will denote the amount of free mRNA in the system.

Note that the notations for the kinetic rate constants are not equivalent in three models. For example, while k_1 notifies the kinetic rate of the cap initiation in the models 1 and 2, it has different measure units in linear and non-linear models. Moreover, in the model 3, k_1 notifies the rate constant for translation initiation (recruiting 40S subunit) of mRNA already in translation process. Hence, the meaning of k_i constants should be considered differently per each model type.

11.4 Simplest Linear Model of Protein Translation

The simplest representation of the translation process has the form of a circular cascade of reactions [12] (see Fig. 11.2). The model contains four chemical species *40S*, *F*, *A* and *P* and three chemical reactions.

The catalytic cycle in which the protein is produced is formed by the following reactions:

1. 40S→*F*, Initiation complex assembly (rate *k1*).
2. *F*→*A*, Some late and cap-independent initiation steps, such as scanning the 5'UTR for the start *A* codon recognition (rate *k2*) and 60S ribosomal unit joining.

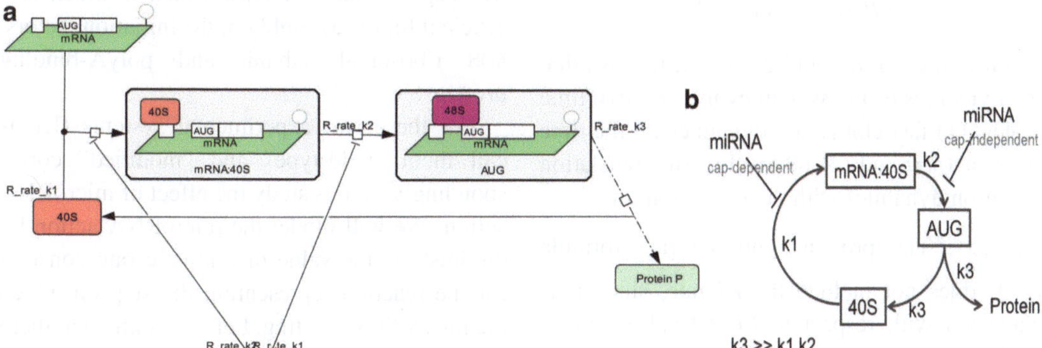

Fig. 11.2 The simplest mathematical model of protein translation which is capable to explain the effect of a miRNA on the very early (rate *k1*) and late (rate *k2*) steps of mRNA initiation; (**a**) graphical presentation of the model in the SBGN standard; (**b**) schematic model presentation. Action of miRNA is modeled by reducing the rate constant of the corresponding translation step

3. $A \to 40S$, combined processes of protein elongation and termination, which leads to production of the protein (rate *k3*), and fall off of the ribosome from mRNA.

The model is described by the following system of equations [12]:

$$\begin{cases} \dfrac{d[40S](t)}{dt} = -k1[40S] + k3[A] \\ \dfrac{d[F](t)}{dt} = k1[40S] - k2[F] \\ \dfrac{d[A](t)}{dt} = k2[F] - k3[A] \\ Psynth(t) = k3[A](t) \end{cases} \quad (11.1)$$

where *Psynth*(*t*) is the rate of protein synthesis.

Following [12], let us assume that *k3* >> *k1*, *k2*. This choice was justified by the following statement: "…The subunit joining and protein production rate (*k3*) is faster than *k1* and *k2* since mRNA-40*S* complexes bound to the *A* without the 60S subunit are generally not observed in translation initiation unless this step is stalled by experimental methods, and elongation is generally thought to not be rate limiting in protein synthesis…" [12].

Under this condition, the Eq. 11.1 have the following approximate solution (which becomes the more exact the smaller the (*k1* + *k2*)/*k3* ratio), suggested earlier in [13]:

$$\begin{bmatrix} 40S(t) \\ F(t) \\ A(t) \end{bmatrix} = \frac{40S_0}{\frac{1}{k1}+\frac{1}{k2}} \left(\begin{bmatrix} 1/k1 \\ 1/k2 \\ 1/k3 \end{bmatrix} + \frac{1}{k3}\begin{bmatrix} -1 \\ 1 \\ 0 \end{bmatrix} e^{-k3t} + \frac{1}{k2}\begin{bmatrix} 0 \\ 1 \\ -1 \end{bmatrix} e^{-(k1+k2)t} \right), \quad (11.2)$$

$$Psynth(\mathrm{t}) = \frac{40S_0}{\frac{1}{k1}+\frac{1}{k2}} \left(1 - \frac{k3}{k2} e^{-(k1+k2)t}\right) \quad (11.3)$$

for the initial condition

$$\begin{bmatrix} 40S(t) \\ F(t) \\ A(t) \\ Psynth \end{bmatrix} = \begin{bmatrix} 40S_0 \\ 0 \\ 0 \\ 0 \end{bmatrix}.$$

From the solution (11.2–11.3) it follows that the dynamics of the system evolves on two time scales: (1) fast elongation dynamics on the time scale ≈ 1/*k3*; and (2) relatively slow translation initiation dynamics with the relaxation time $t_{rel} \approx \frac{1}{k1+k2}$. The protein synthesis rate formula (2–3) does not include the *k3* rate, since it is neglected with respect to *k1*, *k2* values. From (2–3) we can extract the formula for the protein synthesis steady-state rate *Psynth*(*t*) (multiplier before the parentheses) and the relaxation time t_{rel} for it (inverse of the exponent power):

$$Psynth = \frac{40S_0}{\frac{1}{k1}+\frac{1}{k2}}, \; t_{rel} = \frac{1}{k1+k2}. \quad (11.4)$$

Now let us consider two experimental situations: (1) the rate constants for the two translation initiation steps are comparable *k1* ≈ *k2*, and (2) the cap-dependent rate *k1* is limiting: *k1* << *k2*. Accordingly to [12], the second situation can correspond to modified mRNA with an alternative cap-structure (A-cap), which is much less efficient for the assembly of the initiation factors, 40S ribosomal subunit and polyA-binding proteins.

For these two experimental systems (let us call them "wild-type" and "modified" correspondingly), let us study the effect of microRNA action. We will model the microRNA action by diminishing the value of a kinetic rate constant for the reaction representing the step on which the microRNA is acting. Let us assume that there are two alternative mechanisms: (1) microRNA acts in a cap-dependent manner (thus, reducing the *k1* constant) and (2) microRNA acts in a cap-independent manner, for example, through

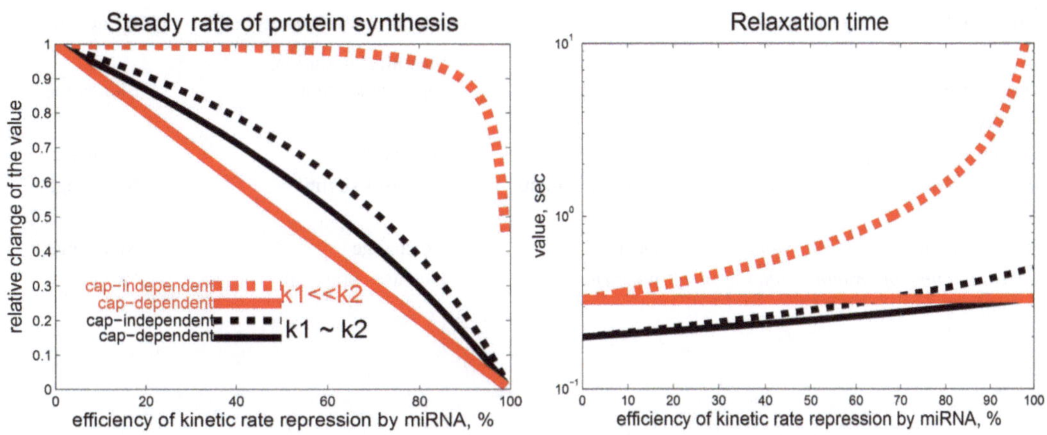

Fig. 11.3 Dependence of the relative change of the protein synthesis steady rate and the relaxation time (time needed to achieve the steady rate) on the efficiency with which microRNA can act at an early cap-dependent (*k1*) or late cap-independent (*k2*) rate of translation. Two scenarios are considered: a wild-type one when *k1* value is similar to *k2* and the case of a modified A-cap structure when *k1*<< *k2* even in the absence of miRNA

interfering with 60S subunit joining (thus, reducing the *k2* constant). The dependence of the steady rate of protein synthesis $Psynth \sim \frac{1}{\frac{1}{k1}+\frac{1}{k2}}$ and the relaxation time $t_{rel} \approx \frac{1}{k1+k2}$ on the efficiency of the microRNA action (i.e., how much it is capable to diminish a rate coefficient) is shown in Fig. 11.3.

Interestingly, experiments with cap structure replacement were made and the effect of microRNA action on the translation was measured [21, 70]. No change in the protein rate synthesis after applying microRNA was observed. From this it was concluded that microRNA in this system should act through a cap-dependent mechanism (i.e., the normal "wild-type" cap is required for microRNA recruitment). It was argued that this could be a misinterpretation [12] since in the "modified" system, cap-dependent translation initiation is a rate limiting process (*k1* << *k2*). Hence, even if microRNA acts in the cap-independent manner (inhibiting *k2*), it will have no effect on the final steady state protein synthesis rate. This was confirmed by the graph similar to the Fig. 11.3a.

From the analytical solution (2–3) we can further develop this idea and claim that it is possible to detect the action of microRNA in the "modified" system if one measures the protein synthesis relaxation time: if it significantly increases then microRNA probably acts in the cap-independent manner despite the fact that the steady state rate of the protein synthesis does not change. This is a simple consequence of the fact that the relaxation time in a cycle of biochemical reactions is limited by the second slowest reaction, see [13, 73]. If the relaxation time does not change in the presence of microRNA then we can conclude that none of the two alternative mechanisms of microRNA-based translation repression is activated in the system, hence, microRNA action is dependent on the structure of the "wild-type" transcript cap.

The observations from the Fig. 11.3 are recapitulated in the Table 11.1. This analysis (of course, over-simplified in many aspects) provides us with an important lesson: observed dynamical features of the translation process with and without presence of microRNA can give clues on the mechanisms of microRNA action and help to distinguish them in a particular experimental situation. Theoretical analysis of the translation dynamics highlights the important characteristics of the

Table 11.1 Modeling two mechanisms of microRNA action on several translation steps in the simplest linear model

Observable value	Initiation	Step after initiation, cap-independent	Elongation
Wild-type cap			
Steady-state rate of protein synthesis	Decreases	Decreases	No change
Relaxation time of protein synthesis	Increases slightly	Increases slightly	No change
A-cap			
Steady-state rate of protein synthesis	Decreases	No change	No change
Relaxation time of protein synthesis	no change	Increases drastically	No change

dynamics which should be measured in order to infer the possible microRNA mechanism.

This conclusion suggests the notion of a **kinetic signature of microRNA action mechanism** which we define as *a set of measurable characteristics of the translational machinery dynamics (features of time series for protein, mRNA, ribosomal subunits concentrations) and the predicted tendencies of their changes as a response to microRNA action through a particular biochemical mechanism.*

11.5 Non-linear Nissan and Parker's Model of Protein Translation

To explain the effect of microRNA interference with translation initiation factors, a non-linear version of the translation model was proposed in [12] which explicitly takes into account recycling of initiation factors (eIF4F) and ribosomal subunits (40S and 60S).

11.5.1 Model Equations and the Steady State Solutions

The model contains the following list of chemical species (Fig. 11.4): *40S*, *60S*, *eIF4F*, *F*, *A*, and *R* and four reactions, all considered to be irreversible:

1. $40S + eIF4F \rightarrow F$, assembly of the initiation complex (rate *k1*).
2. $F \rightarrow A$, some late and cap-independent initiation steps, such as scanning the 5'UTR for the start codon *A* (rate *k2*).
3. $A \rightarrow R$, assembly of ribosomes and protein translation (rate *k3*).
4. $80S \rightarrow 60S + 40S$, recycling of ribosomal subunits (rate *k4*).

The model is described by the following system of equations [12]:

$$\begin{cases} \frac{d[40S](t)}{dt} = -k1[40S][eIF4F] + k4[R] \\ \frac{d[eIF4F]}{dt} = -k1[40S][eIF4F] + k2[F] \\ \frac{d[F](t)}{dt} = k1[40S][eIF4F] - k2[F] \\ \frac{d[A](t)}{dt} = k2[F] - k3[A][60S] \\ \frac{d[60S]}{dt} = -k3[A][60S] + k4[R] \\ \frac{d[R]}{dt} = k3[A][60S] - k4[R] \\ Psynth(t) = k3[A](t) \end{cases} \quad (11.5)$$

The model (11.5) contains three independent conservations laws:

$$\begin{aligned} &[F]+[40S]+[A]+[R]=[40S]_0, \\ &[F]+[eIF4F]=[eIF4F]_0, \\ &[60S]+[R]=[60S]_0, \end{aligned} \quad (11.6)$$

where $[40S]_0$, $[60S]_0$ and $[eIF4F]_0$ are total amounts of available small, big ribosomal subunits and the initiation factor respectively.

The following assumptions on the model parameters were suggested [12]:

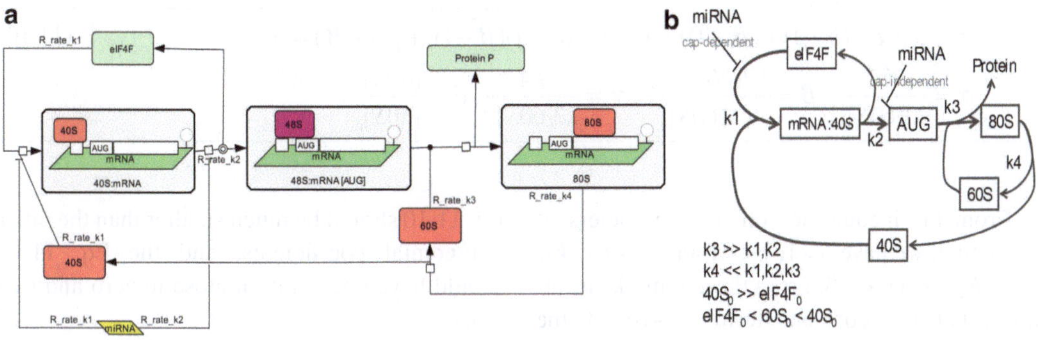

Fig. 11.4 The mathematical model of protein translation which explicitly takes into account recycling of ribosomes and initiation factors; (**a**) graphical presentation of the model in the SBGN standard; (**b**) schematic model presentation

$$\begin{aligned}
&k4 << k1, k2, k3,\\
&k3 >> k1, k2,\\
&[eIF4F]_0 << [40S]_0,\\
&[eIF4F]_0 < [60S]_0 < [40S]_0,
\end{aligned} \tag{11.7}$$

with the following justification: "…The amount 40S ribosomal subunit was set arbitrarily high … as it is thought to generally not be a limiting factor for translation initiation. In contrast, the level of eIF4F, as the canonical limiting factor, was set significantly lower so translation would be dependent on its concentration as observed experimentally… Finally, the amount of subunit joining factors for the 60S large ribosomal subunit were estimated to be more abundant than eIF4F but still substoichiometric when compared to 40S levels, consistent with in vivo levels… The *k4* rate is relatively slower than the other rates in the model; nevertheless, the simulation's overall protein production was not altered by changes of several orders of magnitude around its value…" [12].

The last statement about the value of *k4* is needed to be made more precise: in the model by Nissan and Parker, *k4* is a sensitive control parameter. It does not affect the steady state protein synthesis rate only in one of the possible scenarios (*inefficient initiation*, deficit of the initiation factors, see below).

The final steady state of the system can be calculated from the conservation laws and the balance equations among all the reaction fluxes:

$$\begin{aligned}
k2[F]_s &= k3[A]_s[60S]_s\\
&= k4[R]_s = k1[40S]_s[eIF4F]_s
\end{aligned} \tag{11.8}$$

where "*s*" index stands for the steady state value. Let us designate a fraction of the free [60*S*] ribosomal subunit in the steady state as $x = \frac{[60S]_s}{[60S]_0}$. Then we have

$$\begin{aligned}
&[F]_s = \frac{k4}{k2}[60S]_0(1-x),\ [A]_s = \frac{k4}{k3}\frac{1-x}{x},\ [eIF4F]_s = [eIF4F]_s - \frac{k4}{k2}[60S]_0(1-x),\\
&[60S]_s = [60S]_0 x,\ [R]_s = [60S]_0(1-x),\ [40S]_s = [40S]_0 - [60S]_0(1-x)\left(1+\frac{k4}{k2}\right) - \frac{k4}{k3}\frac{1-x}{x}
\end{aligned} \tag{11.9}$$

and the equation to determine *x*, in which we have neglected the terms of smaller order of magnitude, based on conditions 11.7:

$$x^3 + x^2(\alpha + (\delta - 1) + (\beta - 1)) + x(-\alpha + (\delta - 1)(\beta - 1)) + \gamma(1 - \beta) = 0, \tag{11.10}$$
$$\alpha = \frac{k2}{k1[60S]_0}, \beta = \frac{k2[eIF4F]_0}{k4[60S]_0}, \gamma = \frac{k4}{k3[60S]_0}, \delta = \frac{[40S]_0}{[60S]_0}.$$

From the inequalities on the parameters of the model, we have $\delta > 1$, $\gamma << 1$ and, if $k1 >> k4/[eIF4F]_0$ then $\alpha << \beta$. From these remarks it follows that the constant term $\gamma(1-\beta)$ of the Eq. 11.10 should be much smaller than the other polynomial coefficients, and the Eq. 11.10 should have one solution close to zero and two others:

$$x_0 \approx \frac{k4}{k3([40S]_0 - [60S]_0)}, x_1 \approx 1 - \frac{k2[eIF4F]_0}{k4[60S]_0} + \frac{k2^2[eIF4F]_0}{k1 \cdot k4 \cdot [40S]_0} \frac{1}{1 - \frac{k2[eIF4F]_0}{k4[40S]_0}}, x_2 \approx 1 - \frac{[40S]_0}{[60S]_0} \tag{11.11}$$

provided that $\alpha << |1-\delta|$ or $\alpha << |1-\beta|$. In the expression for $x1$ we cannot neglect the term proportional to α, to avoid zero values in 11.10.

The solution x_2 is always negative, which means that one can have one positive solution $x_0 << 1$, if $\frac{k2[eIF4F]_0}{k4[60S]_0} \geq 1$, and two positive solutions x_0 and x_1, if $\frac{k2[eIF4F]_0}{k4[60S]_0} \leq 1$. However, it is easy to check that if $x_1 > 0$ then x_0 does not correspond to a positive value of $[eIF4F]_s$. This means that for a given combination of parameters satisfying (11.7) we can have only one steady state (either x_0 or x_1).

The two values $x = x_0$ and $x = x_1$ correspond to **two different modes of translation**. When, for example, the amount of the initiation factors $[eIF4F]_0$ is **not enough to provide efficient initiation** ($[eIF4F]_0 < \frac{k2}{k4[60S]_0}$, $x = x_1$) then most of the 40S and 60S subunits remain in the free form, the initiation factor eIF4F being always the limiting factor. If the **initiation is efficient** enough ($[eIF4F]_0 > \frac{k2}{k4[60S]_0}$), then we have $x = x_0 << 1$ when almost all 60S ribosomal subunits are engaged in the protein elongation, and $[eIF4F]$ being a limiting factor at the early stage. However, it is liberated after and ribosomal subunits recycling becomes limiting in the initiation (see the next section for the analysis of the dynamics).

Let us notice that the steady state protein synthesis rate under these assumptions is

$$Psynth = k4[60S]_0(1 - x) = \begin{cases} k4[60S]_0, \; if \; \frac{k2[eIF4F]_0}{k4[60S]_0} > 1 \\ k2[eIF4F]_0, \; else \end{cases}. \tag{11.12}$$

This explains the numerical results obtained in [12]: with low concentrations of $[eIF4F]_0$ microRNA action would be efficient only if it affects $k2$ or if it competes with eIF4F for binding to the mRNA cap structure (thus, effectively further reducing the level $[eIF4F]_0$). With higher concentrations of $[eIF4F]_0$, other limiting factors become dominant: $[60S]_0$ (availability of the heavy ribosomal subunit) and $k4$ (speed of ribosomal subunits recycling which is the slowest reaction rate in the system). Interestingly, in any situation the protein translation rate does not depend on the value of $k1$ directly (of course, unless it does not become "globally" rate limiting), but only through competing with eIF4F (which makes the difference with the simplest linear protein translation model).

Equation 11.12 explains also some experimental results reported in [70]: increasing the concentration of [eIF4F] translation initiation factor enhances protein synthesis but its effect is abruptly saturated above a certain level.

11.5.2 Analysis of the Model Dynamics

It was proposed to use the following model parameters: $k1 = k2 = 2, k3 = 5, k4 = 1, [40S]_0 = 100$, $[60S]_0 = 25$, $[eIF4F]_0 = 6$ [12]. As we have shown in the previous section, there are two scenarios of translation possible in the Nissan and Parker's model which we called "efficient" and "inefficient" initiation. The choice between these two scenarios is determined by the combination of parameters $\beta = \frac{k2[eIF4F]_0}{k4[60S]_0}$. For the original parameters from [12], $\beta = 0.48 < 1$ and this corresponds to the simple one-stage "inefficient" initiation scenario. To illustrate the alternative situation, we changed the value of $k4$ parameter, putting it to 0.1, which makes $\beta = 4.8 > 1$. The latter case corresponds to the "efficient" initiation scenario, the dynamics is more complex and goes in three stages (see below).

Simulations of the protein translation model with these parameters and the initial conditions

$$\begin{bmatrix} [40S] \\ [eIF4F] \\ [F] \\ [A] \\ [R] \\ [60S] \end{bmatrix} = \begin{bmatrix} [40S]_0 \\ [eIF4F]_0 \\ 0 \\ 0 \\ 0 \\ [60S]_0 \end{bmatrix}$$

are shown in Fig. 11.5. The system shows non-trivial relaxation process which takes place in several epochs. Qualitatively we can distinguish the following stages:

1. Stage 1: Relatively fast relaxation with conditions $[40S] \gg [eIF4F]$, $[60S] \gg [A]$. During this stage, the two non-linear reactions $40S + eIF4F \rightarrow F$ and $A + 60S \rightarrow R$ can be considered as pseudo-monomolecular ones: $eIF4F \rightarrow F$ and $A \rightarrow R$ with rate constants dependent on $[40S]$ and $[60S]$ respectively. This stage is characterized by rapidly establishing quasiequilibrium of three first reactions (R1, R2 and R3 with $k1$, $k2$ and $k3$ constants). Biologically, this stage corresponds to the assembling of the translation initiation machinery, scanning for the start codon and assembly of the first full ribosome at the start codon position.
2. Transition between Stage 1 and Stage 2.
3. Stage 2: Relaxation with the conditions $[40S] \gg [eIF4F]$, $[60S] \ll [A]$. During this stage, the reactions $40S + eIF4F \rightarrow F$ and $A + 60S \rightarrow R$ can be considered as pseudo-monomolecular $eIF4F \rightarrow F$ and $60S \rightarrow 80S$. This stage is characterized by two local quasi-steady states established in the two network reaction cycles (formed from R1–R2 and R3–R4 reactions). Biologically, this stage corresponds to the first round of elongation, when first ribosomes move along the coding region of mRNA. The small ribosomal subunit 40S is still in excess which keeps the initiation stage (reaction R1–R2 fluxes) relatively fast.
4. Transition between Stage 2 and Stage 3.
5. Stage 3: Relaxation with the conditions $[40S] \ll [eIF4F]$, $[60S] \ll [A]$. During this stage, the reactions $40S + eIF4F \rightarrow F$ and $A + 60S \rightarrow R$ can be considered as pseudo-monomolecular $40S \rightarrow F$ and $60S \rightarrow R$. During this stage all reaction fluxes are balanced. Biologically, this stage corresponds to the stable production of the protein with constant recycling of the ribosomal subunits. Most of ribosomal subunits 40S are involved in protein elongation, so the initiation process should wait the end of elongation for that they would be recycled.

Our analysis of the non-linear Nissan and Parker's model showed that the protein translation machinery can function in two qualitatively different modes, determined by the ratio $\beta = \frac{k2[eIF4F]_0}{k4[60S]_0}$ [13]. We call these two modes "efficient initiation" ($\beta > 1$) and "inefficient initiation" ($\beta < 1$) scenarios. Very roughly, this ratio determines the balance between the overall

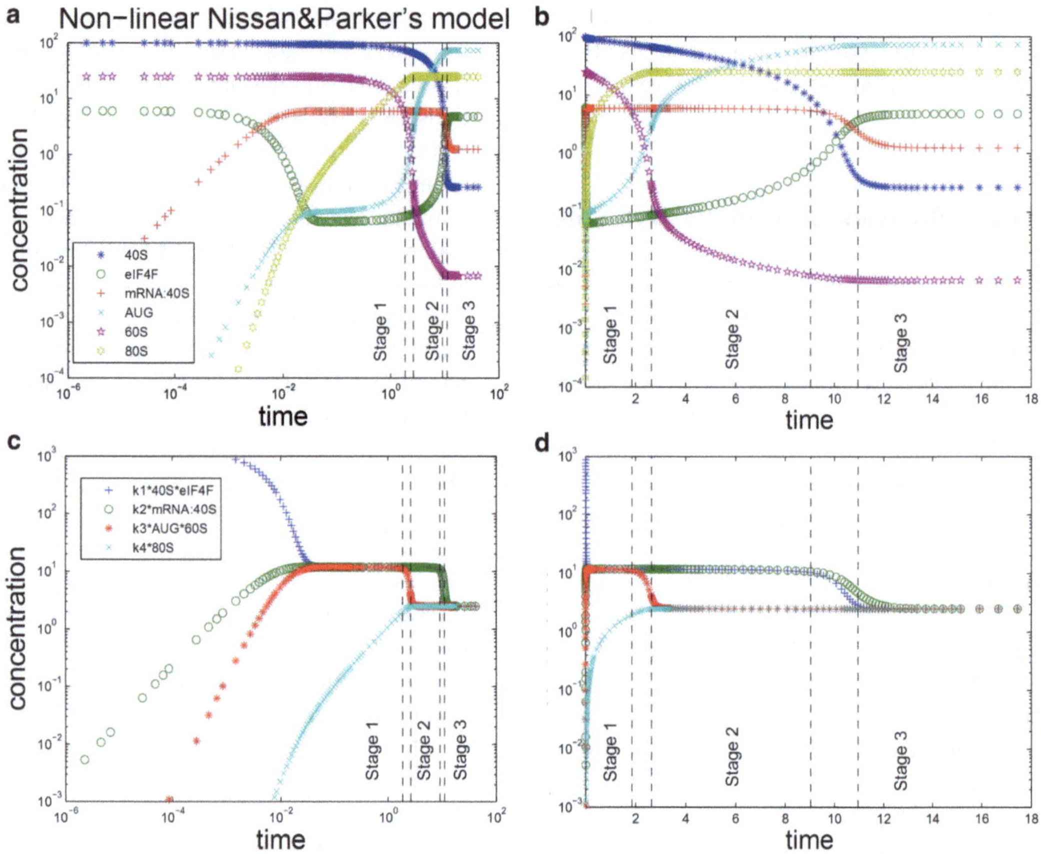

Fig. 11.5 Numerical simulations of the species concentrations and fluxes of the non-linear translation model. (**a**) and (**c**) log-log scale; (**b**) and (**d**) log-scale in values, and linear scale for the time axis (the time units are arbitrary, since the dimensionality of the parameters in [12] was not specified)

speeds of initiation and elongation processes. In the case of "efficient initiation" the rate of protein synthesis is limited by the speed of recycling of the ribosomal components (60S). In the case of "inefficient initiation" the rate of protein synthesis is limited by the speed of recycling of the initiation factors (eIF4F). Switching between two modes of translation can be achieved by changing the availability of the corresponding molecules ($[60S]_0$ or $[eIF4F]_0$) or by changing the sensitive kinetic parameters (*k2* or *k4*).

As a result of the dynamical analysis, we assembled an approximate solution of the non-linear system under assumptions (11.7) about the parameters. The detailed description of this solution is given in [13]. The advantage of such a semi-analytical solution is that one can predict the effect of changing the system parameters.

One of the obvious predictions is that the dynamics of the system is not sensitive to variations of *k3*, so if microRNA acts on the translation stage controlled by *k3* then no microRNA effect could be observed looking at the system dynamics (being the fastest one, *k3* is not a sensitive parameter in any scenario).

If microRNA acts on the translation stage controlled by *k4* (for example, by ribosome stalling mechanism) then we should consider two cases of efficient ($\beta > 1$) and inefficient ($\beta < 1$) initiation. In the first case the steady state protein synthesis rate is controlled by *k4* (as the slowest, limiting step) and any effect on *k4* would lead to the proportional change in the steady state of protein production. By contrast, in the case of inefficient initiation, the steady state protein synthesis is not affected by *k4*. Instead, the relaxation time is affected, being $\sim \frac{1}{k4}$. However, diminishing *k4* increases the β parameter: hence, this changes "inefficient initiation" scenario for the opposite, making *k4* sensitive for the steady state protein synthesis anyway, when *k4* becomes smaller than $\frac{k2[eIF4F]_0}{[60S]_0}$. For example, for the default parameters of the model, decreasing *k4* value firstly leads to no change in the steady state rate of protein synthesis, whereas the relaxation time increases and, secondly, after the threshold value $\frac{k2[eIF4F]_0}{[60S]_0}$ starts to affect the steady state protein synthesis rate directly. This is in contradiction to the message from [12] that the change in *k4* by several orders of magnitude does not change the steady state rate of protein synthesis.

Analogously, decreasing the value of *k2* can convert the "efficient" initiation scenario into the opposite after the threshold value $\frac{k4[60S]_0}{[eIF4F]_0}$. We can recapitulate the effect of decreasing *k2* in the following way: (1) in the case of the "efficient" initiation *k2* does not affect the steady state protein synthesis rate up to the threshold value after which it affects it in a linear manner. The relaxation time drastically increases, because decreasing *k2* leads to elongation of all dynamical stages duration (for example, we have estimated the time of the end of the dynamical Stage 2 as $t''' = \frac{[40S]_0}{k2[eIF4F]_0}$. However, after the threshold value the relaxation time decreases together with *k2*, quickly dropping to its unperturbed value; (2) in the case of "inefficient" initiation the steady state protein synthesis rate depends proportionally on the value of *k2* (12), while the relaxation time is not affected.

MicroRNA action on *k1* directly does not produce any strong effect neither on the relaxation time nor on the steady state protein synthesis rate. This is why in the original work [12] cap-dependent mechanism of microRNA action was taken into account through effective change of the $[eIF4F]_0$ value (total concentration of the translation initiation factors), which is a sensitive parameter of the model (5).

The effect of microRNA through various mechanisms and in various experimental settings (excess or deficit of eIF4F, normal cap or A-cap) is recapitulated in Table 11.2. The conclusion that can be made from this table is that all four mechanisms show clearly different patterns of behavior in various experimental settings. From the simulations one can make a conclusion that it is still not possible to distinguish between the situation when microRNA does not have any effect on protein translation and the situation when it acts on the step which is neither rate limiting nor "second rate limiting" in any experimental setting (*k3* in our case). Nevertheless, if any change in the steady-state protein synthesis or the relaxation time is observed, theoretically, it is possible to specify the mechanism responsible for it.

11.6 General Model of miRNA-Mediated Translation Regulation

Nine distinct mechanisms of microRNA action have been described in the literature: the main experimental data supporting each proposed mechanism are summarized in the review section of this chapter. The complete model containing all known microRNA action mechanisms is shown in Fig. 11.6a using an SBGN standard diagram.

The principal differences between the Nissan and Parker's model and the model described in this section are (1) the complete model describes all nine known mechanisms of miRNA action; (2) mRNA amount is a dynamical variable, i.e. it is modelled explicitly, taking into account its synthesis and degradation; (3) we explicitly model binding of miRNA at various stages of translation, i.e. in our model both mRNA in free

Table 11.2 Modeling of four mechanisms of microRNA action in the non-linear protein translation model

Observable value	Initiation	Step after initiation	Ribosome assembly	Elongation
Wild-type cap, inefficient initiation				
Steady-state rate	Slightly decreases	Decreases	No change	Decreases after threshold
Relaxation time	No change	No change	No change	Goes up and down
Wild-type cap, efficient initiation				
Steady-state rate	No change	Slightly decreases after strong inhibition	No change	Decreases
Relaxation time	No change	Goes up and down	No change	No change
A-cap, inefficient initiation				
Steady-state rate	Decreases	Decreases	No change	Slightly decreases after strong inhibition
Relaxation time	No change	No change	No change	Goes up and down
A-cap, efficient initiation				
Steady-state rate	Decreases after threshold	Slightly decreases after strong inhibition	No change	Decreases
Relaxation time	Goes up and down	Goes up and down	No change	Increases

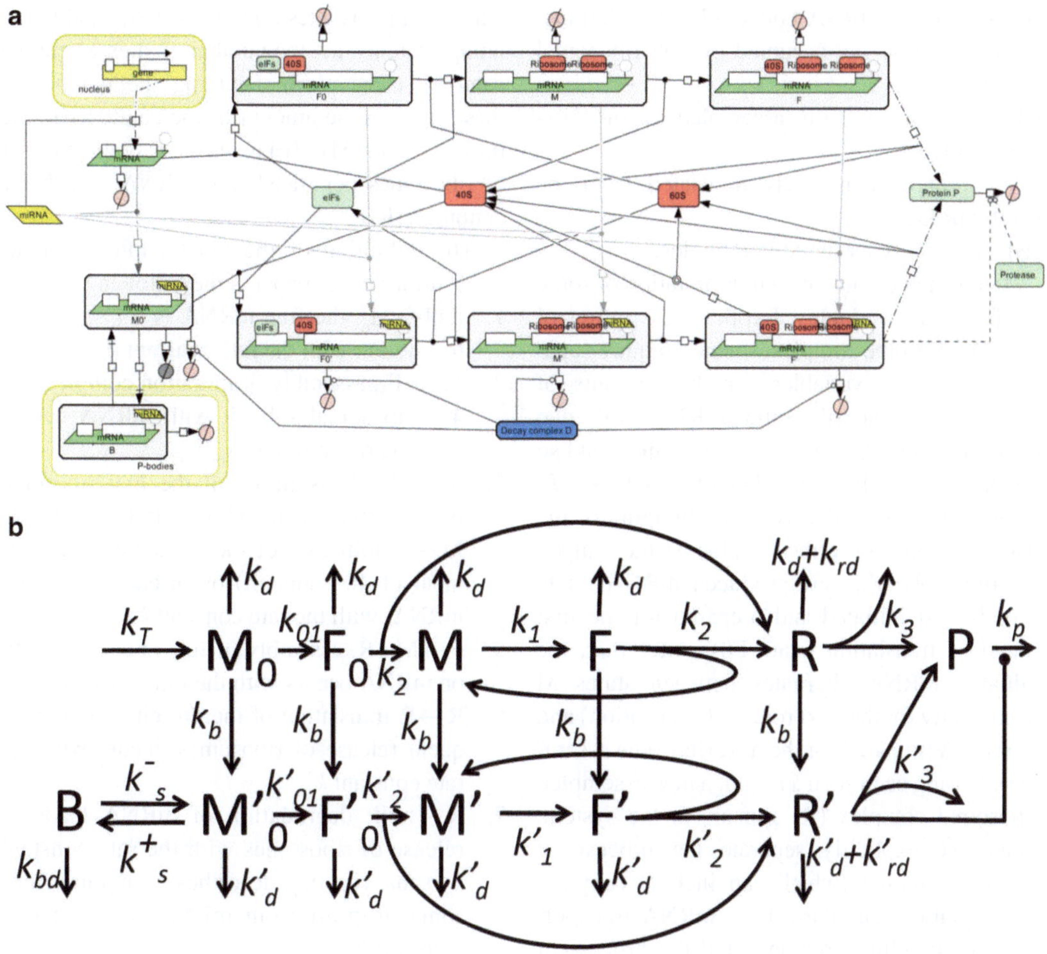

Fig. 11.6 Mathematical model taking into account all nine mechanisms of miRNA action; (**a**) graphical presentation of the model in the SBGN standard; (**b**) schematic model presentation in the assumption that ribosomal subunits and initiation factors are present in excess

and miRNA-bound forms are present; (4) we assume concentration of eIF4F and ribosomal subunits present in excess, as in the simplest model.

For modelling, we assumed that the initiation factors and ribosomal subunits are always available in excess. This allowed us to simplify the model to 12 chemical species and 20 reactions, as described below and schematically shown in Fig. 11.6b:

[M0] – new synthesized and not yet initiated mRNA

[F0] – new initiated mRNA, with initiation complex, including 40S ribosomal subunit

[M] – initiated mRNA with free translation initiation site

[F] – initiated mRNA with translation initiation site occupied by 40S ribosomal subunit

[R] – number of ribosomes fully assembled on miRNA-free mRNA

[M0'] – new synthesized not initiated mRNA with one or more miRNAs bound

[F0'] – new mRNA with initiation complex, including 40S ribosomal subunit, with miRNA(s) bound to mRNA

[M'] – initiated miRNA-bound mRNA with free translation initiation site

[F'] – initiated miRNA-bound mRNA with translation initiation site occupied by 40S ribosomal subunit

[R'] – ribosomes fully assembled on miRNA-bound mRNA

[P] – protein, completely translated from the given mRNA

[B] – mRNA sequestered in P bodies.

Let us make a notice on interpretation of some of the model variables. Explicit description of mRNA:ribosome complexes would require separate dynamical variables for the amounts of mRNA with one ribosome, mRNA with two ribosomes, mRNA with three ribosomes, and so on (potentially, large number of variables). To avoid this complexity, we apply lumping of the detailed model, described in [14]. In the lumped reaction network, new produced mRNA (state M0) is first initiated and prepared for the first round of translation (state F0). After that, the initiated mRNA alternates between states M (state ready for the next round of translation) and F (mRNA prepared for the next ribosome assembly). During each such a round, a new assembled ribosomal complex (R) appears in the system. Thus, we explicitly separate the process of mRNA initiation (which can include capping, adenylylation, circularization, mRNA transport to specific cellular regions) and the process of recruiting 40S ribosomal subunit at already initiated mRNA for the next round of translation. In our model, these two processes proceed with different speeds.

In our interpretation, we consider mRNAs as places for a catalytic reaction (protein translation). These places (amount of catalyzer) in our model can be synthesized or destroyed and present in four states (non-initiated, initiated, in 'translating' state ready for assembling new ribosome and in 'translating' state with a new assembling ribosome). To take into account miRNA, we say that there are two types of catalyzer: miRNA-free and miRNA-bounded, with different rate constants of degradation. miRNA-free catalyzer can be irreversibly transformed into miRNA-bounded type of catalyzer.

Importantly, [R] in our interpretation is not the amount of mRNA translating proteins but the amount of ribosomes bound to mRNA and translating proteins, i.e. the number of sites where the catalysis takes place. Dividing the number of these sites on the amount of the catalyzer in the initiated state [M] + [F] gives the average number of ribosomes per translating mRNA, which we denote as [RB].

The definition of the kinetic rate constants used further in the paper is the following:

1. $null \rightarrow M_0$, the free mRNA is transcribed in the system with the rate constant k_t.
2. $M_0 \rightarrow F_0$, assembly of initiation complex and 40S ribosomal subunit with mRNA occurs with the rate constant k_{01}
3. $F_0 \rightarrow M + R$, assembly of the first ribosome on the initiation site with the rate constant k_2
4. $M \rightarrow F$, initiation of the translation (recruitment of 40S subunit) on already translated mRNA, with the rate constant k_1
5. $F \rightarrow M + R$, assembly of full ribosome (S80) on mRNA occurs with the rate constant k_2
6. $R \rightarrow P$, translation of the protein with consequent release of ribosomes occurs with the rate constant k_3
7. $R \rightarrow null$, degradation of mRNA leads to release of ribosomes with the rate constant k_d, same reaction describes premature ribosome drop-off from mRNA with the rate constant k_{rd}

   We will assume that the process of microRNA binding to mRNA can occur at various stages of translation and that its rate k_b will be the same in each of the following reactions:

8. $M_0 \rightarrow M_0'$
9. $F_0 \rightarrow F_0'$
10. $M \rightarrow M'$
11. $F \rightarrow F'$
12. $R \rightarrow R'$

    In the same way we will assume that the rate of degradation of mRNA not driven by microRNA action (k_d) can be considered as the same one at all stages of translation:

13. $M_0 \rightarrow null$
14. $F_0 \rightarrow null$
15. $M \rightarrow null$
16. $F \rightarrow null$

    The degradation rate of mRNA bound to microRNA could occur with or without direct action of microRNA on its degradation. For the beginning we will assume that this rate constant (k_d') is different from the free mRNA degradation and it is the same one for all stages of translation:

17. $M'_0 \rightarrow null$
18. $F'_0 \rightarrow null$
19. $M' \rightarrow null$
20. $F' \rightarrow null$
21. $R' \rightarrow null$

Next we assume that the reaction corresponding to the assembly of the initiation complex and 40S ribosomal subunit with mRNA in the presence of miRNA ($M'_0 \rightarrow F'_0$) will occur with the rate constant k_{01}'.

22. $M'_0 \rightarrow F'_0$

Recruitment of 40S subunit on already translating miRNA-bound mRNA occurs with the rate constant k_1':

23. $M' \rightarrow F'$

Reactions of assembly of the full ribosome (S80) on mRNA in the presence of microRNA occur with the rate constant k_2':

24. $F'_0 \rightarrow M'+R$
25. $F' \rightarrow M'+R$

The rate of protein production in the case of microRNA action is described by the following reaction:

26. $R' \rightarrow P$, with the rate constant k_3'
27. $R' \rightarrow null$, describes possible mechanism of ribosomal drop-off (without protein production), with the rate constant k_{rd}', and mRNA degradation with ribosome release, with the rate constant k_d'.

Reactions 26 and 27 describe the reverse process of mRNA sequestration in P-bodies, with rates k_{+s} and k_{-s} correspondingly:

28. $M_0' \rightarrow B$,
29. $B \rightarrow M_0'$

The mRNA in P-bodies is degraded with specific rate k_{bd}'

30. $B \rightarrow null$
31. $P+R' \rightarrow null$, the rate of protein degradation by microRNA-independent mechanisms is k_p, while it can be increased in the presence of miRNA by $k_r \times R'$.

The system of equations $d\boldsymbol{x}/dt = \mathbf{K}_0 + \mathbf{K}\boldsymbol{x}$ (where $\boldsymbol{x}$ is the vector of 12 dynamic variables, $\mathbf{K}$ is the kinetic matrix, and $\mathbf{K}_0$ is the vector of production with only one non-zero component corresponding to the transcriptional synthesis of mRNA) has the following form:

Thus, for this simplified linear model we need to define the proper values for 18 coefficients corresponding to the rates of reactions.

For simulations, we needed the numerical values of 18 kinetic coefficients, which were estimated from published reports and are provided in Table 11.3. Although it is obvious that all rates diverge considerably for different mRNAs, experimental data mining allowed us to make a plausible assumption for almost all of the kinetic rates used in the model. For example, mRNA half-lives vary from a few minutes to more than 24 h, with a mean at 10 h [74], which we selected as the corresponding rate. It is nevertheless possible that highly regulated mRNAs, such as most miRNA targets, have shorter half-lives. The same reasoning also applies to protein half-lives.

$$
\begin{aligned}
\frac{d[M_0]}{dt} &= k_t - (k_d + k_{01} + k_b)[M_0] \\
\frac{d[F_0]}{dt} &= k_{01}[M_0] - (k_d + k_2 + k_b)[F_0] \\
\frac{d[M]}{dt} &= k_2([F_0]+[F]) - (k_d + k_1 + k_b)[M] \\
\frac{d[F]}{dt} &= k_1[M] - (k_d + k_2 + k_b)[F] \\
\frac{d[R]}{dt} &= k_2([F_0]+[F]) - (k_d + k_{rd} + k_3 + k_b)[R] \\
\frac{d[M_0']}{dt} &= k_b[M_0] - (k_d' + k_{01}')[M_0'] \\
&\quad - (k_{+s}[M] - k_{-s}[B]) \\
\frac{d[F_0']}{dt} &= k_b[F_0] + k_{01}'[M_0'] - (k_d' + k_2')[F_0'] \\
\frac{d[M']}{dt} &= k_b[M] + k_2'([F_0'] + [F']) - (k_d' + k_1')[M'] \\
\frac{d[F']}{dt} &= k_b[F] + k_1'[M] - (k_d' + k_2')[F] \\
\frac{d[R']}{dt} &= k_b[R] + k_2'([F_0] + [F]) \\
&\quad - (k_d' + k_{rd}' + k_3')[R] \\
\frac{d[P]}{dt} &= k_3[R] + k_3'[R'] - (k_p + k_r[R'])[P] \\
\frac{d[B]}{dt} &= k_{+s}[M'] - k_{-s}[B] - k_{bd}[B]
\end{aligned}
\qquad (11.13)
$$

Table 11.3 Reference set of parameters for the model and their changes according to the action of various miRNA-mediated mechanisms of translation repression

Kinetic rate constant	Reference value or interval	Comment
Parameters of transcription and translation without miRNA action		
k_t	10^{-3}	*Transcription kinetic rate*. If **Transcriptional Inhibition** mechanism is active then this constant is proportionally reduced from k_t (0% efficiency of the mechanism) to zero (100% efficiency of the mechanism).
k_{01}	$2{\cdot}10^{-4}$	*mRNA early initiation rate in the absence of miRNA.*
k_1	1	*Rate of 40S recruitement at already translated mRNA, considered to be fast and not rate-limiting*
k_2	$6{\cdot}10^{-2}$	*60S unit joining and assembly of the full ribosome on mRNA rate in the absence of miRNA.*
k_3	10^{-2}	*Rate including elongation and termination of translation in the absence of miRNA*. In all simulations of translation without miRNA, we assume that $k_3=k_2/6$, which gives six ribosomes sitting on one translated mRNA in average.
k_d	10^{-5}	*mRNA degradation rate in the absence of miRNA*. In all simulations of translation without miRNA, we assume that $k_d<<k_1, k_2, k_3$. Otherwise mRNA will be degraded much faster than it will be initiated and translated.
k_{rd}	0	*Rate of ribosome drop-off.* We neglect the ribosome drop-off in the absence of miRNA
k_p	$5{\cdot}10^{-6}$	*Rate of protein degradation in the absence of miRNA.*
Parameters of various mechanisms of miRNA action		
k_b	10^{-3} (strong) 10^{-4} (medium) 10^{-5} (weak)	*Rate of miRNA binding to mRNA*. This rate depends on many factors including the complementarity of miRNA sequence to the sequence of the binding site. We assume that depending on these factors, the rate can vary in the range of several orders of magnitude. When $k_b<<\min(k_1, k_2, k_3)$, we consider the binding as weak, because it does not considerably influence the rate of translation.
k_{01}'	$[0; k_{01}]$	*mRNA initiation rate with miRNA*. If **Cap Inhibition** mechanism is active then this constant can be proportionally reduced from k_1 to zero.
k_1'	k_{01}	*40S recruitement at already translated miRNA-bound mRNA, we do not consider the corresponding hypothetical mechanism in the model*
k_2'	$[0; k_2]$	*60S unit joining and assembly of the full ribosome on mRNA rate with miRNA*. If **60S Unit Joining Inhibition** mechanism is active then this constant can be proportionally reduced from k_2 to zero.
k_3'	$[0; k_3]$	*Rate including elongation and termination of translation with miRNA*. If **Elongation Inhibition** mechanism is active then this constant can be proportionally reduced from k_3 to zero.
k_d'	$[k_d; 10^2{\cdot}k_d]$	*Rate of mRNA degradation with miRNA*. If **Decay** mechanism is active then this constant can increase ten-fold at 100% mechanism efficiency. If **Cleavage** mechanism is active then this constant can increase by 100-fold.

(continued)

Table 11.3 (continued)

Kinetic rate constant	Reference value or interval	Comment
$k_{\pm s}$	$[0; 5\cdot10^{-2}]$	*Rate of reversible capturing of mRNA to P-bodies.* If **P-bodies Sequestration** mechanism is active, this constant can be proportionally increased from zero to k_{+s}. The reverse rate constant k_{-s} is assumed to be $k_{-s}=5\cdot k_{+s}$. We assume that mRNA can be degraded in P-bodies with the rate k_d'.
k'_{rd}	$[0; 5\cdot k_3']$	*Rate of ribosome drop-off.* If **Ribosome Drop-Off** mechanism is active then this constant is proportionally increased from 0 to $5\cdot k_3'$.
k_r	$[0; 5\cdot10^{-5}]$	*Rate of co-translational protein degradation catalysis.* If **Co-Translational Protein Degradation** mechanism is active then this constant is proportionally increased from zero to $5\cdot10^{-5}$, and the protein degradation rate is increased as $k_p^{miRNA}=k_p+k_r\cdot\mathbf{R'}$.

Similarly, we estimated the elongation time for mRNA translation as 1–2 min [75–77], even though it depends on the mRNA length: at 10 aa/s [78], 1–2 min corresponds to a mean length of 1.8–3.6 kb [79]. Likewise, the numbers of ribosomes per mRNA molecule are highly variable, from 4–5 to more than 10 [29, 75]. We considered six ribosomes per mRNA as being a reasonable assumption. We therefore postulated that six initiation events occur during a cycle of elongation, which leads to an estimate of 6 initiations/min, and is of the same order of magnitude as what has been proposed previously [75]. All information concerning the kinetic coefficients we used for our modelling is summarized in Table 11.3.

11.7 Distinct Dynamical Types of miRNA Action and Kinetic Signatures of miRNA Mechanisms

11.7.1 Analytical Solution of Model Equations for the Case of Normal Translation (No miRNA)

The dynamical variables that can be observed and measured in the experiment are

$$\text{Total amount of mRNA}: \mathbf{MT} = \mathrm{M_0 + F_0 + M + F + M_0' + F_0' + M' + F' + B}$$

$$\text{Total amount of protein}: \mathbf{PR} = \mathrm{P}$$

$$\text{Average number of ribosomes, translating one mRNA}: \mathbf{RB} = \mathrm{(R + R')/(M + M' + F + F')}$$

The solution of model equations and expression for [MT], [PR], [RB] for the trivial case without miRNA in the system were obtained. This can be modelled by putting to zero the binding constant $k_b=0$.

For this case the steady state values for the measurable quantities are

$$MT^{SS} = \frac{k_t}{k_d},$$
$$RB^{SS} = \frac{k_2+k_d}{k_3+k_d+k_{rd}},$$
$$PT^{SS} = \frac{k_3}{k_p}\frac{k_t}{k_d}\frac{k_{01}k_2}{(k_{01}+k_d)(k_3+k_d)} \quad (11.14)$$

and the relaxation times are

$$MT^{RT} = \frac{1}{k_d},$$
$$RB^{RT} = \frac{1}{\min(k_{01}+k_d, k_2+k_d, k_3+k_d+k_{rd})},$$
$$PT^{RT} = \frac{1}{\min(k_d, k_p)} \quad (11.15)$$

where we have assumed that $k_1 >> k_{01}, k_2, k_3$.

These formulas allow qualitative understanding of the effect of miRNA on various steps of translation and the corresponding kinetic signatures. They can also help to decipher experimentally observed kinetic signatures when multiple mechanisms are present simultaneously and the translation parameters are not known. Exact recipe on doing this will be a subject of our future work.

11.7.2 Dominant Paths of the Model and Their Relations to the miRNA Mechanisms

According to the methodology of asymptotology [73, 81], let us consider the case of well separated constants, i.e. when any two kinetic constants in the graph in the Fig. 11.6b have different orders of magnitude at each fork (i.e., a node with several outgoing reactions). Each such a (partial) ordering of kinetic constants will generate a path on the graph (possibly, cyclic), starting at M_0 node. We will call it **the dominant path**. Each path corresponds to one (if it does not contain cycles) or several (if it contains a cycle) dominant systems and to a distinguishable biochemical scenario. For example, the partial ordering ($k_b >> k_1, k_d$; $k_{01} >> k_{-s}, k'_d$; $k'_2 << k'_d$) corresponds to the dominant path describing the process of translation inhibition via 60S subunit joining repression (see Table 11.4, path $M_0M'_0F'_0$).

A dominant path is connected to a dominant system (whose solution of the corresponding dynamics equations provides an asymptotic approximation of the whole system dynamics) in the following way. If the path does not contain cycles, then it represents the dominant system. If the path contains cycles then the cycles should be glued and represented by single nodes (which will represent quasistationary distribution of chemical species concentrations inside the cycle). Then one should find the dominant path for the new graph with glued cycles and continue until an acyclic dominant path will be found. Depending on the ordering of kinetic rates inside each cycle, one cyclic dominant path can lead to several different dominant systems. The dominant system in general represents a hierarchy of glued cycles. The details of constructing dominant systems are provided in [73, 74, 81].

It is convenient to designate each dominant path by nodes through which it passes. There are many possible reaction graph traversals leading to multiple possible dominant paths, if one considers all partial orderings of the constants in the reaction forks. However, some of them are biologically non-relevant. For example, the ordering $k_d >> k_{01}$ (dominant path M_0) will not lead to any translation (the mRNA will be degraded before it will be initiated). In the same way, $k_d >> k_2$ (dominant path M_0F_0) will terminate the normal translation prematurely. Thus, we postulate $k_d << k_{01}, k_2, k_3$. Also for simplicity we assume that binding of miRNA to mRNA is more rapid than normal initiation, i.e., $k_b >> k_{01}, k_2, k_3$ if there is miRNA in the system, and $k_b = 0$, if not. Also we assume that $k_{01} << k_1$, because k_1 corresponds to recruiting 40S subunit on already initiated and translated mRNA (which we assume never be rate-limiting), while k_{01} includes both mRNA initiation and 40S subunit recruiting. This leads to six biologically relevant dominant paths, all of which are listed in Table 11.4.

Table 11.4 shows that the types of dynamical behavior (dominant paths) can be mapped onto the biologically characterized mechanisms of miRNA action, but this mapping is not one-to-one: several biological mechanisms can correspond to one dynamical type (for example, $M_0M'_0$ dominant path corresponds to M1, M7 and M8 biological mechanisms and, conversely, biological mechanism M1 can correspond to $M_0M'_0$ or $M_0M'_0F'_0M'F'R'P$ dominant paths).

11.7.3 Kinetic Signatures of miRNA-Mediated Mechanisms of Protein Translation Inhibition

In order to provide a practical recipe to distinguish between nine different mechanisms of miRNA action, we studied the dynamical behaviour of the model for the reference set of parameters for weak, medium and strong miRNA binding

Table 11.4 Dominant paths of the unified model of microRNA action mechanisms

Dominant path	Biological interpretation	Corresponding miRNA-mediated translation repression mechanism(s)
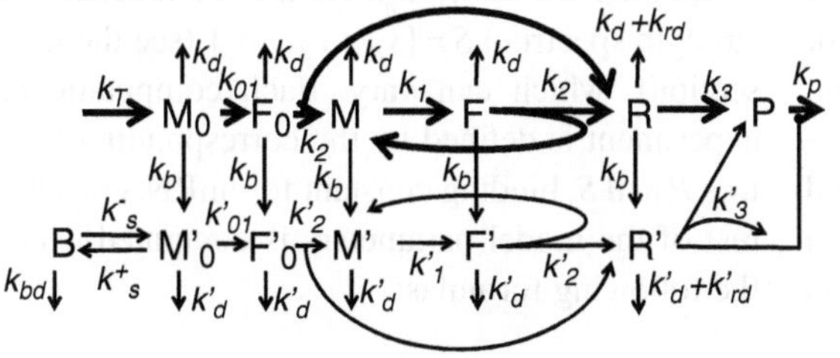	$\mathbf{M_0F_0MFRP}$ normal translation with negligible effect of miRNA	None
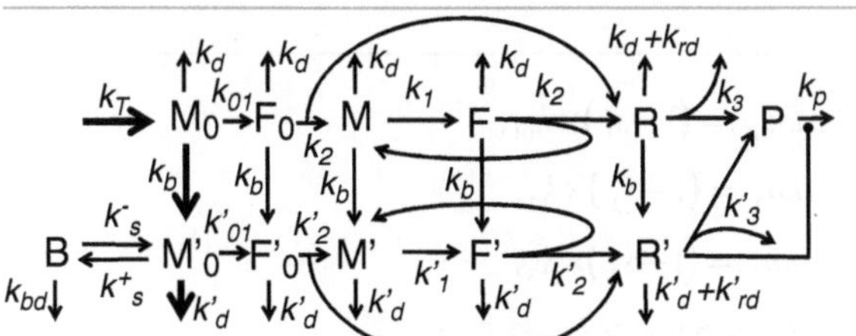	$\mathbf{M_0M'_0}$ the dominant effect is degradation of mRNA by miRNA	M1: Cap inhibition M7: Decay M8: Cleavage
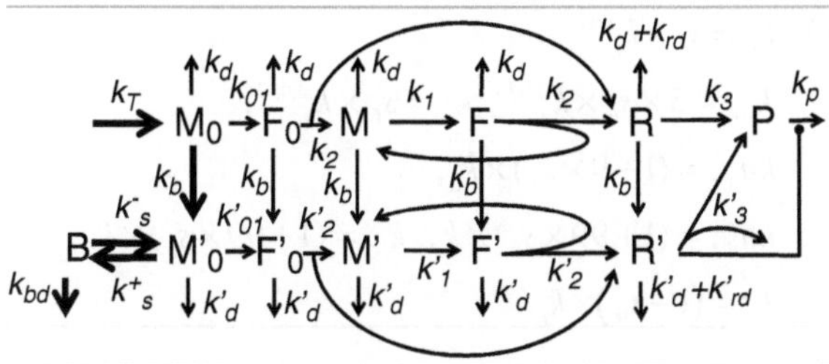	$\mathbf{M_0M'_0B}$ mRNA is captured in P-bodies	M6: Sequestration of mRNA in P-Bodies
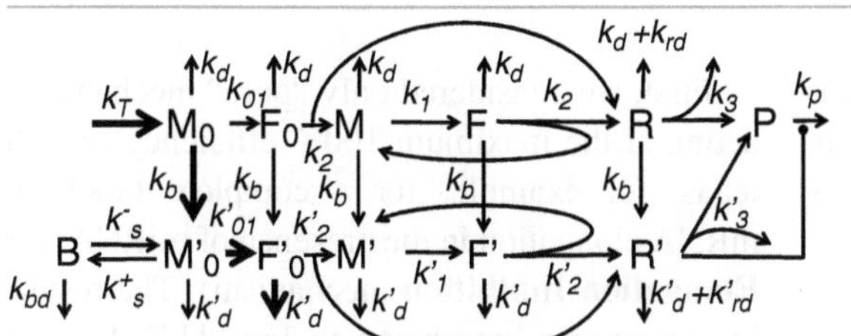	$\mathbf{M_0M'_0F'_0}$ mRNA translation is stuck after initiation, before the assembly of the ribosome	M2: 60S subunit joining inhibition
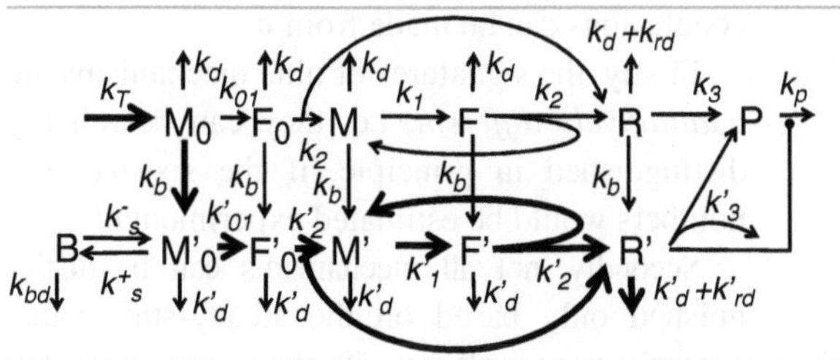	$\mathbf{M_0M'_0F'_0M'F'R'}$ mRNA is stuck with ribosomes on it and destroyed, or mRNA translation is prematurely aborted	M3: Elongation inhibition M4: Ribosome drop-off
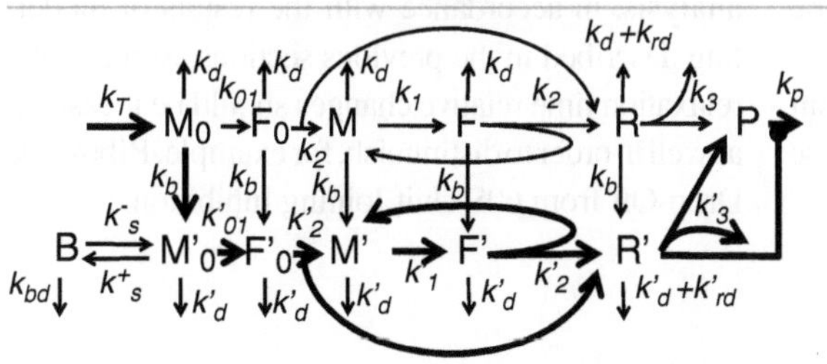	$\mathbf{M_0M'_0F'_0M'F'R'P}$ protein synthesis in the presence of miRNA with low mRNA degradation	M1: Cap inhibition M2: 60S subunit joining inhibition M3: Elongation inhibition M5: Co-translational protein degradation mechanisms

strengths. The simulation was performed in the following way:

(1) First, the system was simulated from zero initial conditions without presence of miRNA (k_b=0) in the time interval [0; 20/k_d]. The steady state and relaxation time values for MT, RB and PR values were estimated from the simulation.
(2) The miRNA binding constant was changed to the corresponding value and the simulation was continued from the steady state obtained before in the time interval [20/*kd*; 40/*kd*]. New steady state and relaxation time values were estimated from the simulation.

The model includes a vector of parameters $P=\{k_t, k_{01}, k_1, k_2, k_3, k_d, k_p\}$ and of mechanism strength spectrum $S=\{s_1, s_2, \ldots, s_9\}$ (see the next section), which can vary. Each computational experiment is defined by the corresponding vectors P and S, binding constant for miRNA (k_b) the rest of the model parameters is computed using the following formulas:

M1 (Cap Inhibition):	$kn_{01}' := (1-s_1)\times k_{01}$,
M2 (60S Unit Joining Inhibition):	$kn_2' := (1-s_2)\times k_2$,
M3 (Elongation Inhibition):	$kn_3' := (1-s_3)\times k_3$,
M4 (Ribosome Drop – Off):	$k_{rd} := 5\times s_4 \times kn_3'$,
M5 (Co – translational protein degradation):	$k_r := s_5 \times k_r^{(ref)}$.
M6 (Sequestration in P – bodies):	$k_{+s} := 5\times s_6 \times k_s^{(ref)},\ k_{-s} = s_6 \times k_s^{(ref)}$,
M7 (Decay of mRNA):	$kn_d' := (1+9\times s_7)\times k_d$,
M8 (Cleavage of mRNA):	$kn_d' := (1+99\times s_8)\times k_d,\ k_b' := (1+99\times s_8)\times k_b$
M9 (Transcriptional Inhibition):	$k_t := (1-s_9)\times k_t$,

The result of the simulation is a kinetic signature for a mixed mechanism of miRNA action, characterized by six numbers: relative changes of the steady states $MT^{SS} = \frac{MT^{ss}_{miRNA}}{MT^{ss}_{no\ miRNA}}$, $RB^{SS} = \frac{RB^{ss}_{miRNA}}{RB^{ss}_{no\ miRNA}}$, $PR^{SS} = \frac{PR^{ss}_{miRNA}}{PR^{ss}_{no\ miRNA}}$ and relative changes of relaxation times $MT^{RT} = \frac{MT^{RT}_{miRNA}}{MT^{RT}_{no\ miRNA}}$, $RB^{RT} = \frac{RB^{RT}_{miRNA}}{RB^{RT}_{no\ miRNA}}$, $PR^{RT} = \frac{PR^{RT}_{miRNA}}{PR^{RT}_{no\ miRNA}}$. For the further analysis, using Principal Component Analysis (PCA) we use the logarithms of these ratios.

First, we considered only "pure" mechanisms acting at the maximum 100% efficiency (which leads, for example, for a complete block of mRNA elongation in the presence of miRNA, for **Elongation Inhibition** mechanism). The resulting signatures are shown in Fig. 11.7. Several conclusions can be made from it.

Firstly, the signatures of nine mechanisms are *qualitatively different*, i.e. they can be reliably distinguished in principle, if the six required numbers would be estimated experimentally.

Secondly, not all mechanisms can be distinguished only based on the steady-state value analysis, in accordance with the results of modelling described in the previous sections. Some of the relaxation time relative changes should be measured as well in order to distinguish, for example, Ribosome Drop-Off from 60S Unit Joining Inhibition.

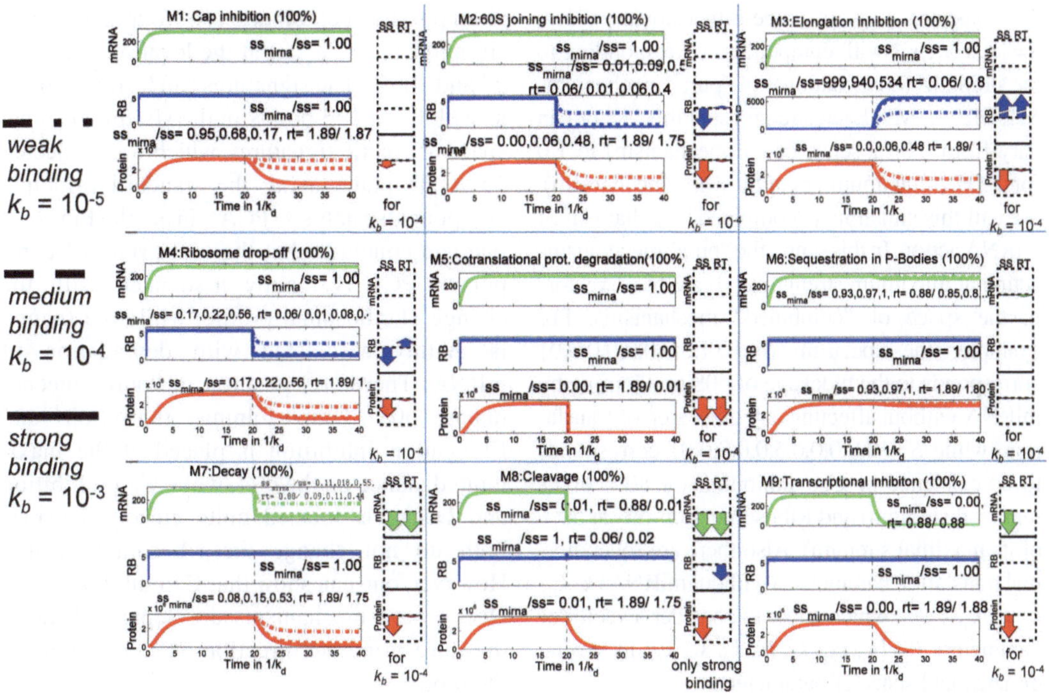

Fig. 11.7 Kinetic signatures of the mechanisms of miRNA action. There are nine signatures corresponding to nine mechanisms. Each plot shows dynamics of three quantities: amount of mRNA (*mRNA*), average number of ribosomes per translated mRNA (*RB*), total amount of protein (*Protein*) in the time units measured in $1/k_d$. The dynamics on the left from the dashed line shows translation without miRNA which is added at the time point 20. Three scenarios are simulated for each signature: strong, medium and weak binding strength of miRNA to mRNA. The numbers on the graphs shows relative change in the steady state (ss_{miRNA}/ss) and change in the relaxation time (*rt*, measured in $1/k_d$). If three numbers are shown separated by comma, they correspond to weak, medium and strong miRNA binding. If only one number is shown, it means that the binding strength does not affect this quantity significantly. The diagrams on the right from the dynamics plot visualize values of six numbers (relative changes of steady state (SS) and relaxation time (RT) for three measurable quantities) for the case of medium binding strength

Thirdly, one can observe that some of the signature components strongly depend in the quantitative fashion on the order of the miRNA binding constant, and some are completely insensitive. This suggest an experiment in which several sequences of miRNA would be utilised having different (weak, medium, tight) affinities to the target mRNA binding site. Observing how the dynamics of observable quantities are changing with the binding affinity, one can distinguish the mechanisms more reliably. For example, in the case of Ribosome Drop-Off the ribosomal profile should be more sensitive to changing miRNA affinity compared to 60S Unit Joining.

11.8 Coexistence of Multiple Mechanisms of miRNA Action

One of the most debated questions on the action of miRNA on translation is the possibility of co-existence of several mechanisms of miRNA action. Let us study formally to what consequences it can lead from the point of view of translation dynamics and kinetic signatures.

We formalize co-existence of several miRNA action mechanisms in the following way. We characterized a situation when a miRNA can interfere with several steps of translation (and transcription) by a *strength spectrum* of nine “pure” mechanisms.

The spectrum is a nine-dimensional vector $S=\{s_1,s_2,\ldots,s_9\}$ with components corresponding to the strengths (contributions) of "pure" mechanisms M1, M2,…, M9. Each strength s_i of this vector can vary from 0.0 (absence of the mechanism) to 1.0 (or 100%, maximum strength of the mechanism). We call this situation a "combined" mechanism of miRNA action. In this sense, the "pure" mechanisms acting at maximum strength (1.0) are basis vectors in the space of "combined" mechanisms. For example, the spectrum $S=\{0,1.0,0,0,0,0,0,0,0\}$ corresponds to the blockage of 60S unit joining by miRNA without affecting any other step of translation, while $S=\{0.8,0,0,0.5,0,0,0,0,0\}$ corresponds to co-existence of Cap Inhibition (at 80% of its maximal strength) and Ribosome drop-off (at 50% of its maximal strength). Also there are seven normal translation parameters (without miRNA) k_t, k_{01}, k_1, k_2, k_3, k_d, k_p, which allow to consider a vector of parameters $P=\{k_t, k_{01}, k_1, k_2, k_3, k_d, k_p\}$ in seven–dimensional space of parameters.

In this section we make two computational experiments in which we exhaustively study the effect of (1) varying S given P fixed at reference parameters; and (2) varying P given S, for four mostly referenced mechanisms: Cap Inhibition, 60S Unit Joining Inhibition, Elongation Inhibition, Decay. In other words, in the first case we study the effect of co-existence of various mechanisms for a given experimental system, characterized by a given set of normal translation parameters. In the second case, we study the effect of variable experimental (or cellular) conditions on the conclusions one can make for the same mixed mechanism of miRNA action. Thus, the results of this section generalize the results of the previous sections to the case of co-existence of several mechanisms at the same time.

11.8.1 Fixed Set of Translation Parameters and Variable Mixed Mechanisms of miRNA Action

In Fig. 11.8 we present the results of the following computational experiment. For a reference set of parameters (Table 11.3) we computed 625 kinetic signatures corresponding to all possible combinations of four mechanism strengths (s_1, s_2, s_3, s_7) at the level of 0, 25, 50, 75 and 100%. The signatures can be represented as a cloud of 625 points in the six-dimensional space of kinetic signatures, which was projected on a 2D plane using the standard principal components analysis (PCA). From the Fig. 11.8 one can conclude that the first principal component PC1 is mainly associated with the change of ribosomal profile, while the second is mainly associated with degradation of mRNA. Therefore, position of "pure" mechanisms 60S Unit Joining Inhibition and Elongation Inhibition is placed at the maximum distance on the plot, while Cap Inhibition and Decay is located quite closely, because both do not change the ribosomal profile. However, one can show that Cap Inhibition and Decay pure mechanisms are separated along the third principle direction PC3, invisible on the plot.

One of the important conclusions that can be made from the plot in the Fig. 11.8 is that the presence of Decay mechanism in the spectrum ($s_7>0$) can mask the effect of other mechanisms leading to the very early blockage of translation (M_0M_0' dominant path). Indeed, it might not matter that a translation in the presence of miRNA is completely blocked at a later stage, if the increased degradation will destroy mRNA even before it can arrive at this blocked later stage. In some cases (such as the mixed mechanism F on the plot, co-existence of complete Cap Inhibition and Decay), the kinetic signature of the mixed mechanism is indistinguishable from Decay.

The kinetic signature K (mix of 60S Unit Joining Inhibition and Elongation Inhibition) is indistinguishable from the pure signature of Elongation Inhibition. The kinetic signature H (mix of three first mechanisms without Decay) reminds pure 60S Unit Joining Inhibition mechanism. Cases F, K and H are three examples of *kinetic signature masking* (or domination) of one mechanism by another.

In other cases the resulting kinetic signature of a mixed mechanism does not remind any signature of the four pure mechanisms: by

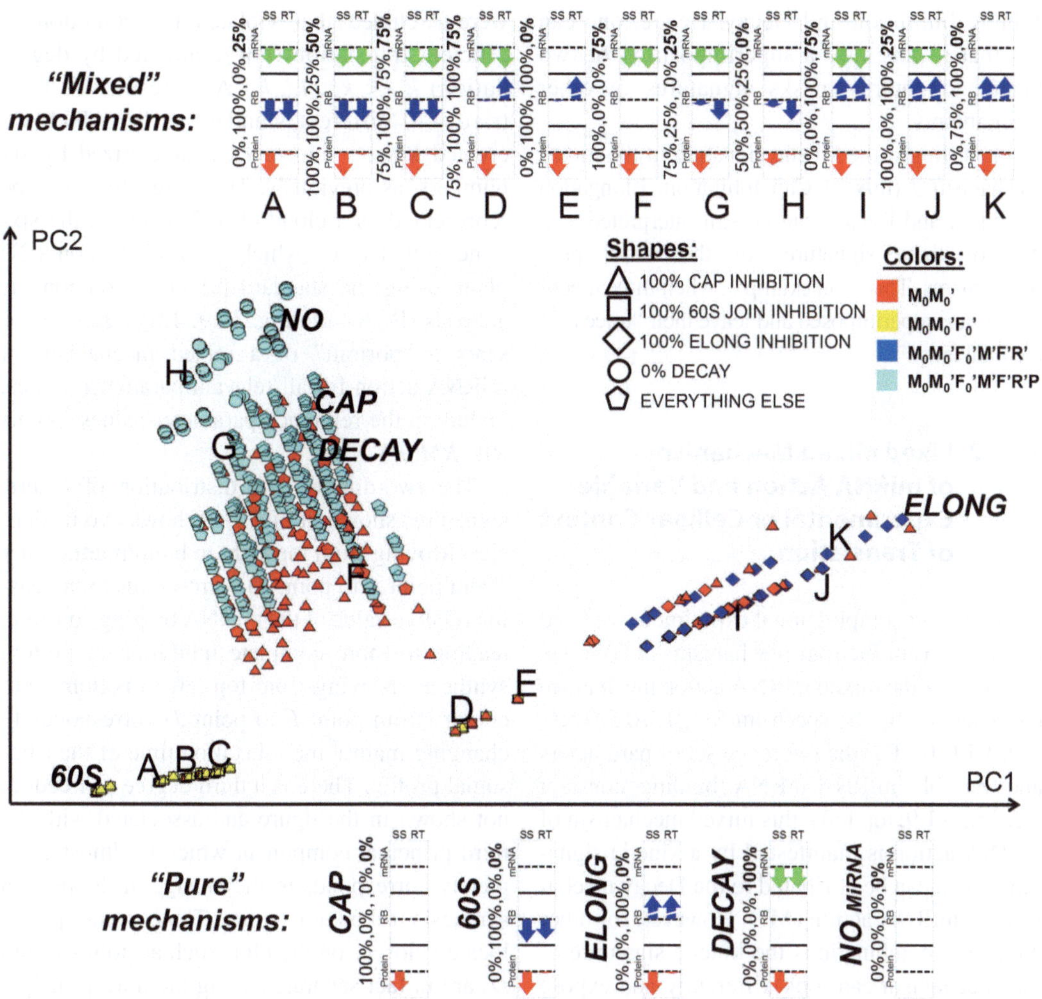

Fig. 11.8 PCA plot for simulations for a fixed set of translation parameters and a variable mixed mechanisms. The mix includes (1) Cap Inhibition, (2) 60S unit joining inhibition, (3) Elongation inhibition and (4) Decay for the reference set of translation parameters and $k_b = 10^{-3}$. The plot represents a projection from a six-dimensional space of measurable quantities: relative changes in steady-state (*SS*) and relaxation time (*RT*) for three quantities: amount of mRNA (*mRNA*), number of ribosomes per mRNA (*RB*) and amount of protein (*Protein*). Each point represents a simulation made for a selected spectrum of strengths of four mechanisms, the colors distinguish the resulting dominant paths and the shapes distinguish spectrums when one of the mechanisms is dominating or when the mRNA decay is not affected by miRNA (circles). For example, red rectangle corresponds to the scenario when 60S unit joining is completely blocked if miRNA is bound, and M_0M_0' dominant path is realized. Few points are annotated with signature diagrams visualizing the numerical values for the six variables. Four numbers on the left of each diagram show the strengths of four miRNA action mechanisms (cap inhibition, 60S unit joining inhibition, elongation inhibition and decay correspondingly). First two principal components explain 86% of data variation

contrast, certain superimposition of the kinetic signatures happens. Thus, the mixed mechanisms *A* (co-existence of complete 60S Unit Joining Inhibition and Decay) and *E* (co-existence of complete 60S Unit Joining Inhibition and Elongation inhibition) give the signature which looks like a superimposition of the kinetic signatures of the initial mechanisms. However, further addition of miRNA action mechanisms does not change the signature qualitatively. Thus, mix of all four mechanisms together (cases *B*, *C*) still looks like a mix of 60S Unit

Joining Inhibition and Decay. Hence, one can say that in this case a superimposition of two kinetic signatures masks signatures of other mechanisms.

Interestingly, the kinetic signature in the mixed mechanism *J* (mix of Cap Inhibition, Elongation Inhibition and Decay) can be still interpreted as a mix of three signatures of the initial pure mechanisms. This is an example, when three mechanisms are superimposed and leave their "traces" in the final mix.

11.8.2 Fixed Mixed Mechanism of miRNA Action and Variable Experimental or Cellular Context of Translation

In the second computational experiment we fixed the strengths of the four mechanisms at 50%, i.e. we consider the mixed miRNA action mechanism characterized by the spectrum $S=\{0.5, 0.5, 0.5, 0, 0, 0, 0.5, 0, 0\}$. For the reference set of parameters and variable miRNA-mRNA binding constant (see Fig. 11.9, top left), this mixed mechanism of miRNA action is manifested by a kinetic signature which can be attributed to the Decay mechanism of miRNA action (M7). However, for other parameter combinations the kinetic signature of this mechanism can look differently and expose features of other mechanisms (see below). The main message of the example shown in Fig. 11.9 is that variation of the parameters of translation mechanism can significantly change the interpretation of the kinetic signature when several mechanisms of miRNA action co-exist.

We study the kinetic signatures of the mixed mechanism $S=\{0.5, 0.5, 0.5, 0, 0, 0, 0.5, 0, 0\}$ when the kinetic parameters of the normal translation are varied in very large intervals (five orders of magnitude). We varied four kinetic rates k_d, k_b, k_{01}, k_2, leaving k_t and k_p fixed at the reference values and putting $k_3=k_2/6$ to provide constant average number of six ribosomes sitting on one mRNA. The parameters took the following range of values: $k_d \in \{10^{-3}, 10^{-4}, 10^{-5}, 10^{-6}, 10^{-7}\}$, k_b, $k_{01}, k_2 \in \{10^{-1}, 10^{-2}, 10^{-3}, 10^{-4}, 10^{-5}\}$ in all possible combinations. From these combinations those were excluded that violated the condition of efficient translation (not dominated by degradation) $k_d \ll k_{01}$, k_2, k_3. As a result, we have tried 440 different simulations for which we created kinetic signatures, characterized by six numbers, as previously. These signatures can be represented as a cloud of 440 points in the six-dimensional space, which was projected on a 2D plane using the standard principal components analysis (PCA), see Fig. 11.9. This figure represents a "portrait" of a mixed mechanism of miRNA action for all relevant parameter values, including the reference parameter values (points *RW*, *RM*, *RS*).

The two-dimensional distribution of kinetic signatures shown in Fig. 11.9 shows two tendencies. Moving from top right to bottom left corner (from point *A* to point *E*) corresponds to increasing relative value of the miRNA binging constant, leading to more complete inhibition of protein synthesis. Moving from top left to bottom right corner (from point *F* to point *J*) corresponds to changing mainly the relaxation time of the ribosomal profile. There is a third degree of freedom not shown in the figure and associated with the third principal component which is almost completely corresponds to the change in the protein synthesis relaxation time. Thus, some points located closely on the plot (such as points *C* and *D*) are in fact separated along the third principal component and have very different protein synthesis relaxation time values.

Several important conclusions can be made from this computational experiment, and the first one concerns the role of miRNA binding strength. Evidently, if k_b is much smaller than the normal translation parameters k_d, k_{01}, k_2 then miRNA binding does not affect the dynamics significantly and the "normal" M_0F_0MFRP dominant path is implemented (case *A*). In the case when the binding is significant but not very strong and comparable to k_d, k_{01}, k_2 parameters (competitive binding), the signature is masked by Decay-like pattern (case *RW*). The Decay mechanism masks all other mechanisms also in those combinations of parameters where k_2 is faster than k_d by several (three) orders of magnitude (cases *RM* and *RS*). In this case, the ribosomal profile is not perturbed by miRNA.

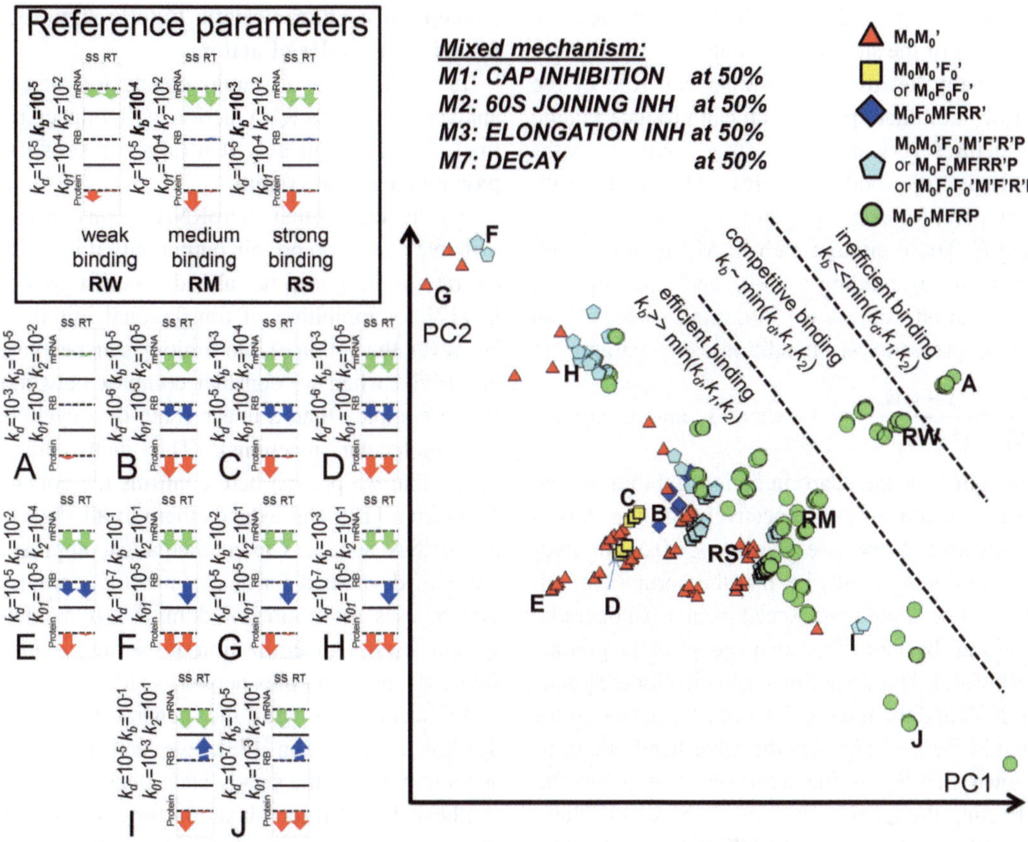

Fig. 11.9 PCA plot (on the *right*) for simulations for one selected mixed mechanism (Cap Initiation Inhibition at 50%, 60S Unit Joining Inhibition at 50%, Elongation Inhibition at 50% and Decay at 50%) and a variable set of internal translation parameters. The plot represents a projection from six-dimensional space of measurable quantities: relative changes in steady-state (*SS*) and relaxation time (*RT*) for three quantities: amount of mRNA (*mRNA*), number of ribosomes per mRNA (*RB*) and amount of protein (*Protein*). Each point represents a simulation made for a combination of k_d, k_b, k_{01} and k_2 parameter values, the color and shape distinguish the resulting dominant paths. Several points are annotated with signature diagrams visualizing the numerical values for the six variables (on the *left*). Cases *RW*, *RM* and *RS* (*top left*) show close to the reference set values for three different miRNA binding rates ($k_b = 10^{-3}$, 10^{-4}, 10^{-5}). First two principal components explain 79% of data variation

The relaxation time of a protein changes in the signatures when the mRNA degradation rate becomes less than the degradation rate of the protein: $k_d \ll k_p = 5 \cdot 10^{-6}$ (see formula 11.15). Notice that for the reference set of parameters the protein is assumed to be more stable than a transcript, and the only "pure" signature where the relaxation time of the protein is affected by miRNA is Cotranslational Protein Degradation. If the protein is less stable than a transcript then this might create confusion in interpreting the signatures and suggesting activation of this mechanism while it is not functional in reality.

The signatures *B*, *C*, *D* and *E* can be interpreted as a superimposition of 60S Unit Joining Inhibition with Decay, with possible role of Cap Inhibition. Elongation Inhibition mechanism leading to the increase of both RB steady state and relaxation time might be suspected in the signatures *I* and *J* even though the RB steady state does not change significantly (this can be attributed to a compensatory effect from mixing Elongation Inhibition with 60S Unit Joining). Signatures *F*, *G* and *H* suggest the role of 60S Unit Joining Inhibition (decreasing the ribosome profile relaxation time) which would be completely missed if one looks at the relative changes of the steady states only.

Finally, let us notice the special role of the M_0M_0' dominant path which can produce kinetic signatures very similar to other dominant paths (this is true both for Figs. 11.8 and 11.9). Compare, for example, pairs of cases *F* and *G*, *D* and *E*. The dominant path M_0M_0' requires relatively strong binding $k_b >> k_{01}$ and relatively fast degradation or slow initiation, which can be expressed as condition on parameters $\frac{k'_{01}}{k'_d} = \frac{(1-s_1)k_{01}}{(1+9s_4)k_d} << 1$, where s_1 and s_7 are the strengths of the Cap Initiation Inhibition and Decay mechanisms respectively in the mixed mechanism. In the case when $s_1 = s_7 = 0.5$, this gives a condition $k_{01} << 9k_d$, i.e. that the normal cap initiation rate should not exceed the normal degradation rate by more than two orders of magnitude (100-fold). This condition is satisfied for the points *B*, *E*, *G* and the reference set of parameters on the plot in the Fig. 11.9. On the other hand, it can be shown that k_{01} is the least sensitive parameter affecting the relative changes of the steady states and relaxation times for the MT, RB and PR values (decreasing k_{01} can affect only the steady state of the protein and not other values, see (14)). Hence, for many kinetic signatures, given relatively strong miRNA binding constant, there is a possibility to implement the M_0M_0' dominant path by slowing down k_{01} without a qualitative signature change. This non-intuitive conclusion can be verified experimentally.

11.9 Concluding Remarks

MicroRNA mode of action is a highly controversial topic. Here, we used mathematical modelling and found that each of the suggested mechanisms has a specific signature (the predicted dynamics of 3 measurable variables of the translational process, namely, the time course of accumulation of protein and mRNA, and of ribososmal loading on the mRNA). These signatures provide a new tool for discriminating between distinct mechanisms. We thus propose the **concept of a characteristic kinetic signature for miRNA modes of action**.

In addition, an essential conclusion of our analysis is that miRNA action will impact the final kinetic output only if it targets a **sensitive parameter of the system**.

The hypothesis that microRNA action can have a visible impact on protein output only if it affects the rate-limiting step has already been suggested in [12] for inhibition of translational initiation. However, the notion of rate-limiting step becomes non-trivial when we consider complex networks (more complex than a linear chain or a cycle of monomolecular reactions). The mathematical model that we present here confirms the conclusions from [12], and extends them to all steps of microRNA action. The mathematical approach we have developed for analysis of this complex system uses the notion of dominant dynamical system, itself a generalization of the rate-limiting step concept to complex networks [13, 73, 80].

In accordance with the general theory of dynamical limitation [73], we can take into account not only the steady-state rates of protein synthesis but also its relaxation time. For example, for a linear chain of reactions, the steady-state rate depends on the slowest kinetic rate parameter (rate-limiting step), whereas the relaxation time of the system depends on the second slowest kinetic rate parameter.

The analysis of our results allowed us to suggest a unifying theory for miRNA modes of action: all proposed modes of action operate simultaneously, and the *apparent* mechanism that will be detected depends on a set of sensitive intrinsic parameters of the individual target mRNA under study. This hypothesis would explain the following set of observations: (1) that the same microRNA apparently uses distinct mechanisms on different targets (e.g. for let7: [18, 24, 29, 52, 70]; for CXCR4: [19, 25, 30, 34]; for miR16: [32, 82]; for miR122: [83]); (2) that microRNA's mode of action depends on the promoter under which the target mRNA is transcribed [68]; and (3) that the status of the cell affects the final observable mode of miRNAs action [37, 39, 54]. Moreover, the possibility of coexistence of two or several mechanisms has already been discussed and proven in the literature [5, 6, 18, 22, 39, 49, 53, 54].

As already stated, our modelling results lead us to propose that, in individual biological systems, the relative abundance and/or activity of some set of intrinsic factors determines the apparent inhibition mechanism that will be detected. These factors are *not related to the miRNA pathways*, but intrinsically determine the **sensitive parameters of the system**. Indeed, RNA-binding proteins not related to the miRNA pathway have been shown to have a strong influence on the final outcome of miRNA regulation [74, 84–86].

A body of studies underscore the importance of intrinsic parameters of mRNAs. Revisiting theses studies in the framework of our model provides an explanation for most of the discrepancies in the literature. Thus, in most of the studies showing initiation inhibition, *in vitro* transcribed mRNAs (transfected into cells or studied directly *in vitro*) were used. In contrast, almost all data supporting elongation inhibition were obtaining in living cells, and thus with physiologically modified target mRNAs [18–21, 25, 70], with only one and very specific exception [33]. Similarly, most of the studies showing IRES-driven mRNAs as being refractory to microRNAs were carried out either *in vitro* [70] or using *in vitro* transcribed mRNAs transfected into cells [18–20], whereas the studies showing IRES-driven mRNAs to be repressed by miRNAs were carried out with mRNAs transcribed *in situ*, inside cells [30, 32]. In all these cases, the difference might come from the status of the target mRNA, rather than from any putative or actual differences in the microRNA machinery.

Another example is the influence of splicing marks attached to mRNAs *in vivo*. The process of mRNA splicing leaves protein marks on mRNAs, which promotes the first round of translation at the initiation step [87, 88]. These marks are dissociated during the first round of translation. Splicing marks, by increasing the initial initiations, would lead to higher initiation rates on intron-containing mRNAs [89]. Elongation would thus become a limiting step. In contrast, *in vitro* transcribed mRNAs lack splicing marks, resulting in a decreased initiation rate, which becomes limiting. Moreover, under *in vitro* conditions, initiation is highly dependent on the concentration of initiation factors, providing another possible explanation for discrepancies between *in vitro* studies.

Another example is the dependence of miRNA effects on codon usage. MicroRNA action has been reported to act on initiation steps when codon usage is optimized for human translation [11, 20], whereas, with non-optimized codons, microRNA was found to act on elongation [27, 30, 33, 35]. This again might have something to do with different rates of elongation, elongation rates being, or not, among the set of limiting (sensitive) parameters for a given mRNA.

Yet another example is the dependence of microRNA mode of action on the experimental procedure for transfection of the mRNA [33]. The transfection procedure is likely to influence the association of the target mRNA with mRNA-binding proteins, which, in turn, changes the sensitive parameters of the system, and hence the final outcome of microRNA action.

All these and some other data clearly support the idea that the observed mode of action of a microRNA depends upon interplay between the intrinsic rates of the different steps of mRNA translation.

In summary, our results provide a mathematical tool to discriminate between different miRNA modes of action. Moreover, we propose a unifying model in which the observed mode of action of a particular miRNA is dictated by the relationships among the intrinsic parameters of its target mRNA. We anticipate that the tool we have developed will promote better analysis of experimental data, and that our model will permit a better understanding of microRNA action. Most importantly, our hypothesis would explain most of the discrepancies in the corresponding literature.

Acknowledgements This work was supported by a grant from the European Commission Sixth Framework Programme (Integrated Project SIROCCO, contract number LSHG-CT-2006-037900) to AHB, and from the Agence Nationale de la Recherche (project ANR-08-SYSC-003 CALAMAR) and from the Projet Incitatif Collaboratif "Bioinformatics and Biostatistics of Cancer" to Institut Curie. AZ is a member of the team "Systems Biology of Cancer", labeled by the Ligue Nationale Contre le Cancer.

References

1. Bartel DP (2009) MicroRNAs: target recognition and regulatory functions. Cell 136:215–233
2. Carthew RW, Sontheimer EJ (2009) Origins and mechanisms of miRNAs and siRNAs. Cell 136:642–655
3. Chekulaeva M, Filipowicz W (2009) Mechanisms of miRNA-mediated post-transcriptional regulation in animal cells. Curr Opin Cell Biol 21:452–460
4. Esquela-Kerscher A, Slack FJ (2006) Oncomirs – microRNAs with a role in cancer. Nat Rev Cancer 6:259–269
5. Eulalio A, Huntzinger E, Izaurralde E (2008) Getting to the root of miRNA-mediated gene silencing. Cell 132:9–14
6. Filipowicz W, Bhattacharyya SN, Sonenberg N (2008) Mechanisms of post-transcriptional regulation by microRNAs: are the answers in sight? Nat Rev Genet 9:102–114
7. Ghildiyal M, Zamore PD (2009) Small silencing RNAs: an expanding universe. Nat Rev Genet 10:94–108
8. Jackson RJ, Standart N (2007) How do microRNAs regulate gene expression? Sci STKE 2007:re1
9. Kloosterman WP, Plasterk RH (2006) The diverse functions of microRNAs in animal development and disease. Dev Cell 11:441–450
10. Moazed D (2009) Small RNAs in transcriptional gene silencing and genome defence. Nature 457:413–420
11. Pillai RS, Artus CG, Filipowicz W (2004) Tethering of human Ago proteins to mRNA mimics the miRNA-mediated repression of protein synthesis. RNA 10:1518–1525
12. Nissan T, Parker R (2008) Computational analysis of miRNA-mediated repression of translation: implications for models of translation initiation inhibition. RNA 14:1480–1491
13. Zinovyev A, Morozova N, Nonne N, Barillot E, Harel-Bellan A, Gorban AN (2010) Dynamical modelling of microRNA action on the protein translation process. BMC Syst Biol 4:13
14. Gorban AN, Zinovyev A, Morozova N, Harel-Bellan A (2012) Modeling coupled transcription, translation and degradation and miRNA-based regulation of this process. E-print arXiv:1204.5941
15. Morozova N, Zinovyev A, Nonne N, Pritchard L-L, Gorban AN, Harel-Bellan A (2012) Kinetic signatures of microRNA modes of action. RNA 18(9):032284
16. Schwarz G (1968) Kinetic analysis by chemical relaxation methods. Rev Mod Phys 40:206–218
17. Hammes GG (1968) Relaxation spectrometry of enzymatic reactions. Accounts Chem Res 1(11):321–329
18. Pillai RS, Bhattacharyya SN, Artus CG, Zoller T, Cougot N, Basyuk E, Bertrand E, Filipowicz W (2005) Inhibition of translational initiation by Let-7 MicroRNA in human cells. Science 309:1573–1576
19. Humphreys DT, Westman BJ, Martin DI, Preiss T (2005) MicroRNAs control translation initiation by inhibiting eukaryotic initiation factor 4E/cap and poly(A) tail function. Proc Natl Acad Sci USA 102:16961–16966
20. Kiriakidou M, Tan GS, Lamprinaki S, De Planell-Saguer M, Nelson PT, Mourelatos Z (2007) An mRNA m7G cap binding-like motif within human Ago2 represses translation. Cell 129:1141–1151
21. Thermann R, Hentze MW (2007) Drosophila miR2 induces pseudo-polysomes and inhibits translation initiation. Nature 447:875–878
22. Zipprich JT, Bhattacharyya S, Mathys H, Filipowicz W (2009) Importance of the C-terminal domain of the human GW182 protein TNRC6C for translational repression. RNA 15(5):781–793
23. Eulalio A, Huntzinger E, Izaurralde E (2008) GW182 interaction with Argonaute is essential for miRNA-mediated translational repression and mRNA decay. Nat Struct Mol Biol 15:346–353
24. Chendrimada TP, Finn KJ, Ji X, Baillat D, Gregory RI, Liebhaber SA, Pasquinelli AE, Shiekhattar R (2007) MicroRNA silencing through RISC recruitment of eIF6. Nature 447:823–828
25. Wang B, Yanez A, Novina CD (2008) MicroRNA-repressed mRNAs contain 40S but not 60S components. Proc Natl Acad Sci USA 105:5343–5348
26. Olsen PH, Ambros V (1999) The lin-4 regulatory RNA controls developmental timing in Caenorhabditis elegans by blocking LIN-14 protein synthesis after the initiation of translation. Dev Biol 216: 671–680
27. Gu S, Jin L, Zhang F, Sarnow P, Kay MA (2009) Biological basis for restriction of microRNA targets to the 3' untranslated region in mammalian mRNAs. Nat Struct Mol Biol 16:144–150
28. Landthaler M, Gaidatzis D, Rothballer A, Chen PY, Soll SJ, Dinic L, Ojo T, Hafner M, Zavolan M, Tuschl T (2008) Molecular characterization of human Argonaute-containing ribonucleoprotein complexes and their bound target mRNAs. RNA 14:2580–2596
29. Maroney PA, Yu Y, Fisher J, Nilsen TW (2006) Evidence that microRNAs are associated with translating messenger RNAs in human cells. Nat Struct Mol Biol 13:1102–1107
30. Petersen CP, Bordeleau ME, Pelletier J, Sharp PA (2006) Short RNAs repress translation after initiation in mammalian cells. Mol Cell 21:533–542
31. Baillat D, Shiekhattar R (2009) Functional dissection of the human TNRC6 (GW182-related) family of proteins. Mol Cell Biol 29:4144–4155
32. Karaa ZS, Iacovoni JS, Bastide A, Lacazette E, Touriol C, Prats H (2009) The VEGF IRESes are differentially susceptible to translation inhibition by miR-16. RNA 15:249–254
33. Lytle JR, Yario TA, Steitz JA (2007) Target mRNAs are repressed as efficiently by microRNA-binding sites in the 5'UTR as in the 3'UTR. Proc Natl Acad Sci USA 104:9667–9672

34. Wang B, Love TM, Call ME, Doench JG, Novina CD (2006) Recapitulation of short RNA-directed translational gene silencing in vitro. Mol Cell 22:553–560
35. Nottrott S, Simard MJ, Richter JD (2006) Human let-7a miRNA blocks protein production on actively translating polyribosomes. Nat Struct Mol Biol 13:1108–1114
36. Sen GL, Blau HM (2005) Argonaute 2/RISC resides in sites of mammalian mRNA decay known as cytoplasmic bodies. Nat Cell Biol 7:633–636
37. Bhattacharyya SN, Habermacher R, Martine U, Closs EI, Filipowicz W (2006) Relief of microRNA-mediated translational repression in human cells subjected to stress. Cell 125:1111–1124
38. Jakymiw A, Lian S, Eystathioy T, Li S, Satoh M, Hamel JC, Fritzler MJ, Chan EK (2005) Disruption of GW bodies impairs mammalian RNA interference. Nat Cell Biol 7:1267–1274
39. Leung AK, Calabrese JM, Sharp PA (2006) Quantitative analysis of Argonaute protein reveals microRNA-dependent localization to stress granules. Proc Natl Acad Sci USA 103:18125–18130
40. Liu J, Rivas FV, Wohlschlegel J, Yates JR 3rd, Parker R, Hannon GJ (2005) A role for the P-body component GW182 in microRNA function. Nat Cell Biol 7:1261–1266
41. Liu J, Valencia-Sanchez MA, Hannon GJ, Parker R (2005) MicroRNA-dependent localization of targeted mRNAs to mammalian P-bodies. Nat Cell Biol 7:719–723
42. Pauley KM, Eystathioy T, Jakymiw A, Hamel JC, Fritzler MJ, Chan EK (2006) Formation of GW bodies is a consequence of microRNA genesis. EMBO Rep 7(9):904–910
43. Eulalio A, Behm-Ansmant I, Schweizer D, Izaurralde E (2007) P-body formation is a consequence, not the cause, of RNA-mediated gene silencing. Mol Cell Biol 27:3970–3981
44. Leung AK, Sharp PA (2006) microRNAs: a safeguard against turmoil? Cell 130:581–585
45. Lim LP, Lau NC, Garrett-Engele P, Grimson A, Schelter JM, Castle J, Bartel DP, Linsley PS, Johnson JM (2005) Microarray analysis shows that some microRNAs downregulate large numbers of target mRNAs. Nature 433:769–773
46. Baek D, Villen J, Shin C, Camargo FD, Gygi SP, Bartel DP (2008) The impact of microRNAs on protein output. Nature 455:64–71
47. Bagga S, Bracht J, Hunter S, Massirer K, Holtz J, Eachus R, Pasquinelli AE (2005) Regulation by let-7 and lin-4 miRNAs results in target mRNA degradation. Cell 122:553–563
48. Behm-Ansmant I, Rehwinkel J, Doerks T, Stark A, Bork P, Izaurralde E (2006) mRNA degradation by miRNAs and GW182 requires both CCR4: NOT deadenylase and DCP1: DCP2 decapping complexes. Genes Dev 20:1885–1898
49. Eulalio A, Rehwinkel J, Stricker M, Huntzinger E, Yang SF, Doerks T, Dorner S, Bork P, Boutros M, Izaurralde E (2007) Target-specific requirements for enhancers of decapping in miRNA-mediated gene silencing. Genes Dev 21:2558–2570
50. Jing Q, Huang S, Guth S, Zarubin T, Motoyama A, Chen J, Di Padova F, Lin SC, Gram H, Han J (2005) Involvement of microRNA in AU-rich element-mediated mRNA instability. Cell 120:623–634
51. Selbach M, Schwanhausser B, Thierfelder N, Fang Z, Khanin R, Rajewsky N (2008) Widespread changes in protein synthesis induced by microRNAs. Nature 455:58–63
52. Wakiyama M, Takimoto K, Ohara O, Yokoyama S (2007) let-7 microRNA-mediated mRNA deadenylation and translational repression in a mammalian cell-free system. Genes Dev 21:1857–1862
53. Wu L, Fan J, Belasco JG (2006) MicroRNAs direct rapid deadenylation of mRNA. Proc Natl Acad Sci USA 103:4034–4039
54. Valencia-Sanchez MA, Liu J, Hannon GJ, Parker R (2006) Control of translation and mRNA degradation by miRNAs and siRNAs. Genes Dev 20(5):515–524
55. Coller J, Parker R (2004) Eukaryotic mRNA decapping. Annu Rev Biochem 73:861–890
56. Llave C, Xie Z, Kasschau KD, Carrington JC (2002) Cleavage of Scarecrow-like mRNA targets directed by a class of Arabidopsis miRNA. Science 297(5589):2053–2056
57. Rhoades MW, Reinhart BJ, Lim LP, Burge CB, Bartel B, Bartel DP (2002) Prediction of plant microRNA targets. Cell 110(4):513–520
58. Hutvagner G, Zamore PD (2002) A microRNA in a multiple-turnover RNAi enzyme complex. Science 297:2056–2060
59. Yekta S, Shih IH, Bartel DP (2004) MicroRNA-directed cleavage of HOXB8 mRNA. Science 304:594–596
60. Aleman LM, Doench J, Sharp PA (2007) Comparison of siRNA-induced off-target RNA and protein effects. RNA 13:385–395
61. Morris KV, Chan SW, Jacobsen SE, Looney DJ (2004) Small interfering RNA-induced transcriptional gene silencing in human cells. Science 305(5688):1289–92
62. Kim DH, Saetrom P, Snove O Jr, Rossi JJ (2008) MicroRNA-directed transcriptional gene silencing in mammalian cells. Proc Natl Acad Sci USA 105:16230–16235
63. Place RF, Li LC, Pookot D, Noonan EJ, Dahiya R (2008) MicroRNA-373 induces expression of genes with complementary promoter sequences. Proc Natl Acad Sci USA 105:1608–1613
64. Weinberg MS, Villeneuve LM, Ehsani A, Amarzguioui M, Aagaard L, Chen ZX, Riggs AD, Rossi JJ, Morris KV (2006) The antisense strand of small interfering RNAs directs histone methylation and transcriptional gene silencing in human cells. RNA 12(2):256–62
65. Orom UA, Nielsen FC, Lund AH (2008) MicroRNA-10a binds the 5'UTR of ribosomal protein mRNAs and enhances their translation. Mol Cell 30:460–471

66. Vasudevan S, Steitz JA (2007) AU-rich-element-mediated upregulation of translation by FXR1 and Argonaute 2. Cell 128:1105–1118
67. Kinch LN, Grishin NV (2009) The human Ago2 MC region does not contain an eIF4E-like mRNA cap binding motif. Biol Direct 4:2
68. Kong YW, Cannell IG, de Moor CH, Hill K, Garside PG, Hamilton TL, Meijer HA, Dobbyn HC, Stoneley M, Spriggs KA et al (2008) The mechanism of micro-RNA-mediated translation repression is determined by the promoter of the target gene. Proc Natl Acad Sci USA 105:8866–8871
69. Kozak M (2008) Faulty old ideas about translational regulation paved the way for current confusion about how microRNAs function. Gene 423(2):108–115
70. Mathonnet G, Fabian MR, Svitkin YV, Parsyan A, Huck L, Murata T, Biffo S, Merrick WC, Darzynkiewicz E, Pillai RS, Filipowicz W, Duchaine TF, Sonenberg N (2007) MicroRNA inhibition of translation initiation in vitro by targeting the cap-binding complex eIF4F. Science 317:1764–1767
71. Eulalio A, Huntzinger E, Nishihara T, Rehwinkel J, Fauser M, Izaurralde E (2009) Deadenylation is a widespread effect of miRNA regulation. RNA 15:21–32
72. Standart N, Jackson RJ (2007) MicroRNAs repress translation of m7Gppp-capped target mRNAs in vitro by inhibiting initiation and promoting deadenylation. Genes Dev 21(16):1975–1982
73. Gorban AN, Radulescu O (2008) Dynamic and static limitation in Multiscale reaction networks, Revisited. Adv Chem Eng 34:103–173, E-print arXiv:physics/0703278 [physics.chem-ph]
74. Radulescu O, Gorban AN, Zinovyev A, Lilienbaum A (2008) Robust simplifications of multiscale biochemical networks. BMC Syst Biol 2:86
75. Yang E, van Nimwegen E, Zavolan M, Rajewsky N, Schroeder M, Magnasco M, Darnell JE Jr (2003) Decay rates of human mRNAs: correlation with functional characteristics and sequence attributes. Genome Res 13:1863–1872
76. Bergmann JE, Lodish HF (1979) A kinetic model of protein synthesis. Application to hemoglobin synthesis and translational control. J Biol Chem 254:11927–11937
77. Hunt T, Hunter T, Munro A (1969) Control of haemoglobin synthesis: rate of translation of the messenger RNA for the alpha and beta chains. J Mol Biol 43:123–133
78. Scornik OA (1974) In vivo rate of translation by ribosomes of normal and regenerating liver. J Biol Chem 249:3876–3883
79. Gilchrist MA, Wagner A (2006) A model of protein translation including codon bias, nonsense errors, and ribosome recycling. J Theor Biol 239:417–434
80. Hartl DL, Jones EW (2005) Genetics: analysis of genes and genomes, 6th edn. Jones & Bartlett Publishers, Sudbury
81. Gorban AN, Radulescu O, Zinovyev A (2010) Asymptotology of chemical reaction networks. Chem Eng Sci 65:2310–2324, E-print arXiv:0903.5072 [physics.chem-ph]
82. Huang J, Liang Z, Yang B, Tian H, Ma J, Zhang H (2007) Derepression of microRNA-mediated protein translation inhibition by apolipoprotein B mRNA-editing enzyme catalytic polypeptide-like 3G (APOBEC3G) and its family members. J Biol Chem 282:33632–33640
83. Jopling CL, Schütz S, Sarnow P (2008) Position-dependent function for a tandem microRNA miR-122-binding site located in the Hepatitis C Virus RNA genome. Cell Host Microbe 4(1):77–85
84. Mayr C, Bartel DP (2009) Widespread shortening of 3'UTRs by alternative cleavage and polyadenylation activates oncogenes in cancer cells. Cell 138:673–684
85. Moore MJ (2005) From birth to death: the complex lives of eukaryotic mRNAs. Science 309:1514–1518
86. Sandberg R, Neilson JR, Sarma A, Sharp PA, Burge CB (2008) Proliferating cells express mRNAs with shortened 3' untranslated regions and fewer microRNA target sites. Science 320:1643–1647
87. Le Hir H, Seraphin B (2008) EJCs at the heart of translational control. Cell 133:213–216
88. Moore MJ, Proudfoot NJ (2009) Pre-mRNA processing reaches back to transcription and ahead to translation. Cell 136:688–700
89. Kapp LD, Lorsch JR (2004) The molecular mechanics of eukaryotic translation. Annu Rev Biochem 73:657–704

Web Resources for microRNA Research

12

Ulf Schmitz and Olaf Wolkenhauer

Abstract

Over the last decade thousands of microRNAs (miRNAs) have been discovered in all kinds of taxa. The ever growing number of identified miRNA genes required ordered cataloging and annotation. This has led to the development of miRNA web resources.

MiRNA web resources can be referred to either as web accessible databases (repositories) or web applications that provide a defined computational task upon user request. Today, more than three dozen web accessible resources exist that gather, organize and annotate all kinds of miRNA related data. According to the type of data or data processing method, these miRNA web resources can be classified as miRNA sequence and annotation databases, resources and tools for predicted as well as experimentally validated targets, databases of miRNA regulation and expression, functional annotation and mapping databases and a number of other tools and resources that are species-specific or focus on particular phenotypes.

This chapter provides an overview of the different types of miRNA web resources and their purpose and gives some examples for each category. Furthermore, some valuable miRNA web applications will be introduced. Finally, strategies for miRNA data retrieval and associated risks and pitfalls will be discussed.

Keywords

microRNA sequence database • miRNA targets • miRNA function • miRNA expression • Data retrieval • Data integration • Data mining

U. Schmitz (✉)
Department of Systems Biology and Bioinformatics, Institute of Computer Science, University of Rostock, Rostock 18051, Germany
e-mail: ulf.schmitz@uni-rostock.de

O. Wolkenhauer
Department of Systems Biology and Bioinformatics, Institute of Computer Science, University of Rostock, Rostock 18051, Germany

Stellenbosch Institute for Advanced Study (STIAS), Wallenberg Research Centre at Stellenbosch University, Stellenbosch, South Africa
e-mail: olaf.wolkenhauer@uni-rostock.de

U. Schmitz et al. (eds.), *MicroRNA Cancer Regulation: Advanced Concepts, Bioinformatics and Systems Biology Tools*, Advances in Experimental Medicine and Biology 774,
DOI 10.1007/978-94-007-5590-1_12,

12.1 Introduction

The first microRNA (miRNA) molecule *lin-4* was discovered in *C. elegans* in 1993 [1]. It was shown that *lin-4* can post-transcriptionally regulate *lin-14* expression, which is involved in larval development of the round worm. This small molecule turned out to be just one entity of a new class of non-coding RNAs (ncRNAs) that are present in many, if not all, animals and plants. The second example, *let-7,* was found in the same organism seven years later [2, 3]. Later on, homologs of these miRNAs were found in human and fly genomes. Shortly after, in year 2001, by conducting cloning experiments of small ncRNAs, more miRNAs have been identified in human, drosophila and *C. elegans* [4–6]. The ever increasing number of identified miRNAs demanded formal registration and naming conventions. Consequently, the miRNA Registry was set up with the aim of a controlled registration, naming and annotation of newly discovered miRNA stem loop- and mature sequences [7]. This platform was the first miRNA-centered web resource and with the advancement of knowledge about the role and function of miRNAs and their occurrence in numerous animals, many more miRNA web resources have sprung up. The Nucleic Acids Research (NAR) journal implemented in its annual special issue on biological databases a section for ncRNA databases. In the year 2012 it listed a total of 84 resources in the category of RNA sequence databases.

MiRNA web resources are designed to satisfy multiple purposes: They can be classified according to the kind of data or data processing that they provide, namely as (i) miRNA sequence and annotation databases, (ii) resources and tools for predicted targets of miRNAs as well as experimentally validated targets, (iii) databases of miRNA regulation and expression, and (iv) functional annotation and mapping databases. A coarser way to classify miRNA web resources would be based on the origin of the data. Towards this end, the classes of primary, secondary and tertiary web resources can be defined (Fig. 12.1). Primary miRNA web resources are those that store or process first-hand data, which are for example miRNA sequences or miRNA expression profiles. Secondary web resources are those that perform computational inferences or predictions based on the primary data, such as the tools and databases for miRNA target predictions. Other databases merge data from various primary and/or secondary resources, e.g. several target prediction repositories. This kind of web resource is referred to as tertiary web resource.

Another categorization of miRNA web resources was used in Jacobsen et al. [8]. The authors differentiated between resources of (i) genomic contexts and evolutionary conservation of miRNAs (e.g., miROrtho [9], miRGen [10], miRfunc [11], microTranspoGene [12]); (ii) prediction and validation of targets (e.g., TargetScan [13], miRNAMap [14], microRNA.org [15], miRDB [16], miRecords [17], TarBase [18]); and (iii) biological functions and phenotypes of individual miRNAs (e.g., miR2Disease [19], DIANA-mirPath [20], MMIA [21]).

In addition to databases and repositories, miRNA devoted tools and services were made available via the web, which enable users to predict and visualize binding sites of miRNAs in putative target genes (e.g., microRNA.org [15]), the structure of miRNA precursors and miRNA:mRNA hybrids (e.g., RNA22 [22]) or to perform functional analyses based on miRNA expression data and target predictions (e.g., miRGator [23]).

In this chapter we provide an overview of different classes of web resources for animal miRNAs, with a focus on human miRNAs. Furthermore, we introduce the concept behind each class and discuss how the data is generated (experiments, literature search, *in silico* methods etc.). In addition, we provide some examples for each category. Finally, we discuss things to consider when making use of miRNA web resources and potential pitfalls.

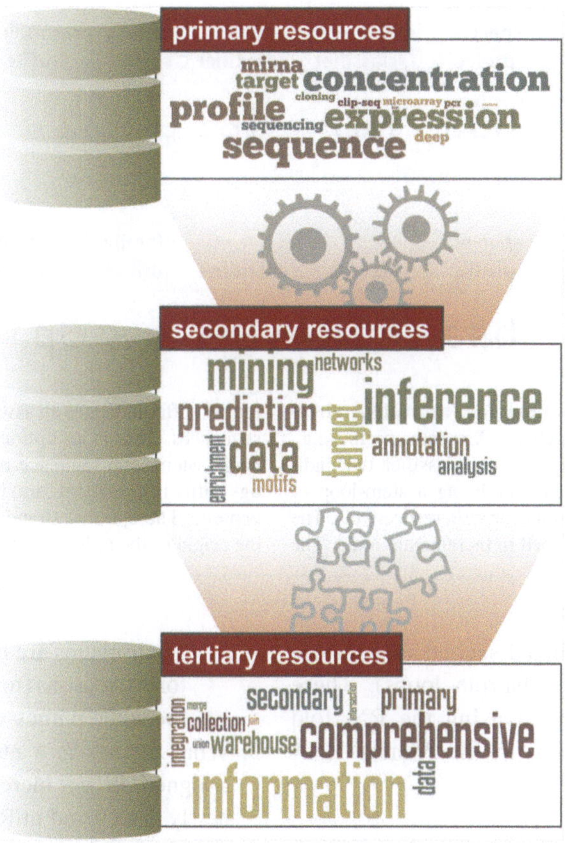

Fig. 12.1 Classification of miRNA web resources

12.2 MicroRNA Sequence Databases

The first developed miRNA web resource was called miRNA Registry [7], today better known as the **miRBase database** (http://mirbase.org). This database was a logical consequence of the ever growing number of newly discovered miRNAs. Its purpose was to register and categorize pre-miRNA and mature miRNA sequences, to assign names to miRNA sequences following a defined system [24] and to provide further annotations like experimental evidences, references to publications, etc. The first release of the miRNA Registry contained records of 218 miRNA precursors from human, mouse, fruit fly, *C. elegans* and A*rabidopsis thaliana*. As of today (release 18) the miRBase database provides sequence and structure information for more than 18,000 miRNA precursors and their maturated forms in 168 species including animals, plants, algae and viruses.

Sequences in miRBase are typically derived from cloning, sequencing or northern blotting experiments. In addition to that, every miRNA gene is put in a genomic context by indicating relationships to neighboring transcripts (intergenic, intragenic) and other putative miRNA members of a co-regulated stretch of DNA commonly referred to as miRNA cluster [25]. Sequences with homologs in other organisms are assigned to a family and information about other family members is cross-linked. The miRBase database

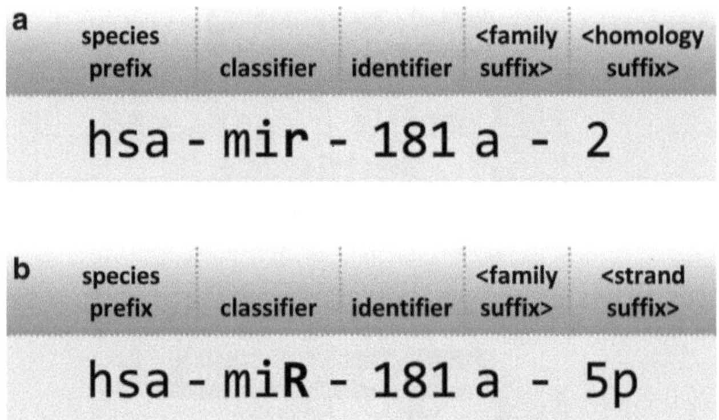

Fig. 12.2 Naming conventions for (a) miRNA stem-loop and (b) mature sequences. A species prefix (e.g., *hsa* for homo sapiens) precedes the classifier that indicates the designated sequence as being a stem-loop or mature miRNA sequence by *mir* or *miR*, respectively. The numeric identifiers are assigned in incremental order. The family suffix indicates an association to a group of similar sequences. This is an optional element of the sequence name. Stem-loop sequence names can contain a homology suffix in numerical form if homologs exist in the same genome. The optional strand suffix (*-3p* or *-5p*) indicates the origin of the miRNA in the stem-loop precursor

also provides the predicted secondary structure for miRNA precursors (hairpin loops). These predictions are generated using the RNAfold software, which belongs to the Vienna RNA Package [26].

12.2.1 MiRNA Sequence Names and Identifiers

After the first two miRNAs (*let-7* and *lin-4*) and their homologues have been discovered, it became clear that there might be many more of the same kind of non-coding RNAs. It was then time for a common nomenclature for the annotation of miRNA sequences [24]. To all miRNA stem-loop precursor and mature sequences deposited in the miRNA Registry (now miRBase) names and identifiers were assigned following a well-defined nomenclature with the following rules:

1. MiRNA stem-loop and mature sequences have a three letter prefix representing the species (e.g. *mmu* for mouse or *hsa* for human miRNAs).
2. The class of the molecule is denoted by either *mir* or *miR* for pre-miRNAs or mature miRNAs, respectively.
   (a) The only exceptions to this rule are the members of the *let-7* and *lin-4* family. Their names are being conserved for historical reasons (for newly found homologs the same names will be assigned).
3. What follows is a numeric identifier that is assigned in an incremental order for each newly discovered miRNA.
   (a) Exceptions are homologs to known miRNAs. When possible the same numeric identifier as for the original sequence will be assigned (e.g., for *hsa-mir-121* existing in human a newly discovered homolog in Xenopus would be named as *xla-mir-121*).
   (b) For homologous stem-loop sequences within the same genome an additional numeric suffix is assigned (e.g., *hsa-let-7a-1* and *hsa-let-7a-2*).
   (c) For hairpins expressing similar mature miRNAs an alphabetic character is appended (e.g. *mmu-mir-181a* and *mmu-mir-181b*).
4. For mature sequences an additional suffix is applied (*-3p* or *-5p*) denoting the location in the precursor sequence that they originate from.

The rules described above are applied for naming miRNA precursor and mature sequences in animals (see also Fig. 12.2). These names are used as common names. Additionally, an unambiguous identifier of the form ‘MI0000xxx’ and ‘MIMAT000xxx’ (xxx denotes an incremental numeric identifier) is assigned to each precursor and mature miRNA sequence. Homologs across species are associated with miRNA families that also have common names and unique identifiers (e.g., common name: *mir-21* family; ID: 'MIPF0000060').

Sequences of miRNA genes and their primary transcripts can also be found in other databases which are not exclusively dedicated to miRNAs. The Ensembl database (www.ensembl.org), for example, provides miRNA gene- and transcript sequences along with information about splice variants, phylogenetic information, orthologs and paralogs [27]. NCBIs GenBank stores miRNA gene-, transcript- and mature sequences along with some annotation and links to other specialized resources [28].

miRNA sequence data can be accessed in all the databases mentioned above via a web interface in HTML representation. Additionally, the download of the sequence data in FASTA formatted files is facilitated. GenBank provides sequence representations in its own well annotated format, whereas ensembl presents miRNA genes in the genomic context inside its genome browser and provides the export to several file formats (including flat file representation).

12.3 Resources for Predicted and Validated miRNA Targets

MiRNAs regulate the expression of a large number of genes at the post-transcriptional level [29, 30]. Different mechanisms, how this regulation is conducted, are discussed in the literature and in detail in the Chap. 11 'Mathematical modeling of microRNA-mediated mechanisms of translation repression' in this book. MiRNA regulation of its targets can basically lead to two different effects: (i) translation repression (the translation initiation or elongation is blocked or early translation termination is enforced) and (ii) target mRNA destabilization by decapping or deadenylation. Based on the observed patterns in early discovered miRNA:target interactions [2, 31, 32], the first computational miRNA target prediction algorithms were developed that predict putative binding sites in gene transcripts that mature miRNAs might be able to hybridize to and thus induce target repression [33–38]. To this day, many more approaches have been published and every new detail that has been uncovered in the process of target regulation led to refinements and adjustments in the prediction approaches. Figure 12.3 shows a timeline of the emergence of target prediction tools. In general, many of the available tools construct their algorithms on the basis of the following features: (a) the sequence complementarity between a miRNA and its target site, especially at the seed region (nucleotides 2–8 of the miRNA), (b) hybridization energy of the miRNA/mRNA duplex, (c) evolutionary conservation of target sites, and (d) other diverse compositional and sequence features of the target site. The predictions of many tools and algorithms for *homo sapiens* and also for other animals, plants, fungi, etc. have been deposited in publically accessible web resources. In the following paragraphs we will introduce some of these resources and group them according to the approaches that are used in the different target prediction algorithms.

12.3.1 Sequence Complementarity, Conservation and Thermodynamic Stability

The early miRNA target prediction algorithms identified subsequences within the 3′ UTR of gene transcripts that exhibit a certain degree of complementarity to a given miRNA and are conserved in related genomes. For these algorithms, the seed region (nucleotides 2–8 of the miRNA) is of special importance [34, 38, 39]. Additionally, the predicted thermodynamic stability of putative miRNA target duplexes is another criterion used since the early algorithms [37, 40].

miRanda – www.microrna.org [15]: On this website, target sites predicted by the miRanda algorithm (including mirSVR scores) [41] are shown and directly mapped to the 3′ UTR of the designated target mRNA. The users can retrieve

2003	2004	2005	2006	2007	2008	2009	2010	2011
TargetScan	RNAhybrid	TargetBoost	RNA22	NBmiRTar	MirTarget2	DIANA microT 3.0	T-Rex	RepTar
miRanda	DIANA-microT	TargetScanS	MicroTar	GenMir++/3	FindTar	TargetMiner	TargetSpy	mirSOM
		PicTar	MiTarget	PITA	mirWIP		SVMicrO	
		MovingTargets	MirTarget	TargetRank	MirTif		mirSVR	
		MicroInspector		EIMMo				
		GenMir		StarMir				
				mRTP				

Fig. 12.3 Timeline of miRNA target prediction algorithms released in the last decade

predictions for five species. Two separate interfaces facilitate the search for either targets of a specific miRNA or miRNAs that putatively regulate a specific target gene. On top of that, precompiled predictions, expression profiles and a stand-alone target prediction tool can be downloaded.

TargetScan – www.targetscan.org [13]: This website provides a search interface to TargetScan predictions for ten animal species. The user can search for targets of miRNAs, miRNAs targeting a gene of interest and for common targets of conserved miRNA families. Predictions can be filtered for different degrees of conservation.

DIANA-microT – http://diana.cslab.ece.ntua.gr/microT [42]: The DIANA LAB website hosts several projects including the database of validated miRNA targets *TarBase*; a tool for the association of miRNAs with KEGG pathways DIANA-*mirPath*; a tool for the analysis of expression data for miRNA function *DIANA-mirExTra* and the target prediction algorithm *DIANA-microT*. The latter can be queried by miRNA identifiers to search for the corresponding targets, by gene names to search for its miRNA regulators or both to look for specific miRNA:target pairs. Additionally, miRNA sequences can be submitted to search for putative target genes in human and mouse. Results are annotated and interlinked and can be filtered by defining a threshold for the so called miTG score or for KEGG pathways.

12.3.2 Target Site and Flanking Features (Site Accessibility)

Results of the first generation of miRNA target prediction algorithms were subjected to experimental validation, and soon it has been realized that the performance of the early algorithms was not satisfactory as they produce a high fraction of false positive predictions. Further refinements were necessary and as the body of validated miRNA:target interactions grew, new patterns have emerged that led to the development of new algorithms. For example, the flanking regions of predicted binding sites have come into focus as they are supposed to determine the site accessibility. Features, like the flanking AU content and the site position and proximity to co-operating target sites, have emerged as relevant factors [13, 43]. Two algorithms that considered these new patterns in their approach are:

PITA – http://genie.weizmann.ac.il/pubs/mir07 [44]: Target predictions for human, mouse, fly and worm can be downloaded or queried. The results of such queries can be filtered by minimum seed size, allowance of G:U pairs, allowance of single mismatch, minimum seed conservation and inclusion of site flanks into the prediction. UTR sequences provided by the user can be scanned for binding sites of known miRNAs or user-provided miRNA sequences. Alternatively, a stand-alone executable program can be downloaded.

PicTar – http://pictar.mdc-berlin.de [45]: This database provides an interface to PicTar predictions for different taxa. The users can search for either targets of a specified miRNA or miRNAs putatively regulating the expression of a specified gene. Results are annotated and link out to supplementary/continuative information on external web sites. Loci of the predicted target sites are highlighted in the target 3′ UTR which is shown in a multiple sequence alignment with related species and the structure of the predicted duplex is shown.

12.3.3 Machine Learning Approaches

Increasing availability of experimental evidences that prove microRNA:target interactions and translation repression efficacy have provided a good source of data to train classifiers, based on machine learning algorithms, that can be used to predict new miRNA:target interactions. It is now possible to statistically exploit the features described above (binding energy, sequence and conformational features), and thus rank them by their importance and subsequently predict target sites with higher accuracy. Machine learning enables us to find rules for the classification of miRNA targets and non-targets based on observations from the training data. This idea has been implemented in tools like TargetBoost (genetic programming with boosting) [46], miTarget (support vector machine – SVM) [47], mRTP (ensemble algorithm) [48], and MiRTif (SVM) [49]. It is however important to carefully choose strong discriminative features which should be extracted from a good and representative pool of positive and negative samples. Machine learning based target prediction algorithms employ a great number of different features for (i) seed conformation (ii) outer-seed conformation (iii) structural and (iv) position specific features. The most recent algorithms using a machine learning approach are TargetSpy (Multiboost) [50], SVMicrO (support vector machine) [51], mirSVR (support vector regression) [41] and RepTar (hidden markov model – HMM) [52].

TargetSpy – www.targetspy.org [50]: On this website, a simple search interface can be used to query TargetSpy predictions for human, mouse, rat, chicken and fly miRNA targets. The user can select between two predefined cut-offs score, 'sensitive' and 'specific', and filter target sites with seed complementarity. All predictions as well as the TargetSpy stand-alone version can be downloaded.

miRDB – http://mirdb.org [16]: Results of expression based target predictions from MirTarget2 [53] are deposited in the miRDB database. MirTarget2 is a SVM classifier that was trained with microarray data from miRNA transfection experiments that were extracted from the NCBI GEO database [54] (accession numbers GSE6838 and GSE6207). Downregulated genes were mapped to predictions from TargetScan [13], PicTar [45], miRanda [36] and MirTarget [55], and features of putative binding sites were extracted and used for training. The miRDB database stores target predictions for five different species (human, mouse, rat and chicken) which can be downloaded as tab delimited text files for the current and previous releases of miRDB.

12.3.4 Composite Resources of Target Predictions

With the growing number of prediction tools, different sets of putative targets for a miRNA have been proposed. However, it has been shown that these sets are only partially overlapping and that each target prediction tool may produce a significant amount of false positive and false negative predictions. Therefore users put more confidence in miRNA:target interactions that are predicted by several algorithms. Some web resources collect predictions from several algorithms and make these available. Two examples are:

miRecords – http://mirecords.biolead.org [17]: This web resource contains two databases, one for experimentally validated targets of miRNA regulation and the other for miRNA target predictions. The latter integrates predictions from 11 miRNA target prediction algorithms (DIANA-microT, MicroInspector, miRanda, MirTarget2, miTarget, NBmiRTar, PicTar, PITA, RNA22, RNAhybrid, and TargetScan/TargertScanS) and nine different species. The results can be filtered for those with a minimum support by *n* algorithms and can be downloaded as Excel files.

miRWalk – http://mirwalk.uni-hd.de [56]: This database also provides data about predicted and validated miRNA:target interactions for human, mouse and rat (including the regulation of mitochondrial transcripts). Predictions include that of eight established algorithms (DIANA-microT, miRanda, MirTarget2, RNAhybrid, PicTar, PITA, RNA22 and TargetScan) plus a newly developed algorithm called miRWalk. Furthermore, predicted targets are mapped to pathways and diseases, and validated targets are additionally mapped to corresponding organs and cell lines. All results can be downloaded as Excel files.

12.3.5 Resources of Validated miRNA: Target Interactions

A widely used approach to support computationally predicted targets of miRNAs is to perform expression experiments (e.g. microarrays) for both miRNAs and mRNAs. From these experiments inverse correlations in the expression profiles of miRNAs and mRNAs indicate putative regulatory interdependencies, e.g. a miRNA-induced repression of a potential mRNA target [29, 57]. This approach has already been considered in some data-driven prediction algorithms (e.g. MirTarget2 [53], GenMir3 [58], T-REX [59]). Recently, however, a new approach has been developed for the detection of binding sites of RNA-binding proteins (RBP) and miRNA-containing ribonucleoprotein complexes (miRNPs, e.g. Argonaute or TNRC6). The approach is based on cross-linking and immoprecipitation followed by high-throughput sequencing of the cross-linked RNA sequences and is referred to as HITS-CLIP [60, 61]. Databases of miRNA:target interactions with experimental support are:

starBase – http://starbase.sysu.edu.cn [62]: The starBase database (sRNA target Base) maps reads from CLIP-Seq (HITS-CLIP, PAR-CLIP) and degradome sequencing (Degradome-Seq, PARE) experiments to miRNA binding sites in mRNA targets predicted by six different algorithms in six different species (including *homo sapiens*). In this way, the database provides information about *semi-validated* miRNA target interactions. starBase can be queried with a miRNA or gene symbol and the results can be exported as comma separated value file (csv). Beyond that starBase provides tools for functional enrichment analysis of miRNA targets in GO terms and KEGG pathways.

TarBase – www.microrna.gr/tarbase [63], **miRecords** – http://mirecords.umn.edu/mirecords [17] and **miRTarBase** – http://mirtarbase.mbc.nctu.edu.tw [64]: Apart from the integrated approaches of high-throughput experiments with miRNA target predictions, a number of experimental techniques are being used to unravel new or confirm predicted miRNA:target pairs. These range from mRNA-based (microarrays, quantitative real-time PCR assays, reporter assays [29, 65]) and immunoprecipitation-based target analyses [66] to protein-based target analyses [30, 67, 68]. There exist several databases that collect information about experimentally validated miRNA target interactions for a growing number of organisms, usually by manual literature curation. The data collections of the three databases, TarBase [63], miRecords [17] and miRTarBase [64], can be explored through a web interface, while miRecords and miRTarBase provide their data for download as well. The data records contain annotations for the miRNAs as well as the mRNA targets, information about the experimental evidence, and links to the original publication and other external resources. On top of that TarBase allows a fine grained filtering of the search results (for species, method, regulation type, year etc.) and miRecords shows the algorithms that correctly predict a validated interaction.

12.4 Functional Annotation and Mapping Databases

MicroRNAs are post transcriptional regulators of gene expression and their biogenesis pathway and regulatory mechanisms have been well studied (see Filipowicz et al. [69] and Krol et al. [70] for reference). MiRNAs can suppress protein synthesis of a large number of targets [30] by either repressing their translation or by causing mRNA decay, e.g. through target deadenylation. They have been shown to play a role in almost every cellular process. Additionally, many miRNA:disease relationships have been described. In some cases deregulated miRNAs even have a causal relationship to pathogenesis or the progression of diseases [71]. The role of miRNAs in cancer for example has been studied in great detail. It has been found that miRNAs can become oncogenic by inhibiting tumor suppressors or can themselves act as tumor suppressors by targeting oncogenes [72, 73]. Both, deregulated tumor suppressor miRNAs and oncogenic miRNAs (*oncomir*), can induce or mediate tumorigenesis when being silenced and overexpressed, respectively. Moreover, some miRNAs that are referred to as *metastamir* have been associated with tumor progression and metastasis development [74–77].

But, what is the designated function of a particular miRNA? This question cannot be easily answered, and often miRNAs do not have a distinct function but may drive, catalyze or suppress different cellular processes in dependence of the physiological and/or environmental conditions [78]. Furthermore, some miRNAs exhibit a pathogenic potential when being overexpressed or silenced, mutated or post-transcriptionally edited [70, 79–82]. It can be concluded that studies of miRNA function have to consider a larger context, as their function is only exhibited in an interrelation with other molecules. Still, many miRNAs have been linked to specific cell functions as a mediator [83] or even as the driving force [84]. It has been shown that information about miRNA expression profiles in different pathologies could be used to define novel diagnostic markers [85]. On the other hand, knowledge about the role of miRNAs in diseases is of vital importance and may give rise of new therapeutic targets. The miR2Disease database, for example, is a valuable resource for miRNA:disease associations.

miR2Disease – www.mir2disease.org [19]: Data in miR2Disease is literature curated and consists of binary relationships (causal/non-causal) between deregulated miRNAs and human diseases. This information is complemented with the specific miRNA expression pattern in the disease state and literature references. The miR2Disease database contains more than 3,000 entries (beginning of 2012) that link miRNAs with 163 human diseases (including many cancer types). Information from the miR2Disease database can be downloaded as a tab delimited text file where each record is associated to a disease and being composed of the miRNA, its mRNA target, the date of publication and a literature reference.

Examples like this show that there is a high demand for the functional characterization of miRNAs. Different approaches have been developed to narrow down functions of miRNAs *in silico*. One such approach uses already known, as well as predicted target genes. Many genes have been functionally characterized and therefore functional enrichment analyses, performed by determining statistical prevalence for associations with GO terms, pathways or diseases in sets of miRNA targets, is used to infer the biological function of miRNAs. One miRNA web resource that is dedicated to the functional characterization of miRNAs by using this approach is the miRGator database.

miRGator 2.0 – http://mirgator.kobic.re.kr [23]: The second release of the miRGator database is a composite collection of miRNA related data that pursues the aim of a functional characterization of miRNAs through the combined miRNA and target mRNA expression analysis. This web resource hosts miRNA expression profiles for various experimental conditions (data extracted from PhenomiR, GEO and ArrayExpress) and gene expression data of miRNA transfection or knockdown experiments (data extracted from GEO and ArrayExpress). These data sets are complemented by information about known and predicted mRNA targets (algorithms used are miBridge, TargetScan, miRanda and PITA) as well as miRNA-disease relationships (extracted from PhenomiR). miRGator provides three interfaces that allow the user to browse miRNA and gene expression experiments, perform association analyses (e.g. miRNA-disease; miRNA-phenotype) and conduct customized *gene set analyses* (GSA) and *miRNA set analyses* (miRSA).

In line with the functional enrichment approach is the analysis of target involvement in signaling, metabolic and disease related pathways to refine the functional characterization of miRNAs [86]. On the other hand there are experimental strategies to determine miRNA function, e.g. expression profiling and functional assays. One database that links miRNA expression to human diseases is the PhenomiR database.

PhenomiR 2.0–http://mips.helmholtz-muenchen.de/phenomir [87]: This database primarily maps miRNA expression to human pathology. Furthermore, it links biological processes to miRNA activity profiles. Experimental data were extracted from the literature and manually curated. Each entry contains general information about the bioprocess or disease (such as PubMed ID, Taxon ID, GO ID and disease), details about the study that uncovered this relationship (such as study design, patients, control, samples information and tissue/cell line), information about the miRNAs (such as name, method used to determine expression and pattern/fold change), information about genes associated with the specific miRNA and additional relevant information. PhenomiR, at the beginning of 2012, contained 675 entries derived from 365 publications, 145 diseases and 98 bioprocesses. The entire database can be downloaded as a tab delimited '.tbl' file.

miRó database – http://ferrolab.dmi.unict.it/miro [88]: miRó is another database that links miRNAs with diseases, biological processes and functions. Here, in contrast to miR2Disease and PhenomiR, human miRNA-phenotype associations are not extracted from the literature but predicted based on putative as well as validated miRNA:target interactions. Target predictions come from three established algorithms (TargetScan, PicTar and miRanda). This database primarily integrates data from several web resources of miRNAs, targets, diseases and ontologies. Furthermore, miRó provides detailed information about the miRNAs and their target genes.

Other non-miRNA centralized mapping databases that contain information for the functional characterization of genes including miRNAs are the Online Mendelian Inheritance in Man® (OMIM®; http://omim.org) and WikiGenes which is a community driven project for collaborative publishing [89].

12.5 Resources of microRNA Expression and Gene Regulation

Deregulation of miRNA expression can cause phenotypic changes in the cell and thus may, in the worst case, be a cause for human diseases. MiRNA expression profiles have been demonstrated to be useful biomarkers [90, 91]. Furthermore, target gene expression is typically negatively correlated to that of the regulatory miRNA. Consequently, such observations may give rise to new regulatory relationships between miRNAs and putative target genes. This shows that transcriptomic as well as proteomic (high-throughput) data is of great interest for miRNA research. During the last decade a tremendous amount of such data has been generated for different cell types, disease states, tissues and organisms. As a consequence, some databases that collect and organize such data and make it publically accessible have been developed. Apart from the popular expression databases like NCBIs Gene Expression Omnibus (GEO) [54], EBIs ArrayExpress [92] and the cancer focused Oncomine [93], several specialized miRNA web resources collect miRNA and target expression data; some of those also facilitating some functional analyses.

mirZ – www.mirz.unibas.ch [94]: Developers of the mirZ database gathered miRNA expression data from sequencing experiments carried out in human, mouse, rat, zebrafish, worm and fruitfly (most of the data originates from a large scale miRNA expression profiling study conducted by

Landgraf and co-authors [95]). The data were integrated into a web resource, together with target prediction and data mining tools. Target predictions are based on the ElMMo algorithm [96], which uses a Bayesian approach for target site detection.

microRNA.org – http://microRNA.org [15]: We have introduced this database above, in the section about target prediction resources, for its miRanda predictions. Additionally, this database collects tissue-based miRNA expression profiles from mouse, rat and human, and shares them with the community. MiRNA expression data can be visualized as heatmap or bar graph and can be directly compared across different tissue types.

dbDEMC – http://159.226.118.44/dbDEMC/index.html [97]**:** This is the only database so far that is fully dedicated to miRNAs in cancer, namely a **d**ata**b**ase of **D**ifferentially **E**xpressed **M**iRNAs in human **C**ancers (dbDEMC). The data of miRNA profiles in cancer was extracted from the literature (48 microarray experiments in 14 cancer types). Then Significance Analysis of Microarrays (SAM) was applied to identify significantly differentially expressed miRNAs in cancer as compared to normal tissue. The database can be queried by miRNA (name or accession number) or by a specific cancer type. For the former a list of cancer types will be returned were this miRNA is up- or down regulated, plus a heatmap representation and more details on the tissue or cell line used. For the latter a list of all miRNAs up- or downregulated in this cancer type is returned. Finally, information about differentially expressed miRNAs can be downloaded as a tab delimited text file.

There is no doubt that knowledge about miRNA expression can improve our understanding of miRNA function and their involvement in human pathologies. Today, miRNAs are being accepted as important regulators of gene expression. But the question how the miRNA expression itself is regulated still remains to be clarified. To do so, a number of studies have been undertaken to identify promoter regions and transcription start sites of miRNA genes, *cis*-regulatory and *trans*-acting elements respectively transcription factors (TF) that can promote or inhibit miRNA expression [98–101].

miRStart – http://mirstart.mbc.nctu.edu.tw [102]: This web resource indicates putative transcription start sites (TSS) for miRNA genes based on experimental evidence. This evidence includes cap analysis of gene expression (CAGE) tags, derived from the FANTOM4 web resource [103], TSS Seq tags derived from the DataBase of Transcriptional Start Sites (DBTSS) [104] and H3K4me3 chromatin signatures (enriched around putative TSSs) derived from Barski et al. [100]. Expressed sequence tags (EST) data derived from the UCSC browser was used to underpin experimental evidences of miRNA TSSs. Here, authors looked for conserved ESTs upstream of the pre-miRNA sequence. Finally, a SVM classifier was trained to predict high-confidence TSSs from the available data. The miRStart database provides details on TSSs for human miRNAs. For each miRNA the distribution of experimental evidences (CAGE tags, TSS tags and H3K4me3) along the 50 kb upstream region of the miRNA precursor sequence is displayed together with sequence conservation patterns and ESTs. Below a ranked list of putative TSSs is given along with the genomics location, offset to the precursor and a score.

TransmiR – http://cmbi.bjmu.edu.cn/transmir [105]: The TF-miRNA regulatory database (TransmiR) collects information on experimentally validated TF-miRNA interactions. Therefore the authors conducted a literature survey to identify TF–miRNA regulatory relationships. The database can be queried for miRNAs or TFs. The results include information about miRNA function, disease associations, the type of regulation (activation/repression) and a link to the original publication. The whole TransmiR dataset can be downloaded as tab delimited text file or Excel table.

PuTmiR – www.isical.ac.in/~bioinfo_miu/TF-miRNA/TF-miRNA.html [106]: This web resource in contrast to TransmiR collects predicted TF-miRNA interactions. PuTmiR lists TFs that putatively bind to sites within the 10kb upstream and downstream regions of human miRNA precursor sequences. These data have been extracted from the track of conserved transcription factor binding sites of the UCSC genome browser. The results are presented in tabular form including information about the genomic location and a prediction score. The data can be downloaded in a tabular text format.

From the previous paragraphs it becomes apparent that miRNAs are involved in complex regulatory structures that are composed of TFs, miRNAs and target genes. Closer inspection often reveals regulatory motifs like feedback and feed forward loops that can exhibit non-linear dynamics and thus play critical roles in various biological processes. The accumulation of TF-miRNA and miRNA-target regulations in small regulatory modules sometimes scales up to a complex regulatory network [107]. There are two web resources that were developed based on this idea.

CircuitsDB – http://biocluster.di.unito.it/circuits [108]: This database focuses on feed-forward loops (FFLs) composed of a TF that regulates the expression of a miRNA and common targets of both. Authors performed an *ab-initio* bioinformatics analysis of regulatory regions (promoter, 3′ UTRs) of protein coding genes and miRNA precursor sequences to search for conserved putative transcription factor binding sites (TFBS) in promoter regions and conserved seed complementary sites in the 3′ UTRs of protein coding genes [101]. The obtained results were complemented with miRNA-target predictions from TargetScan and TargetMiner [13, 109]. The database can be queried for TFs, precursor and mature miRNAs as well as joint target genes. A table where each row represents a mixed FFL, containing its well annotated components, will be returned. Furthermore, the database has two sections dedicated to MYC and ER FFLs. Additionally; users can search for binary transcriptional relationships (TF->gene, TF->miRNA) and binary post-transcriptional relationships (miRNA-gene). All data in CircuitsDB can be downloaded as tab delimited text files.

MIR@NT@N – http://mironton.uni.lu [110]: This web resource predicts regulatory networks involving TFs, miRNAs and target genes at genome level by integrating data from several primary resources: TF regulation data from PAZAR, JASPAR and oPOSSUM [111–113]; target predictions from MicroCosm (former miRBase Targets database) and microRNA.org [15, 114]; protein-protein interaction data from UniHI [115]; and gene-annotations from Ensembl [116]. MIR@NT@N (MIRna @Nd Transcription factor @nalysis Network) can be queried with lists of TFs, miRNAs and target genes, and it returns predictions of TF-miRNA, miRNA-TF and miRNA-gene regulations. Additionally, a regulatory network including all these interaction types will be constructed which can then be used for the automated detection of FBLs and FFLs. For single entities (TF, miR or gene) some annotations can be retrieved including information about regulators and/or targets. Regulatory network analyses with MIR@NT@N can be integrated into workflows of the data mining framework M@IA [117].

12.6 Web Tools

In Chap. 3 of this book algorithms for miRNA target prediction were introduced. Results of most target prediction algorithms can be accessed from dedicated web databases. Some of these databases allow filtering predictions by adjusting prediction parameters and to tune the sensitivity by defining more or less stringent cutoffs for the prediction score. Others facilitate the direct search for target genes of a submitted (mock or newly discovered) miRNA sequence.

Beyond that there exists a range of other miRNA-devoted web tools and collections of predictive computations. One example for such a predictive computation is the text-mining based identification of miRNA-target interactions. Others tools provide access to functional annotations of miRNAs or information about targeted pathways. Still others deal with miRNA expression profiles and realize an expression based inference of miRNA function. In contrast some tools study the effect of miRNA regulation in gene expression data. In a similar way DNA microarray or high-throughput sequencing data are used to identify the potential influence of miRNAs on biological processes or phenotypic changes. Some of these tools visualize regulatory networks that have been constructed computationally by mining target predictions and miRNA/mRNA expression profiles. Very popular for the functional characterization of miRNAs are enrichment analyses that search for enriched associations of miRNAs or their predicted/validated targets in the GO terms, and pathway and disease databases.

In the following paragraphs a few of these tools will be introduced. It has to be noted that this is only a small selection of available tools and upon publication of this book, more web-based miRNA analysis tools will have emerged.

MiRror – www.proto.cs.huji.ac.il/mirror [118]: Given a set of miRNAs miRror computes a ranked list of targets based on 11 prediction algorithms that complement each other in search criteria. Top ranked targets are those that are regulated by multiple miRNAs of the input set. The principal underlying the miRror approach is that targets are regulated in coordination by multiple miRNAs [45, 119]. Additionally, the reverse scenario has also been implemented that allows querying the miRNA regulation of a set of genes.

MMIA – http://cancer.informatics.indiana.edu/mmia [21]: This web service is devoted to the functional characterization of miRNAs. MMIA (microRNA and mRNA integrated analysis) integrates miRNA and mRNA expression data with miRNA target predictions for analyzing miRNA-associated phenotypes and biological functions by gene set analysis (GSA). Therefore, the users go through an analysis pipeline starting with the upload of miRNA expression data, followed by the selection of one target prediction algorithm or the intersection of two algorithms (TargetScan, PITA and/or PicTar), and the upload of mRNA expression data. Next, the users can select gene sets for the GSA (KEGG [120], MSigDB [121] and G2D [122]). MMIA offers some optional analyses, e.g. the identification of a prevalence of transcription factor binding sites (TFBS) in miRNA promoter regions, enrichment in disease associations, and gene set enrichment analysis with the GSEA-P software tool [121]. The output informs the users about differentially expressed miRNAs (including a heatmap representation), gene sets enriched in regulated miRNAs and TFBS enriched in the promoter regions of the regulated miRNAs. Finally, predicted mRNA targets of the regulated miRNAs are shown.

SylArray – www.ebi.ac.uk/enright-srv/sylarray [123]: This EMBL-EBI hosted web-service identifies enriched miRNA signatures in targets from high-throughput miRNA perturbation experiments. By using the Sylamer algorithm [124] SylArray scans the 3′ UTR sequences of an ordered gene list (e.g. by fold-change), submitted by the user, for potential miRNA binding sites. It then calculates the significance of enriched binding sites based on hypergeometric *p*-values. The graphical output shows a landscape plot of the enrichment *p*-values for miRNA seed complementary sites in 3′ UTRs, which is calculated in incremental parts of the submitted ranked gene list.

MiRonTop – www.microarray.fr:8080/miRonTop/index [125]: This is another web service that makes inferences on miRNA regulation based on high-throughput transcriptomic data. It uses several target prediction algorithms to predict genes that are under the regulatory control of miRNAs. Based on that, miRNAs with strongest regulatory impact on the given state data are determined. Furthermore, a functional enrichment analysis for up and downregulated genes is performed for the three Gene Ontologies (molecular function, biological process and cellular component).

TAM – http://cmbi.bjmu.edu.cn/tam [126]: This web service named as TAM (Tool for Annotations of human MiRNAs) is also devoted to the functional characterization of a given list of miRNAs. The idea is to provide meaning to results of miRNA medium and high-throughput experiments like microarray, HT-qPCR or deep sequencing. These results are usually lists of up- or down regulated miRNAs (e.g. in disease state or under treatment). TAM can be used to identify over- or underrepresentation of miRNA families and clusters, of functional or disease/tissue associations against a user defined background or one that is provided by the database. Predefined sets of miRNAs as well as the source code can be downloaded.

MAGIA – http://gencomp.bio.unipd.it/magia [127]: In this web resource the integration of miRNA target prediction with mRNA and miRNA expression data analysis is provided as an interactive web service. MAGIA (MiRNA And Genes Integrated Analysis) has a target prediction interface that can be used to query the database for the union or intersection of miRanda, Pita and/or TargetScan predictions to find either target genes for a group of miRNAs or regulatory miRNAs for a group of genes. On the other hand, MAGIA provides a web service for the analysis of mRNA and miRNA expression data for the detection of inverse correlated miRNA:mRNA pairs. Therefore, after uploading miRNA as well as mRNA expression data the user can choose an appropriate measure (Spearman correlation, Pearson correlation, mutual information, Genmir and meta-analysis) and select target prediction algorithm(s) to be used for the integrated analysis.

miRó – http://ferrolab.dmi.unict.it/miro [88]: This web resource is in the first place a collection of data that integrates information from various resources about miRNAs, their predicted targets, and associated gene ontology terms and diseases. Thus, it can be considered as a composite or tertiary miRNA web resource. All data are interlinked and thus by e.g. selecting a miRNA, a list of diseases and biological processes will be returned based on predicted targets of the miRNA. Additionally, expression profiles for miRNAs in a number of tissues are given. Furthermore, miRó can be used to predict new functional associations by analyzing common targets of miRNAs involved in different processes. The data mining module of miRó can be used to cluster miRNAs that are associated with common biological processes or diseases.

It becomes apparent that the aim of most miRNA web tools is the functional characterization of single miRNAs, miRNA clusters and coregulated miRNAs. What scientists bother about are the regulatory interactions by which miRNAs are involved in the regulation of almost all biological processes and how they are involved in the emergence of a large number of diseases. One way to the functional categorization of miRNAs leads through their mRNA targets. That's why most miRNA web tools incorporate predictions of miRNA targets and information about validated targets into their analyses. In many cases functional enrichment analyses are conducted to identify overrepresentations of miRNAs or their targets in gene regulatory networks, signaling and metabolic pathways or gene ontology terms.

In the next section we focus on the design principles of miRNA databases. We will describe the underlying technologies and the interfaces through which the users can access or download the information provided.

12.7 MicroRNA Data Retrieval

Most of the miRNA web resources are designed in a way that the users can browse the information and can extract data via a web interface. This is typically realized by CGI or JavaScript applications that query a backend database upon users' requests and represent results in HTML format. This is sufficient in many scenarios and is a user-friendly way to explore data, structure and search information about a particular miRNA or a set of miRNAs. However, sometimes this is not enough when for example users want to get hold of the entire database, query the database in batch mode, extract a specific subset of information or apply customized filters. Therefore, more and more web resources provide other means to extract data from their repositories.

12.7.1 Data Download

Many web resources make their collections of experimental data, literature derived knowledge or computationally generated data accessible via download. For example, sequence and structure information of miRNAs, lists of predicted or validated targets, or functional associations are being provided in custom file formats (e.g. fasta, gff and bed), as spreadsheets (like in miRecords) or simply as comma or tab delimited files. Files of smaller size are usually uncompressed, while larger files (e.g. more than 100 MB) are typically compressed in zip, tar or gz archives (e.g. in miRDB).

Some other resources provide XML formatted files for download which are basically text files that provide hierarchically structured data and can be used for exchange over the internet and between (web-) services (e.g. in miRMaid).

12.7.2 Programmatic Access

So far, only few resources facilitate a programmatic access to their data. One example is the miRMaid web service that provides, via a programming interface, HTML, XML and FASTA representations of data that were previously extracted from the miRBase database [8]. Actually, the miRMaid framework and database can be used as local installation and can be accessed by Ruby or Perl programs, while the RESTful[1] web API (application interface) of miRMaid can be accessed via any internet browser or programmatically via a uniform URL for every record in miRMaid that is uniquely addressable.

Another example is the fRNAdb (functional RNA database; [128]) that hosts information about known and predicted non-coding RNA (ncRNA) sequences (including miRNAs) assembled from a number of other databases. The fRNAdb API can be used to query and retrieve data via SOAP (previously known as Simple Object Access Protocol) or REST (Representational state transfer) interface. Using the REST interface one can post queries from simple to complex via a URL used in a standard web browser or the HTTP GET method in a computer script. Results are returned XML formatted, for sequences other formats can be requested (gff, bed or fasta). The SOAP interface of the fRNAdb supports several programming languages (Perl, Python, Ruby and Java). In this way very specific or complex requests are facilitated and can be processed in an automated manner. Responses to SOAP requests are always XML formatted.

Despite these two exceptions it has to be noted that this field lacks services for programmatic data retrieval and automatic access to miRNA web resources. The BioCatalogue, a popular registry for web services in the biosciences, lists more than 2,000 services (as of 2012) but it still lacks miRNA-related tools and resources [129]. The only hits when searching the database for miRNA related services are (i) 'miRMaid', (ii) ASRP – a web service for the retrieval of small RNAs from the *Arabidopsis small RNA Project* database [130], and (iii) RNAhybrid – a tool for finding minimum free energy hybridizations between long and short RNAs (used for miRNA target prediction; [40]). This situation has to be improved in future in order to facilitate quick, automated access to RNA specific information for reference or more interesting for data mining investigations. Furthermore, web accessible tools for the computational prediction of regulatory interactions involving miRNAs or predictions of functional, phenotype and pathology associations of miRNAs should be equipped with a programmatic interface.

12.8 Things to Consider When Using miRNA Web Resources

Nowadays, the use of diverse information and services that are provided in miRNA web resources seem indispensable in any kind of miRNA related analysis or experimental effort conducted [131]. For example, expression profiles of miRNAs, designated targets, sequence information of miRNA precursors or maturated miRNAs are frequently being requested. The scenarios in which miRNA-associated information is demanded are countless. However, one has to be

[1] RESTful characterizes a web service that meets constrains defined in the Representational State Transfer (REST) architectural style principles.

aware of some pitfalls and risks that coincide with the use of these resources and tools.

miRNA/target expression data. Expression profiles of miRNAs can be retrieved from databases like microRNA.org [15] and MirZ [94]. Although it has been shown that a cross-platform comparison of high-throughput miRNA expression experiments reveals high correlation [132], one has to keep in mind that these data were often generated under different experimental conditions with different experimental set-ups. Consequently, a quantitative comparison of miRNA expression data carries the risk of being unreliable. This is in line with mRNA expression data from miRNA perturbation experiments that are often used for the validation of predicted miRNA:mRNA interactions. Here again, quantitative conclusions about miRNA:target regulation tend to be very imprecise. Additionally, does this approach for target validation lack a significant point, namely that many targets are regulated by translation repression (e.g. by inhibition of translation initiation, prolongation or early translation termination), which means that their mRNA concentration stays constant while only their protein concentration drops upon miRNA induction. It has to be kept in mind that targets' transcriptomics analysis captures only the regulatory mechanism of target degradation (e.g. caused by mRNA deadenylation). Unfortunately, resources for protein based target expression analysis do not exist, except for the results of the recently introduced pSILAC method (stable isotope labeling with amino acids in cell culture, [30]) that have been made accessible in the web (http://psilac.mdc-berlin.de/).

Sequence data. In terms of sequence information for miRNA genes, precursors and mature sequences it has to be differentiated between predicted (e.g. via homology analysis) and validated sequences (e.g. via cloning or sequencing). Additionally, one has to be aware that some miRNA resources contain entries with obsolete names/identifiers which might have changed (through an adaptation of the naming convention) or been deleted from the primary miRNA sequence registry (miRBase). One example is the previously used way to indicate that a mature miRNA sequence originates from the passenger strand (less active) of the hairpin precursor by appending an asterisk '*' to the name. This notion has been discarded and now the strands, that maturated miRNAs originate from, are indicated by the strand-suffix (*-3p* or *-5p*) in the miRNA name (see also 'miRNA sequence databases' section).

Target prediction data. Most of the existing miRNA target prediction algorithms provide their prediction results in web resources. These include but are not limited to DIANA-microT, TargetScan, miRanda and PicTar. Beyond that, there exist some web resources that collect and integrate target prediction data from various target prediction algorithms and in some cases add some analytical facilities to it (e.g. miRWalk [56], miRGen [10] and miRecords [17]). It is important to keep in mind that even the most recent target prediction algorithms contain many false positive predictions and that there might be a bias towards patterns found in previously identified miRNA-target interactions that have been used as template for these algorithms. A common strategy to extract high confidence targets is to retrieve results of several algorithms and consider only commonly predicted targets for further investigation [50]. However, it has been shown by Ritchie et al. [133] that this strategy does not always hold its promises. The authors have shown that the overlapping predictions by one group of algorithms did not constitute a majority in the overlapping predictions of another group of algorithms.

Target validation data. The experimental identification of miRNA target genes can be achieved in different ways. However, there are several resources that provide data on experimentally validated targets, like Tarbase, miRecords and miRTarBase [17, 63, 64]. It has to be noted that some experimental procedures are more reliable in the detection or validation of miRNA targets as compared to others. Here we list the most commonly used methods in the order of their reliability: reporter assays, CLIP (crosslinking and immunoprecipitation combined with target predictions), western blots, qRT-PCR and microarrays. One exception to the common resources for validated miRNA targets is the starBase database which maps data from HITS-CLIP, PAR-CLIP and degradome sequencing experiments [60, 62] with predictions from four established target prediction algorithms. The result is a collection of *high-confidence* miRNA:target interactions inferred from experimental indications that match computational predictions.

Phenotype associations. There are several databases that provide information about miRNA disease or phenotype associations (e.g. miR2disease; PhenomiR 2.0 and miR6). This information is often derived by a manual literature curation process [19] or through computational predictions [88]. In any case these data can contain errors or unspecific information. For example, it has to be clarified which organism a miRNA:phenotype association refers to and how the experimental prove has been conducted (type of experiment, *in vivo*/*in vitro*, treatment, etc.).

Names, identifiers and accession numbers. A widely spread problem in miRNA resources is the incompleteness in terms of miRNAs and the use of obsolete identifiers. Often only those miRNAs that have been known at the time of the database release were considered and never updated. Additionally, many miRNA names have changed in newer versions of the miRBase database. This can lead to confusion, if other databases do not update their resources accordingly – miRNAs do not seem to exist in other databases and newly discovered miRNAs are missing. This problem can often be observed in databases of miRNA target predictions. In some cases predictions are not repeated for newly discovered miRNAs. This is also problematic for the automated data retrieval and integration.

Another issue that database designers should take care of is the use of a controlled and common vocabulary with respect to functional annotations. This can be realized by the use of ontology (e.g. GO) or MeSH (Medical Subject Headings®) terms that ensure a consistent way to retrieve and exchange information.

Interconnectedness. It can be observed that newer miRNA web resources make use of and incorporate information from already existing miRNA databases (e.g. [65]). Here, it is important that data from third party resources have to be updated according to their origins. Others prefer the supposedly saver way by linking to complementary information hosted in other resources. However, it has to be ensured that URLs are valid and that interlinked information does not become obsolete.

In summary. When making use of data from a miRNA web resource one should get aware of the release date or the time of the latest update and what has been changed compared to the previous release. Predicted sequences change upon experimental verification, likewise predicted miRNA targets can be proven to be false, new regulatory relationships are being discovered, improved computational algorithms lead to better predictions in terms of sensitivity and/or accuracy, and new phenotypic association are being revealed. No one likes to build up an analysis on outdated information, but rather wants to make sure to be on top of the latest developments. It can for example be very frustrating when, say a regulatory relationship (e.g. miRNA-mRNA or TF-miRNA) could not be found in a certain web resource only because in this resource it was registered with an obsolete identifier. Sometimes this leads to false assumptions and to misguided subsequent investigations.

Further, it is important to differentiate between literature derived and curated or experimentally supported information and computational predictions. The confidence of the latter can often be inferred from the applied scoring system. In this case it is important to be aware of the range and distribution of scores. Additional aid for the estimation of the confidence in computational predictions can be provided by (independent) benchmarks as can sometimes be found in review articles. Sethupathy and coauthors [134] for example compared the performance of several miRNA target prediction algorithms. Although not always a guarantee for more reliable predictions but at least an increase in confidence can be gained by comparing the results from different algorithms (ideally based on complementary approaches) and considering only the intersection for further analyses. For example, different resources that predict transcriptional regulators of miRNA genes exist (e.g., PutmiR and MIR@NT@N). Their predictions often differ a lot. So far no independent benchmark exists that could prove the superiority of one approach over the other. Therefore, in case an experimental validation of predicted TF-miRNA regulation is planned it is only wise to consider commonly predicted regulatory relationships in order to save time and resources. On the other hand, this approach is certainly at the cost of sensitivity. If highest sensitivity is the goal it makes sense to consider the union of computational predictions, rather than the intersection, for further investigation. Likewise the experimental approach used to determine or validate an interaction can result in more or less confident assumptions (see paragraph about Target validation data). In a recent publication we have collected predicted and validated interactions and have combined them in a regulatory map for the miRNA target hub p21 [107]. In our analysis, for the first time, a system was established that computes and assigns confidence scores to binary relationships of TF-miRNA and miRNA-mRNA pairs in regulatory networks.

Another criterion, that should not be underestimated when selecting web resources for data retrieval, is the quantity and quality of annotations that each data set is equipped with. It can be very convenient when all the information required can be obtained from one hand rather than through a time consuming and error prone journey through a number of web resources. This bears the risk of non-conformity (e.g. due to different identifiers) and an extra effort to bring the different formats used to present data into a common data format. Therefore, resources that make use of standardized data exchange formats and provide a programmatic access are to be preferred.

12.9 Conclusion

Since the launch of the first miRNA web resource ~10 years ago, several dozens of new resources have been released. Lots of experimental, categorical and computationally generated data that give a deeper insight into all aspects of miRNA biology and function can be accessed through these databases and web tools. It is not easy to keep track of all available resources, their quality (reliability) and how up-to-date they are. Though in a natural selection process; only good resources will be used and cited by the community.

In this chapter we have introduced concepts of miRNA web resources and gave a number of examples for each category. However, it has to be mentioned that we did not discuss every type of miRNA web resources and introduced a few but by far not all web resources. One category of tools that we did not discuss is that of applications for miRNA gene or hairpin precursor prediction. For the interested reader we would like to refer to the review article by Mendes et al. [135]. In this chapter we focused on resources for human miRNAs and we here assume that most of the existing human miRNAs have already been detected by deep sequencing analyses.

In Table 12.1 we list miRNA web databases and services which are most crucial for current miRNA studies.

Acknowledgements The work of US and OW was supported by the German research foundation (DFG, Project: WO 991/4-1) and the German Federal Ministry of Education and Research (BMBF) as part of the project e:Bio-Metsys. We would like to thank Michael Hecker for fruitful discussions on this topic as well as Julio Vera and Xin Lai for continuous mutual encouragement and for proof reading the manuscript.

Table 12.1 Overview of current miRNA web resources and services

Resource	Description	URL	PMID
Primary microRNA web resources			
mimiRNA	Web resource of miRNA expression data and tool for the discovery of functional relations between miRNAs and mRNAs	http://mimirna.centenary.org.au	19933167
miRBase	The miRNA sequence and annotation database	www.mirbase.org	17991681
miRecords	Database of validated animal miRNA-target interactions (also predicted targets from 11 algorithms)	http://mirecords.biolead.org	18996891
miRex	Database of miRNA gene expression	http://miracle.igib.res.in/mirex/	14681370
miRStart	A database of human miRNA transcription start sites (TSSs)	http://mirstart.mbc.nctu.edu.tw	21821656
miRTarBase	Databases of experimentally validated miRNA-target interactions	http://mirtarbase.mbc.nctu.edu.tw/	21071411
miRvar	Database for genomic variations in miRNAs	http://genome.igib.res.in/mirlovd	21618345
smirnaDB/mirZ	Databases of cloning-based miRNA expression profiles and of predicted miRNA target sites	www.mirz.unibas.ch/cloningprofiles/	19468042
TransmiR	Manually curated database of TF-miRNA regulations	http://202.38.126.151/hmdd/mirna/tf/	19786497
Secondary microRNA web resources			
CID-miRNA	Web service for the prediction of miRNA precursors	http://mirna.jnu.ac.in/cidmirna/	18522801
CircuitsDB	A database of miRNA-TF regulatory circuits in human and mouse	http://biocluster.di.unito.it/circuits	20731828
dbDEMC	Database of differentially expressed miRNAs in human cancers	http://159.226.118.44/dbDEMC/index.html	21143814
DIANA-microT	Predicted targets of miRNA regulation based on the DIANA-microT algorithm	www.microrna.gr/microT	19765283
FAME	Database that provides functional associations of human miRNAs derived with the FAME algorithm	http://acgt.cs.tau.ac.il/fame/	20576699
HMDD	Manually curated database of human miRNA-disease associations	http://202.38.126.151/hmdd/mirna/md	18923704
Magia	Interactive web service that integrates miRNA target predictions with mRNA and miRNA expression data analysis	http://gencomp.bio.unipd.it/magia/start	20484379
MapMi	Web service for mapping miRNA sequences to genomic loci across many species	www.ebi.ac.uk/enright-srv/MapMi	20233390
MicroCosm	Web resource of computationally predicted targets for miRNAs based on the miRanda algorithm (former miRBase Targets database)	www.ebi.ac.uk/enright-srv/microcosm	17991681
MicroInspector	Web application for searching miRNA binding sites in a target gene sequences	http://bioinfo.uni-plovdiv.bg/microinspector/	15980566
microRNA.org	A resource for predicted miRNA targets (miRanda) and miRNA expression profiles	www.microrna.org	18158296
microTranspo Gene	Database of miRNAs derived from transposable elements (TE)	http://microtranspogene.tau.ac.il	17986453
miR2Disease	Literature curated database of human miRNA-diseases relationships	www.mir2disease.org	18927107
miRDB	Database for miRNA target predictions (SVM classifier MirTarget2) and functional annotations	http://mirdb.org/miRDB	18426918

(continued)

Table 12.1 (continued)

Resource	Description	URL	PMID
miRdSNP	Database of disease associated SNPs and their distance from miRNA target sites on the 3′ UTRs of human genes	http://mirdsnp.ccr.buffalo.edu/	22276777
mirEval	Web service for the prediction of miRNA precursors	http://tagc.univ-mrs.fr/mireval	18453555
miRNAminer	Web service that searches for homologous miRNA genes using BLAST	http://pag.csail.mit.edu/mirnaminer	18215311
miRNEST	Collection of animal, plant and virus microRNA data with predicted miRNAs based on EST data	http://mirnest.amu.edu.pl	22135287
MiRonTop	Web service that makes inferences on miRNA regulation based on high-throughput transcriptomic data	www.microarray.fr:8080/miRonTop/index	20959382
miROrtho	Database of miRNA gene candidates and homologs of known miRNAs	http://cegg.unige.ch/mirortho	18927110
MiRror	Web tool for the analysis of cooperative regulation by ensembles of miRNAs on gene sets and pathways	http://www.proto.cs.huji.ac.il/mirror/	20529892
MMIA	Web tool for associating miRNA and mRNA expression with phenotypes and biological functions	http://cancer.informatics.indiana.edu/mmia	19420067
PhenomiR	Database of literature curated miRNA phenotype relationships based on miRNA expression profiles in diseases and biological processes	http://mips.helmholtz-muenchen.de/phenomir/	20089154
PuTmiR	Resource for putative miRNA transcription factors	http://www.isical.ac.in/~bioinfo_miu/TF-miRNA/TF-miRNA.html	20398296
RNA22	Web service for the prediction of miRNA target sites based on the RNA22 algorithm	http://cbcsrv.watson.ibm.com/rna22.html	16990141
SSCprofiler	Web service for the HMM based prediction of miRNA precursors	www.imbb.forth.gr/SSCprofiler.html	19324892
starBase	Database that maps reads from CLIP-Seq and degradome sequencing experiments to predicted miRNA binding sites in mRNA targets	http://starbase.sysu.edu.cn/	21037263
SylArray	Web-server for the prediction of miRNA effects on mRNA expression data	www.ebi.ac.uk/enright-srv/sylarray/	20871108
TAM	Web service for the functional characterization of miRNA sets	http://cmbi.bjmu.edu.cn/tam	20696049
TargetRank	Target prediction database with an integrated ranking of conserved and nonconserved miRNA targets	http://genes.mit.edu/targetrank/	17872505
TargetScan	Predicted targets of miRNA regulation based on the TargetScan algorithm	www.targetscan.org	15652477
Tertiary microRNA web resources			
CoGemiR	Collection of information on miRNA genomic location, conservation and expression data	http://cogemir.tigem.it/	18837977
mirDIP	Collection of miRNA-target predictions from several prediction databases	http://ophid.utoronto.ca/mirDIP	21364759

(continued)

Table 12.1 (continued)

Resource	Description	URL	PMID
miRGator	miRGator is an integrated resource for miRNA and mRNA expression profiles from diff. tissues, diseases and miRNA perturbation experiments	http://mirgator.kobic.re.kr	21062822
miRGen	miRGen provides an overview about the genomic context of miRNAs (incl. clusters), target predictions from six algorithms and experimentally supported targets	www.diana.pcbi.upenn.edu/miRGen.html	17108354
miRMaid	Web framework that integrates miRNA data resources and provides an interface for programmable computer access	www.mirmaid.org	20074352
miRNAMap	Resource that collects information about known miRNAs and their targets, expression profiles, predicted targets and miRNA tissue specificity	http://mirnamap.mbc.nctu.edu.tw	18029362
miRò	Data collection that integrates information from various resources about miRNAs, predicted targets, associated gene ontologies terms and diseases	http://ferrolab.dmi.unict.it/miro	20157481
miRSel	Text-mining based identification of miRNA-target interactions with links to PubMed	http://services.bio.ifi.lmu.de/mirsel	20233441
miRWalk	Database that integrates miRNA target predictions from various resources and information on validated targets	http://mirwalk.uni-hd.de/	21605702

The resources are ordered alphabetically and with respect to the defined classes, although some resources might fit two or more classes. We note that we cannot guarantee that this list is complete and that all these resources are still accessible after the publication of this book. Note that the table is only up-to-date the time of writing. In this list we considered only databases dedicated to human and animal miRNAs

References

1. Lee RC, Feinbaum RL, Ambros V (1993) The C. elegans heterochronic gene lin-4 encodes small RNAs with antisense complementarity to lin-14. Cell 75:843–854
2. Reinhart BJ, Slack FJ, Basson M, Pasquinelli AE, Bettinger JC, Rougvie AE, Horvitz HR, Ruvkun G (2000) The 21-nucleotide let-7 RNA regulates developmental timing in Caenorhabditis elegans. Nature 403:901–906
3. Slack FJ, Basson M, Liu Z, Ambros V, Horvitz HR, Ruvkun G (2000) The lin-41 RBCC gene acts in the C. elegans heterochronic pathway between the let-7 regulatory RNA and the LIN-29 transcription factor. Mol Cell 5:659–669
4. Lagos-Quintana M, Rauhut R, Lendeckel W, Tuschl T (2001) Identification of novel genes coding for small expressed RNAs. Science 294:853–858
5. Lau NC, Lim LP, Weinstein EG, Bartel DP (2001) An abundant class of tiny RNAs with probable regulatory roles in Caenorhabditis elegans. Science 294:858–862
6. Lee RC, Ambros V (2001) An extensive class of small RNAs in Caenorhabditis elegans. Science 294:862–864
7. Griffiths-Jones S (2004) The microRNA registry. Nucleic Acids Res 32:D109–D111
8. Jacobsen A, Krogh A, Kauppinen S, Lindow M (2010) miRMaid: a unified programming interface for microRNA data resources. BMC Bioinform 11:29
9. Gerlach D, Kriventseva EV, Rahman N, Vejnar CE, Zdobnov EM (2009) miROrtho: computational survey of microRNA genes. Nucleic Acids Res 37:D111–D117
10. Megraw M, Sethupathy P, Corda B, Hatzigeorgiou AG (2007) miRGen: a database for the study of animal microRNA genomic organization and function. Nucleic Acids Res 35:D149–D155
11. Taccioli C, Fabbri E, Visone R, Volinia S, Calin GA, Fong LY, Gambari R, Bottoni A, Acunzo M, Hagan J, Iorio MV, Piovan C, Romano G, Croce CM (2009) UCbase & miRfunc: a database of ultraconserved sequences and microRNA function. Nucleic Acids Res 37:D41–D48

12. Levy A, Sela N, Ast G (2008) TranspoGene and microTranspoGene: transposed elements influence on the transcriptome of seven vertebrates and invertebrates. Nucleic Acids Res 36:D47–D52
13. Grimson A, Farh KK-H, Johnston WK, Garrett-Engele P, Lim LP, Bartel DP (2007) MicroRNA targeting specificity in mammals: determinants beyond seed pairing. Mol Cell 27:91–105
14. Hsu PWC, Huang H-D, Hsu S-D, Lin L-Z, Tsou A-P, Tseng C-P, Stadler PF, Washietl S, Hofacker IL (2006) miRNAMap: genomic maps of microRNA genes and their target genes in mammalian genomes. Nucleic Acids Res 34:D135–D139
15. Betel D, Wilson M, Gabow A, Marks DS, Sander C (2008) The microRNA.org resource: targets and expression. Nucleic Acids Res 36:D149–D153
16. Wang X (2008) miRDB: a microRNA target prediction and functional annotation database with a wiki interface. RNA 14:1012–1017
17. Xiao F, Zuo Z, Cai G, Kang S, Gao X, Li T (2009) miRecords: an integrated resource for microRNA-target interactions. Nucleic Acids Res 37:D105–D110
18. Sethupathy P, Corda B, Hatzigeorgiou AG (2006) TarBase: a comprehensive database of experimentally supported animal microRNA targets. RNA 12:192–197
19. Jiang Q, Wang Y, Hao Y, Juan L, Teng M, Zhang X, Li M, Wang G, Liu Y (2009) miR2Disease: a manually curated database for microRNA deregulation in human disease. Nucleic Acids Res 37:D98–D104
20. Papadopoulos GL, Alexiou P, Maragkakis M, Reczko M, Hatzigeorgiou AG (2009) DIANA-mirPath: integrating human and mouse microRNAs in pathways. Bioinformatics 25:1991–1993
21. Nam S, Li M, Choi K, Balch C, Kim S, Nephew KP (2009) MicroRNA and mRNA integrated analysis (MMIA): a web tool for examining biological functions of microRNA expression. Nucleic Acids Res 37:W356–W362
22. Miranda KC, Huynh T, Tay Y, Ang Y-S, Tam W-L, Thomson AM, Lim B, Rigoutsos I (2006) A pattern-based method for the identification of MicroRNA binding sites and their corresponding heteroduplexes. Cell 126:1203–1217
23. Cho S, Jun Y, Lee S, Choi H-S, Jung S, Jang Y, Park C, Kim S, Lee S, Kim W (2011) miRGator v2.0: an integrated system for functional investigation of microRNAs. Nucleic Acids Res 39:D158–D162
24. Ambros V, Bartel B, Bartel DP, Burge CB, Carrington JC, Chen X, Dreyfuss G, Eddy SR, Griffiths-Jones S, Marshall M, Matzke M, Ruvkun G, Tuschl T (2003) A uniform system for microRNA annotation. RNA 9:277–279
25. Tanzer A, Stadler PF (2004) Molecular evolution of a MicroRNA cluster. J Mol Biol 339:327–335
26. Hofacker IL (2003) Vienna RNA secondary structure server. Nucleic Acids Res 31:3429–3431
27. Flicek P, Amode MR, Barrell D, Beal K, Brent S, Chen Y, Clapham P, Coates G, Fairley S, Fitzgerald S, Gordon L, Hendrix M, Hourlier T, Johnson N, Kahari A, Keefe D, Keenan S, Kinsella R, Kokocinski F, Kulesha E, Larsson P, Longden I, McLaren W, Overduin B, Pritchard B, Riat HS, Rios D, Ritchie GRS, Ruffier M, Schuster M, Sobral D, Spudich G, Tang YA, Trevanion S, Vandrovcova J, Vilella AJ, White S, Wilder SP, Zadissa A, Zamora J, Aken BL, Birney E, Cunningham F, Dunham I, Durbin R, Fernandez-Suarez XM, Herrero J, Hubbard TJP, Parker A, Proctor G, Vogel J, Searle SMJ (2010) Ensembl 2011. Nucleic Acids Res 39:D800–D806
28. Benson DA, Karsch-Mizrachi I, Clark K, Lipman DJ, Ostell J, Sayers EW (2011) GenBank. Nucleic Acids Res 40:D48–D53
29. Lim LP, Lau NC, Garrett-Engele P, Grimson A, Schelter JM, Castle J, Bartel DP, Linsley PS, Johnson JM (2005) Microarray analysis shows that some microRNAs downregulate large numbers of target mRNAs. Nature 433:769–773
30. Selbach M, Schwanhäusser B, Thierfelder N, Fang Z, Khanin R, Rajewsky N (2008) Widespread changes in protein synthesis induced by microRNAs. Nature 455:58–63
31. Wightman B, Ha I, Ruvkun G (1993) Posttranscriptional regulation of the heterochronic gene lin-14 by lin-4 mediates temporal pattern formation in C. elegans. Cell 75:855–862
32. Brennecke J, Hipfner DR, Stark A, Russell RB, Cohen SM (2003) bantam encodes a developmentally regulated microRNA that controls cell proliferation and regulates the proapoptotic gene hid in Drosophila. Cell 113:25–36
33. Stark A, Brennecke J, Russell RB, Cohen SM (2003) Identification of Drosophila MicroRNA targets. PLoS Biol 1:E60
34. Enright AJ, John B, Gaul U, Tuschl T, Sander C, Marks DS (2003) MicroRNA targets in Drosophila. Genome Biol 5:R1
35. Rajewsky N, Socci ND (2004) Computational identification of microRNA targets. Dev Biol 267:529–535
36. John B, Enright AJ, Aravin A, Tuschl T, Sander C, Marks DS (2004) Human MicroRNA targets. PLoS Biol 2:e363
37. Kiriakidou M, Nelson PT, Kouranov A, Fitziev P, Bouyioukos C, Mourelatos Z, Hatzigeorgiou A (2004) A combined computational-experimental approach predicts human microRNA targets. Genes Dev 18:1165–1178
38. Lewis BP, Shih I, Jones-Rhoades MW, Bartel DP, Burge CB (2003) Prediction of mammalian microRNA targets. Cell 115:787–798
39. Lewis BP, Burge CB, Bartel DP (2005) Conserved seed pairing, often flanked by adenosines, indicates that thousands of human genes are microRNA targets. Cell 120:15–20
40. Rehmsmeier M, Steffen P, Hochsmann M, Giegerich R (2004) Fast and effective prediction of microRNA/target duplexes. RNA 10:1507–1517
41. Betel D, Koppal A, Agius P, Sander C, Leslie C (2010) Comprehensive modeling of microRNA

targets predicts functional non-conserved and non-canonical sites. Genome Biol 11:R90
42. Maragkakis M, Reczko M, Simossis VA, Alexiou P, Papadopoulos GL, Dalamagas T, Giannopoulos G, Goumas G, Koukis E, Kourtis K, Vergoulis T, Koziris N, Sellis T, Tsanakas P, Hatzigeorgiou AG (2009) DIANA-microT web server: elucidating microRNA functions through target prediction. Nucleic Acids Res 37:W273–W276
43. Nielsen CB, Shomron N, Sandberg R, Hornstein E, Kitzman J, Burge CB (2007) Determinants of targeting by endogenous and exogenous microRNAs and siRNAs. RNA 13:1894–1910
44. Kertesz M, Iovino N, Unnerstall U, Gaul U, Segal E (2007) The role of site accessibility in microRNA target recognition. Nat Genet 39:1278–1284
45. Krek A, Grün D, Poy MN, Wolf R, Rosenberg L, Epstein EJ, MacMenamin P, da Piedade I, Gunsalus KC, Stoffel M, Rajewsky N (2005) Combinatorial microRNA target predictions. Nat Genet 37:495–500
46. Saetrom O, Snøve O, Saetrom P (2005) Weighted sequence motifs as an improved seeding step in microRNA target prediction algorithms. RNA 11:995–1003
47. Kim S-K, Nam J-W, Rhee J-K, Lee W-J, Zhang B-T (2006) miTarget: microRNA target gene prediction using a support vector machine. BMC Bioinformatics 7:411
48. Yan X, Chao T, Tu K, Zhang Y, Xie L, Gong Y, Yuan J, Qiang B, Peng X (2007) Improving the prediction of human microRNA target genes by using ensemble algorithm. FEBS Lett 581:1587–1593
49. Yang Y, Wang Y-P, Li K-B (2008) MiRTif: a support vector machine-based microRNA target interaction filter. BMC Bioinform 9(Suppl 12):S4
50. Sturm M, Hackenberg M, Langenberger D, Frishman D (2010) TargetSpy: a supervised machine learning approach for microRNA target prediction. BMC Bioinform 11:292
51. Liu H, Yue D, Chen Y, Gao S-J, Huang Y (2010) Improving performance of mammalian microRNA target prediction. BMC Bioinform 11:476
52. Elefant N, Berger A, Shein H, Hofree M, Margalit H, Altuvia Y (2011) RepTar: a database of predicted cellular targets of host and viral miRNAs. Nucleic Acids Res 39:D188–D194
53. Wang X, Naqa IME (2008) Prediction of both conserved and nonconserved microRNA targets in animals. Bioinformatics 24:325–332
54. Barrett T, Troup DB, Wilhite SE, Ledoux P, Evangelista C, Kim IF, Tomashevsky M, Marshall KA, Phillippy KH, Sherman PM, Muertter RN, Holko M, Ayanbule O, Yefanov A, Soboleva A (2010) NCBI GEO: archive for functional genomics data sets–10 years on. Nucleic Acids Res 39: D1005–D1010
55. Wang X, Wang X (2006) Systematic identification of microRNA functions by combining target prediction and expression profiling. Nucleic Acids Res 34: 1646–1652
56. Dweep H, Sticht C, Pandey P, Gretz N (2011) miR-Walk–database: prediction of possible miRNA binding sites by "walking" the genes of three genomes. J Biomed Inform 44:839–847
57. Farh KK-H, Grimson A, Jan C, Lewis BP, Johnston WK, Lim LP, Burge CB, Bartel DP (2005) The widespread impact of mammalian MicroRNAs on mRNA repression and evolution. Science 310:1817–1821
58. Huang JC, Morris QD, Frey BJ (2007) Bayesian inference of MicroRNA targets from sequence and expression data. J Comput Biol 14:550–563
59. Volinia S, Visone R, Galasso M, Rossi E, Croce CM (2009) Identification of microRNA activity by Targets' Reverse EXpression. Bioinformatics 26:91–97
60. Hafner M, Landthaler M, Burger L, Khorshid M, Hausser J, Berninger P, Rothballer A, Ascano M, Jungkamp A-C, Munschauer M, Ulrich A, Wardle GS, Dewell S, Zavolan M, Tuschl T (2010) Transcriptome-wide identification of RNA-binding protein and microRNA target sites by PAR-CLIP. Cell 141:129–141
61. Chi SW, Zang JB, Mele A, Darnell RB (2009) Argonaute HITS-CLIP decodes microRNA-mRNA interaction maps. Nature 460:479–486
62. Yang J-H, Li J-H, Shao P, Zhou H, Chen Y-Q, Qu L-H (2011) StarBase: a database for exploring microRNA-mRNA interaction maps from Argonaute CLIP-Seq and Degradome-Seq data. Nucleic Acids Res 39:D202–D209
63. Vergoulis T, Vlachos IS, Alexiou P, Georgakilas G, Maragkakis M, Reczko M, Gerangelos S, Koziris N, Dalamagas T, Hatzigeorgiou AG (2011) TarBase 6.0: capturing the exponential growth of miRNA targets with experimental support. Nucleic Acids Res 40:D222–D229
64. Hsu S-D, Lin F-M, Wu W-Y, Liang C, Huang W-C, Chan W-L, Tsai W-T, Chen G-Z, Lee C-J, Chiu C-M, Chien C-H, Wu M-C, Huang C-Y, Tsou A-P, Huang H-D (2011) miRTarBase: a database curates experimentally validated microRNA-target interactions. Nucleic Acids Res 39:D163–D169
65. Krützfeldt J, Rajewsky N, Braich R, Rajeev KG, Tuschl T, Manoharan M, Stoffel M (2005) Silencing of microRNAs in vivo with "antagomirs". Nature 438:685–689
66. Karginov FV, Conaco C, Xuan Z, Schmidt BH, Parker JS, Mandel G, Hannon GJ (2007) A biochemical approach to identifying microRNA targets. Proc Natl Acad Sci 104:19291–19296
67. Vinther J, Hedegaard MM, Gardner PP, Andersen JS, Arctander P (2006) Identification of miRNA targets with stable isotope labeling by amino acids in cell culture. Nucleic Acids Res 34:e107
68. Jovanovic M, Reiter L, Picotti P, Lange V, Bogan E, Hurschler BA, Blenkiron C, Lehrbach NJ, Ding XC, Weiss M, Schrimpf SP, Miska EA, Grosshans H, Aebersold R, Hengartner MO (2010) A quantitative targeted proteomics approach to validate predicted

microRNA targets in C. elegans. Nat Methods 7:837–842
69. Filipowicz W, Bhattacharyya SN, Sonenberg N (2008) Mechanisms of post-transcriptional regulation by microRNAs: are the answers in sight? Nat Rev Genet 9:102–114
70. Krol J, Loedige I, Filipowicz W (2010) The widespread regulation of microRNA biogenesis, function and decay. Nat Rev Genet 11:597–610
71. Croce CM (2009) Causes and consequences of microRNA dysregulation in cancer. Nat Rev Genet 10:704–714
72. Hammond SM (2006) MicroRNAs as oncogenes. Curr Opin Genet Dev 16:4–9
73. Hammond SM (2007) MicroRNAs as tumor suppressors. Nat Genet 39:582–583
74. Kloosterman WP, Plasterk RHA (2006) The diverse functions of microRNAs in animal development and disease. Dev Cell 11:441–450
75. Garzon R, Calin GA, Croce CM (2009) MicroRNAs in cancer. Annu Rev Med 60:167–179
76. Hurst DR, Edmonds MD, Welch DR (2009) Metastamir: the field of metastasis-regulatory microRNA is spreading. Cancer Res 69:7495–7498
77. Farazi TA, Spitzer JI, Morozov P, Tuschl T (2011) miRNAs in human cancer. J Pathol 223:102–115
78. Bushati N, Cohen SM (2007) microRNA functions. Annu Rev Cell Dev Biol 23:175–205
79. Calin GA, Croce CM (2006) MicroRNA signatures in human cancers. Nat Rev Cancer 6:857–866
80. Calin GA, Croce CM (2006) MicroRNA-cancer connection: the beginning of a new tale. Cancer Res 66:7390–7394
81. Ryan BM, Robles AI, Harris CC (2010) Genetic variation in microRNA networks: the implications for cancer research. Nat Rev Cancer 10:389–402
82. Gallo A, Locatelli F (2012) ADARs: allies or enemies? The importance of A-to-I RNA editing in human disease: from cancer to HIV-1. Biol Rev Camb Philos Soc 87:95–110
83. Gu J, Iyer VR (2006) PI3K signaling and miRNA expression during the response of quiescent human fibroblasts to distinct proliferative stimuli. Genome Biol 7:R42
84. Yu Z, Li Z, Jolicoeur N, Zhang L, Fortin Y, Wang E, Wu M, Shen S-H (2007) Aberrant allele frequencies of the SNPs located in microRNA target sites are potentially associated with human cancers. Nucleic Acids Res 35:4535–4541
85. Heneghan HM, Miller N, Lowery AJ, Sweeney KJ, Newell J, Kerin MJ (2010) Circulating microRNAs as novel minimally invasive biomarkers for breast cancer. Ann Surg 251:499–505
86. Xu J, Wong C (2008) A computational screen for mouse signaling pathways targeted by microRNA clusters. RNA 14:1276–1283
87. Ruepp A, Kowarsch A, Schmidl D, Buggenthin F, Brauner B, Dunger I, Fobo G, Frishman G, Montrone C, Theis FJ (2010) PhenomiR: a knowledgebase for microRNA expression in diseases and biological processes. Genome Biol 11:R6
88. Laganà A, Forte S, Giudice A, Arena MR, Puglisi PL, Giugno R, Pulvirenti A, Shasha D, Ferro A (2009) miRò: a miRNA knowledge base. Database (Oxford) 2009:bap008
89. Hoffmann R (2008) A wiki for the life sciences where authorship matters. Nat Genet 40:1047–1051
90. Barbarotto E, Schmittgen TD, Calin GA (2008) MicroRNAs and cancer: profile, profile, profile. Int J Cancer 122:969–977
91. Volinia S, Calin GA, Liu C-G, Ambs S, Cimmino A, Petrocca F, Visone R, Iorio M, Roldo C, Ferracin M, Prueitt RL, Yanaihara N, Lanza G, Scarpa A, Vecchione A, Negrini M, Harris CC, Croce CM (2006) A microRNA expression signature of human solid tumors defines cancer gene targets. Proc Natl Acad Sci U S A 103:2257–2261
92. Parkinson H, Kapushesky M, Shojatalab M, Abeygunawardena N, Coulson R, Farne A, Holloway E, Kolesnykov N, Lilja P, Lukk M, Mani R, Rayner T, Sharma A, William E, Sarkans U, Brazma A (2007) ArrayExpress–a public database of microarray experiments and gene expression profiles. Nucleic Acids Res 35:D747–D750
93. Rhodes DR, Kalyana-Sundaram S, Mahavisno V, Varambally R, Yu J, Briggs BB, Barrette TR, Anstet MJ, Kincead-Beal C, Kulkarni P, Varambally S, Ghosh D, Chinnaiyan AM (2007) Oncomine 3.0: genes, pathways, and networks in a collection of 18,000 cancer gene expression profiles. Neoplasia 9:166–180
94. Hausser J, Berninger P, Rodak C, Jantscher Y, Wirth S, Zavolan M (2009) MirZ: an integrated microRNA expression atlas and target prediction resource. Nucleic Acids Res 37:W266–W272
95. Landgraf P, Rusu M, Sheridan R, Sewer A, Iovino N, Aravin A, Pfeffer S, Rice A, Kamphorst AO, Landthaler M, Lin C, Socci ND, Hermida L, Fulci V, Chiaretti S, Foà R, Schliwka J, Fuchs U, Novosel A, Müller R-U, Schermer B, Bissels U, Inman J, Phan Q, Chien M, Weir DB, Choksi R, Vita GD, Frezzetti D, Trompeter H-I, Hornung V, Teng G, Hartmann G, Palkovits M, Lauro RD, Wernet P, Macino G, Rogler CE, Nagle JW, Ju J, Papavasiliou FN, Benzing T, Lichter P, Tam W, Brownstein MJ, Bosio A, Borkhardt A, Russo JJ, Sander C, Zavolan M, Tuschl T (2007) A mammalian microRNA expression atlas based on small RNA library sequencing. Cell 129:1401–1414
96. Gaidatzis D, van Nimwegen E, Hausser J, Zavolan M (2007) Inference of miRNA targets using evolutionary conservation and pathway analysis. BMC Bioinformatics 8:69
97. Yang Z, Ren F, Liu C, He S, Sun G, Gao Q, Yao L, Zhang Y, Miao R, Cao Y, Zhao Y, Zhong Y, Zhao H (2010) dbDEMC: a database of differentially expressed miRNAs in human cancers. BMC Genomics 11(Suppl 4):S5

98. Ozsolak F, Poling LL, Wang Z, Liu H, Liu XS, Roeder RG, Zhang X, Song JS, Fisher DE (2008) Chromatin structure analyses identify miRNA promoters. Genes Dev 22:3172–3183
99. Shalgi R, Lieber D, Oren M, Pilpel Y (2007) Global and local architecture of the mammalian microRNA-transcription factor regulatory network. PLoS Comput Biol 3:e131
100. Barski A, Cuddapah S, Cui K, Roh T-Y, Schones DE, Wang Z, Wei G, Chepelev I, Zhao K (2007) High-resolution profiling of histone methylations in the human genome. Cell 129:823–837
101. Re A, Corá D, Taverna D, Caselle M (2009) Genome-wide survey of microRNA-transcription factor feed-forward regulatory circuits in human. Mol Biosyst 5:854–867
102. Chien C-H, Sun Y-M, Chang W-C, Chiang-Hsieh P-Y, Lee T-Y, Tsai W-C, Horng J-T, Tsou A-P, Huang H-D (2011) Identifying transcriptional start sites of human microRNAs based on high-throughput sequencing data. Nucleic Acids Res 39:9345–9356
103. Kawaji H, Severin J, Lizio M, Waterhouse A, Katayama S, Irvine KM, Hume DA, Forrest AR, Suzuki H, Carninci P, Hayashizaki Y, Daub CO (2009) The FANTOM web resource: from mammalian transcriptional landscape to its dynamic regulation. Genome Biol 10:R40
104. Yamashita R, Wakaguri H, Sugano S, Suzuki Y, Nakai K (2009) DBTSS provides a tissue specific dynamic view of Transcription Start Sites. Nucleic Acids Res 38:D98–D104
105. Wang J, Lu M, Qiu C, Cui Q (2010) TransmiR: a transcription factor-microRNA regulation database. Nucleic Acids Res 38:D119–D122
106. Bandyopadhyay S, Bhattacharyya M (2010) PuTmiR: a database for extracting neighboring transcription factors of human microRNAs. BMC Bioinform 11:190
107. Lai X, Schmitz U, Gupta S, Bhattacharya A, Kunz M, Wolkenhauer O, Vera J (2012) Computational analysis of target hub gene repression regulated by multiple and cooperative miRNAs. Nucleic Acids Res 40:8818–8834
108. Friard O, Re A, Taverna D, Bortoli MD, Cora' D (2010) CircuitsDB: a database of mixed microRNA/transcription factor feed-forward regulatory circuits in human and mouse. BMC Bioinform 11:435
109. Bandyopadhyay S, Mitra R (2009) TargetMiner: microRNA target prediction with systematic identification of tissue-specific negative examples. Bioinformatics 25:2625–2631
110. Le Bechec A, Portales-Casamar E, Vettter G, Moes M, Zindy P-J, Saumet A, Arenillas D, Theillet C, Wasserman W, Lecellier C-H, Friederich E (2011) MIR@NT@N: a framework integrating transcription factors, microRNAs and their targets to identify sub-network motifs in meta-regulation network model. BMC Bioinform 12:67
111. Portales-Casamar E, Arenillas D, Lim J, Swanson MI, Jiang S, McCallum A, Kirov S, Wasserman WW (2009) The PAZAR database of gene regulatory information coupled to the ORCA toolkit for the study of regulatory sequences. Nucleic Acids Res 37:D54–D60
112. Portales-Casamar E, Thongjuea S, Kwon AT, Arenillas D, Zhao X, Valen E, Yusuf D, Lenhard B, Wasserman WW, Sandelin A (2009) JASPAR 2010: the greatly expanded open-access database of transcription factor binding profiles. Nucleic Acids Res 38:D105–D110
113. Ho Sui SJ, Fulton DL, Arenillas DJ, Kwon AT, Wasserman WW (2007) oPOSSUM: integrated tools for analysis of regulatory motif over-representation. Nucleic Acids Res 35:W245–W252
114. Griffiths-Jones S, Grocock RJ, van Dongen S, Bateman A, Enright AJ (2006) miRBase: microRNA sequences, targets and gene nomenclature. Nucleic Acids Res 34:D140–D144
115. Chaurasia G, Malhotra S, Russ J, Schnoegl S, Hanig C, Wanker EE, Futschik ME (2009) UniHI 4: new tools for query, analysis and visualization of the human protein-protein interactome. Nucleic Acids Res 37:D657–D660
116. Hubbard TJP, Aken BL, Ayling S, Ballester B, Beal K, Bragin E, Brent S, Chen Y, Clapham P, Clarke L, Coates G, Fairley S, Fitzgerald S, Fernandez-Banet J, Gordon L, Graf S, Haider S, Hammond M, Holland R, Howe K, Jenkinson A, Johnson N, Kahari A, Keefe D, Keenan S, Kinsella R, Kokocinski F, Kulesha E, Lawson D, Longden I, Megy K, Meidl P, Overduin B, Parker A, Pritchard B, Rios D, Schuster M, Slater G, Smedley D, Spooner W, Spudich G, Trevanion S, Vilella A, Vogel J, White S, Wilder S, Zadissa A, Birney E, Cunningham F, Curwen V, Durbin R, Fernandez-Suarez XM, Herrero J, Kasprzyk A, Proctor G, Smith J, Searle S, Flicek P (2009) Ensembl 2009. Nucleic Acids Res 37:D690–D697
117. Le Béchec A, Zindy P, Sierocinski T, Petritis D, Bihouée A, Le Meur N, Léger J, Théret N (2008) M@IA: a modular open-source application for microarray workflow and integrative datamining. In Silico Biol 8:63–69
118. Friedman Y, Naamati G, Linial M (2010) MiRror: a combinatorial analysis web tool for ensembles of microRNAs and their targets. Bioinformatics 26:1920–1921
119. Ivanovska I, Ball AS, Diaz RL, Magnus JF, Kibukawa M, Schelter JM, Kobayashi SV, Lim L, Burchard J, Jackson AL, Linsley PS, Cleary MA (2008) MicroRNAs in the miR-106b family regulate p21/CDKN1A and promote cell cycle progression. Mol Cell Biol 28:2167–2174
120. Kanehisa M, Araki M, Goto S, Hattori M, Hirakawa M, Itoh M, Katayama T, Kawashima S, Okuda S, Tokimatsu T, Yamanishi Y (2008) KEGG for linking genomes to life and the environment. Nucleic Acids Res 36:D480–D484

121. Subramanian A (2005) From the cover: gene set enrichment analysis: a knowledge-based approach for interpreting genome-wide expression profiles. Proc Natl Acad Sci 102:15545–15550
122. Perez-Iratxeta C, Bork P, Andrade-Navarro MA (2007) Update of the G2D tool for prioritization of gene candidates to inherited diseases. Nucleic Acids Res 35:W212–W216
123. Bartonicek N, Enright AJ (2010) SylArray: a web server for automated detection of miRNA effects from expression data. Bioinformatics 26:2900–2901
124. van Dongen S, Abreu-Goodger C, Enright AJ (2008) Detecting microRNA binding and siRNA off-target effects from expression data. Nat Methods 5:1023–1025
125. Brigand KL, Robbe-Sermesant K, Mari B, Barbry P (2010) MiRonTop: mining microRNAs targets across large scale gene expression studies. Bioinformatics 26:3131–3132
126. Lu M, Shi B, Wang J, Cao Q, Cui Q (2010) TAM: a method for enrichment and depletion analysis of a microRNA category in a list of microRNAs. BMC Bioinform 11:419
127. Sales G, Coppe A, Bisognin A, Biasiolo M, Bortoluzzi S, Romualdi C (2010) MAGIA, a web-based tool for miRNA and Genes Integrated Analysis. Nucleic Acids Res 38:W352–W359
128. Mituyama T, Yamada K, Hattori E, Okida H, Ono Y, Terai G, Yoshizawa A, Komori T, Asai K (2009) The Functional RNA Database 3.0: databases to support mining and annotation of functional RNAs. Nucleic Acids Res 37:D89–D92
129. Bhagat J, Tanoh F, Nzuobontane E, Laurent T, Orlowski J, Roos M, Wolstencroft K, Aleksejevs S, Stevens R, Pettifer S, Lopez R, Goble CA (2010) BioCatalogue: a universal catalogue of web services for the life sciences. Nucleic Acids Res 38:W689–W694
130. Gustafson AM, Allen E, Givan S, Smith D, Carrington JC, Kasschau KD (2005) ASRP: the Arabidopsis Small RNA Project Database. Nucleic Acids Res 33:D637–D640
131. Angerstein C, Hecker M, Paap BK, Koczan D, Thamilarasan M, Thiesen H-J, Zettl UK (2012) Integration of MicroRNA databases to study MicroRNAs associated with multiple sclerosis. Mol Neurobiol 45:520–535
132. Yauk CL, Rowan-Carroll A, Stead JD, Williams A (2010) Cross-platform analysis of global microRNA expression technologies. BMC Genomics 11:330
133. Ritchie W, Flamant S, Rasko JEJ (2009) Predicting microRNA targets and functions: traps for the unwary. Nat Methods 6:397–398
134. Sethupathy P, Megraw M, Hatzigeorgiou AG (2006) A guide through present computational approaches for the identification of mammalian microRNA targets. Nat Methods 3:881–886
135. Mendes ND, Freitas AT, Sagot M-F (2009) Current tools for the identification of miRNA genes and their targets. Nucleic Acids Res 37:2419–2433

13 Discovery of microRNA Regulatory Networks by Integrating Multidimensional High-Throughput Data

Jian-Hua Yang and Liang-Hu Qu

Abstract

MicroRNAs (miRNAs) are endogenous non-coding RNAs (ncRNAs) of approximately 22 nt that regulate the expression of a large fraction of genes by targeting messenger RNAs (mRNAs). However, determining the biologically significant targets of miRNAs is an ongoing challenge. In this chapter, we describe how to identify miRNA-target interactions and miRNA regulatory networks from high-throughput deep sequencing, CLIP-Seq (HITS-CLIP, PAR-CLIP) and degradome sequencing data using starBase platforms. In starBase, several web-based and stand-alone computational tools were developed to discover Argonaute (Ago) binding and cleavage sites, miRNA-target interactions, perform enrichment analysis of miRNA target genes in Gene Ontology (GO) categories and biological pathways, and identify combinatorial effects between Ago and other RNA-binding proteins (RBPs). Investigating target pathways of miRNAs in human CLIP-Seq data, we found that many cancer-associated miRNAs modulate cancer pathways. Performing an enrichment analysis of genes targeted by highly expressed miRNAs in the mouse brain showed that many miRNAs are involved in cancer-associated MAPK signaling and glioma pathways, as well as neuron-associated neurotrophin signaling and axon guidance pathways. Moreover, thousands of combinatorial binding sites between Ago and RBPs were identified from CLIP-Seq data suggesting RBPs and miRNAs coordinately regulate mRNA transcripts. As a means of comprehensively integrating CLIP-Seq and Degradome-Seq data, the starBase platform is expected to identify clinically relevant

J.-H. Yang
RNA Information Center, Key Laboratory of Gene Engineering of the Ministry of Education, State Key Laboratory for Biocontrol, Sun Yat-sen University, Guangzhou 510275, P. R. China

L.-H. Qu (✉)
RNA Information Center, Key Laboratory of Gene Engineering of the Ministry of Education, State Key Laboratory for Biocontrol, Sun Yat-sen University, Guangzhou 510275, P. R. China

Biotechnology Research Center, Sun Yat-sen University, Guangzhou 510275, P. R. China
e-mail: lssqlh@mail.sysu.edu.cn

U. Schmitz et al. (eds.), *MicroRNA Cancer Regulation: Advanced Concepts, Bioinformatics and Systems Biology Tools*, Advances in Experimental Medicine and Biology 774,
DOI 10.1007/978-94-007-5590-1_13,

miRNA-target regulatory relationships, and reveal multi-dimensional post-transcriptional regulatory networks involving miRNAs and RBPs. starBase is available at http://starbase.sysu.edu.cn/.

Keywords
microRNAs • Ago • CLIP-Seq • starBase • RNA-binding proteins • miRNA-target interactions • Cancer-associated miRNAs • Degradome-Seq • Post-transcriptional regulation • deepView

13.1 Introduction

MicroRNAs (miRNAs) represent a class of small non-coding RNAs (ncRNAs) that play an important role in various biological processes by targeting mRNAs [1, 2]. Emerging evidence is revealing that miRNAs serve as nodes of signaling networks that regulate cancer, apoptosis, proliferation, differentiation and stem cell biology [2, 3].

After the identification of hundreds of miRNAs, the best way to understand their biological function and regulatory mechanism is to identify the genes they regulate [1, 4]. To date, various miRNA target prediction programs have been developed [5]. However, the different target prediction programs produce rather different lists of predicted targets and all have a high false-positive rate [1, 2, 5, 6]. Therefore, we must spend substantial time and effort choosing targets from these predicted lists for experimental validation. High-throughput sequencing of Ago-immunoprecipitated RNAs after crosslinking (CLIP-Seq, HITS-CLIP, PAR-CLIP) and mRNA degradome (Degradome-Seq) now provide powerful ways to identify biologically relevant miRNA-target interactions [7–11]. The application of the CLIP-Seq and Degradome-Seq methods has significantly reduced the rate of false-positive predictions of miRNA binding sites [7–11].

Numerous candidate miRNA targets have been generated by the above-mentioned computational algorithms and high-throughput experimental methods, but choosing the biologically significant targets from these candidate miRNA targets is daunting [1, 4]. Many studies have shown that miRNAs act as critical regulators of genes that function at different steps in complex biological pathways [1, 4, 5]. Therefore, enrichment analyses of candidate target genes in Gene Ontology (GO) categories and biological pathways can be useful methods to identify targets that have important biological functions.

To identify biological miRNA-target interactions and miRNA regulatory networks, our starBase platform [12] integrated all published CLIP-Seq (PAR-CLIP, HITS-CLIP) data from various tissues or cell lines to identify miRNA targets. Multiple web-based computational tools were developed to explore these data and perform enrichment analyses of target genes in GO and Kyoto Encyclopedia of Genes and Genomes (KEGG) data. In addition, we explore combinatorial effects between Ago protein and other RBPs that might affect the efficiency of microRNA-mediated regulation. starBase provides a variety of interfaces and graphical visualizations to facilitate analysis of the massive and heterogeneous CLIP-Seq, Degradome-Seq data, miRNA targets and regulatory networks in normal and cancer cells (Fig. 13.1).

13.2 Materials

13.2.1 Data Sources

All known miRNAs were downloaded from miRBase [13]. Known non-coding RNA and protein-coding genes were downloaded from Ensembl [14] or UCSC [15]. miRNA targets predicted by five programs (TargetScan [16, 17], PicTar [18], miRanda [19], PITA [20] and RNA22 [21]) were downloaded from their corresponding websites.

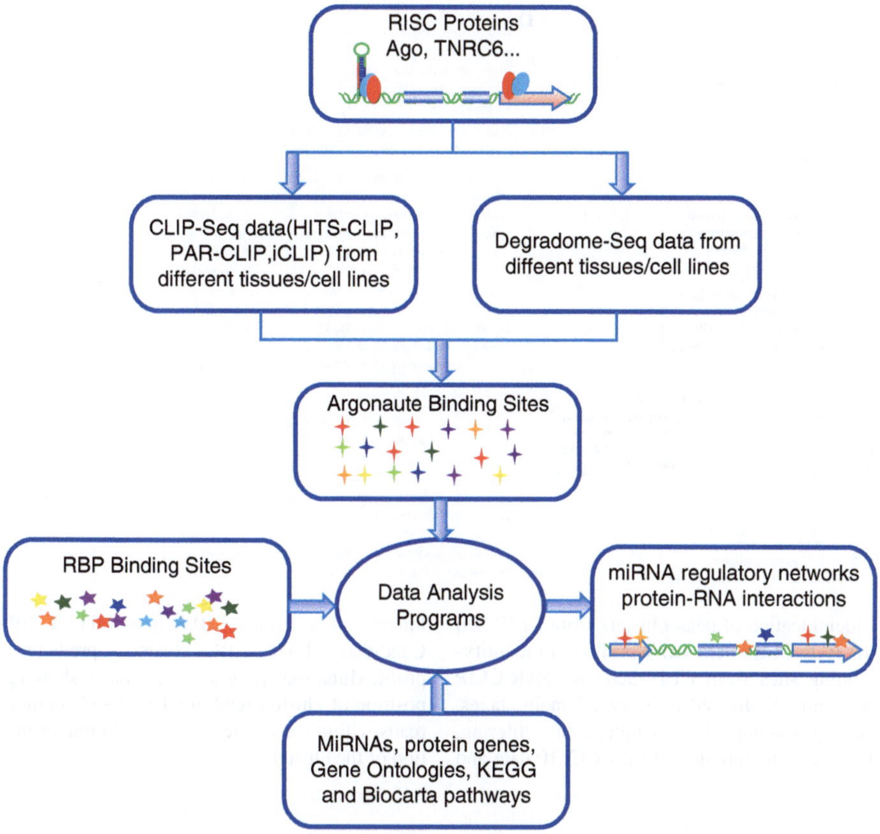

Fig. 13.1 Basic framework for identifying Ago binding sites and miRNA functional networks from CLIP-Seq and Degradome-Seq data. All results generated by this framework are displayed in the visual browser and web page. Web-based and stand-alone tools for data analysis are provided in the starBase platform. RISC is an abbreviation of RNA-Induced Silencing Complex. RBP is an abbreviation RNA-Binding Protein. RBSs is an abbreviation of RBP binding sites

Experimentally validated miRNA target sites were downloaded from the miRecords database [22]. CLIP-Seq data and Degradome-Seq data were downloaded from NCBI GEO database [23] or obtained from the supplementary material of the original articles [7–11, 24–26].

13.2.2 Gene Ontology and Pathways

GO ontology data [27] for the NCBI RefSeq genes were downloaded from the NCBI ftp site (ftp://ftp.ncbi.nih.gov/gene/DATA/), the Kyoto Encyclopedia of Genes and Genomes (KEGG) pathways [28] were downloaded from the KEGG database (http://www.genome.jp/kegg/), and the BioCarta pathways were downloaded from the CGAP ftp site (ftp://ftp1.nci.nih.gov/pub/CGAP). For the network or pathway analysis, we imported the above information into our starBase database. Enrichment analysis for the GO, KEGG and BioCarta pathways in the dataset was determined using a hypergeometric test and Bonferroni correction [29]. A cutoff of 0.1 on the Bonferroni-corrected P value was applied.

13.3 Methods and Applications

The methods presented in this chapter describe how to discover Ago binding and cleavage sites, miRNA-target regulatory relationships,

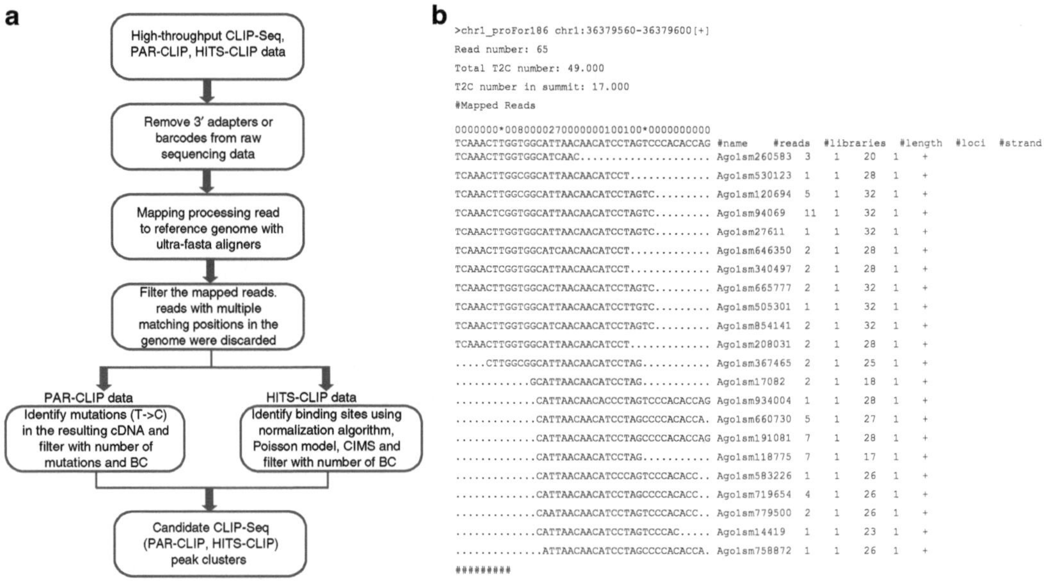

Fig. 13.2 Identification of peak clusters from CLIP-Seq (HITS-CLIP, PAR-CLIP) data. (**a**) Workflow for identifying Ago binding sites from CLIP-Seq and PAR-CLIP data. The workflow is divided into several main stages, including data processing, data mapping, data filtering, and identification of mutations from PAR-CLIP data and binding sites from CLIP-Seq or HITS-CLIP data. (**b**) Candidate PAR-CLIP clusters predicted by the profileMutations program. The page showing genomic position of cluster, read number, T→C number, mapped reads within the cluster and T→C number in each position of the cluster

miRNA functional networks and combinatorial effects between Ago and other RBPs from CLIP-Seq and Degradome-Seq data using the starBase web interfaces and tools. Further, we also describe how to run a comparative analysis of these data using the deepView genome browser [30].

13.3.1 Discovering Argonaute Binding Sites from CLIP-Seq Data

The high-throughput CLIP-Seq method provides a powerful way to identify the sites of Argonaute interaction. In this section, we summarize the features and workflow used to identify Argonaute binding sites from CLIP-Seq data and describe how to identify/filter PAR-CLIP clusters using our profileMutations tools.

13.3.1.1 Workflow for Identifying Argonaute Binding Sites from CLIP-Seq and PAR-CLIP Data

The general computational workflow summary from a series of recent publications [7–9, 31] is as follows (Fig. 13.2a):

1. Use next-generation sequencing and data processing. Remove 3′ adapters or barcodes from raw sequencing data. Identical reads are collapsed into unique reads. Users can remove adapters using our ClipRead program (http://deepbase.sysu.edu.cn/clipReadSearch.php), as described in our deepBase paper [30].
2. Map processed reads to the reference genome. According to different CLIP-Seq experimental methods, the user may select various ultrafast aligners. Previous studies often select aligners that only allow mismatches. However, recent studies found that the reverse transcriptase used in CLIP frequently skips the cross-linked amino acid-RNA adduct, resulting

in a nucleotide deletion [31]. Thus, it is better to select an aligner that allows mismatches and indels, such as novoalign or BWA [32].

3. Filter the mapped reads. To avoid repeat associated sequences, reads with multiple matching positions in the genome are often discarded. To remove potential duplicates resulting from PCR amplifications, all reads with identical sequences are treated as a single read [7].
4. Group overlapping reads into clusters to facilitate the analysis of large numbers of CLIP-Seq reads.
5. Filter candidate sites using mutation features and biological complexities (a measure of reproducibility between biologic replicates) [7]. (i) For PAR-CLIP, the actual crosslink sites often include transitions from thymidine to cytidine (T→C) in the resulting cDNA, the clusters therefore must include one or more (T→C) mutations [9]. (ii) For CLIP-Seq (HITS-CLIP), various methods have been used to identify binding sites of RBPs. Yeo et al. identified binding sites using a Poisson distribution model and filtered the candidates using biological complexity [8]. The Darnell lab identified binding sites by combining biological complexity and a normalization algorithm called in silico random CLIP [7]. Recently, they found that cross-linking induced mutation sites (CIMS) can be used to identify high-confidence binding sites from HITS-CLIP data [31].

13.3.1.2 Identify and Filter CLIP-Seq Clusters Using profile Mutations Tool

The methods mentioned above have been suggested for identifying Ago binding sites. However, there is no tool to analyze these CLIP-Seq data and identify clusters. Our profileMutations tool was developed to filter and identify PAR-CLIP clusters based on the number of PAR-CLIP reads and T→C mutations. In this section, we describe how to filter PAR-CLIP clusters using our profileMutations program, a new tool provided in the starBase platform [12].

1. Genome preparation. The genome can be downloaded from public genome centers, such as the UCSC Bioinformatics web site.
2. Mapping CLIP-Seq reads (the format of read is described in Readme file of program) from one or multiple experiments to the genome using an aligner, such as Bowtie [33] and BWA [32]. Alignment reads must be saved in Bowtie [33] or SAM [34] format.
3. Search. Start profileMutations with the following options: profileMutations -m 3 -n 5 genome.fasta alignments.bwt. The user can set the option to output different results. For example, the option -m 3 -n 5 tells profileMutations to output only the clusters with mutations (T→C) ≥3 and a number of reads ≥5. You can use option -h to view more options for profileMutations. After a short time, the program will return the results (Fig. 13.2b).

13.3.2 Exploring miRNA-Target Regulatory Relationships from CLIP-Seq Data

To examine the relationship between Ago binding clusters and potential sites of miRNA action, animal miRNA target sites predicted by five prediction programs (TargetScan [16], PicTar [18], miRanda [19], PITA [20] and RNA22 [21]) were intersected with all CLIP-Seq clusters.

13.3.2.1 Evaluating Predicted miRNA Target Sites

The following steps describe how to use the starBase website to evaluate predicted miRNA target sites overlapping with CLIP-Seq data.

1. Click "miRNA-target->miRNA-target relationship" to open the miRNA-target interaction page.
2. Select the clade, genome, and database of interest. For example, choose "mammal->human->hg19". If you want to search miRNA target genes of interest, you can enter the gene name in the text box.
3. Because some miRNAs may not be expressed in the tissue/cell lines used for CLIP-Seq, you should select the miRNA read number to ensure they are expressed in the tissue/cell lines. For example, select read number >=100.

If you want to search all genes targeted by a miRNA of interest, you can select one miRNA in dropdown menu.

4. Some CLIP-Seq clusters with small read numbers may simply represent experimental or biological noise. Therefore, we provide an option to allow users to filter these target sites further by limiting the number of biological complexities, T→C changes or read numbers of CLIP-CLIP data.
5. Finally, you can click "Search" to see a list of all miRNA-target interactions in the human genome. The miRNA name, the official gene name, the number of deep-sequencing reads, number of target sites and CLIP-Seq reads are indicated in a table. The user can click on the title of the table to sort miRNA-target interactions according to various features, such as the number of reads, miRNA names, gene names and target sites.
6. (a) Click on any non-zero number within the table to launch a detailed page providing further information on that miRNA-target interaction. The detailed information for an interaction includes a description of the target gene, the CLIP-Seq cluster overlapping with target site, and the number of CLIP-Seq reads or the alignment between a miRNA and the target gene. (b) Click the target coordinates to view the genomic context of the target site in our deepView genome browsér. (c) By clicking the CLIP-Seq cluster, the user can view the cluster location, number of library and T→C mutations, gene ontology categories and pathways of the target gene, and reads of clusters. The user can further filter the candidate target sites using this information. See Fig. 13.3a for a sample screenshot. (d) The "references" section enables the retrieval of the primary articles yielding the CLIP-Seq sequencing data or predicted miRNA targets. Click the article title link to visit the NCBI PUBMED website.

One target site predicted by multiple different programs could be viewed as high confidence candidate site. Thus, the user can search the intersections of different programs by selecting "miRNA-target->target site intersection". The above-mentioned workflow is also applicable for the "target site intersection" website.

Studies have revealed that miRNAs are often deregulated in cancer [2, 3]. We used the "miRNA-target relationship" website to examine the known target genes of oncomiRs from human CLIP-Seq data. Our starBase platform identified 12 of 20 known oncomiR target genes [35], such as BCL2 (miR-21,miR15/16) [36], HMGA2 (let-7) [37] and E2F1 (miR-17/20) [38] (Fig. 13.3b). The results demonstrate that even though the CLIP-Seq data used in this study are from a specific cell type (HEK293 cell), the starBase platform can be used to identify most known targets of oncomiRs. Thus, as CLIP-Seq technology is applied to a broader set of cell lines, tissues and conditions, our starBase will continuously be updated to facilitate the comprehensive exploration of clinically relevant miRNA-target regulatory relationships.

13.3.2.2 Predicting miRNA Target Sites from CLIP-Seq Data

To identify target sites of novel small RNAs from CLIP-Seq data, the ClipSearch program [12] was developed to search for 6-8-mers (8-mer, 7-mer-m8 and 7-mer-A1) in CLIP-Seq data. This tool starts by scanning peak clusters overlapping the 3′-UTR for potential miRNA targets (6mer-8mer) and then outputs the detailed information. In this section, we introduce how to use the ClipSearch program to predict target sites of small RNAs from CLIP-Seq data.

Click "targetTools->Animal Target" to open the ClipSearch page. The user is required to select an intended organism and then enter nucleotides 2–8 of a mature miRNA sequence. To reduce false positives in the predicted targets from the ClipSearch program, the user can filter the candidate targets by selecting site types, which are classified into 8-mer, 7-mer-m8 and 7-mer-A1; minimum number of CLIP-Seq reads; and biological complexity. Click "Search ClipSeq" to submit a task; a typical run may take several minutes to finish. The output of the ClipSearch program includes site type, information about the target gene and visual sequence alignments matched to a specific CLIP-Seq cluster.

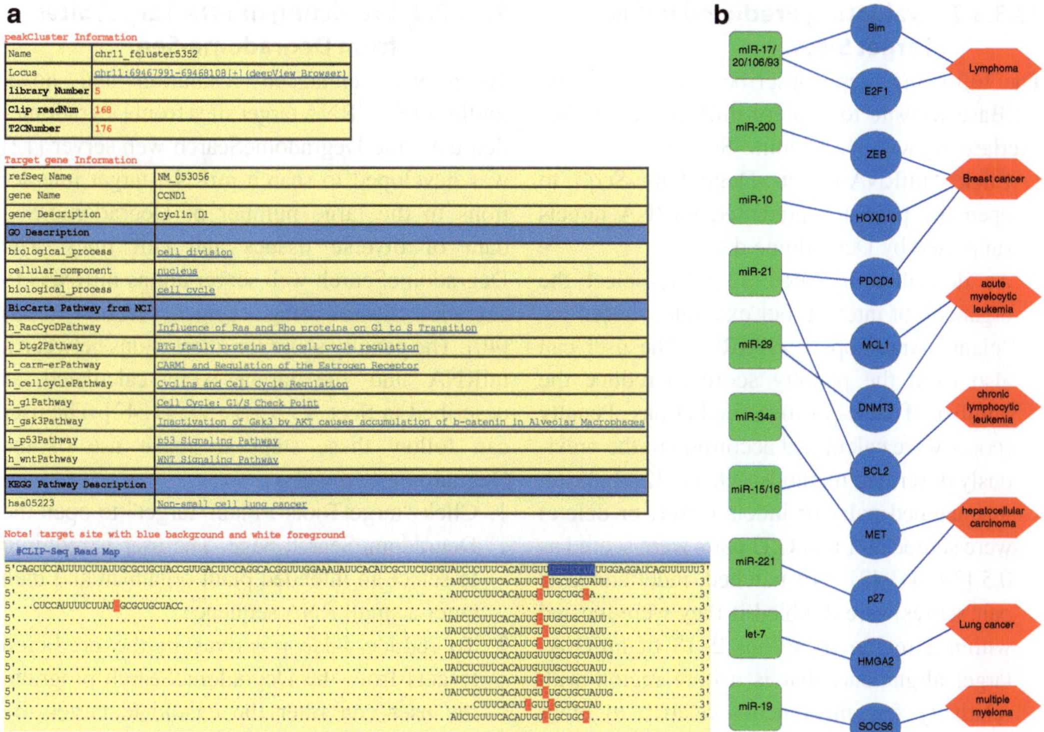

Fig. 13.3 miRNA target gene overlap with CLIP-Seq data. (**a**) Information for the candidate miRNA target gene. The page for a candidate miRNA target gene generated by starBase platform, showing peak cluster information, target gene information that includes gene description, GO description, KEGG and Biocarta pathways, and CLIP-Seq reads mapped to the potential peak cluster. (**b**) Regulatory relationships among some known oncomiRs, target genes and cancers in our starBase. The regulatory relationships were drawn using the Cytoscape program [53]. *Rectangles* represent miRNAs, *ellipses* represent target genes and hexagons represent cancers

13.3.3 Discovering Argonaute Cleavage Sites and miRNA-Target Interactions from Degradome-Seq Data

In this section, we summarize the features and workflow used to identify Argonaute cleavage sites and miRNA-target interactions from Degradome-Seq data.

13.3.3.1 Workflow for Identifying Argonaute Cleavage Sites from Degradome-Seq Data

The general computational workflow summary from a series of recent publications [10–12, 39] is as follows:

(a) Deep sequencing data processing. As described in Sect. 13.3.1.1, remove 3′ adapters or barcodes from raw deep sequencing data. Identical reads are collapsed into unique reads.

(b) Mapping processed reads to the reference genome and cDNA sequences. The user can select among various ultra-fast aligners, such as Bowtie [33] and segemehl [40]. To avoid repeat-associated sequences, reads with >50 hits in the genome were discarded.

(c) Identifying cleavage sites from mapping reads in genomic or cDNA sequences. Reads with the same 5′ start position were merged as one candidate cleavage site and their number were calculated as the abundance of each cleavage signature. If candidate cleavage site located within in the central region (10–11 nt region from 5′-end of miRNA) of an miRNA complementary site, it were taken as cleavage site of corresponding miRNA.

13.3.3.2 Evaluating Predicted miRNA Target Sites

The following steps describe how to use the starBase website to explore miRNA target sites overlapping with Degradome-Seq data.

1. Click "miRNA-target->Degradome-Seq" to open the page for predicted miRNA targets supported by Degradome data.
2. As described in Sect. 13.3.2.1, select the organism of interest. For example, choose the "plant->Arabidopsis->TAIR9". The user can also limit the penalty score to reduce the number of false-positive predictions. Penalty scores were calculated according to the previously described methods [39, 41, 42]. In brief, mismatched pairs or indels (insert or delete) were scored as 1 and G:U pairs were scored as 0.5 [39, 41, 42]. Mismatched, indel and G:U pair scores were doubled if they were located within a region (positions 2–13 of miRNA-target alignment) that is a core region with relatively few mismatches relative to other positions [39, 41, 42].
3. You may search all genes targeted by one miRNA of interest by selecting one miRNA in dropdown menu.
4. Finally, you can click "Search" to see a list of all miRNA-target interactions in the Arabidopsis genome. The miRNA name, the official gene name, the penalty score (cleaveLand score) [39], the abundance of each cleavage signature and genome coordinates are all indicated in a table. The user can click on the title of the table to sort miRNA-target interactions according to various features, such as cleaveLand score, miRNA names, gene names and the abundance of each cleavage signature.
5. Click on the "Target gene" within the table to launch a detailed page providing further information on that miRNA-target interaction. The detailed information for an interaction includes a description of the target gene, miRNA-mRNA alignments and Degradome-Seq information. See Fig. 13.4a for a sample screenshot. The user can further filter the candidate targets by the number of reads mapped to the cleavage site and the number of Degradome-Seq experiments including the cleavage site.

13.3.3.3 Predicting miRNA Target Sites from Degradome-Seq

To provide enhanced resolution and novel findings for miRNA target sites from Degradome-Seq data, the DegradomeSearch web server [12] was developed to search miRNA-target interactions in the large number of Degradome-Seq data of diverse tissues and cell lines. The DegradomeSearch web server aligns miRNAs to extended clusters using segemehl (version 0.093) [40]. The penalty score for interactions between a miRNA and the target were calculated as described in Sect. 13.3.1.1 [39, 41, 42]. The user can follow these steps to make use of the Degradome-Seq web server:

1. Click "targetTools->Plant Target" to open the DegradomeSearch page. The user is required to select an intended plant organism and then enter a small RNA sequence.
2. To reduce false positives in the predicted targets from the DegradomeSearch program, the user can filter the candidate targets by selecting a cutoff penalty score (e.g. default is less than 4.5) and minimum number of cleavage tags.
3. Click "Search DegradomeSeq" to submit the task; a typical run may take several minutes to finish. The output of the Degradome program includes the penalty score, miRNA-mRNA alignments, the genome locus, a link to the deepView genome browser, sample information and number of cleavage tags. See Fig. 13.4b for a sample screenshot.

13.3.4 Predicting miRNA Functional Networks

CLIP-Seq and Degradome-Seq technologies have provided an elegant way to study biologically relevant miRNA–target interactions at the transcriptome-wide level. Thus, these miRNA targets overlapping with CLIP-Seq and Degradome-Seq data are expected to precisely define functional networks by combining the analysis of Gene Ontology (GO) and biological pathway. In the starBase platform, two novel computational tools, miRGO and miRPathway, were developed to

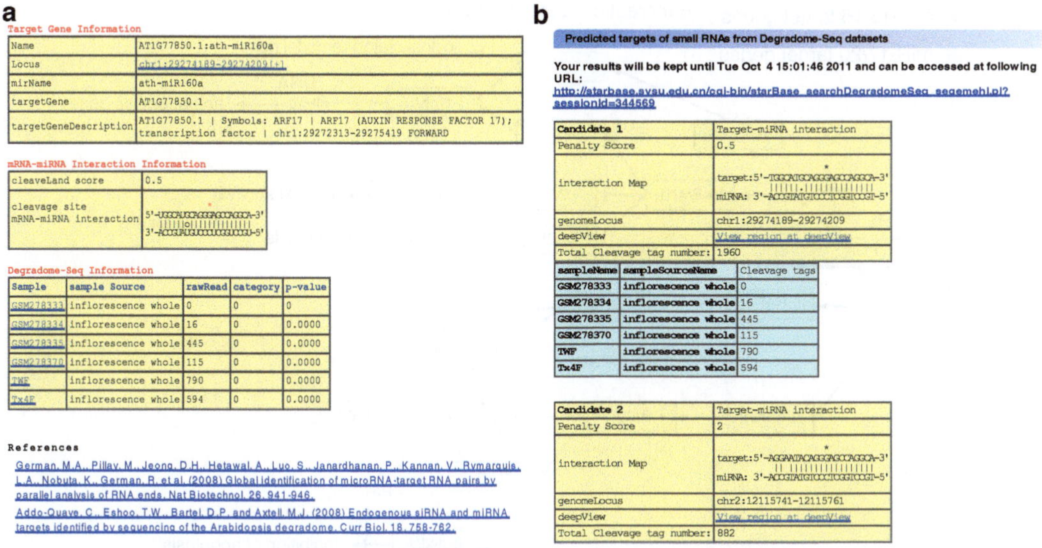

Fig. 13.4 miRNA-target regulatory relationships overlapping with Degradome-Seq. (**a**) miRNA-target interaction information in starBase. The page for an ath-miR160a target gene generated by starBase platform, showing target gene description, interaction between ath-miR160a and AT1G77850.1, and Degradome-Seq information. (**b**) The output of the DegradomeSearch program. The information for each candidate includes the penalty score, duplex between miRNA and target gene, genomic locus of target gene, total cleavage tag number and sample information

investigate the biological function of miRNAs that are expressed in tissues or cell lines for CLIP-Seq experiments. These two tools provide a comprehensive enrichment analysis of miRNA targets in >10,000 categories and biological pathways, including Gene Ontology terms and KEGG and Biocarta pathways, by using all genes that have at least one GO term as a background set. In addition, these two tools can also perform an enrichment analysis of genes targeted by multiple miRNAs and thus can be used for elucidating functional targets that are affected by multiple co-expressed miRNAs.

We used miRPathway to examine the pathways that contain the target genes of oncomiRs, such as miR-21, miR-17/20, miR-15/16. The set of genes regulated by these oncomiRs was enriched for proteins that have key roles in cancer biology. For example, miR-21, is involved in colorectal cancer, chronic myeloid leukemia and small cell lung cancer pathways; miR-15/16 is involved in cell cycle, the p53 signaling pathway, and pancreatic cancer; miR-17/20 is involved in pancreatic cancer and chronic myeloid leukemia. The above information shows that miR-15/16 and miR-17/20 may combinatorially regulate pancreatic cancer (Fig. 13.5). Moreover, we also applied the miRPathway tool to miRNA targets overlapping with HITS-CLIP data from the mouse brain. We found these targets to be most significantly enriched in the MAPK signaling pathway, the neurotrophin signaling pathway and the axon guidance pathway (Fig. 13.6a). Many miRNAs in these datasets also target genes that have key roles in the glioma pathway and other cancer-associated pathways (Fig. 13.6a). Our results suggest that applying bioinformatics approaches to high-throughput experimental datasets may help to precisely define miRNA function and pinpoint important target genes.

The following steps describe how to use the miRPathway website (this workflow is also applicable for the miRGO tool) to predict miRNA functional networks.

1. Click "miRFunction->miRPathway" to open the "KEGG Analysis for miRNA target genes" page.
2. As described in Sect. 13.3.2.1, select the clade, genome, and database of interest.

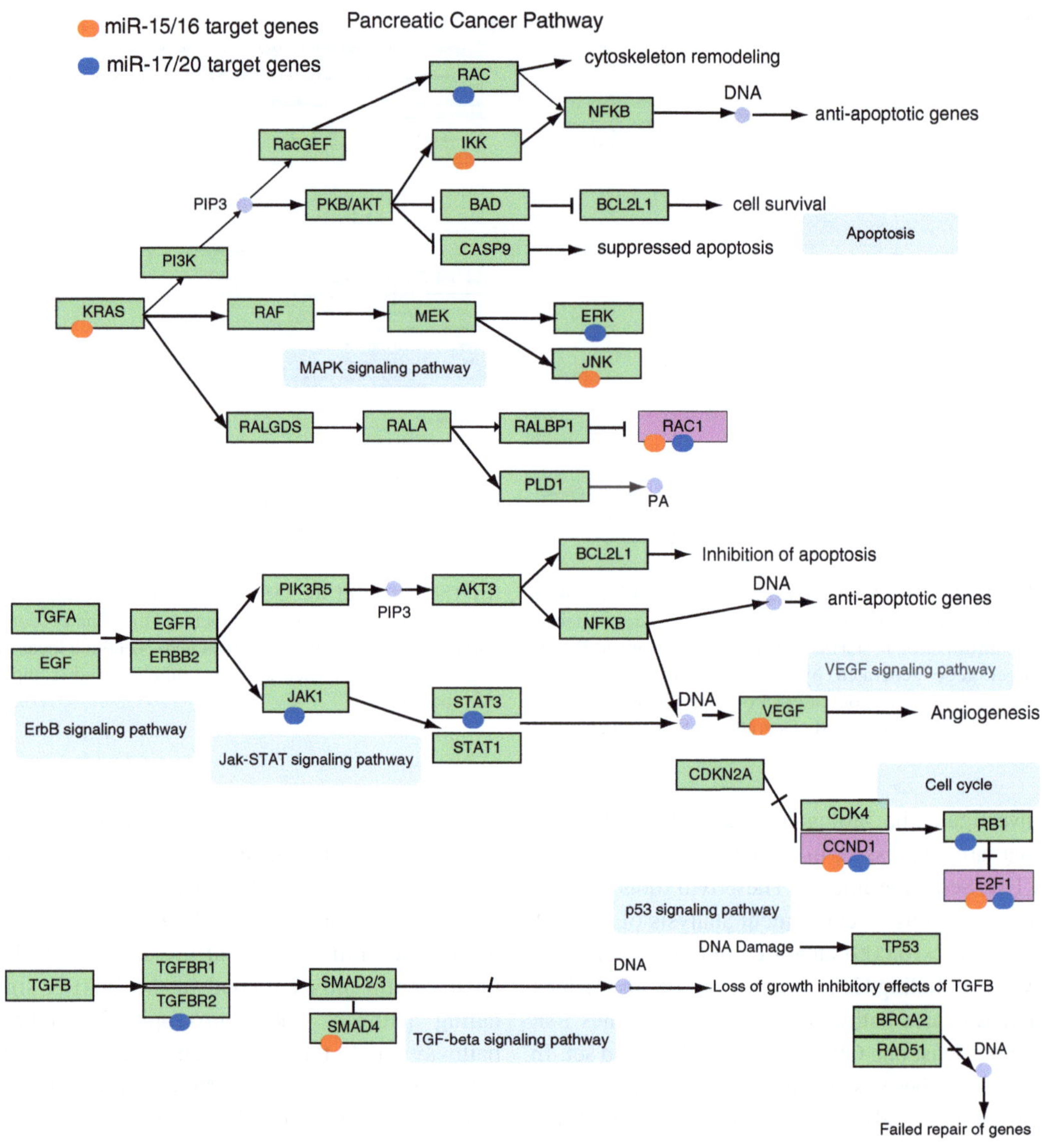

Fig. 13.5 Pancreatic cancer pathway targeted by miR-15/16 and miR-17/20. *Yellow circles* represent miR-15/16 targets and *blue circles* represent miR-17/20 targets. The Pancreatic cancer pathway shown is based on the KEGG pathway

3. You must select one or multiple miRNAs of interest to perform an enrichment analyses for miRNA target genes. You should select the minimum read number that the miRNAs should have in the CLIP-Seq experiment to be considered in your analysis.
4. Some CLIP-Seq clusters with small read numbers may simply represent experimental or biological noises. Thus, target sites may be further limited by the read number in the CLIP-Seq experiment (e.g., T→C number or read number ≥3).
5. You can select a P-value to set the significance threshold for the results. To balance sensitivity versus specificity, we considered several significance levels, including 0.1, 0.05, 0.01, 0.005, and 0.001.
6. Finally, you can click "Search" to see a list of all pathways for miRNA targets (Fig. 13.6b).

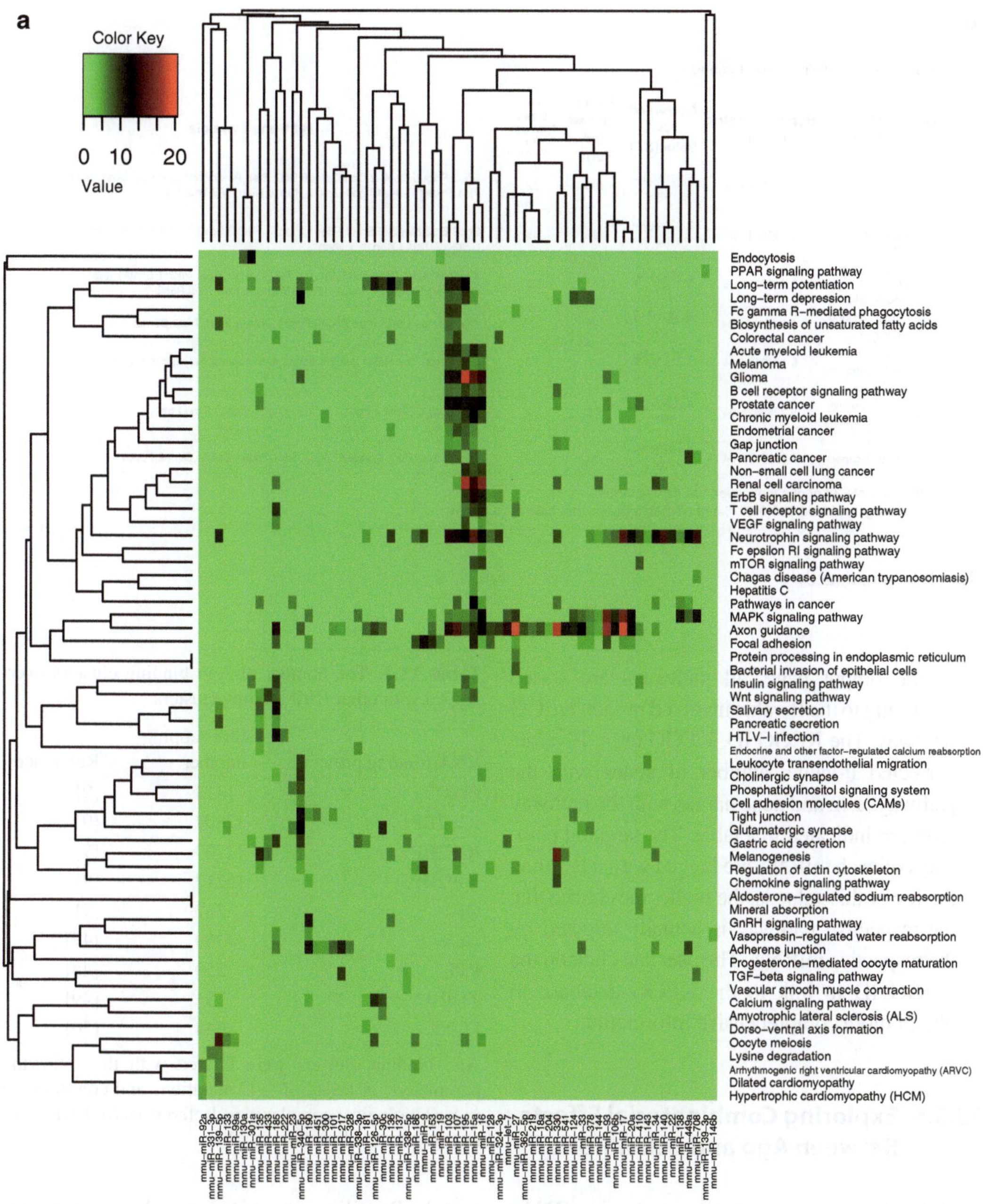

Fig. 13.6 Enrichment analysis of miRNA functional networks. (**a**) A heat map derived from KEGG-enrichment analysis of genes targeted by miRNAs with >=5,000 reads. The tree shows the hierarchical clustering of miRNAs based on their involvement in KEGG pathways. All target genes are overlapping with HITS-CLIP (mouse brain) clusters with BC >=2 and >=5 reads. Bonferroni-corrected P-value <0.05 was used to denote the significance of the enrichment in a pathway. We performed a 2-way hierarchical clustering with -log2(P-value) using the heatmap.2 function with the Pearson correlation coefficient and average linkage clustering in the R package [54]. The *color key* in the heatmap represent the -log2(p-value). For miRNAs from the same family, only one was selected for the hierarchical clustering. (**b**) miRPathway output page. The page for enrichment analysis of mmu-miR-9, showing KEGG IDs, KEGG terms, the Bonferroni-corrected P-values and target genes

b

KEGG Analysis for miRNA target genes

KEGG ID	KEGG Term	Hypergeometric P-value	Corrected P-value(BF)	Gene Number for Term	Hit Gene Number for Term	Hit Gene Symbols
mmu04510	Focal adhesion	3.613024e-07	5.925360e-05	196	22	Shc2,Pdgfrb,Col4a2,Ccnd2,Itga6,Bcl2,Pten,Pak6,Ppp1cb,Pik3r3,Sos1,Itga1,Tln1,Pik3r1,Col4a1,Itgb1,Gsk3b,Rap1b,Diap1,Vcl,Tnc,Pak2
mmu04722	Neurotrophin signaling pathway	1.539884e-05	2.525410e-03	128	15	Shc2,Bcl2,Map3k3,Sort1,Psen1,Ptpn11,Pik3r3,Sos1,Pik3r1,Map3k1,Pdk1,Gsk3b,Arhgdia,Rap1b,Ywhab
mmu04810	Regulation of actin cytoskeleton	1.611026e-05	2.642082e-03	211	20	Arpc1a,Pdgfrb,Itga6,Pak6,Ppp1cb,Pik3r3,Sos1,Itga1,Pik3r1,Slc9a1,Itgb1,Wasf2,Myh9,Iqgap1,Diap1,Arhgef12,Gng12,Fgf10,Vcl,Pak2
mmu05222	Small cell lung cancer	3.858123e-04	6.327321e-02	85	10	Col4a2,Itga6,Bcl2,Pten,Pik3r3,Pik3r1,Col4a1,Itgb1,Traf3,Ccne2
mmu05414	Dilated cardiomyopathy	5.107815e-04	8.376817e-02	88	10	Dag1,Itga1,Cacna2d1,Adcy5,Slc8a1,Itgb1,Lmna,Dmd,Itga6,Cacnb4
mmu05412	Arrhythmogenic right ventricular cardiomyopathy (ARVC)	5.285682e-04	8.668518e-02	73	9	Itga6,Dag1,Cacnb4,Itga1,Cacna2d1,Itgb1,Slc8a1,Dmd,Lmna
mmu05215	Prostate cancer	5.592315e-04	9.171397e-02	89	10	Pdgfrb,Bcl2,Pten,Creb3l2,Pik3r3,Sos1,Pik3r1,Gsk3b,Ccne2,Foxo1

mmu-miR-9 target 638 genes that cover 164 KEGG terms.
mmu-miR-9 targets picked 638(3.207%) of all 19891 genes that having GO terms.

Fig. 13.6 (continued)

In the output page, all pathways are sorted according to the P-value involved in each KEGG Pathway. The KEGG ID, KEGG term, P-value, corrected P-value, number of genes with the pathway term and gene names with the pathway term are indicated in a table. The targeted pathways of each miRNA are tagged with a P-value; the lower this value is, the higher the chance that the respective pathway is actually targeted by the respective miRNA. The user can click on the KEGG ID to link to the KEGG database to obtain more comprehensive information.

13.3.5 Exploring Combinatorial Effects Between Ago and Other RBPs

RNA-binding proteins (RBPs) and microRNAs (miRNAs) often stimulate and inhibit gene expression by binding to the 3′-UTR regions of target mRNAs [43, 44]. Their combinatorial effects are often linked to differentiation or oncogenesis [43–45]. Recent studies have shown that many RBPs can affect the efficiency of miRNA regulation by controlling miRNA accessibility [45, 46]. In particular, recent studies have shown that the Pumilio protein is required for the regulation of cel-let-7 [46] and hsa-miR-221/222.

To explore the possible relationships among binding sites of Ago and other RBPs, we intersected 39,514 stringent Ago binding clusters (>=3 T→C mutations and BC>=2) with other RBP clusters obtained from original papers (Table 13.1). We identified thousands of binding sites that bind both Ago and other RBPs, including

Table 13.1 The number of Ago binding clusters overlapped with other RBP binding clusters

RNA-binding protein	Overlapping number	References
PUM2	1313	[9]
IGF2BP1	12802	[9]
IGF2BP2	12709	[9]
IGF2BP3	16212	[9]
QKI	367	[9]
HuR	3360	[47]
PTB	550	[48]
TDP-43	751	[49]
FOX2	734	[50]

Ago binding clusters were filtered with the following parameters: at least 3 T→C mutations and conservation in >=2 biological complexities. References for RBP binding clusters are listed in the table

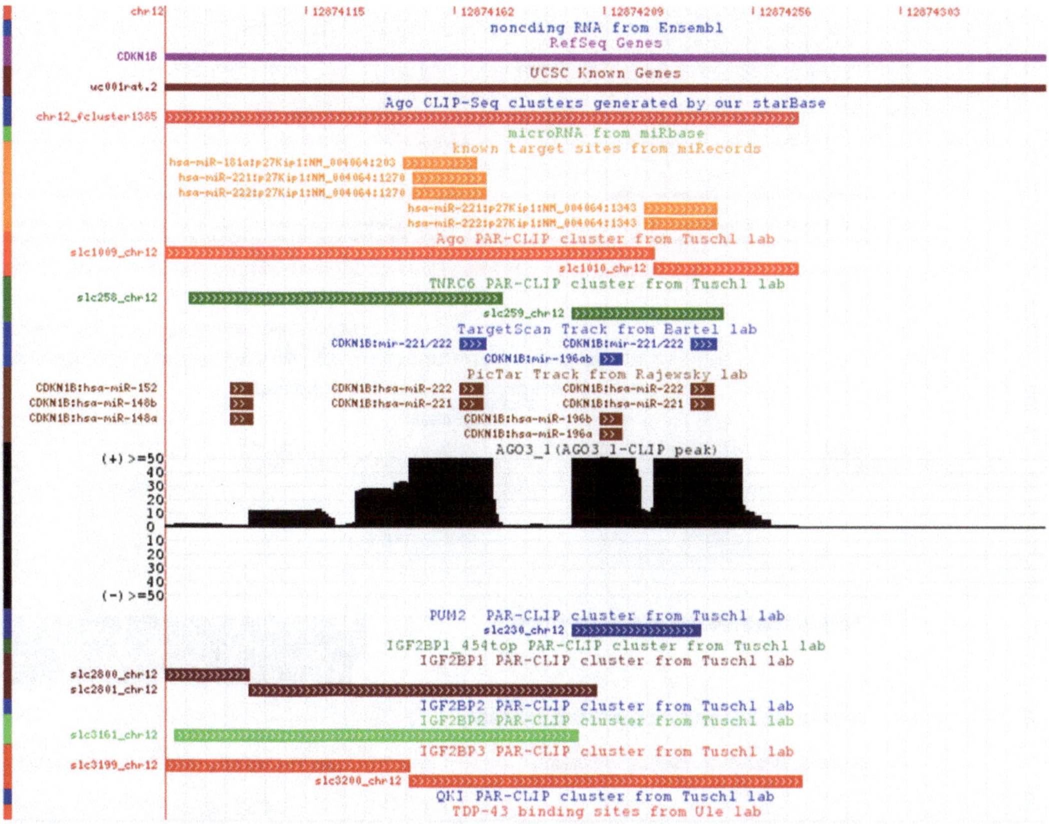

Fig. 13.7 Combinatorial effects among Ago, Pum2 and IGF2BP. Pum2 binding cluster overlaps with the known miR-221/222 target site and IGFBP1-3 binding clusters. The binding maps of various RNA Binding Proteins (RBPs) and their combinatorial effects can be simultaneously and visually inspected in our deepView genome browser by selecting the RBP binding track of interest

PUM2 [9], IGF2BP1 [9], IGF2BP2 [9], IGF2BP3 [9], QKI [9], HuR [47], PTB [48], TDP-43 [49], FOX2 [50] (Table 13.1). Approximately 30% of Ago/TNRC6 binding clusters overlap with binding sites of the IGF2BP family. The majority of RBPs have >500 clusters that overlap with AGO clusters. These data show that miRNA binding sites can themselves be the target of multiple RBPs. Importantly, the above-mentioned Pumilio binding sites can also be found in our data (Fig. 13.7). Interestingly, we found that IGF2BP1-3 proteins also bind to target sites of miR-221/222, suggesting that combinatorial effects of Ago, Pum and IGF2BP proteins may control miR-221/222 accessibility (Fig. 13.7).

In starBase, we also provide protein-RNA interaction maps and a website to explore relationships between Ago and other RBPs. One can click "protein-RNA->AgoRBPs" to open the web page for the interplays between Ago and other RBPs. As described in the above-mentioned tutorials, you can very easily access these data and select the Ago-RBP binding sites of interest.

13.3.6 Comparative Analysis of microRNA Targets Using the deepView Genome Browser

The large amount of candidate miRNA targets and high-throughput CLIP-Seq and Degradome-Seq data has increased the demand for visual tools that allow rapid visual correlation of different types of information. To enable the user to browse seamlessly along the genome and to zoom effortlessly in a very large set of CLIP-Seq and

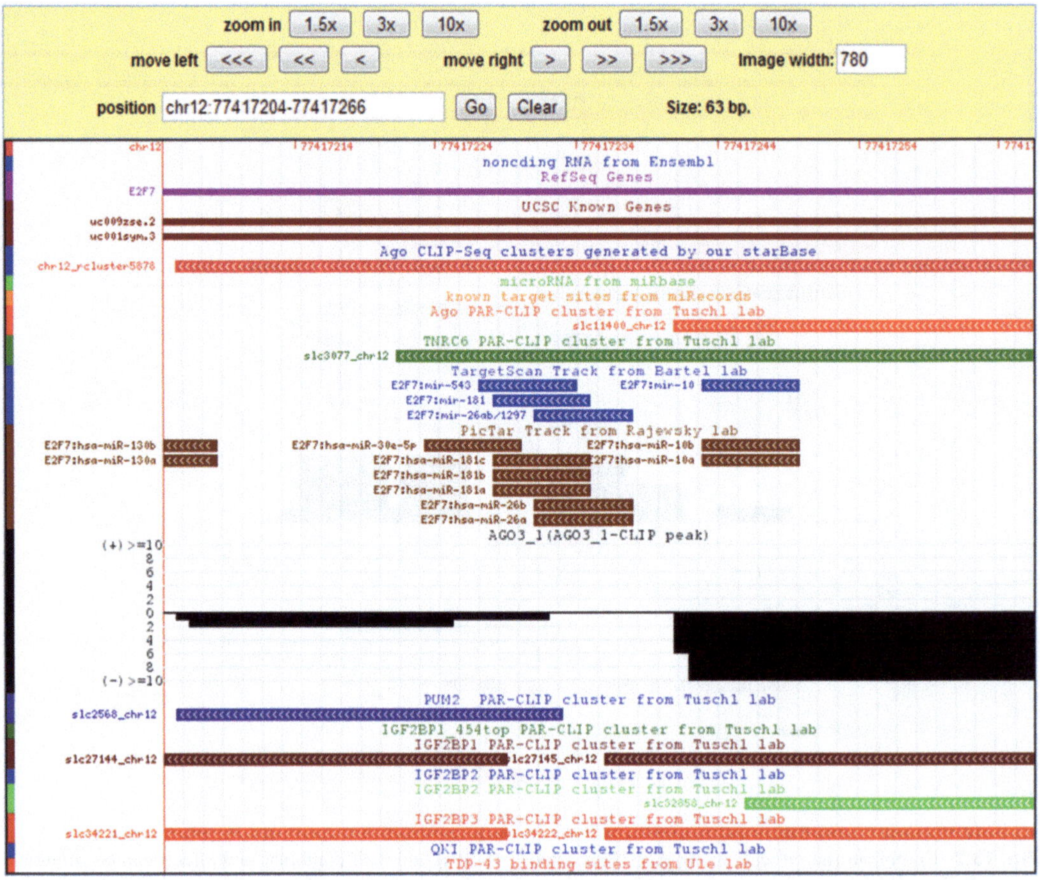

Fig. 13.8 The deepView browser page displaying the chr12:77417204–77417266 region in the human genome (UCSC hg19). The navigation buttons are visible at the *top* of the image. The deepView browser provides an integrated view of mapped reads, predicted and known miRNA targets, ncRNAs, protein-coding genes, target clusters, target-peaks, target-plots and RBP binding clusters. This page also displays combinatorial effects of Ago, Pum, HuR and IGF2BP proteins that may combinatorially control miRNA accessibility

Degradome-Seq data, the improved deepView genome browser was developed to provide an integrated view of mapped reads, predicted and known miRNA targets, ncRNAs, protein-coding genes, target clusters, target-peaks, target-plots and RBP binding clusters. See Fig. 13.8 for a sample screenshot for the output of the deepView browser. The user can zoom into a region of interest and proceed to a detailed view of the track item within the browser by clicking on it. Moreover, the "zoom out" or "zoom in" button can be used to extend or shrink the width of the displayed coordinate range.

To determine miRNA target sites, the user can type the symbol or name of the target gene in the position text-box, and then click the "Go" button to change the deepView genome browser display image completely. In the image, the user can view the peak patterns generated from different CLIP-Seq and Degradome-Seq experiments to determine the bona fide binding site or distinguish true miRNA cleavage sites from background noise. Moreover, the user can further filter the candidate target site by considering only those site that were predicted by other programs as well. In addition, the user can click a check box to display the RBP binding track of interest and then determine whether this site can be regulated by other RBPs.

13.4 Conclusions

In this chapter, we have comprehensively explored how to analyze CLIP-Seq and Degradome-Seq data, miRNA regulatory networks, and interplay between Ago and other RBPs using our starBase platform. Unlike other databases or tools that predict miRNA regulatory networks using computationally predicted miRNA targets [51], starBase provides enhanced resolution to determine miRNA functional networks based on miRNA-target interactions overlapping with high-throughput CLIP-Seq and Degradome-Seq data. Moreover, the numerous combinatorial binding sites between Ago and other RBPs identified in this study have shown a complex post-transcriptional operon system, in which RBPs and miRNAs coordinately regulate mRNA transcripts [52].

The dataset and tools can be freely downloaded from starBase platform. The dataset can also be viewed, along with additional annotation information, with our deepView genome browser. These tools and diverse resources should serve as important materials for future studies to elucidate the miRNA regulatory networks and for investigations into the biological functions of genes and ncRNAs whose expression is controlled by miRNAs and RBPs.

Acknowledgments This research is supported by the National Natural Science Foundation of China (No. 30830066, 30900820); Ministry of Science and Technology of China, National Basic Research Program (No. 2011CB811300); the funds from Guangdong Province (No. S2012010010510); The project of Science and Technology New Star in ZhuJiang Guangzhou city (No. 2012J2200025); Fundamental Research Funds for the Central Universities (No. 2011330003161070); China Postdoctoral Science Foundation (No. 200902348).

References

1. Bartel DP (2009) MicroRNAs: target recognition and regulatory functions. Cell 136:215–233
2. Filipowicz W, Bhattacharyya SN, Sonenberg N (2008) Mechanisms of post-transcriptional regulation by microRNAs: are the answers in sight? Nat Rev Genet 9:102–114
3. Inui M, Martello G, Piccolo S (2010) MicroRNA control of signal transduction. Nat Rev Mol Cell Biol 11:252–263
4. Thomas M, Lieberman J, Lal A (2010) Desperately seeking microRNA targets. Nat Struct Mol Biol 17:1169–1174
5. Rajewsky N (2006) microRNA target predictions in animals. Nat Genet 38(Suppl):S8–S13
6. Ambros V (2004) The functions of animal microRNAs. Nature 431:350–355
7. Chi SW, Zang JB, Mele A et al (2009) Argonaute HITS-CLIP decodes microRNA-mRNA interaction maps. Nature 460:479–486
8. Zisoulis DG, Lovci MT, Wilbert ML et al (2010) Comprehensive discovery of endogenous Argonaute binding sites in Caenorhabditis elegans. Nat Struct Mol Biol 17:173–179
9. Hafner M, Landthaler M, Burger L et al (2010) Transcriptome-wide identification of RNA-binding protein and microRNA target sites by PAR-CLIP. Cell 141:129–141
10. German MA, Pillay M, Jeong DH et al (2008) Global identification of microRNA-target RNA pairs by parallel analysis of RNA ends. Nat Biotechnol 26:941–946
11. Addo-Quaye C, Eshoo TW, Bartel DP et al (2008) Endogenous siRNA and miRNA targets identified by sequencing of the Arabidopsis degradome. Curr Biol 18:758–762
12. Yang JH, Li JH, Shao P et al (2011) starBase: a database for exploring microRNA-mRNA interaction maps from Argonaute CLIP-Seq and Degradome-Seq data. Nucleic Acids Res 39:D202–D209
13. Griffiths-Jones S, Saini HK, van Dongen S et al (2008) miRBase: tools for microRNA genomics. Nucleic Acids Res 36:D154–D158
14. Kersey PJ, Lawson D, Birney E et al (2010) Ensembl Genomes: extending Ensembl across the taxonomic space. Nucleic Acids Res 38:D563–D569
15. Rhead B, Karolchik D, Kuhn RM et al (2010) The UCSC Genome Browser database: update 2010. Nucleic Acids Res 38:D613–D619
16. Lewis BP, Burge CB, Bartel DP (2005) Conserved seed pairing, often flanked by adenosines, indicates that thousands of human genes are microRNA targets. Cell 120:15–20
17. Lewis BP, Shih IH, Jones-Rhoades MW et al (2003) Prediction of mammalian microRNA targets. Cell 115:787–798
18. Krek A, Grun D, Poy MN et al (2005) Combinatorial microRNA target predictions. Nat Genet 37:495–500
19. John B, Enright AJ, Aravin A et al (2004) Human MicroRNA targets. PLoS Biol 2:e363
20. Kertesz M, Iovino N, Unnerstall U et al (2007) The role of site accessibility in microRNA target recognition. Nat Genet 39:1278–1284
21. Miranda KC, Huynh T, Tay Y et al (2006) A pattern-based method for the identification of MicroRNA binding sites and their corresponding heteroduplexes. Cell 126:1203–1217
22. Xiao F, Zuo Z, Cai G et al (2009) miRecords: an integrated resource for microRNA-target interactions. Nucleic Acids Res 37:D105–D110

23. Barrett T, Troup DB, Wilhite SE et al (2009) NCBI GEO: archive for high-throughput functional genomic data. Nucleic Acids Res 37:D885–D890
24. Wu L, Zhang Q, Zhou H et al (2009) Rice MicroRNA effector complexes and targets. Plant Cell 21:3421–3435
25. Pantaleo V, Szittya G, Moxon S et al (2010) Identification of grapevine microRNAs and their targets using high throughput sequencing and degradome analysis. Plant J 62:960–976
26. Zhou MGL, Li P, Song X, Wei L, Chen Z, Cao X (2010) Degradome sequencing reveals endogenous small RNA targets in rice (Oryza sativa L. ssp. indica). Front Biol 5:67–90
27. Ashburner M, Ball CA, Blake JA et al (2000) Gene ontology: tool for the unification of biology. The Gene Ontology Consortium. Nat Genet 25:25–29
28. Kanehisa M, Goto S (2000) KEGG: kyoto encyclopedia of genes and genomes. Nucleic Acids Res 28:27–30
29. Castillo-Davis CI, Hartl DL (2003) GeneMerge–post-genomic analysis, data mining, and hypothesis testing. Bioinformatics 19:891–892
30. Yang JH, Shao P, Zhou H et al (2010) deepBase: a database for deeply annotating and mining deep sequencing data. Nucleic Acids Res 38:D123–D130
31. Zhang C, Darnell RB (2011) Mapping in vivo protein-RNA interactions at single-nucleotide resolution from HITS-CLIP data. Nat Biotechnol 29:607–614
32. Li H, Durbin R (2009) Fast and accurate short read alignment with Burrows-Wheeler transform. Bioinformatics 25:1754–1760
33. Langmead B, Trapnell C, Pop M et al (2009) Ultrafast and memory-efficient alignment of short DNA sequences to the human genome. Genome Biol 10:R25
34. Li H, Handsaker B, Wysoker A et al (2009) The Sequence Alignment/Map format and SAMtools. Bioinformatics 25:2078–2079
35. Iorio MV, Croce CM (2009) MicroRNAs in cancer: small molecules with a huge impact. J Clin Oncol 27:5848–5856
36. Cimmino A, Calin GA, Fabbri M et al (2005) miR-15 and miR-16 induce apoptosis by targeting BCL2. Proc Natl Acad Sci U S A 102:13944–13949
37. Mayr C, Hemann MT, Bartel DP (2007) Disrupting the pairing between let-7 and Hmga2 enhances oncogenic transformation. Science 315:1576–1579
38. O'Donnell KA, Wentzel EA, Zeller KI et al (2005) c-Myc-regulated microRNAs modulate E2F1 expression. Nature 435:839–843
39. Addo-Quaye C, Miller W, Axtell MJ (2009) CleaveLand: a pipeline for using degradome data to find cleaved small RNA targets. Bioinformatics 25:130–131
40. Hoffmann S, Otto C, Kurtz S et al (2009) Fast mapping of short sequences with mismatches, insertions and deletions using index structures. PLoS Comput Biol 5:e1000502
41. Jones-Rhoades MW, Bartel DP (2004) Computational identification of plant microRNAs and their targets, including a stress-induced miRNA. Mol Cell 14:787–799
42. Allen E, Xie Z, Gustafson AM et al (2005) microRNA-directed phasing during trans-acting siRNA biogenesis in plants. Cell 121:207–221
43. George AD, Tenenbaum SA (2006) MicroRNA modulation of RNA-binding protein regulatory elements. RNA Biol 3:57–59
44. Kishore S, Luber S, Zavolan M (2010) Deciphering the role of RNA-binding proteins in the post-transcriptional control of gene expression. Brief Funct Genomics 9:391–404
45. Kedde M, van Kouwenhove M, Zwart W et al (2010) A Pumilio-induced RNA structure switch in p27-3′ UTR controls miR-221 and miR-222 accessibility. Nat Cell Biol 12:1014–1020
46. Nolde MJ, Saka N, Reinert KL et al (2007) The Caenorhabditis elegans pumilio homolog, puf-9, is required for the 3′UTR-mediated repression of the let-7 microRNA target gene, hbl-1. Dev Biol 305:551–563
47. Lebedeva S, Jens M, Theil K et al (2011) Transcriptome-wide analysis of regulatory interactions of the RNA-binding protein HuR. Mol Cell 43:340–352
48. Xue Y, Zhou Y, Wu T et al (2009) Genome-wide analysis of PTB-RNA interactions reveals a strategy used by the general splicing repressor to modulate exon inclusion or skipping. Mol Cell 36:996–1006
49. Tollervey JR, Curk T, Rogelj B et al (2011) Characterizing the RNA targets and position-dependent splicing regulation by TDP-43. Nat Neurosci 14:452–458
50. Yeo GW, Coufal NG, Liang TY et al (2009) An RNA code for the FOX2 splicing regulator revealed by mapping RNA-protein interactions in stem cells. Nat Struct Mol Biol 16:130–137
51. Papadopoulos GL, Alexiou P, Maragkakis M et al (2009) DIANA-mirPath: integrating human and mouse microRNAs in pathways. Bioinformatics 25:1991–1993
52. Keene JD (2007) RNA regulons: coordination of post-transcriptional events. Nat Rev Genet 8:533–543
53. Shannon P, Markiel A, Ozier O et al (2003) Cytoscape: a software environment for integrated models of biomolecular interaction networks. Genome Res 13:2498–2504
54. Gentleman RC, Carey VJ, Bates DM et al (2004) Bioconductor: open software development for computational biology and bioinformatics. Genome Biol 5:R80

Discovering Functional microRNA-mRNA Regulatory Modules in Heterogeneous Data

14

Bing Liu, Lin Liu, Anna Tsykin, Gregory J. Goodall, Murray J. Cairns, and Jiuyong Li

Abstract

microRNAs (miRNAs) are small non-coding RNAs that cause mRNA degradation and translation inhibition. They are pivotal regulators of development and cellular homeostasis through their control of diverse processes. Recently, great efforts have been made to elucidate many targets that are affected by miRNAs, but the functions of most miRNAs and their precise regulatory mechanisms remain elusive. With more and more matched expression profiles of miRNAs and mRNAs having been made available, it is of great interest to utilize both expression profiles and sequence information to discover the functional regulatory networks of miRNAs and their target mRNAs for potential biological processes that they may participate in. In this chapter, we first briefly review the computational methods for discovering miRNA targets and miRNA-mRNA regulatory modules, and then focus on a method of identifying functional miRNA-mRNA regulatory modules by integrating multiple data sets from different sources.

Keywords

miRNAs • Functional miRNA-mRNA regulatory modules (FMRMs) • Corr-LDA • Cancer • Microarray

B. Liu (✉)
Children's Cancer Institute Australia for Medical Research, Lowy Cancer Research Centre, University of New South Wales, High St., Randwick, NSW 2031, Australia
e-mail: BLiu@ccia.unsw.edu.au

L. Liu • J. Li (✉)
School of Computer and Information Science, University of South Australia, Adelaide, SA, Australia
e-mail: Jiuyong.Li@unisa.edu.au

A. Tsykin
Centre for Cancer Biology, SA Pathology, Adelaide, SA, Australia

G.J. Goodall
Centre for Cancer Biology, SA Pathology, Adelaide, SA, Australia

School of Molecular and Biomedical Science and Department of Medicine, University of Adelaide, Adelaide, SA, Australia

M.J. Cairns
School of Biomedical Sciences and Pharmacy, Faculty of Health, University of Newcastle, Newcastle, NSW, Australia

U. Schmitz et al. (eds.), *MicroRNA Cancer Regulation: Advanced Concepts, Bioinformatics and Systems Biology Tools*, Advances in Experimental Medicine and Biology 774,
DOI 10.1007/978-94-007-5590-1_14,

14.1 Introduction

The genetic material of an organism, or genome [1], plays a central role in encoding both the cellular fabric and the regulatory machinery that controls cell homeostasis and internal functions, such as DNA replication and response to environmental signals. While the genome is encoded by DNA, the complex biological processes derived from the genome involve a myriad of interacting and co-functioning RNA molecules and diverse proteins. These co-functioning groups of molecules, described as gene regulatory modules, are essential components in biological systems. In order to understand the composition of these modules and their roles in an organism, detailed investigation of gene structures, functions, and activities must be determined within individual cells and in various tissues throughout development. However, since gene structures and functions are relatively constant from one cell to another or from one species to another, it is the patterns of gene expression and its regulation or dysregulation that have the greatest consequence in normal biology and diseases.

While gene expression can be influenced by many factors, post-transcriptional gene regulation involving a type of small non-coding RNAs known as microRNAs (miRNAs) is particularly fascinating because of the breadth of their interactions and synergistic/combinatorial relationships to target genes. Increasing evidence suggests that miRNAs are pivotal regulators of development and cellular homeostasis through their control of diverse biological processes. miRNAs regulate target mRNAs causing mRNA degradation and translational inhibition to a large extend through the logic of complementary base pairing [2], thereby making fine-scale adjustments to protein outputs. Consequently, dysregulation of miRNAs may lead to human diseases.

Recent studies have reported differentially regulated miRNAs in diverse cancer types, such as breast cancer [3], lung cancer [4], prostate cancer [5], colon cancer [6], ovarian cancer [7] and head and neck cancer [8]. Thus, identifying miRNAs, their target mRNAs, and building their regulatory networks are critical for understanding normal biological processes and their roles in the development of diseases. Here, we define the functional miRNA-mRNA regulatory modules (FMRMs) as groups of interactional miRNAs and mRNAs which are believed to participate in specific biological processes. The identification of FMRMs will potentially make significant contribution to the development of gene-based therapeutic treatments and miRNA based drugs [9].

In recent years, a large number of studies have been conducted to analyze miRNA-mRNA interactions in cell culture and animal models using both low throughput and high throughput techniques. These endeavours have led to an increase in the amount of miRNA and mRNA data at both expression and sequence levels. While some validated miRNAs and their target genes have been collected in databases, such as TarBase [10] and miRecords [11], these contain only a fraction of the diversity and abundance of potential miRNA regulatory influences. Indeed, it is impractical to explore empirically all the possibilities in this combinatorial matrix. As such, a complete understanding of miRNA functions and their precise regulatory mechanisms remains elusive.

High-throughput technologies, such as microarray, mass spectrometry, and especially the newly developed next generation sequencing, have provided tremendous potential for profiling RNAs and proteins at several levels with unprecedented resolution, depth, and speed. These features of the new technologies present major bioinformatics challenges, particularly for discovering and modelling the regulatory mechanisms of molecules involved in specific biological functions by integrating heterogeneous data.

In this chapter, we first briefly review the computational methods for discovering miRNA targets and miRNA-mRNA regulatory modules, and then focus on a method of identifying FMRMs by integrating multiple data sets from different sources.

14.2 Computational Methods for miRNA Discovery

Computational approaches provide efficient ways to identify putative miRNA targets as well as their regulatory mechanisms. They facilitate experimental validation by producing statistically significant hypotheses from biological measurements. We propose the following categories of computational approaches for miRNA research: (i) miRNA target prediction [12–15], that is, to identify the targeted mRNAs of miRNAs based on intrinsic sequence homology and conservation; (ii) discovering miRNA regulatory modules (MRMs), that is, to identify a group of co-expressed miRNAs and mRNAs, either at sequence level [16], or by integrating sequence and expression profiles of miRNAs and mRNAs [17–20]; and (iii) prediction of functional miRNA regulatory modules (FMRMs), which are regulatory networks of miRNAs and their target miRNAs for specific biological processes [21–23].

14.2.1 Predicting miRNA Targets

Like most other non-protein-coding RNAs, miRNAs target mRNAs through complementary base pairing, in either complete or incomplete fashion. The preliminary task of understanding the regulatory mechanisms of miRNAs is to identify miRNAs and their target mRNAs in different species. Therefore, previous work largely focuses on the genome-wide discovery of miRNAs [12] and the prediction of putative target mRNAs [24] at the sequence level.

It has been generally believed that miRNAs bind to the 3′ untranslated regions (3′UTRs) of the target transcripts in at least one of two classes of binding patterns [25]. One class of target sites have perfect Watson-Crick complementarity to bases 2–7 at the 5′ end of miRNAs, referred to as the 'seed region'. The 'seed region' has been shown to be sufficient for miRNAs to suppress their targets without requiring significant further base pairings at the 3′ end of the miRNA. The second class of target sites has imperfect complementary base pairing at the 5′ end of the miRNA, but it is compensated via additional base pairings in the 3′ end of the miRNA. However, the 3′UTR boundaries of many mRNAs are not clearly defined in many species and it is still a undergoing project to characterize the location, extent, or splice variation of 3′UTRs in a variety of species [26]. In addition, it has been demonstrated that a transcript can contain multiple target sites for a single miRNA and a transcript can have target sites for several different miRNAs. The multiple-to-multiple relations between miRNAs and mRNAs lead to further complexity in miRNA regulatory mechanisms.

Recognition of miRNA binding sites on the basis of sequence alone is unreliable, because many potential sites are non-functional, perhaps due to incorporation into RNA secondary structure or occlusion by RNA binding proteins. Various web-based algorithms, such as miRanda [27], Pictar [28], RNA22 [29], and TargetScan [30], have been developed to predict miRNA targets, using various "rules" of base pairing, target accessibility, and evolutionary conservation of target site. However, they produce widely different lists of predictions and all suffer from high false positive and false negative rates [31]. Most algorithms apply a cross-species conservation requirement to reduce the number of false positives although this does increase the number of false negatives because not all miRNA targets are conserved [32]. Furthermore, in a study that used mass spectrometry to measure the global impact of deletion of a single microRNA, hundreds of proteins were found to respond, but the best performing target prediction algorithms, TargetScan [30] and PicTar [28], which restrict their predictions primarily to conserved sites in 3′UTRs, were reported to nevertheless have false positive rates of about 67% [33].

Another assessment was conducted by Alexiou et al. [34] who compared the performance for eight widely used target prediction programs, including EIMMo [35], miRanda [27], miRBase [15], PicTar [28], PITA [36], RNA22 [29], and TargetScan 5.0 [30], for the human and mouse genome, using experimentally validated targets in Selbach et al. [37]. They found those programs have a precision of ~50% with a sensitivity that ranges from 6 to 12%. Overall, the complex features of miRNA pose great challenges on the computational approaches for miRNA target prediction.

14.2.2 Discovering miRNA Regulatory Modules

Despite the significant increase in the number of both experimentally validated and computationally predicated miRNA targets, the majority of the miRNA targets and their responses to the miRNAs remain largely unknown. Thus, great interests have been moved to discover functions of miRNAs by identifying miRNA-mRNA regulatory modules (MRMs).

At the sequence level, Yoon and De Micheli [16] proposed a prediction method for MRMs. In their method, miRNA-mRNA regulatory relationships are first modelled as a weighted bipartite graph based on sequence binding information. With the bipartite graph, nodes represent miRNAs and mRNAs, edges stand for weights corresponding to the miRNA-mRNA binding strength. Candidate MRMs are defined as bicliques in which all the edges have similar weights. Therefore, a graph mining method is proposed to discover such bicliques in the given bipartite graphs. This is the first method that explicitly searches for the multiple to multiple relationships among miRNAs and their target genes. However, predictions based on sequence only may not be sufficient to determine the complex interactions of miRNA-mRNA pairs. The modules identified at sequence level do not necessarily function at biological levels.

In order to minimize false positive and to effectively detect MRMs, recently developed methods have integrated the analysis of expression profiles of miRNAs and mRNAs in conjunction with the predicted miRNA targets. Most of the integrative methods of MRM discovery are based on the assumption that miRNA negatively regulate their target mRNAs to the effect that an inverse relationship should exist between the expression a specific miRNA and its targets.

Huang et al. [17, 38] applied Bayesian network parameter learning to infer miRNA-mRNA interactions using both miRNA-mRNA sequence binding information and expression profiles of miRNAs and mRNAs. An initial network representing the putative target relationships between miRNAs and mRNAs are first constructed according to the target information predicted from sequence binding. Then inverse patterns of expression values between miRNAs and mRNAs are encoded in this network where the changes of mRNA expression follow a Gaussian distribution, and they are a summation of negatively weighted changes of the expression of their regulator miRNAs. The Gaussian Bayesian network parameter learning is used to infer the probabilities of miRNAs relating to their target mRNAs at the expression level. This method explicitly encodes the inverse expression patterns between miRNAs and their target mRNAs in the network. It reflects the findings of miRNA regulatory mechanisms in the early stage of miRNA research. Furthermore, this model searches co-expressed miRNAs and mRNAs which are presumed to function together. Thus, this model detects not only miRNA targets but also co-functional miRNAs and mRNAs in certain biological processes.

Joung et al. [18] improved Yoon's method in [16] by integrating sequence and expression profiles of miRNAs and mRNAs to discover MRMs. They proposed a population-based probabilistic learning model to identify synergistic miRNAs involved in the regulation of their targets. More specifically, this method employs a genetic algorithm to search for a subset of miRNAs and mRNAs with a best fitness score. The fitness score is a balanced aggregation of the binding strengths of miRNAs and their targets at the sequence level, the expression coherence score of miRNA in the miRNA subsets, and the expression coherence score of mRNA in the selected mRNA subsets.

Tran et al. [19] proposed a rule based method for identifying MRMs. This method is based on an assumption that genes regulated by the same miRNAs show similar expression profiles. This method first utilizes the putative targets of miRNAs predicated by PicTar to construct the miRNA-mRNA relationships at the sequence level. Then, the method calculates the Pearson's

Correlations of miRNAs using their pair-wise expression values. The correlation calculation is also applied to mRNAs. A correlation table denoting similar or dissimilar miRNAs and their targets is constructed for each miRNA constrained by the miRNA and mRNA target relationships predicted at the sequence level. Finally, the CN2-SD rule induction method [39] is applied to search for groups of miRNAs and mRNAs that have similar expression patterns.

Peng et al. [20] used simultaneously profiled expression of cellular miRNAs and mRNAs to construct a miRNA-mRNA regulatory network. This network comprises of two disjoint sets of miRNAs and mRNAs. A connection is made between a pair of miRNA and predicted target mRNA if there is a significant inverse correlation between their expression profiles. A network structure, named biclique where every miRNA is connected to every mRNA in the given set of miRNAs and mRNAs, is considered to be a candidate regulatory module. The statistical significance of all bicliques is systematically assessed through a permutation test in the given network, and the statistically significant bicliques are the final MRMs.

Recently, Bonnet et al. [40] proposed to combine several techniques to discover MRMs using expression profiles of miRNAs and mRNAs only. Their method involves two steps. In the first step, multiple clusters of co-expressed genes are identified. A Gibbs sampling approach is used for a two-way clustering of both genes and conditions in order to avoid the local optima. In the second step, a set of regulator genes are identified for each cluster using a fuzzy decision tree model. In the decision trees, each node defines a split between two sets of conditions. The major regulators are assigned to those genes with the most significant counts in the multiple clusters estimated by a probabilistic model. This method, however, considers the expression profiles of miRNAs and mRNAs as a single data matrix, thus ignores the nature of their differences.

The above methods aim at exploring general miRNA-mRNA regulatory modules by integrating expression information of miRNAs and mRNAs with or without considering sequence information. They have archived variant successes on different trail data sets. However, they identify groups of co-expressed miRNAs and mRNAs without considering the biological conditions of the samples. Therefore, no information regarding the functions of MRMs can be identified by the afore-described methods directly. Thus, the functions of MRMs usually are unclear until a functional enrichment analysis is conducted by querying the identified target genes against the Gene Ontology (GO) or other similar annotation databases [16, 19]. Those biological conditions are very important in biological experimental design, and hence, some conditionally related MRMs may be missed out if we do not take into account the conditions. This question, however, is of great interests in understanding the biological pathways of MRMs.

14.2.3 Inferring Functional miRNA Regulatory Modules

In order to resolve some of the limitations of MRM outlined above, we proposed the concept of functional miRNA-mRNA regulatory modules (FMRMs) [21]. FMRMs explicitly indicate how groups of miRNAs regulate their target mRNAs and how they co-act together to form pathways in complex regulatory networks with significance for specific conditions.

In this work, we proposed a putative miRNA regulatory network model based on a bipartite graph, where an unbiased connection is made between a specific miRNA and its predicated target mRNAs independent of the inverse expression relationship assumption [21]. The FMRMs are defined as a subset of the bipartite graph in which the expression profiles of both miRNAs and mRNAs are associated with specific biological conditions. In order to discover the FMRMs, firstly, this method searches for a set of maximal bicliques in the given bipartite graphs defined by the predicated target information. Then association rule mining is applied to the maximal bicliques but using the expression data to discover the association between specific biological conditions and the inverse expression pattern of miRNAs and mRNAs. The association relationships among miRNAs, mRNAs, and conditions are merged to form the final FMRMs. This work is the first published work to explicitly discover FMRMs.

Joung and Fei [22] proposed a probabilistic graphical model to identify FMRMs. It is a generative model directly adopted from the Author-Topic model [41] used in information retrieval. It models the miRNA-mRNA regulatory mechanism as hierarchical steps in which the FMRMs are defined as functional clusters of miRNAs with their target mRNAs involved in the same biological processes. In this model, each mRNA has events of its expression in a specific condition that is likely to be associated with the expression events of miRNAs. A hierarchical generative process hypothesizes that a miRNAs is sampled from a multinomial distribution over FMRMs, and then the sampled miRNA is used to sample a mRNA which has a multinomial distribution over conditions. An approximate method, Gibbs sampling, is used to infer the FMRMs because the exact inference of the model is intractable. This method integrates datasets from diverse sources including miRNA target information and expression profiles of mRNAs. The drawback of this method is that it does not use the expression profiles of miRNAs. Thus, the regulatory relationships of miRNAs and mRNAs largely rely on the miRNA target information predicated at the sequence level, instead of at the expression level.

We also targeted this problem and proposed a Bayesian network (BN) based method to discover FMRMs with complex miRNA-mRNA interactions [23]. It is designed to explore all possible miRNA-mRNA interactions by integrating miRNA target information, expression profiles of miRNAs and mRNAs, and sample categories. In order to capture all possible interactions, it splits expression profiles of miRNAs and mRNAs according to sample categories, and then builds BNs on separate data sets. Interaction networks identified using individual data sets are then integrated by BN averaging procedure. To avoid statistically insignificant results due to small data sets, it employs bootstrapping to achieve reliable inference and integration.

In order to capture the correspondence between miRNAs and mRNAs, we proposed another probabilistic graphical model based on Correspondence Latent Dirichlet Allocation (Corr-LDA) [42] to discover FMRMs [43]. This approach enables the integration of heterogeneous data sets, including expression profiles of miRNAs and mRNAs, with or without the prior target binding information. In this method, FMRMs are dependent groups of miRNAs and mRNAs linked by the assumption that they participate in the same latent functions. The following sections will focus on the details of this method and its application for inferring FMRMs.

14.3 Data and Problem Definition

A large amount of data for miRNAs and mRNAs at both sequence and expression levels has been accumulated and pending for analysis. On one hand, miRNA target prediction programs usually produce hundreds to thousands of putative targets for each miRNA according to the base pairing at the sequence level. Among them, some have been validated by biological experiments, but the majority of targets remain uncertain. The predicted miRNA target information usually is organized as a table where each row indicates a target pair of a miRNA and mRNA. Various other information may also be available in the row such as the binding strength between the miRNA and its target mRNA. It provides a potential relationship between miRNAs and mRNAs. On the other hand, microarray experiments have been able to profile not only mRNAs but also miRNAs at the expression level. The expression profiles of miRNAs and mRNAs are usually organized as two dimensional tables as well where columns stand for samples and rows denote miRNAs/mRNAs. Each cell of the expression table is the expression value of the specific miRNA/mRNA in a sample. Each sample may belong to different categories, such as cancers or normal tissues. It is of great interest to integrate the miRNA putative target information, paired expression profiles of miRNAs and mRNAs, and the sample information to investigate the miRNA regulatory mechanisms,

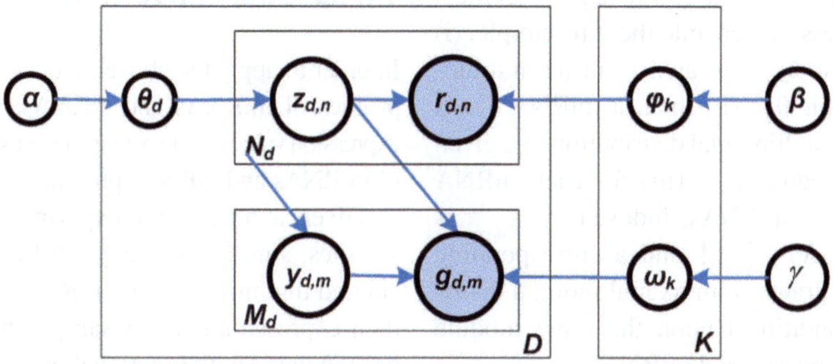

Fig. 14.1 Generative model of FMRM discovery. This model is illustrated with a plate notation (details in context). Given expression data of miRNAs and mRNAs of D samples, each sample d is a mixture of random miRNAs and mRNAs. Each miRNA $r_{d,n}$ and mRNA $g_{d,m}$ are generated from one of the K latent functional modules, indexed by $z_{d,n}$

that is, to understand how miRNAs regulate their target mRNAs and identify groups of miRNAs and their target mRNAs that contribute to specific biological processes.

The following section will introduce the techniques of integrating these heterogeneous data to discover FMRMs with Corr-LDA, and also demonstrate how to interpret the results in terms of biological perspective.

14.4 Inferring FMRMs with Corr-LDA

Given the expression profiles of miRNAs and mRNAs for matched samples and putative target information linking miRNA and mRNA, we assume that there are functional modules (FMRMs) governing miRNA and mRNA expression under the prevailing biological conditions. We model FMRMs with latent random variables which act as a bridge between miRNAs and mRNAs. By inferring the latent variables, we can identify FMRMs.

14.4.1 Modelling FMRMs

Specifically, we model FMRMs with a probabilistic generative process. Given the K latent functions presented in the samples, our method considers miRNAs and mRNAs as observations generated from a probabilistic process over these K functions. Thus, each sample is a random mixture of miRNAs and mRNAs associated with K functional modules. By inferring the probability distributions of the latent variables, we are able to obtain the probabilities of how samples, miRNAs, and mRNAs are related to functional modules.

We depict the model in Fig. 14.1 with a plate notation. In this notation, nodes stand for random variables (observed variables are shaded and latent ones are unshaded); edges denote conditional dependency between random variables; and plates denote replications of a substructure with the number of repetitions given in the bottom corner (either right or left side).

In Fig. 14.1, the D samples were profiled with a set of miRNAs V and a set of mRNAs T. Random variable $r_{d,n}$ and $g_{d,m}$ denote the indexes of a miRNA and mRNA expressed in the d-th sample, respectively, where $d \in \{1,\ldots,D\}, n \in \{1,\ldots,N_d\}, m \in \{1,\ldots,M_d\}$. N_d and M_d are the total numbers of times the miRNAs and mRNAs which are expressed in the d-th sample. Random variable $z_{d,n}$ stands for the latent functional module associating with the n-th miRNA in the d-th sample. We assume that $z_{d,n}$, $r_{d,n}$, and $g_{d,m}$ all have multinomial distributions with parameters θ_d, φ_k, and ω_k, respectively. Each parameter has a Dirichlet prior with hyperparameters α, β, and γ, correspondingly.

Without considering the putative target constraints, the generative procedure for each sample

d can be illustrated by the following hierarchical sampling process: to generate the d-th sample, (i) a latent module $z_{d,n}$ is drawn from its multinomial distribution θ_d; (ii) then, a miRNA $r_{d,n}$ is drawn from its multinomial distribution φ_k, given the selected module $z_{d,n}$; (iii) for each mRNA $g_{d,m}$, one of the miRNAs, indexed by $y_{d,m}$, is selected from $R_d = \{r_{d,n}\}$ and a corresponding mRNA $g_{d,m}$ is drawn from its multinomial distribution ω_k, conditional upon the same module that generates the selected miRNA $r_{d,n}$.

When the constraint of the putative target relationship between miRNAs and mRNAs is preferred, for each mRNA, one of the miRNAs from the set of hosting miRNAs of that mRNAs is selected, and a corresponding mRNA is drawn from the multinomial distribution of mRNAs, conditional upon the same module that generates the selected miRNA.

The generative procedure without putative target constraint for each sample d is illustrated by the following sampling process:

1. Choose $\theta_d \mid \alpha \sim Dir(\alpha)$
2. Choose $\varphi_k \mid \beta \sim Dir(\beta)$
3. Choose $\omega_k \mid \gamma \sim Dir(\gamma)$
4. For each $r_{d,n}$, $n \in \{1,\ldots,N_d\}$:
   (a) Choose module $z_{d,n} \mid \theta_d \sim Mult(\theta_d)$
   (b) Choose a $r_{d,n} \mid \{z_{d,n}, \varphi_{1:K}\} \sim Mult(\varphi_{z_{d,n}})$
5. For each $g_{d,m}, m \in \{1,\cdots,M_d\}$:
   (a) Choose miRNA index $y_{d,m} \mid N \sim Unif\{1,\cdots,N_d\}$
   (b) Choose $g_{d,m} \mid \{y_{d,m}, \overline{z}, \omega_{1:K}\} \sim Mult(\omega_{z_{y_{d,m}}})$

From the above generative process, we see that the parameter $= \{\theta_d\}$ associates samples with modules, $= \{\varphi_k\}$ assigns the probability of miRNAs expressed in module $Z = \{z_{d,n}\}$, and $= \{\omega_k\}$ indicates the probability of mRNAs expressed in Z corresponding to the miRNAs. Therefore, by estimating Θ, Φ, and π, we can identify FMRMs (details in Sects. 3.3, 3.4 and 3.5).

Under this model, miRNAs can associate with any modules, but mRNAs may only associate with the modules that produce the miRNAs. In effect, this model captures the hierarchical notion that miRNAs are generated under specific FMRMs, and mRNAs are regulated by the miRNAs.

14.4.2 Data Conversion

In order to apply the above model to the expression profiles of miRNAs and mRNAs, we convert the expression values to the counts of the expression events of miRNAs and mRNAs present in the samples.

Given a microarray experiment profiled D' samples, similar to Joung and Fei [22], we considered that miRNAs and mRNAs have events of their expression in every sample that are likely to be associated with functional modules. Therefore, each miRNA or mRNA can be represented as a vector of variables, $\{s_1^+, s_2^-, \cdots, s_{D-1}^+, s_D^-\}$. It corresponds to the expression events of a miRNA or mRNA in all samples, where duplex $\{s_{2d-1}^+, s_{2d}^-\}$ indicates an over- and under- expressed miRNA or mRNA of sample d, $d \in 1,\cdots,D'$, thus, $D = 2D'$. To get the integer counts $(\sigma_{2d-1,i}, \sigma_{2d,i})$ for the duplex expression status, we convert the expression value of a miRNA or mRNA of sample d with,

$$\sigma_{2d-1,i}, \sigma_{2d,i} = \begin{cases} \lceil \varepsilon \cdot |e_{d,i} - med_d| \rceil, 0, & if\, e_{d,i} \le med_d \\ 0, \lceil \varepsilon \cdot |e_{d,i} - med_d| \rceil, & if\, e_{d,i} < med_d \end{cases} \tag{14.1}$$

where, $e_{d,i}$ is the expression value of a miRNA or mRNA in the d-th sample, ε is a scaling constant, and med_d denotes the expression median of all miRNAs or mRNAs in the d-th sample.

Then, the counts of miRNAs and mRNAs are replaced by the indexes from the set of miRNAs, V and the set of mRNAs, T. The indexes, therefore, are the random variables $r_{d,n}$ and $g_{d,m}$ used in the model (Fig. 14.1).

14.4.3 Estimating Model Parameters

The exact inference for the parameters of this model is intractable, we used the collapsed Gibbs sampling method [44] to estimate the parameters.

This method iteratively generates samples that converge to draws from a target distribution of random variables Z through integrating out the parameters Θ, Φ, and Ω for each sampling. For the d-th sample and the n-th miRNA, the sampling is expressed as a conditional probability:

$$p\left(z_{d,n} = k \mid Z_{-(d,n)}, \boldsymbol{Y}_d, \boldsymbol{R}_d, \boldsymbol{G}_d\right) \propto p\left(z_{d,n} \mid Z_{-(d,n)}\right) p\left(r_{d,n} \mid z_{d,n}\right) \prod_{m=1}^{M_d} p\left(g_{d,m} \mid y_{d,m}, Z_d\right) \propto \frac{n_{d,-(d,n)}^{(k)} + \alpha_k}{\left(\sum k = 1 K n_d^{(k)} + \alpha_k\right) - 1} \cdot \frac{n_{k,-(d,n)}^{(v)} + \beta_v}{\left(\sum v = v_1 v_P n_k^{(v)} + \beta_v\right) - 1} \cdot \frac{m_{k,-(d,n)}^{(t)} + \gamma_t}{\left(\sum t = t_1 t_Q m_k^{(t)} + \gamma_t\right) - 1} \tag{14.2}$$

where $z_{d,n}$ is the current module assignment of the n-th miRNA of the d-th sample. $Z_{-(d,n)}$ is the current module assignment of all miRNAs in all samples excluding that of the n-th miRNA of d-th sample. $n_{d,-(d,n)}^{(k)}$ is the number of times that the k-th FMRM has been observed with miRNAs across samples excluding that of the n-th miRNA of the d-th sample. $n_{k,-(d,n)}^{(v)}$ is the number of times that miRNA v is assigned to the k-th FMRM excluding that of the n-th miRNA of d-th sample. $m_{k,-(d,n)}^{(t)}$ is the number of times that mRNA t is assigned to the k-th FMRM excluding the current assignment.

After sufficient sampling, the distribution of $z_{d,n}$ converges to the target distribution of $\boldsymbol{Z}$, then we estimate the parameters Θ, Φ, and Ω based on the values of the module assignments produced from the sampling:

$$\theta_{d,k} = \frac{n_d^{(k)} + \alpha_k}{\sum_{K-1}^{K} n_d^{(k)} + \alpha_k} \tag{14.3}$$

$$\varphi_{k,v} = \frac{n_k^{(v)} + \beta_v}{\sum_{v-v_1}^{v_P} n_k^{(v)} + \beta_v} \tag{14.4}$$

$$\omega_{k,t} = \frac{m_k^{(t)} + \gamma_t}{\sum_{t-t_1}^{t_Q} m_k^{(t)} + \gamma_t} \tag{14.5}$$

14.4.4 Algorithm of Estimating Model Parameters

Unlike the preceding sampling procedure, here $n_d^{(k)}$, $n_k^{(v)}$, and $m_k^{(t)}$ are calculated from the assignment results for all data without excluding the current module. Using Eqs. (14.2), (14.3), (14.4), and (14.5), the Gibbs sampling procedure can be designed. The algorithm includes three stages: initialization, sampling, and reading out of parameters.

Algorithm: Gibbs Sampling for FMRM Discovery

***Initialization**
Assign zeros to all count variables, $n_d^{(k)}$, n_d, $n_k^{(v)}$, n_k, $m_k^{(t)}$, m_k
foreach $d \in \{1,\ldots,D\}$ do
  foreach miRNA $r_{d,n}$, $n \in \{1,\ldots,N_d\}$do
    sample FMRM index $z_{d,n} = k \sim Mult(1/K)$
      increment sample-FMRM count: $n_d^{(k)} + 1$
      increment sample-FMRM sum: $n_d + 1$
      increment FMRM-miRNA count: $n_k^{(v)} + 1$
      increment FMRM-miRNA sum: $n_k + 1$
  end for
  foreach mRNA $g_{d,m}$ $m \in [1, M_d]$ do
      sample index for FMRM index $y_{d,m} = x \sim Uniform\{1,\cdots,N_d\}$
      assign the FMRM $k = z_{d,y_{d,m}}$ to mRNA $g_{d,m}$
      increment FMRM-mRNA count: $m_k^{(t)} + 1$
      increment FMRM-mRNA sum: $m_k + 1$
  end for
***Gibbs sampling over burn-in period and sampling period**
while not converge or not reach iteration limit do
    foreach $d \in \{1,\cdots,D\}$ do
      for each miRNA $r_{d,n}$, $n \in \{1,\cdots,N_d\}$do
        *for the current assignment k to a miRNA term v indexed by miRNA $r_{d,n}$: decrement counts and sums: $n_d^{(k)} - 1$, $n_d - 1$, $n_k^{(v)} - 1$, $n_k - 1$
        *sample index $y_{d,m} = x \sim Uniform\{1,\ldots M_d\}$ for mRNA $g_{d,x_{d,n}}$, which corresponding to miRNA $r_{d,n}$

*for the current assignment of k to a term t for mRNA $g_{d,x_{d,n}}$: decrement counts and sums: $m_k^{(t)}-1$, m_k-1
*multinomial sampling according to Eq. (14.2), sample topic index $\hat{k} \sim p\left(z_{d,n} \mid Z_{-(d,n)}, Y_d, R_d, G_d\right)$
*use the new assignment of $z_{d,n}=\hat{k}$ to the miRNA term v for miRNA $r_{d,n}$ and increment counts and sum: $n_d^{(k)}+1$, n_d+1, $n_{\hat{k}}^{(v)}+1$, $n_{\hat{k}}+1$
*use the new assignment of $z_{d,n}=\hat{k}$ to the term t for mRNA $g_{d,x_{d,n}}$ and increment counts and sum: $m_{\hat{k}}^{(t)}+1$, $m_{\hat{k}}+1$
end for
end for
***Check convergence and read out parameters**
If converged and L sampling iterations since last read out then
read out parameter set Θ, Φ, and Ω according to Eqs. (14.3), (14.4), and (14.5)
end if
end while

14.4.5 Assigning Biological Conditions to Modules

The parameters inferred from this model provide insights into the data sets at several levels. Θ clusters samples into modules that should relate to the biological conditions of the experiments.

We conceive a statistical model to identify the connection between biological conditions and modules. Let C be the number of biological conditions of the D samples in the data set, and c_i be the number of samples belonging to condition i, where $\Sigma_{i=1}^{C} c_i = D$. For each module, assume there are x samples among the n highest probability samples that belong to the same condition i. The random variable x follows a hypergeometric distribution with parameters D, c_i, and n, denoted as

$$p(x) \sim hypergeometric\left(x; D, c_i, n\right) \quad (14.6)$$

We assign biological condition i to module k when x is at a statistically significant level, for example, p-value < 0.05.

14.4.6 Identifying miRNAs and mRNAs for Modules

The parameters Φ and Ω indicate the probabilities of each miRNA and mRNA participating in a FMRM. For a K-FMRMs involving P miRNAs, Φ is a $K \times P$ probability matrix where the element $\varphi_{k,v}$ indicates the likelihood that miRNA v belongs to the k-th FMRM. Similarly, Ω is a $K \times Q$ probability matrix where the element $\omega_{k,t}$ indicates the belief of mRNA t participating in the k-th FMRM, and Q is the number of mRNAs under investigation.

For each FMRM, we consider the top ranked miRNAs and mRNAs with the highest probabilities to be the participants of the FMRM.

14.5 Reconstructing miRNA-mRNA Target Relationships

We query a miRNA target database to reconstruct the target relationship of the miRNAs and mRNAs in each module. Hypothesis tests are conducted on the identified miRNAs and mRNAs to evaluate whether they are likely to have been identified by chance or not.

14.5.1 Functional and Pathway Analysis of FMRMs

Function and pathway analysis of the identified FMRMs is conducted by reviewing literature and querying the Ingenuity Pathway Analysis (IPA, www.ingenuity.com) database of functional biological pathways to identify the significantly enriched functions and pathways.

14.6 Experimental Validation

To validate this method, we apply it to a mouse mammary data set for breast cancer research. In this section, we first briefly introduce the original data set and how we prepare it for our experiment, and then we focus on presenting the analysis and results of applying our model to the data.

14.6.1 Materials and Experiments Data Sets

The data set were profiled with 46 samples derived from 9 classes of mouse models, representing one normal type and two breast cancer subtypes: basal and luminal. The expression data were screened with 1,336 probes of miRNAs (corresponding to 334 unique miRNAs) and 22,626 probes of mRNAs. For each type of the conditions, 3–7 samples were profiled with miRNAs and mRNAs. The sample name and their corresponding mouse model class and tumor subtypes are listed in Table 14.1.

The mRNA expression data were profiled with mouse genome 430A 2.0 GeneChip (Affymetrix) and scanned on Affymetrix GeneChip scanner 3,000. The microRNA microarray chip (LMT_miRNA_v2 microarray) was designed using the Sanger miR9.0 database (http://microrna.sanger.ac.uk) and manufactured by Agilent Technologies as custom-synthesized 8 × 15 k microarrays. The array contains 1,667 unique mature miRNA sequences across all species, among them, 334 unique miRNAs were for mouse. Each mature miRNA is represented by + and – (reverse complement) strand sequences, and each with four replicate probes.

14.6.2 miRNA Expression Data Normalization

The gProcessSignal values of probes designed for mouse miRNAs were feature extracted using the GE2 protocol (www.agilent.com) with exclusion of internal control probes, non-mouse probes, and all negative strand probes. A global median normalization procedure was applied to the gProcessSignal values of the selected probes across all arrays.

14.6.3 mRNA Expression Data Normalization

mRNA array data were normalized using GC-RMA of Partek Genomic Suite (www.partek.com). The normalized data were further filtered using MAS5 detection calls for probes designated as “P” (present) or “M” (Marginal) in less than 3 samples of all data population.

In order to compare with the target prediction, the expression data sets of miRNAs and mRNAs were further filtered with MicroCosm Targets V5.0 [15], and only those in MicroCosm were maintained for analysis. Consequently, 1,112 probes of miRNAs and 19,223 probes of mRNAs were used in our experiment.

14.6.4 Implementation

Given the above expression data of miRNAs and mRNAs, the input data for our model include a 1,112 × 46 matrix of miRNA expression values and a 19,223 × 46 matrix of mRNA expression values. In the following discussion, we do not consider the putative target information to avoid the bias probably incurred by the prior prediction [45].

In the experiment, the constant ε for converting the expression values was 30. After the data conversion, the number of samples D is 92. We set the number of FMRMs, K, to 20. This value is determined by the number of sample types. Our data sets were profiled with nine classes of mouse models. miRNAs and mRNAs could be over- or under- expressed in the samples so the number of sample types is 18. In addition, two extra types were added to allow the redundancy as our model could discover subtypes of classes. We set the hyperparameters α, β, and γ to ten. The number of iterations of Gibbs sampling is 2,500. These value settings are based on empirical experiments.

14.6.5 Associating FMRMs with Biological Conditions

The parameter Θ obtained with our method is a 92 × 20 probability matrix. Referring to Sect. 3.3, the element $\theta_{d,k}$ of Θ is the belief of sample d belonging to module k. We extracted the top 5% (5) ranked samples with highest probabilities in each module, and assigned the biological conditions to each module as discussed in Sect. 3.5. Figure 14.2 illustrates the probability map for the

Table 14.1 Sample information

Sample name	Mouse model	Model class	Tumor subtype
X503_BT	Brca1–/– p53 503 BT	BRCA-p53	Basal
X4176_BT	Brca1–/– p53 4176 BT	BRCA-p53	Basal
X627_BT	Brca1–/– p53 T627 BT	BRCA-p53	Basal
X572_BT	Brca1–/– p53 572 BT	BRCA-p53	Basal
X53447_BT	Brca1–/–p53447BT	BRCA-p53	Basal
C3Tag_2	C3TAg 2	C3TAg	Basal
C3Tag_4	C3TAg 4	C3TAg	Basal
C3Tag_5	C3TAg 5	C3TAg	Basal
C3Tag_7	C3TAg 7	C3TAg	Basal
C3Tag_8	C3TAg 8	C3TAg	Basal
P53_1570	p53–/– 1570R_PN1b	p53	Basal
P53_2979	p53–/– 2979R_PN1b	p53	Basal
P53_5354	p53–/– 5354L_PN10	p53	Basal
P53_5809	p53–/– 5809R_PN2(254c)	p53	Basal
P53_5817	p53–/– 5817_PN2(254c)	p53	Basal
P53_5851	p53–/– 5851L_PN2(254c)	p53	Basal
P53_8546	p53–/– 8546R_PN1b	p53	Basal
Cmyc_043508	C-Myc Tumor 043508	C-Myc	Luminal
Cmyc_04004022	C-Myc Tumor 04004022	C-Myc	Luminal
Cmyc_04005648	C-Myc Tumor 04005648	C-Myc	Luminal
Cmyc_04004021	C-Myc Tumor 04004021	C-Myc	Luminal
H2N_Founder_A	MMTV-H2N Founder A	H2N	Luminal
H2N_53	MMTV-H2N 53	H2N	Luminal
H2N_1	MMTV-H2N 1	H2N	Luminal
H2N_61	MMTV-H2N 61	H2N	Luminal
H2N_64	MMTV-H2N 64	H2N	Luminal
Hras_1.4	MMTV-Haras An #1-4	Hras	Luminal
Hras_3.4	MMTV-Haras An #3-4	Hras	Luminal
Hras_5.4	MMTV-Haras An #5-4	Hras	Luminal
Hras_4.4	MMTV-Haras An #4-4	Hras	Luminal
Hras_2.4	MMTV-Haras An #2-4	Hras	Luminal
PyMT_436	PyMT 436	MMTV-PyMT	Luminal
PyMT_437	PyMT 437	MMTV-PyMT	Luminal
MMTV.11567	MMTV-PymT #11567	MMTV-PyMT	Luminal
MMTV.11568	MMTV-PymT #11568	MMTV-PyMT	Luminal
MMTV.11570	MMTV-PymT #11570	MMTV-PyMT	Luminal
MMTV.5.1	MMTV-PymT #5-1	MMTV-PyMT	Luminal
Wnt_4675	MMTV-Wnt 4675	MMTV-Wnt	Luminal
Wnt_4676	MMTV-Wnt 4676	MMTV-Wnt	Luminal
Wnt_4635	MMTV-Wnt 4635	MMTV-Wnt	Luminal
Wnt_4677	MMTV-Wnt 4677	MMTV-Wnt	Luminal
FVB_M1_2	FVB pregnant M2-1	NormalMammary	Normal
FVB_M1_4	FVB pregnant M4-1	NormalMammary	Normal
FVB_M1_1	FVB pregnant M1-1	NormalMammary	Normal
FVB_M1_3	FVB pregnant M3-1	NormalMammary	Normal
FVB_M1_5	FVB pregnant M5-1	NormalMammary	Normal

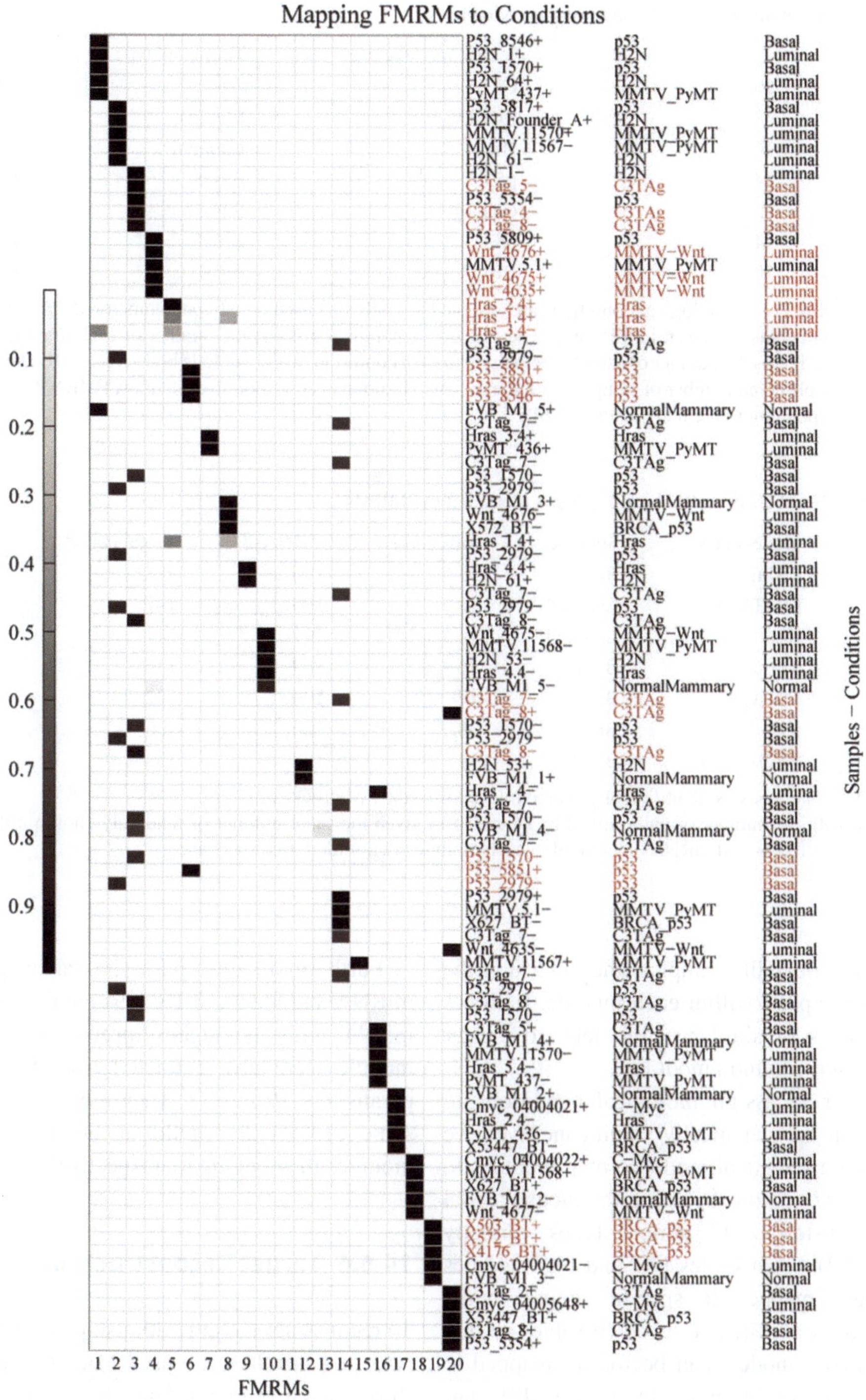

Fig. 14.2 Assigning biological conditions to FMRMs. The y-axis on the *right side* of the figure denotes names, mouse model types, and breast cancer subtypes in three columns. Using the parameter Q, the likelihood that a particular sample is associated with a specific module, the top 5% of samples associated with each module is displayed using *grey* scale to represent the probability that sample is associated with each module. Some samples are significantly associated with more than one module. Some modules, such as module-11, have only rather low probability of association with samples, and so have nearly *white shading* even for their top five samples. Clustering of certain sample types with modules is *highlighted*

Table 14.2 Assigning biological conditions to FMRMs

FMRM#	c_i	x	Mouse model class	Tumor subtype	p-Value
3	10	3	C3TAg	Basal	0.0081
4	8	3	MMTV_Wnt	Luminal	0.004
5	10	3	Hras	Luminal	0.0081
6	14	3	p53	Basal	0.0222
11	10	3	C3TAg	Basal	0.0081
13	14	3	p53	Basal	0.0222
19	10	3	BRCA_p53	Basal	0.0081

According to Eq. (14.6), biological conditions are assigned to FMRMs based on a hypergeometric distribution
The significant results are given in this table. The size of the population is 92, the number of each draw is 5% of the population, i.e. 5. c_i is the number of samples belonging to each condition, which include both over- and under expressed status. x is the observed number of samples with the assigned biological condition in each draw. FMRM# is the module number corresponding to the number in Fig. 14.2

Table 14.3 Numbers of miRNA-mRNA pairs identified in FMRMs

FMRM#	Model class	Subtype	miRNAs#	mRNAs#	Target pairs#	p-Value
3	C3Tag	Basal	33	190	273	1.70E-07
4	MMTV_Wnt	Luminal	18	190	147	3.23E-08
5	Hras	Luminal	16	191	144	2.98E-07
6	p53	Basal	16	189	146	1.48E-06
11	C3TAg	Basal	17	190	122	1.13E-11
13	p53	Basal	18	186	136	1.29E-10
19	BRCA_p53	Basal	18	188	133	2.71E-12

The miRNAs and mRNAs identified in each module are linked by MicroCosm
Compared with the number of pairs linked by MicroCosm given the same number of randomly chosen miRNAs and mRNAs, the miRNAs and mRNAs identified in each module are not identified by chance

5 highest probability samples (the top 5% of the 92 total samples) within each module. It shows that samples of similar nature tend to cluster together into common modules.

In order to assign biological conditions to modules at the statistically significant level, we conceived a statistical model to map modules to biological conditions by using the mouse model classes instead of tumor types directly (Table 14.2). From Tables 14.2 and 14.7 modules have been mapped to specific mouse model classes at a significant level (p-value <0.05). These mouse models can be further mapped to two human breast tumor subtypes [46–48], suggesting that the identified modules are associated with those biological conditions. Other modules are clustered by samples with mixed biological conditions, suggesting that they may participate in several cellular processes.

Furthermore, the top 5% (56) ranked probes of miRNA and the top 0.1% (192) ranked probes of mRNA with the highest probabilities in each module were also extracted from the inferred parameters Φ and Ω. They are assigned to the same biological conditions according to the modules they belong to, respectively.

14.6.6 Target Reconstruction

To reconstruct target relationships between miRNAs and mRNAs, we used MicroCosm Targets (http://www.ebi.ac.uk/enright-srv/microcosm/htdocs/targets/v5/) to link miRNAs and mRNAs identified in each FMRM. The numbers of linked miRNAs and mRNAs are given in Table 14.3.

To investigate whether the miRNAs and mRNAs in each module were identified by

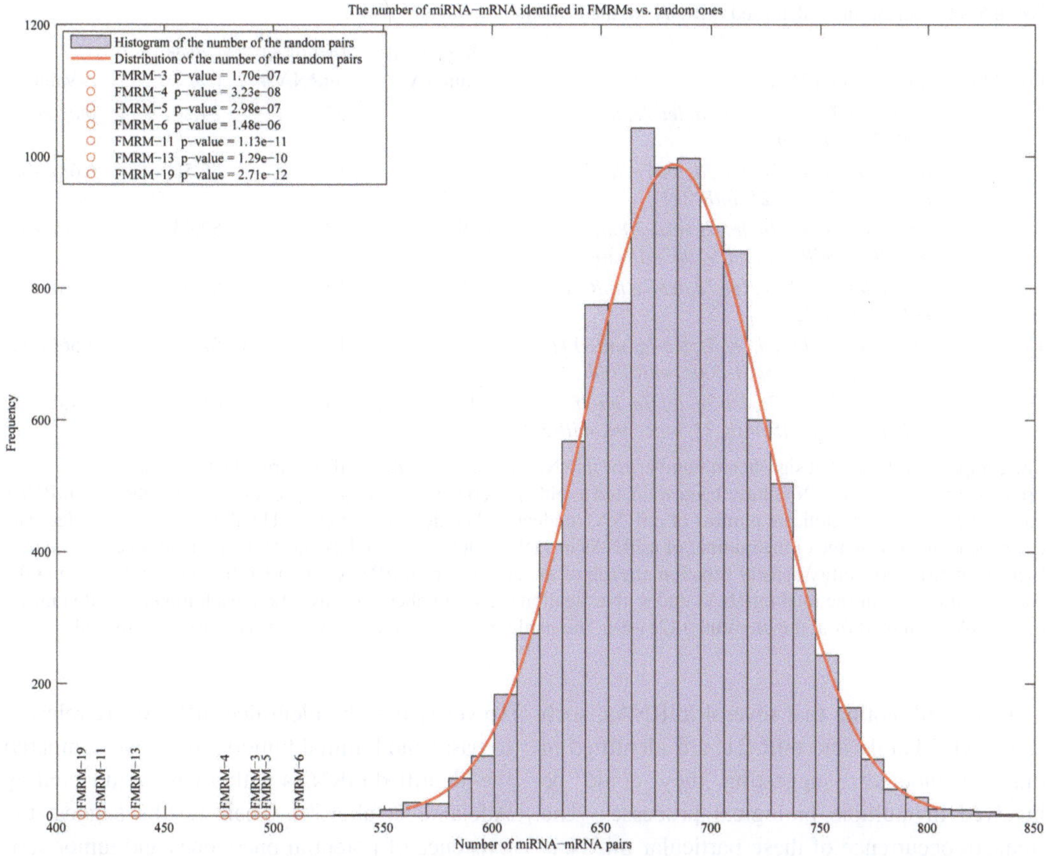

Fig. 14.3 Comparison of the numbers of miRNA-mRNA pairs in the identified modules with the ones from the random matching. The distribution of the number of matched target pairs is estimated by simulation which was executed for 10,000 times. It indicates that the miRNAs and mRNAs identified in each module are not identified by chance. (The numbers here are for probes, not like those in Table 14.3 have been mapped to the unique names of miRNAs and mRNAs)

chance, we randomly selected a group of miRNAs and a group of mRNAs from MicroCosm with the same numbers as those in the identified modules, and queried how many pairs that can be linked by MicroCosm. The distribution of the number of matched pairs was estimated by a simulation which was executed 10,000 times. Illustrated in Fig. 14.3, the estimated distribution shows that the numbers of target relationships of the randomly chosen miRNAs and mRNAs are significantly different from those of the identified miRNAs and mRNAs in each module (p-value <0.05). It indicates that the miRNAs and mRNAs in each module are not identified by chance.

14.6.7 Functional Validation of miRNAs

To further validate that the identified miRNAs are relevant to cancers, we investigated the implications of miRNAs for cancers through literature review. We built a benchmark based on the current knowledge and compared it with the miRNAs identified in the modules.

From the literature [49], 42 miRNAs have been validated to have implications for cancers. We identified a significant number of miRNAs covered by the benchmark shown in Table 14.4. The comparison shows that the miRNAs identified by our method are largely consistent with the current knowledge of miRNAs for cancers.

Table 14.4 Validation of identified miRNAs in the FMRMs

FMRM#	Supported miRNAs	Supported miRNA #	Identified miRNA #	Coverage (%)	*p*-Value
3	*let-7a, let-7b, let-7c, let-7d, let-7e, let-7f, miR-221,miR-29a*	8	33	24.24	0.0264
4	*let-7a, let-7b, let-7c, let-7d, let-7e, let-7f, let-7g, let-7i, miR-21, miR-221*	10	18	55.56	6.68E–06
5	*let-7b, let-7c, let-7d, let-7i, miR-200b, miR-200c, miR-29a, miR-29b, miR-30c*	9	17	52.94	3.56E–05
6	*let-7a, let-7b, let-7c, let-7d, let-7i, miR-103, miR-21, miR-221*	8	16	50.00	1.76E–04
15	*let-7a, let-7c, let-7d, let-7f, let-7g, miR-141, miR-19b, miR-21, miR-200a, miR-200b*	9	17	52.94	3.56E–05
19	*let-7a, let-7b, let-7c, let-7d, let-7e, let-7f, miR-143, miR-145, miR-21, miR-29a, miR-29b*	11	18	61.11	5.45E–07

The comparison shows that significant numbers of miRNAs identified in the FMRMs are relevant to cancers

From the literature, 42 miRNAs have been validated as either oncogenes or tumor suppressors. Among the 334 miRNAs under investigation, a significant number of miRNAs in identified modules are supported by the current knowledge. The coverage is the percentage of the number of miRNAs in each module supported by literature. p-value is calculated by a hypergeometric probability density function at each of the numbers of miRNAs supported by the literature, using the corresponding size of the total miRNAs under investigation (334), numbers of miRNAs in each module, and numbers of miRNAs identified from the literature (42) [49]. The modules with significant supports are given in this table

It is worth noting that several miRNAs, such as the let-7 family and miR-21, are identified in multiple modules, suggesting they could be involved in multiple biological processes. The frequent occurrence of these particular miRNAs is consistent with their known strong association with multiple cancer types, including breast cancers. The identification of multiple modules containing different but overlapping sets of miRNAs is likely to be the consequence of activation of distinct subsets of common gene interaction networks in specific cancer subtypes. For example, Blenkiron et al. [46] identified 31 miRNAs differentially expressed between basal and luminal tumors. Among them, let-7a, b, and f are under-expressed in basal tumors but over-expressed in luminal tumors. These miRNAs were identified in module 3, 4, 5, and 6 using our method and show patterns that are consistent with their reported involvement in these tumors.

14.6.8 Functional Validation of miRNA Target Genes

It is expected that the miRNA target genes are also relevant to the specific biological processes. To verify that the identified mRNAs are relevant to basal and luminal tumors, firstly we compared the identified mRNAs with a work conducted by Adelaide et al. [50]. Their results suggest the existence of potential oncogenes and tumor suppressor genes differentially associated with the basal and luminal subtype. As their results are largely consistent with many previous reports [51–53], we validate our analysis based on their results.

In our results, 18 genes have been identified by Adelaide et al. [50] as in Table 14.5. Among these genes, Ccdc77 identified in FMRM-3 also is targeted by miR-29a and miR-221, Hspa14 identified in FMRM-4 is targeted by miR-21, and Cox4i1 identified in FMRM-19 is targeted by let-7c and let-7e. It further confirms that let-7e, miR-21, miR-29a, and miR-221 may have important regulatory functions towards basal and luminal tumors. In addition, Rbm4b identified in FMRM-3 is targeted by miR-697 and miR-700, Rbx1 identified in FMRM-5 is targeted by miR-709, Gspt1 identified in FMRM-11 is targeted by miR-669c and miR-710, and Cox4i1 identified in FMRM-19 is targeted by miR-709. It suggests that miR-669c, miR-697, miR-709, and miR-710 may also play important roles in regulating basal

Table 14.5 Validation of identified mRNAs in the FMRMs

Gene	Expression	Associated type	Module
Cct3	Over	Basal	1
Upf2	Over	Basal	1
Eif4a1	Under	Luminal	2,18,20
Ccdc77	Over	Basal	3
Rbm4b	Over	Luminal	3
Hspa14	Over	Basal	4
Rbx1	Under	Luminal	5,18
Ppap2a	Under	Basal	7
Tpd52	Over	Luminal	8
Tulp3	Over	Basal	9
Gpm6a	Under	Basal	10
Gdap1	Over	Luminal	10
Gspt1	Over	Luminal	11
Rbx1	Over	Luminal	12
Npy1r	Under	Basal	16
Rpl13	Under	Luminal	18
Cox4i1	Under	Luminal	19
Arfgef1	Over	Luminal	20

Adelaide et al. [50] suggest the existence of potential oncogenes and tumor suppressor genes differentially associated with the basal and luminal subtype
In our results, 18 genes identified in FMRMs are consistent with their results

and luminal tumors. It is worth noting that many previously reported results were not recovered in the current study because the investigated data were profiled with mouse model while the results of Adelaide et al. [50] were produced on breast cancer samples of humans.

Furthermore, we queried the mRNAs identified in each module against the Ingenuity Pathway Analysis (IPA) Database. We specifically focused on human species as we are interested in the networks of human cancers. The networks participated by the mRNAs identified in FMRMs are highly associated with cancers. Many genes are directly related to cancers and genetic disorders. They are co-targeted by a group of miRNAs identified from our method, suggesting the identified miRNAs and their target mRNAs have implications for cancers. For example, a network involving the miRNAs and mRNAs identified by our method is associated with cancer, cellular compromise, DNA replication, and repair (Fig. 14.4). The networks that are explicitly associated with cancers and within the top five networks of each module are given in Table 14.6. The detailed networks are also given in Figs. 14.5, 14.6, 14.7 and 14.8. The identified genes of FMRMs, which are relevant to cancers, are given in Table 14.7. The results indicate that our methods effectively identified many cancer related genes. Those genes are targeted by a group of miRNAs, suggesting those miRNAs also participate in the networks of cancers.

14.7 Discussion and Conclusion

miRNAs have been regarded as one of the most important gene regulators. Identifying their functions and regulatory mechanisms is critical in understanding biological processes of organisms. Great efforts, in both biological experiments and computational methods, have been made to illustrate their functions. However, the precise regulatory functions of most miRNAs remain elusive due to the complexity of the regulatory mechanisms.

In this chapter, we first reviewed the computational methods for miRNA function discovery briefly. We proposed three categories for miRNA research: miRNA target prediction, discovering miRNA regulatory modules (MRMs), and prediction of functional miRNA regulatory modules (FMRMs). Second, we demonstrated a method of discovering FMRMs. This method is inspired by the Corr-LDA, which has been used to extract the correspondence patterns from heterogeneous data. We modified Corr-LDA and derived the solution for FMRM discovery.

This method models FMRMs with a generative process. It makes use of the expression profiles of miRNAs and mRNAs, with or without using the target relationships between miRNAs and mRNAs based on the sequence binding information. It simultaneously identifies groups of interactive miRNAs and mRNAs, which are believed to participate in specific biological functions.

We have applied this method to a mouse model data set for human breast cancer research. The method has effectively identified several modules related to breast cancer subtypes: basal and luminal. Since the data sets used were profiled from mouse

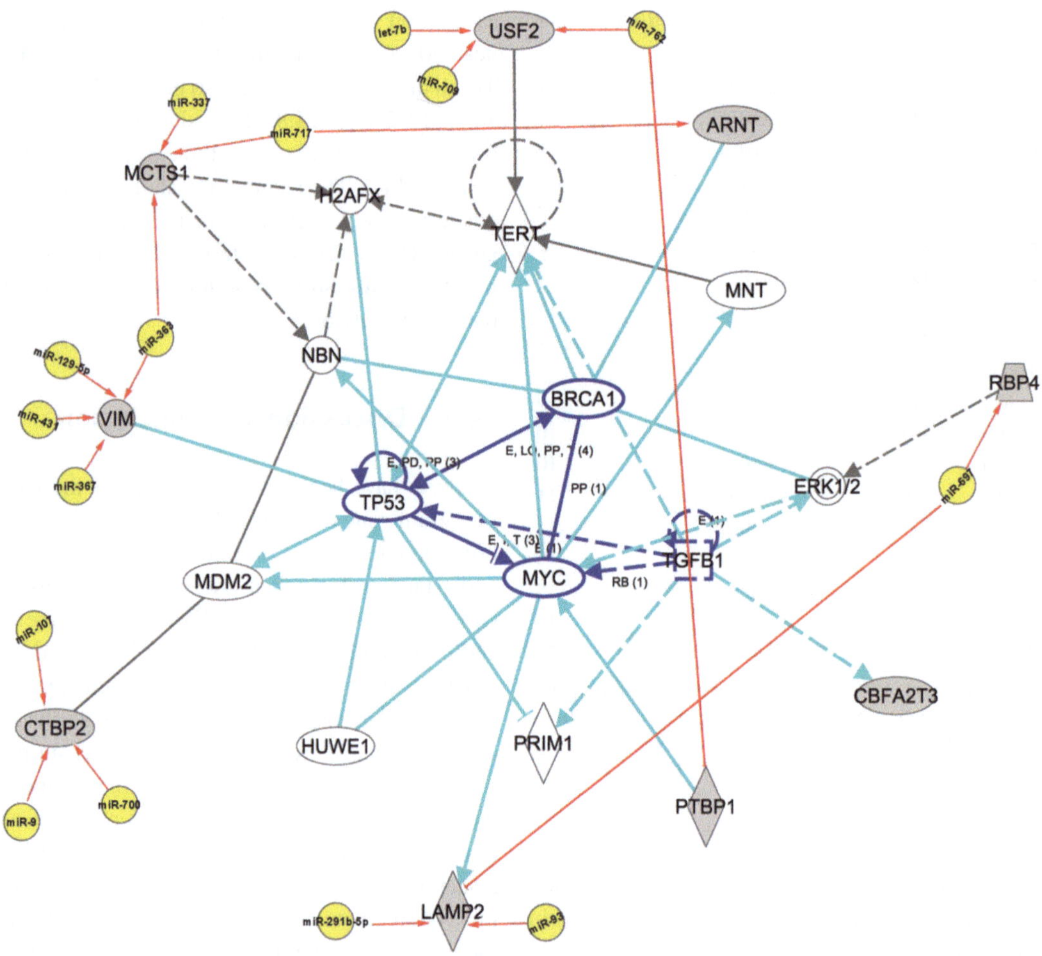

Fig. 14.4 A network associated with cancer, cellular compromise, DNA replication, and repair. It is participated by a group of miRNAs and their target mRNAs identified by our method, suggesting these miRNAs and their target mRNAs have the function of cancers

Table 14.6 Associated network functions of FMRMs

FMRM#	Associated network function	Ref figure	Score
3	Cancer, cellular compromise, DNA replication, and repair	Fig. 14.4	12
3	Gene expression, cancer, immunological disease	Fig. 14.5	12
4	Cellular growth and proliferation, cancer, dermatological diseases and conditions	Fig. 14.6	14
4	Cellular assembly and organization, cancer, cellular development	–	2
13	Cancer, cell cycle, DNA replication, recombination, and repair	Fig. 14.7	18
13	Cancer, cell morphology, cellular development	–	12
19	Cancer, cell-to-cell signalling and interaction, cellular function and maintenance	Fig. 14.8	2
19	Amino acid metabolism, cancer, cell morphology	–	2

The networks participated by the genes identified in FMRMs are highly related to cancers.
The networks associated with cancers that are explicitly within the top five networks of each module are listed. It is worth noting that many mouse genes have been filtered out as we specifically target human breast cancers

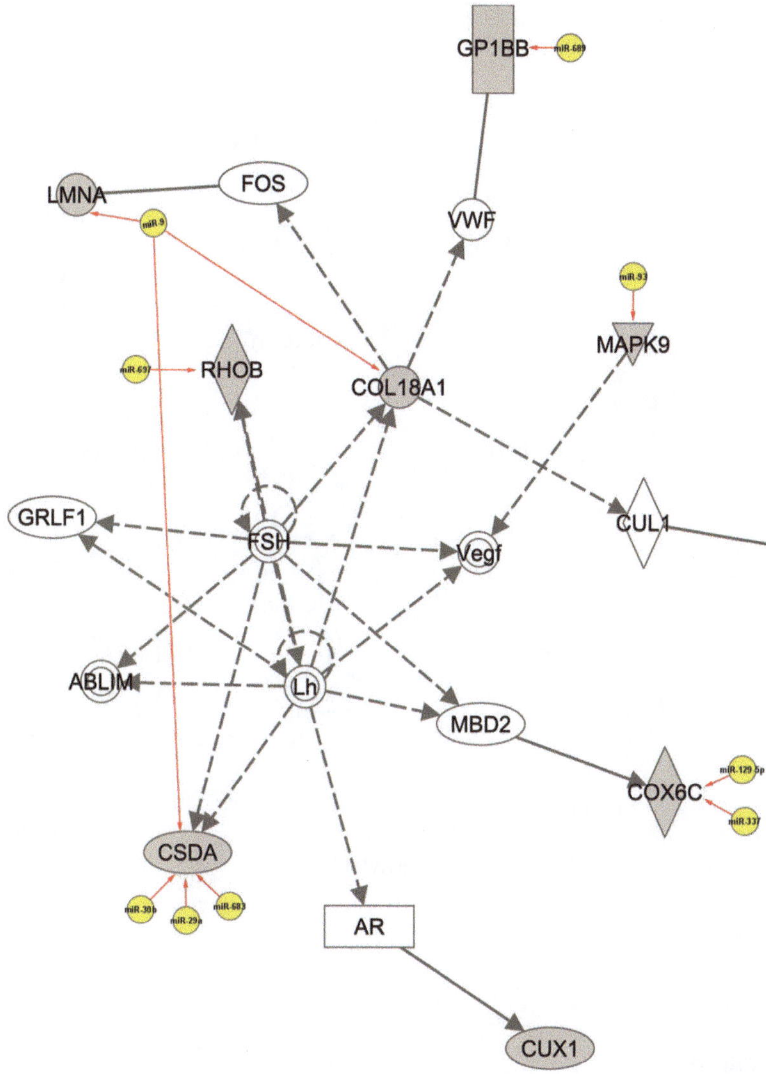

Fig. 14.5 A network associated with gene expression, cancer, immunological disease. It is participated by a group of miRNAs and their target mRNAs identified in FMRM-3, suggesting these miRNAs and their target mRNAs have the function of cancers

tissues, many genes have been filtered out because we focus on human genes. Thus, previously reported results were not fully recovered in this work. However, a large proportion of miRNAs and mRNAs identified in the modules have been reported to have associations with basal and luminal subtypes. Many others have direct indications on cancers and genetic disorders. Furthermore, many novel associations among miRNAs, mRNAs, and biological processes have been predicted by our model. Several miRNAs and mRNAs are highly related to cancers as reported by previous works, suggesting those modules may have roles in the development processes of cancers.

This model allows discovery of FMRMs with or without using the target relationship between miRNAs and mRNAs. Some researchers have suggested that algorithms that do not consider known targets may avoid biases [30, 45, 54]. Bonnet et al. [40] also showed that expression profiles alone can be used to infer miRNA

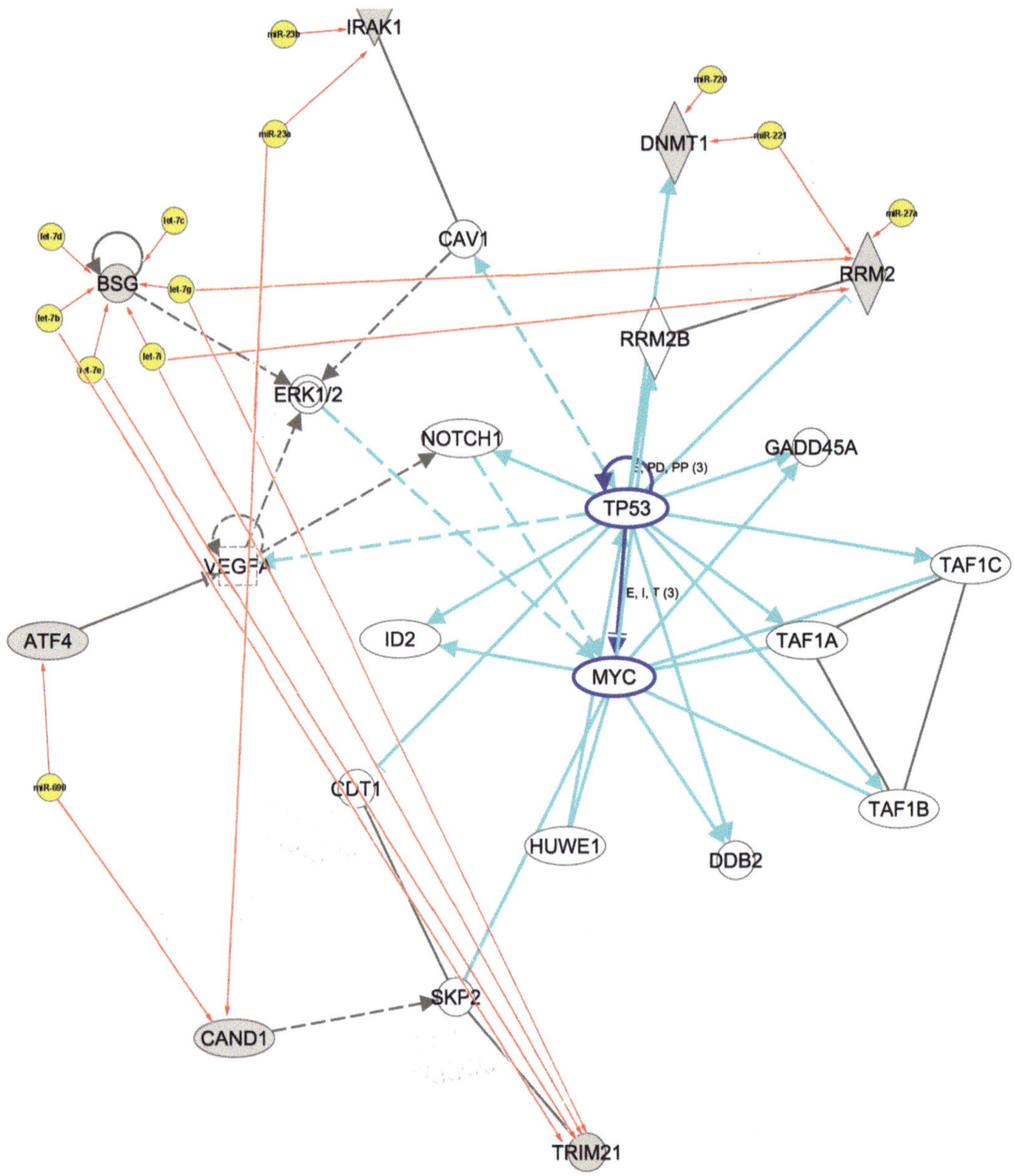

Fig. 14.6 A network associated with of cellular growth and proliferation, cancer, dermatological diseases and conditions. It is participated by a group of miRNAs and their target mRNAs identified in FMRM-4, suggesting these miRNAs and their target mRNAs have the function of cancers

regulatory networks. This method provides the flexibility of inferring FMRMs with or without target relationships of miRNAs and mRNAs. We have demonstrated this model without using the prior target prediction. The results suggest that expression profiles of miRNAs and mRNAs are crucial for both target identification and regulatory module discovery.

With more and more data available at different levels, it is possible to integrate multiple data sets to explore functions of miRNAs. Current methods have been limited in identifying miRNA

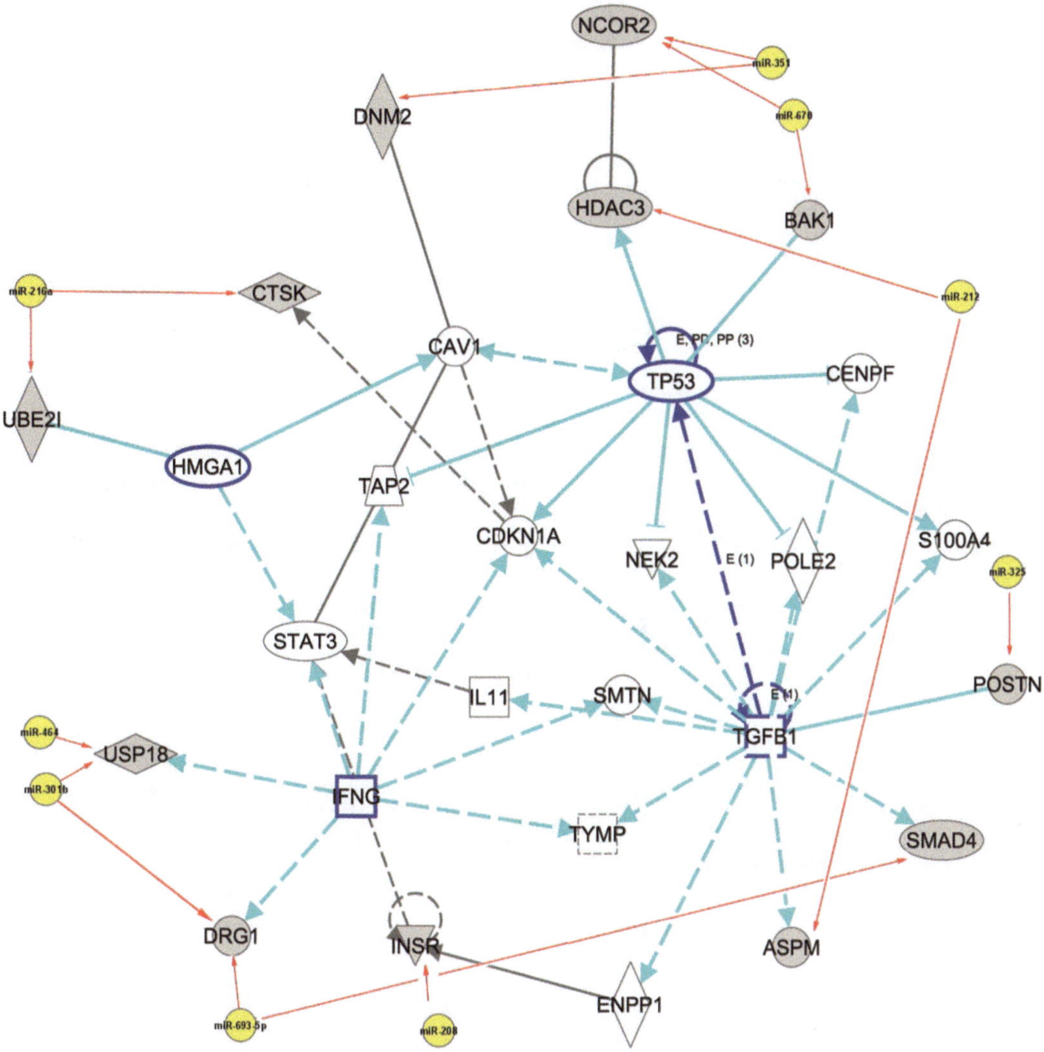

Fig. 14.7 A network associated with cancer, cell cycle, DNA replication, recombination, and repair. It is participated by a group of miRNAs and their target mRNAs identified in FMRM-13, suggesting these miRNAs and their target mRNAs have the function of cancer

targets, MRMs, or FMRMs at miRNA-mRNA levels. Biological discovery has suggested that miRNA regulation can degrade mRNAs as well as inhibit protein translation. Although one third of mRNAs repressed in the translation process display detectable destabilization, more are repressed without detectable changes in mRNA levels [33]. The current computational methods have only analysed at the miRNA-mRNA level. The global impact on protein output remains to be examined. In the future, when matched data sets of miRNA, mRNA, and protein are available, we can involve protein in the above model by adding another layer of latent variable. By involving proteins, it can potentially discover the miRNA regulatory pathway of miRNAs → mRNAs → proteins → conditions.

Acknowledgement We thank Dr Jeffrey E. Green and Dr. Min Zhu for providing the data sets.

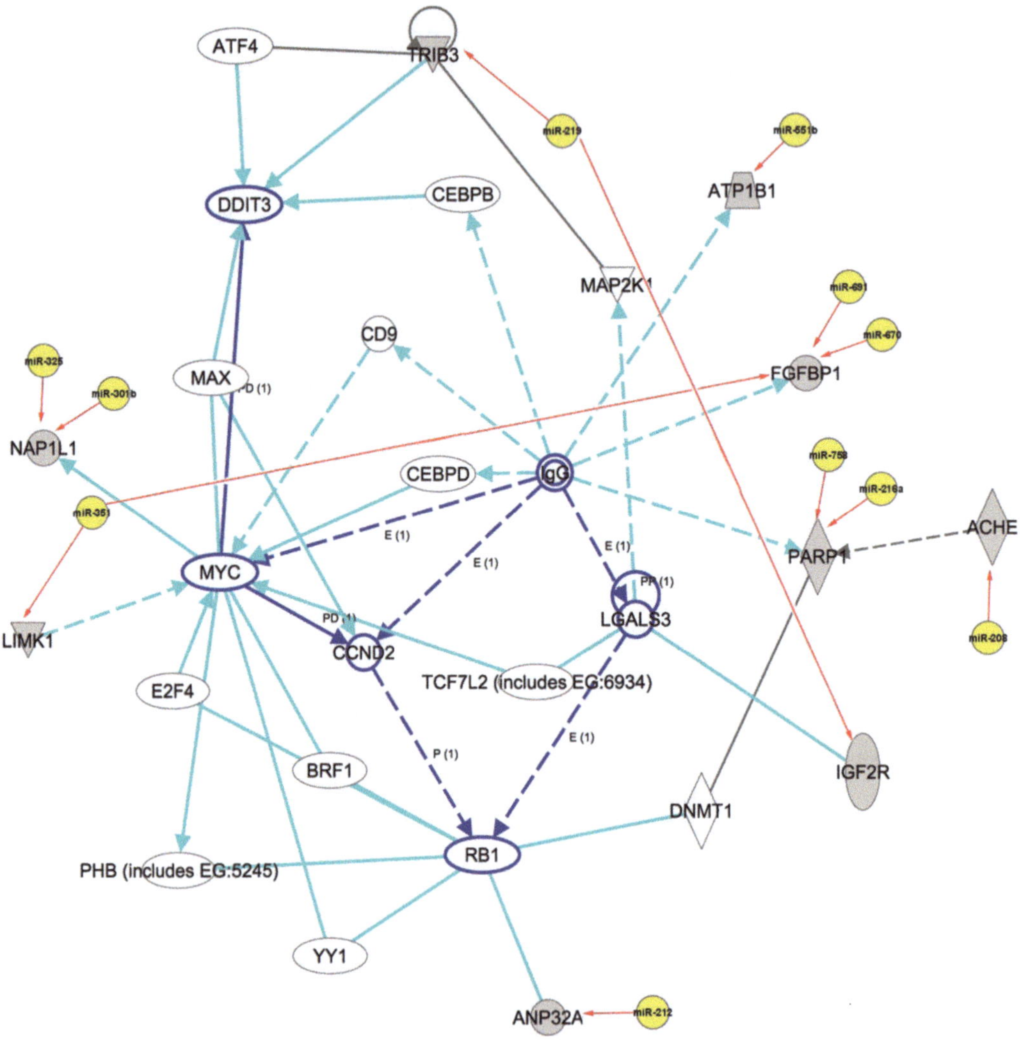

Fig. 14.8 A network associated with cancer, cell-to-cell signaling and interaction, cellular function and maintenance. It is participated by a group of miRNAs and their target mRNAs identified FMRM-19, suggesting these miRNAs and their target mRNAs have the function of cancers

Table 14.7 Cancer associated genes in the FMRMs

FMRM#	mRNAs	# mRNAs	p-Value (adj.)
3	CALR, COL18A1, VIM, SDCBP, MCTS1, AK2, RBP4, AP2S1, ARNT, NDUFV2, PRPF8, RPS15	12	4.89E–03 to 2.54E–02
4	DNMT1, NF2, RRM2	3	2.13E–03 to 3.05E–02
5	CEBPB, DDX39, HSP90AB1, MT2A, NUP62, SQLE, TCP1, TRIO	8	2.01E–03 to 4.84E–02
11	DICER1, ENO1, HSP90B1, RXRB, SPRY2	5	5.26E–03 to 4.84E–02
13	IGF2R, LSM14B, NCOR2, SP110, STX5, TOR2A, ACHE, HDAC3, PARP1, POSTN, SMAD4, UBE2I, RNF6, BAK1	14	6.88E–03 to 4.56E–02

Many genes identified in FMRMs are relevant to cancers

Genes identified in FMRMs are directly assigned to diseases and disorders. The cancer related genes of FMRMs within their top five bio-functions are listed

References

1. Ridley M (2006) Genome: the autobiography of a species in 23 chapters book description. Topeka Bindery, New York
2. He L, Hannon GJ (2004) MicroRNAs: small RNAs with a big role in gene regulation. Nat Rev Genet 5(7): 522–531
3. Iorio MV et al (2005) MicroRNA gene expression deregulation in human breast cancer. Cancer Res 65(16):7065–7070
4. Yanaihara N et al (2006) Unique microRNA molecular profiles in lung cancer diagnosis and prognosis. Cancer Cell 9(3):189–198
5. Porkka KP et al (2007) MicroRNA expression profiling in prostate cancer. Cancer Res 67(13):6130–6135
6. Akao Y, Nakagawa Y, Naoe T (2007) MicroRNA-143 and −145 in colon cancer. DNA Cell Biol 26(5): 311–320
7. Yang H et al (2008) MicroRNA expression profiling in human ovarian cancer: miR-214 induces cell survival and cisplatin resistance by targeting PTEN. Cancer Res 68(2):425–433
8. Zhang X et al (2009) Alterations in miRNA processing and expression in pleomorphic adenomas of the salivary gland. Int J Cancer 124(12):2855–2863
9. Croce CM (2009) Causes and consequences of microRNA dysregulation in cancer. Nat Rev Genet 10(10):704–714
10. Papadopoulos GL et al (2009) The database of experimentally supported targets: a functional update of TarBase. Nucl Acids Res 37(suppl_1):D155–D158
11. Xiao F et al (2009) miRecords: an integrated resource for microRNA-target interactions. Nucl Acids Res 37(suppl_1):D105–D110
12. Bentwich I et al (2005) Identification of hundreds of conserved and nonconserved human microRNAs. Nat Genet 37(7):766–770
13. Krek A et al (2005) Combinatorial microRNA target predictions. Nat Genet 37:495–500
14. Hatzigeorgiou AG (2007) Same computational analysis, different miRNA target predictions. Nat Methods 4(3):191
15. Griffiths-Jones S et al (2008) miRBase: tools for microRNA genomics. Nucl Acids Res 36(suppl_1): D154–D158
16. Yoon S, De Micheli G (2005) Prediction of regulatory modules comprising microRNAs and target genes. Bioinformatics 21(suppl_2):ii93–ii100
17. Huang JC et al (2007) Using expression profiling data to identify human microRNA targets. Nat Method 4(12):1045–1049
18. Joung J-G et al (2007) Discovery of microRNA-mRNA modules via population-based probabilistic learning. Bioinformatics 23(9):1141–1147
19. Tran D, Satou K, Ho T (2008) Finding microRNA regulatory modules in human genome using rule induction. BMC Bioinform 9(Suppl 12):S5
20. Peng X et al (2009) Computational identification of hepatitis C virus associated microRNA-mRNA regulatory modules in human livers. BMC Genomics 10(1):373
21. Liu B, Li J, Tsykin A (2009) Discovery of functional miRNA-mRNA regulatory modules with computational methods. J Biomed Inform 42(4):685–691
22. Joung J-G, Fei Z (2009) Identification of microRNA regulatory modules in Arabidopsis via a probabilistic graphical model. Bioinformatics 25(3):387–393
23. Liu B et al (2009) Exploring complex miRNA-mRNA interactions with Bayesian networks by splitting-averaging strategy. BMC Bioinform 10(1):408
24. Krek A et al (2005) Combinatorial microRNA target predictions. Nat Genet 37(5):495–500
25. Rajewsky N (2006) MicroRNA target predictions in animals. Nat Genet 38:S8–S13
26. Mazière P, Enright AJ (2007) Prediction of microRNA targets. Drug Discov Today 12(11–12):452–458
27. John B et al (2004) Human MicroRNA targets. PLoS Biol 2(11):e363
28. Lall S et al (2006) A genome-wide Map of conserved MicroRNA targets in *C. elegans*. Curr Biol 16(5): 460–471
29. Miranda KC et al (2006) A pattern-based method for the identification of MicroRNA binding sites and their corresponding heteroduplexes. Cell 126(6): 1203–1217
30. Lewis BP et al (2003) Prediction of mammalian MicroRNA targets. Cell 115(7):787–798
31. Ritchie W, Flamant S, Rasko JEJ (2009) Predicting microRNA targets and functions: traps for the unwary. Nat Method 6(6):397–398
32. Grimson A et al (2007) MicroRNA targeting specificity in mammals: determinants beyond seed pairing. Mol Cell 27(1):91–105
33. Baek D et al (2008) The impact of microRNAs on protein output. Nature 455(7209):64–71
34. Alexiou P et al (2009) Lost in translation: an assessment and perspective for computational microRNA target identification. Bioinformatics 25(23):3049–3055
35. Gaidatzis D et al (2007) Inference of miRNA targets using evolutionary conservation and pathway analysis. BMC Bioinform 8(1):69
36. Kertesz M et al (2007) The role of site accessibility in microRNA target recognition. Nat Genet 39(10): 1278–1284
37. Selbach M et al (2008) Widespread changes in protein synthesis induced by microRNAs. Nature 455(7209): 58–63
38. Huang J, Morris Q, Frey B (2006) Detecting microRNA targets by linking sequence, microRNA and gene expression data. Res Comput Mol Biol 3909:114–129
39. Lavrac N et al (2004) Subgroup discovery with CN2-SD. J Mach Learn Res 5:153–188
40. Bonnet E et al (2010) Module network inference from a cancer gene expression data set identifies MicroRNA regulated modules. PLoS One 5(4):e10162
41. Steyvers M et al(2004) Probabilistic author-topic models for information discovery. In: Proceedings of

the tenth ACM SIGKDD international conference on knowledge discovery and data mining, ACM, Seattle, WA

42. Blei DM, Jordan MI(2003) Modeling annotated data. In: Proceedings of the 26th annual international ACM SIGIR conference on research and development in information retrieval, ACM, Toronto, Canada
43. Liu B et al (2010) Identifying functional miRNA–mRNA regulatory modules with correspondence latent dirichlet allocation. Bioinformatics 26(24): 3105–3111
44. Liu JS (1994) The collapsed Gibbs sampler in Bayesian computations with applications to a gene regulation problem. J Am Stat Assoc 89:958–966
45. Bartel DP (2009) MicroRNAs: target recognition and regulatory functions. Cell 136(2):215–233
46. Blenkiron C et al (2007) MicroRNA expression profiling of human breast cancer identifies new markers of tumor subtype. Genome Biol 8(10):R214
47. Desai KV et al (2002) Initiating oncogenic event determines gene-expression patterns of human breast cancer models. Proc Natl Acad Sci USA 99(10): 6967–6972
48. Herschkowitz J et al (2007) Identification of conserved gene expression features between murine mammary carcinoma models and human breast tumors. Genome Biol 8(5):R76
49. Sassen S, Miska EA, Caldas C (2008) MicroRNA – implications for cancer. Virchows Arch 452(1):1–10
50. Adelaide J et al (2007) Integrated profiling of basal and luminal breast cancers. Cancer Res 67(24):11565–11575
51. Bergamaschi A et al (2006) Distinct patterns of DNA copy number alteration are associated with different clinicopathological features and gene-expression subtypes of breast cancer. Genes Chromosomes Cancer 45(11):1033–1040
52. Chin K et al (2006) Genomic and transcriptional aberrations linked to breast cancer pathophysiologies. Cancer Cell 10(6):529–541
53. Neve RM et al (2006) A collection of breast cancer cell lines for the study of functionally distinct cancer subtypes. Cancer Cell 10(6):515–527
54. Lewis BP, Burge CB, Bartel DP (2005) Conserved seed pairing, often flanked by adenosines, indicates that thousands of human genes are MicroRNA targets. Cell 120(1):15–20

Elucidating the Role of microRNAs in Cancer Through Data Mining Techniques

15

Luciano Cascione, Alfredo Ferro, Rosalba Giugno, Alessandro Laganà, Giuseppe Pigola, Alfredo Pulvirenti, and Dario Veneziano

Abstract

microRNAs (miRNAs) have been shown to play a crucial role in the most important biological processes and their dysregulation has been connected to a variety of diseases, including cancer. The number of computational tools for the analysis of miRNA related data is continuously increasing. They range from simple look-up resources to more sophisticated tools for functional analysis of miRNAs. These systems may help to investigate the role of miRNAs in key biological processes and their involvement in diseases. The ultimate goal is to allow the development of regulatory models describing complex processes and the effects of their dysregulation.

Here we review the most important and recent methods for the analysis of miRNA expression profiles and the tools available on the web for target prediction and functional analysis of miRNAs.

Particular emphasis is given to the integration of heterogeneous data, including target predictions and expression profiles, which can be used to infer miRNA/phenotype associations and for the generation of network models of miRNA function.

L. Cascione • R. Giugno • A. Pulvirenti • D. Veneziano
Department of Clinical and Molecular Biomedicine, University of Catania, Catania, Italy
e-mail: cascione@dmi.unict.it; giugno@dmi.unict.it; apulvirenti@dmi.unict.it; veneziano@dmi.unict.it

A. Ferro (✉)
Department of Clinical and Molecular Biomedicine, University of Catania, Catania, Italy

Dipartimento di Matematica e Informatica, Università degli Studi di Catania, Città Universitaria – Viale A. Doria, 6, Catania 95125, Italy
e-mail: ferro@dmi.unict.it

A. Laganà
Department of Molecular Virology, Immunology and Medical Genetics, Comprehensive Cancer Center, The Ohio State University, Columbus, OH, USA
e-mail: alessandro.lagana@osumc.edu

G. Pigola
Research and Development, IGA Technology Services, Udine, Italy
e-mail: gpigola@igatechnology.com

U. Schmitz et al. (eds.), *MicroRNA Cancer Regulation: Advanced Concepts, Bioinformatics and Systems Biology Tools*, Advances in Experimental Medicine and Biology 774,
DOI 10.1007/978-94-007-5590-1_15, © Springer Science+Business Media Dordrecht 2013

Keywords
microRNA • Database • Expression profiles • Functional analysis • Network models

15.1 Introduction

In the past decade, many efforts have been spent to demonstrate the crucial role of miRNAs in the most important biological processes, including apoptosis, development and immune response [1–3]. Moreover, the dysregulation of miRNAs has been connected to a variety of diseases, cancer being probably the most extensively studied one [4–6].

The partial complementarity that most miRNAs exhibit to their targets, especially in animals, is the key to their flexibility. Indeed, a single miRNA is usually able to bind to many targets, often in several sites, and a single gene can be targeted by different miRNAs acting cooperatively. This is a clear indication that the simple miRNA/target interactions are actually part of a more complex regulatory system and should be analyzed in the wider context of expression networks.

The initial focus of bioinformatics miRNA research was primarily on the development of tools for the identification of miRNAs and their targets. The prediction of miRNA binding sites on targets still remains a challenge. Indeed, although several studies have uncovered the basic rules of miRNA/target interactions [7], the target prediction tools currently available still produce a significant number of false positives and are not able to identify some experimentally validated miRNA/target pairs [8, 9].

Nevertheless, target prediction tools constitute the essential basis of functional miRNA analysis, allowing to link miRNAs to processes, diseases and pathways, through their targets. Recently, many bioinformatics tools for functional analysis of miRNAs have been developed. Their ultimate goal is the identification of non trivial relationships between miRNAs and other molecular actors, such as genes and transcription factors, and the development of regulatory models describing complex processes and the effects of their dysregulation. These purposes can be fulfilled thanks to the huge amount of data that are produced daily and made publicly available on the internet, among which miRNA/target matches and miRNA expression profiles are mostly predominant.

In this chapter we review the most important and recent methods for the analysis of miRNA expression profiles and the tools available on the web for functional analysis of miRNAs. In particular, in Sect. 15.2 the most used miRNA profiling technologies are described, together with the computational and statistical methods for the analysis of the related data. Emphasis is particularly given to some aspects such as data normalization, the identification of differentially expressed microRNAs, clustering and the role of miRNAs as biomarkers. Section 15.3 is focused on miRNA target prediction. An overview of the general features is given, together with a brief description of the most popular target prediction tools available on the web. Finally, in Sect. 15.4 we present a series of tools for functional analysis of miRNAs. Particular emphasis is given to the integration of heterogeneous data, including target predictions and expression profiles, which can be used to infer miRNA/phenotype associations and for the generation of network models of miRNA function.

15.2 miRNA Profiling: Technologies and Data Analysis

Several methodological approaches for mRNA expression profiling have been applied to profile miRNA expression. Current methods widely used in the study of miRNA expression include northern blotting with radiolabelled probes, oligonucleotide microarrays, qPCR-based detection of mature miRNAs, single molecule detection in liquid phase, in situ hybridization (ISH) and massively parallel sequencing.

In this section we will review the main technologies used for miRNA profiling as well as the computational and statistical methods used for the normalization and the analysis of the produced data.

15.2.1 Profiling Technologies

In general, all existing profiling methods can be separated into two categories: one that utilizes direct oligo hybridization without sample RNA amplification and the other requiring sample amplification. A caveat to keep in mind is that there are inherent advantages and disadvantages to both approaches.

Nonetheless, three principal methods are currently used more than others to measure the expression levels of miRNAs and genes in general: microarray hybridization [10], real-time reverse transcription- PCR (qPCR) [11, 12] and massively parallel/next-generation sequencing (NGS) [13], all of which face unique challenges compared to their use in mRNA profiling. For example, the existence of miRNA families, the largest encompassing nine variants (hsa-let-7a–i), whose members differ by as little as one nucleotide but nevertheless exhibit differential expression patterns, represents a real challenge in miRNA recognition, regardless of the technology used. Microarray technology is actually based on the Watson–Crick base pairing nature of nucleic acids and thus involves nucleic acid hybridization between target molecules and their corresponding complementary probes. Synthesized antisense probes are spotted and immobilized onto a nylon support platform using a hand held spotting device. This method is relatively low cost and readily available to labs without specialized robotics and equipments dedicated to array fabrication. A disadvantage to this method is its scale. Oligo spots from a hand held device are macroscopic in nature, so the resulting array will be relatively large. About 30 mg of total RNA is commonly used to hybridize an array of this size [14]. To address this issue, automated robots have been employed to spot microscopic oligo dots onto a glass slide [15, 16].

Probes designed to differentiate between mature miRNAs and pre-miRNAs and probes that detect hypothetical miRNAs can all be spotted onto the same array. The isolated microRNAs are labeled with fluorescent dye and then hybridized with the microRNA microarray, resulting in specific binding of the labeled microRNAs to the corresponding probes. The fluorescence emission from labeled microRNAs bound at different positions on the slides can be detected. Consequently, the kinds of microRNAs and their relative quantities in the studied sample can be evaluated by analyzing the fluorescence signal data. The design of the microRNA probes, the preparation of microRNA samples and the labeling of microRNAs are considered the most important procedures in the microRNA microarray platform.

Direct hybridization of miRNA samples onto an oligo array may require a large amount of total RNA; however, some research protocols might have access to a small and limited amount of RNA—such as needle biopsies. A PCR based approach was developed to address this issue. The principle of qPCR is based on the detection, in real-time, of a fluorescent reporter molecule whose signal intensity correlates with amount of DNA present in each cycle of amplification.

In this method, total RNA is isolated as usual. The first step in qPCR of miRNAs is the accurate and complete conversion of RNA into complementary DNA (cDNA) by reverse transcription (RT). The RT reaction first consists of small RNA fractionation, followed by polyadenylation. Then a standard RT protocol is applied where poly(T)s are added to prime the synthesized poly(A) tail so reverse transcriptase can produce cDNAs from the small RNA. Finally, miRNA specific primers will probe for a specific miRNA through PCR amplification [17, 18]. The specificity and sensitivity of qPCR assays are dependent upon primer design. In fact, due to specificity issues and inability to differentiate between mature and pre-miRNA, changes have been made to the

RT step. Instead of a general poly(A) reaction in combination with universal priming through poly(T) adapter molecules, a miRNA specific stem-loop reverse primer is used. This specially designed primer contains a sequence that is antisense to a portion of the 30nt long sequence of the miRNA that is to be amplified. To increase the specificity of the PCR amplification step, the forward primer contains an antisense sequence derived from the mature miRNA, and the reverse primer consists of sequences taken from the stem-loop of the reverse primer. Sensitivity and specificity was found to be dramatically improved. In addition, the nature of specific priming allows this protocol to differentiate between the longer pre-miRNA and the shorter mature active form of the miRNA. Finally, it is claimed that this protocol can discriminate between isoforms of related miRNAs that differ by only one or two base pairs [18, 19].

The major advantages of qPCR over microarrays are (1) the speed and the sensitivity of the qPCR assays, (2) considerably larger dynamic range compared to microarray analysis and (3) a convenient requirement for low amounts of starting material (in the range of nanograms of total RNA).

However, both the RNA ligation [20] and the PCR amplification steps bear inherent biases, the method is laborious and costly, and associated tools for computational analysis are in their infancy. The reliability of miRNA expression profiling depends also on the quality of the total RNA used as input material. Robust, reproducible methods for RNA isolation and estimation of RNA quality should be employed prior to initiating the characterization of miRNA expression levels. The successful outcome of qPCR analysis depends upon a number of interconnected steps that require individual optimization. To perform qPCR that provides meaningful and reproducible results, several parameters such as RNA extraction, RNA integrity control, cDNA synthesis, primer design, amplicon detection, and data normalization must be taken into account.

qPCR is often considered a "gold standard" in the detection and quantitation of gene expression. However, the rapid increase in number of miRNAs renders qPCR inefficient on a genomic scale, and it is probably better used as a validation rather than a discovery tool.

As with genomic DNA and RNA analysis, microarrays are still the best choice for a standardized genome-wide assay that is amenable to high-throughput applications. Whole-genome screening generates a qualitative and quantitative evaluation of how experimental conditions affect miRNA profiles.

High-throughput sequencing of miRNAs, though, is coming into wider use and is unmatched for the discovery and experimental validation of novel or predicted miRNAs. The high demand for low-cost sequencing has driven the development of high-throughput sequencing technologies that parallelize the sequencing process, producing thousands or millions of sequences at once [21, 22]. These technologies are intended to lower the cost of DNA sequencing beyond what is possible with standard dye-terminator methods. In particular, next generation sequencing (NGS) technologies provide a digital expression profiling readout that is fundamentally different than analog measurement systems like microarrays. A variety of different approaches are being used. They generally involve the amplification of DNA templates by PCR and the physical binding of template DNA to a solid surface or to tiny beads called microbeads. These techniques are often referred to as massively parallel DNA sequencing, because thousands or millions of sequencing reactions are run at once to greatly speed up the process. All next generation sequencing systems use clonal cluster sequencing. The process, which begins with a single target molecule, involves creation of a clonal target during an intermediate amplification step. Multiple identical copies are required to produce a high signal-to-noise-ratio.

Finally, the Nanostring technology can be used to detect any type of nucleic acid in solution and could be modified with appropriate recognition probes to detect other biological molecules as well.

Nanostring utilizes a digital technology that is based on direct multiplexed measurement of gene expression that is capable of high level precision and sensitivity at less than one transcript copy per cell [23]. The technology uses molecular "barcodes" and single-molecule imaging to detect and count hundreds of unique transcripts in a single reaction [24]. Each color-coded barcode is attached to a single target-specific probe corresponding to a gene of interest. Mixed together with controls, they form a multiplexed assay. The degree of multiplexing is in the hundreds, which is less than that of microarrays. However, the Nanostring technology has higher throughput, accuracy and sensitivity than microarrays, which makes it preferable for low-multiplex applications, such as biomarker validation or molecular diagnostics [25].

Cancer research and biomarker validation are two of the areas where Nanostring has been most rapidly adopted. Advantages over existing platforms include direct measurement of mRNA expression levels without enzymatic reactions or bias, sensitivity coupled with high multiplex capability, and digital readout. Comparison of the Nanostring gene expression system with microarrays and TaqMan PCR demonstrated that the Nanostring system is more sensitive than microarrays and similar in sensitivity to real-time PCR [24].

Although each of these methods has their own unique advantages, they have not been perfected yet. However, at present, the method chosen for miRNA detection should best fit experience, the experimental conditions in the laboratory, and the goal of research.

15.2.2 miRNA Profiling-Normalization

The signal intensities of miRNA microarray experiments may be biased by differences in sample RNA preparation, dye labelling, hybridization and washing efficiency, peculiarities of print tip, spatial or hybridization specific effects or pre-amplification of extracted RNA. For these reasons normalization is an essential aspect of data processing.

It can minimize systematic, technical or experimental variation and thus has significant impact on the detection of differentially expressed miRNAs between two or more conditions.

Several studies pointed out that the selection of the data pre-processing method can have great impact on the resulting data outcome [26–30].

Inappropriate normalization of the data can lead to incorrect conclusions. Rigorous normalization of miRNA data may even be more critical than that of other RNA functional classes since relatively small changes in miRNA expression may be biologically and clinically significant [31, 32].

At present, there is no consensus normalization method for the three miRNA profiling approaches cited above. Several normalization techniques are similar to mRNA profiling normalization methods while others are specifically modified or developed for miRNA data. Indeed, miRNAs have some unique signatures such as their small total number and short length.

Prior to normalization, data pre-processing of miRNA profiling experiments includes platform and vendor specific steps, such as baseline adjustment and threshold setting for RT-qPCR analyses, background correction for microarray technology, or filtering for small RNA-sequence data. Following these very first steps of raw data pre-processing, one needs to choose the optimal normalization strategy to correct for systematic and technical variation enabling a better estimation of the biological variation.

15.2.2.1 Normalization Approaches for microRNA RT-PCR

RT-PCR is generally accepted as gold standard for miRNA expression measurement and normalized microRNA RT-PCR profiling data is used for the evaluation of the goodness of miRNA microarray normalization methods [27, 33].

Normalization of RT-qPCR miRNA profiling data is needed because signal intensities may depend on reverse transcription and PCR reaction efficiencies.

There are two types of sources of variation in RT-qPCR experiments. The first one is technical: there may be differences in sample procurement, stabilization, RNA extraction, reverse transcription and PCR reaction efficiencies. The second one is biological, there may be sample-to-sample inconsistencies in cellular subpopulations or even differences in bulk transcriptional activity. For these reasons normalization of RT-qPCR miRNA profiling data is needed.

The common normalization methods for microRNA RT-PCR profiling are based on predefined invariant endogenous controls, reference miRNAs [31] or other small non-coding RNAs such as small nuclear and small nucleolar RNA [28, 34, 35].

However, in [36] the authors argued that it is best to normalize genes with reference genes belonging to the same RNA class because the use of small non-coding RNAs other than miRNAs does not mirror the physicochemical properties of miRNA molecules.

Using non-miRNA reference genes for qPCR normalization is not advisable when the overall abundance of miRNA varies, e.g., in experiments affecting the miRNA processing machinery, or in comparisons involving multiple tissues or combinations of tissues and cell lines [37].

Selection of invariant miRNAs identified by algorithms specifically developed for reference gene evaluation and selection was superior over small non-coding RNA based normalization [31, 35]. These algorithms are based on reference gene ranking and stepwise elimination of the least stable gene [36], repeated pairwise correlation and regression analysis [38] or statistical linear mixed-effects modelling [39] of the respective experimental data.

Moreover invariant miRNAs can be selected based on a distinguishable low standard deviation and high-mean population as suggested by Pradervand et al. [28] for miRNA microarray preprocessing and this approach is applicable for RT-qPCR profiling experiments as well. Basically, the use of more than one reference gene increases the accuracy of quantification compared to the use of a single reference gene [36, 39].

Commonly used methods for miRNA raw data processing use median or mean value of the raw readings as normalization factor. However, many miRNAs may not be expressed in a biological sample, and thus median or mean value may be skewed towards the assay readings for lowly expressed miRNAs, which tend to be more variable compared with the readings for more abundantly expressed miRNAs. A scaling method suggested by Wang et al. [40], uses the average expression values of eight selected miRNAs with relatively high expression from a descending sorted list.

For large scale microRNA expression profiling studies the mean expression value normalization outperformed the current normalization strategy that makes use of stable small RNA controls, such as snoRNAs proposed by manufacturers, in terms of better reduction of technical variation [35].

However, the selection of a limited number of miRNAs or small RNA controls that resemble the mean expression value can be successfully used for normalization in follow-up studies where only a limited number of miRNA molecules are profiled to allow a more accurate assessment of relevant biological variation from a miRNA RT-qPCR profiling experiment [32, 35].

15.2.2.2 Normalization Methods for miRNA Microarray Experiments

Different normalization methods have been used on miRNA microarray expression profiling data sets, but there is currently no clear consensus about their relative performances [28].

Some have even chosen to omit normalization [41–43] but comparative studies on the relative performance of different normalization methods within a miRNA microarray platform have emphasized the need for evaluating and identifying appropriate normalization methods [27, 28, 44]. miRNA microarrays can be single-color or dual-color systems calling for different normalization approaches. Single-color miRNA microarrays have been predominately used, while dual-color hybridization systems are less frequently prevalent [44].

Both can be observed with respect to intra-array normalization for the correction of dye effects and inter-array approaches for the balance of the distribution differences among experiments [45].

The first normalization methods to be used with miRNA array data employed centring to median values [46–48] or scaling based on total array intensities [49, 50].

Certain methodologies currently used for large-scale genome arrays have been adapted to and modified for miRNA arrays such as Quantile [51] and LOESS (Locally Weighted Regression and Smoothing Scatterplots) [52]. Various assumptions are often taken by several normalization methods. Scaling, LOESS and Quantile [26, 27] are based on two assumptions, (i) only a small portion of spots is differentially expressed, and (ii) differentially expressed spots are homogeneously distributed with respect to both, over- and under-expressed miRNAs [29].

However, these assumptions could fail for miRNA platforms as they are printed with a relatively small number of selected sequences [27, 29]. Moreover, the number of expressed miRNAs in a miRNA microarray profiling is small (typically in the order of hundreds) compared with a few thousands of genes [30]. Hence, among the expressed microRNAs the proportion of those that are differentially expressed is much larger than that observed in mRNA expression profiling [30]. Experiments with most miRNAs differentially expressed predominantly in one direction, that is only up- or down-regulated, are not unusual.

Thus, it must be verified whether these assumptions hold true for the respective datasets and one should choose a normalization method that makes only minimal assumption about the presence of a set of constant miRNAs, like invariant-based normalization [28]. Alternatively, a normalization method free of assumption, the majority of algorithms for variance stabilization normalization [53] or even an assumption free approach [54] can be utilized instead.

Quantile normalization is a transformation method originally proposed by Bolstad et al. [51] for oligonucleotide arrays. It is now widely used for one-color miRNA microarrays as well and was confirmed as one of the most robust methods [27, 28, 44, 55]. It is an inter-array approach and equalizes the distributions of expression intensities across arrays. Thus, quantile normalization assumes that the overall distribution of signal intensity does not change. While this assumption likely holds true for the comparison of p53 overexpressing versus control cells [28] or even for brain–heart comparisons according to Rao et al. [44] where only 5% of miRNAs were differentially expressed, it may not hold true in case large numbers of miRNAs are differentially expressed in only one direction.

Such cases may be, for example, the knockout of proteins essential for miRNA biogenesis, which lead to a dramatic reduction in steady state miRNA levels by blocking production of mature miRNAs [44].

Rao et al. [44] compare the performance of several normalization methods on miRNA single channel microarray profiling, showing the better performance of quantile normalization.

Quantile normalization can be applied to dual-labeled array data if red and green channels are treated as two independent single-labeled array data. On the contrary two single-labeled array data can be considered as a dual-labeled data and LOESS normalization may be used in this case [56].

For the two colors microarray data, normalization is usually applied to the log-ratios of green channel signal (Cy3) and red channel (Cy5) signal, which will be written as M and A .

The LOESS normalization and its variants [27, 29, 44] are the most used transformation based methods. They use local regression via locally weighted scatter plot smoothing. M is defined as the log transformation of Cy3/Cy5 and A as the log transformation of the squared root of Cy3*Cy5 (as used in the MA-plot). It is advisable to introduce weights that penalize outliers because outlier values can strongly influence the local regression curve (LOWESS). However, Lowess and Loess are treated as synonyms. Local regression via LOESS uses a quadratic polynomial weighted regression function with Tukey's biweight function [52] of the log ratios of the Cy3 and Cy5 signals on overall spot intensity of the two signals (Cy3*Cy5).

In addition to intensity-dependent variation in log ratios, spatial bias could also be a significant source of systematic error (print-tip effect). It is possible to correct for both print-tip and intensity-dependent bias by performing a within-print-tip-group normalization using LOESS.

Print-tip LOESS normalizes each M value by subtracting from it the corresponding value on the tip-group LOESS curve [27]. Finally, the normalized log-ratios (N) are:

$$N = M - loess_i(A)$$

where $loess_i(A)$ is the loess curve as a function of A for the *ith* tip group.

However, Sarkar et al. [30] did not find significant differences between print-tip LOESS and other normalizations.

Hua et al. [27] compared 15 normalization methods using microarray data and RT-PCR data. It was found that microRNA normalized data by print-tip LOESS method were most consistent with the RT-PCR results.

A variant of LOESS normalization called LOESSM was proposed by Risso et al. [29]. This non-parametric normalization scales the expression data on the global median expression rather than on zero. This modification relaxes the assumption of symmetry among up- and down-regulated genes and it was shown that LOESSM, in case of absence of channel-effect, has better performance. In addition, LOESS combined with Generalized Procrustes Analysis (GPA), an assumption free inter-array normalization [54], improved its results and outperformed the other normalizations in terms of sensitivity and specificity [29].

LOESS normalizations and its variants emerged as being robust in the reduction of non-biological bias.

Variance stabilization normalization (VSN), an inter-array transformation method, is widely used for microRNA microarray data [28, 30]. It was developed for mRNA arrays and is based on a parameterized arcsinh transformation instead of a logarithmic transformation that calibrates sample-to-sample variations and renders variance approximately independent of the mean intensity [53].

Spike-in VSN normalization restricts the model fit to spike-in spots. These spots recognize specific RNA transcripts that can be added as internal controls in the experiments. Normalization intensities for all miRNAs are then obtained by applying the resulting transformation to all spots of interest on the array [30].

One limitation of this approach is that reliable results can only be obtained for intensities within the range covered by the spike-in used and that excludes targets that are not expressed.

Pradervand et al. [30] proposed a linear regression method to select a set of miRNAs with constant expression (invariants) and used these invariants to calculate VSN parameter (VSN-INV). The invariant probes are those that have medium-high mean intensity and low variance across samples. VSN used with default parameter settings assumes that most genes are not differentially expressed whereas the invariant-based regression only assumes that a subpopulation of expressed genes does not change. So, VSN-INV is appropriate only if a significant fraction of miRNAs is expected to be differentially expressed.

Based on theirs comparisons, Pradervand et al. found that VSN-INV and quantile normalization were the most robust normalization methods compared to VSN with default parameter or scaling. In general, one should note that VSN strongly affects the distribution of the large fraction of miRNAs whose expression is near or at background, resulting in the large increase of variability for those microRNAs.

15.2.2.3 Scaling Normalization

The first normalization methods for mRNA microarray were based on the selections of predefined and stably expressed housekeeping genes, as described by Garzon et al. and Perkins at al. [57, 58] that uses all probes. These methods have been applied to one- or two-channel miRNA microarray profiling. Most commercially available miRNA microarrays do not have controls for endogenous RNAs that have been shown to be robustly invariant between various different tissue samples or conditions [44]. To date, there is no consensus on the existence and reliability of reference gene miRNAs. The selection of reference genes to normalize miRNA levels depends on the bioinformatics analysis of the respective data (as shown for mRNA in [36, 39]) and is otherwise still rather empirical due to the lack of robust reference miRNAs [34], although a universal reference miRNA reagent set has been proposed [30].

Bargaje et al. [55] identified constitutively expressed miRNAs across tissues. A mean of expression levels of a set of 16 microRNAs showing minimum variability, was reasonably successful as a normalization factor for comparing datasets generated by the same platforms. However, normalization using constitutive microRNAs was ineffective when comparing bead-based and microarray-based datasets. In these cases quantile and Z-score normalization were both successful in transforming the data sets generating comparable means and scale.

The scaling methods like Z-score, mean, median, or 75th percentile assume that different sets of intensities differ by a constant global factor and all raw intensity values are multiplied with one common (i.e., global) scaling factor [26, 27, 55]. The Z-score provides a mean-centered rank for the expression level in units of standard deviation. Z-scores thus provide an index of the expression level of the miRNA with respect to the cellular pool of miRNA. Unlike other normalization methods, Z-scores are not influenced by the addition of new datasets allowing flexible cross-platform validation of miRNA microarray profiling experiments [55].

Recently, Wang et al. [40] suggested the pre-evaluation of the overall miRNA expression pattern by a panel of miRNAs using RT-qPCR assays to build a logistic regression model based on these results. The personalized logistic regression model based on 29 miRNAs efficiently calibrated the variance across arrays and improved miRNA microarray discovery accuracy compared with different scaling methods, LOESS or quantile normalization [40].

15.2.3 Identification of Differentially Expressed Genes and miRNA

Several methods have been applied to the identification of differentially expressed genes and microRNA in microarray data.

The simplest method is to evaluate the log ratio between two conditions (or the average of ratios when there are replicates) and consider all the genes that differ by more than an arbitrary cut-off value to be differentially expressed. This is not a statistical test, and there is no associated value that can indicate the level of confidence in the designation of genes as differentially or not differentially expressed.

It is considered to be unreliable [59] because statistical variability is not taken into account and is susceptible to outliers.

More sophisticated statistical methods have been proposed. The classification success is affected by the choice of the method, the number of genes in the genelist, the number of cases (samples) and the noise in the dataset.

Different methods produce dissimilar gene lists, which can produce dramatically different discrimination performance when trained as gene classifiers.

The gene lists produced by the feature selection methods can be grouped broadly according to the manner in which they treat gene variance.

15.2.3.1 t-Statistic

The simplest statistical method for detecting differential expression is *t* test. It can be used to compare two conditions when there is replication of samples. With more than two conditions, analysis of variance (ANOVA) can be used.

The t-test calculates the observed t-statistic for each gene. The idea is to compare between-group difference and within-group difference and then to calculate the probability value (p-value) of t-statistic for each gene from t-distribution.

The output of the analysis is a p-value for each gene. It represents the chance of getting the t-statistic as large as, or larger than the observed one, under the hypothesis of no differential expression (null hypothesis). A small p-value indicates that the hypothesis of no differential expression is not true and the gene is differentially expressed.

15.2.3.2 SAM

Several modified t-statistics have been proposed to address this problem. SAM [60] is one of the most popular. It performs moderately well except when applied to data with low sample size and to the noisy datasets.

SAM uses a moderated t-statistic, whereby a constant is added to the denominator of the t-statistic. The addition of this constant reduces the chance of detecting genes which have a low standard deviation by chance. The constant is estimated from the sum of the global standard error of the genes [61–63].

15.2.3.3 Empirical Bayes Method (Limma)

The empirical bayes method provides a more complex model of the gene variance. The gene standard error is estimated as a representative value of the variance of the genes at the same level of expression as the gene of interest [64]. In training sets with a large number of cases, the empirical bayes method performed comparably with ANOVA. Importantly, unlike most other methods, the empirical bayes t-statistic proved equally robust with low numbers of cases. The Bayesian statistic also provides p-values and has the advantage that it can be expanded to deal with datasets that have more then two classes.

Limma provides advanced statistical methods for linear modelling of microarray data and for identifying differentially expressed genes. It fits a linear model to the data and uses an empirical Bayes method for assessing differential expression [65]. One or two experiment definition matrices need to be specified during the analysis: a *design matrix* defining the RNA samples and a *contrast matrix* (optional for simple experiments) defining the comparisons to be performed.

When there are more than two conditions in an experiment, a more general concept of relative expression is needed. One approach that can be applied to cDNA microarray data from any experimental design is to use an analysis of variance model (ANOVA) to obtain estimates of the relative expression (*VG*) for each gene in each sample [66, 67]. In the ANOVA model, the expression level of a gene in a given sample is computed relative to the weighted average expression of that gene over all samples in the experiment.

The microarray ANOVA model is not based on ratios but it is applied directly to intensity data; the difference between two relative expression values can be interpreted as the mean log ratio for comparing two samples (as $\log A - \log B = \log(A/B)$, where $\log A$ and $\log B$ are two relative expression values). Alternatively, if each sample is compared with a common reference sample, one can use normalized ratios directly. This is an intuitive but less efficient approach to obtain relative expression values than using the ANOVA estimates. Direct estimates of relative expression can also be obtained from single-color expression assays [68].

The set of estimated relative expression values, one for each gene in each RNA sample, is a derived data set that can be subjected to a second level of analysis. There should be one relative expression value for each gene in each independent sample. The distinction between technical replication and biological replication should be kept in mind when interpreting results from the analysis of a derived data. If inference is being made on the basis of biological replicates and there is also technical replication in the experiment, the technical replicates should be averaged to yield a single value for each independent biological unit. The derived data can be analyzed on a gene-by-gene basis using standard ANOVA methods to test for differences among conditions.

15.2.3.4 ROC

Classifiers built using gene lists from the ROC method outperform all other methods when applied to large datasets. High RCI scores are observed even when only a few of the most highly ranked genes are examined. These high RCI scores are maintained when the number of genes examined is increased. It is possible to obtain p-values using this method [69]. ROC, like the t-statistic methods, loses power when the number of samples is reduced. It ranks a gene based on its power to discriminate between the groups given a threshold false positive rate. This means that it ignores the level of expression of the gene in the two groups. Therefore as the training size decreases, the likelihood of a gene with low variance and no biological meaning being a good discriminator by chance increases. ROC is an unsuitable method when the sample size is below 30 (class size of 15).

15.2.3.5 Rank Product

The Rank Product [70] package contains functions for the identification of differentially expressed genes using the rank product non-parametric method described in [63]. It generates a list of up- or down-regulated genes based on the estimated percentage of false positive predictions (pfp), which is also known as false discovery rate (FDR). The attractiveness of this method is its ability to analyse data sets from different origins (e.g. laboratories) or variable environments.

Rank product assumes constant variance across all samples. It compares the product of the ranks of genes in a class with the product of the ranks of genes in the second class. For each gene in the dataset, rank products sorts the genes according to the likelihood of observing their ranked positions on the lists of differentially expressed genes just by chance.

15.2.4 Clustering

Clustering algorithms are widely used in the analysis of microRNA profiling data. In clinical studies, they are not only used to cluster microRNA into groups of co-regulated miRNA, but also for clustering patients, and thereby defying novel disease entities based on miRNA expression profiles.

A reliable and precise classification of tumors is essential for successful diagnosis and treatment of cancer.

Current methods for classifying human malignancies rely on a variety of morphological, clinical, and molecular variables. In spite of recent progress, there are still uncertainties in diagnosis. Also, it is likely that the existing classes are heterogeneous and comprise diseases which are molecularly distinct and follow different clinical courses. microRNA microarray datasets have been used to characterize the molecular variations among tumors by monitoring microRNA expression profiles on a genomic scale. This led to more reliable classification of tumors and to the identification of marker miRNA that distinguish among these classes. Eventual clinical implications include an improved ability to understand and predict cancer survival. However, there are three main types of statistical problems associated with tumor classification:

- The identification of new tumor classes using microRNA expression profiles – unsupervised learning;
- The classification of malignancies into known classes – supervised learning
- The identification of marker microRNA that characterize the different tumor classes – feature selection.

Clustering can answer these problems. It is possible to cluster rows, columns or both. Rows (miRNA) clustering can identify groups of co-regulated miRNA, spatial or temporal expression patterns, reduce redundancy (cf. feature selection) in prediction, and detect experimental artefacts. On the other hand columns clustering allows to identify new classes of biological samples, new tumor classes or new cell types. Moreover, it allows to detect experimental artefacts.

In order to perform clustering, a way to measure how similar or dissimilar two objects are is needed. The feature data are often transformed to an n×n distance or similarity matrix, D=(dij), for the n objects to be clustered. Features correspond to expression levels of different microRNAs and possible classes include tumor types or clinical outcomes (survival, non-survival). Other information such as age and sex may also be important and can be included in the analysis. The most popular distances are Euclidean distance and Manhattan distance. Hamming distance is used for ordinal, binary or categorical data.

Clustering procedures can be divided into three categories: Hierarchical, Partitioning (K-means K-medoids/partitioning around medoids) and Model based approaches. The first one is either divisive or agglomerative and provides a hierarchy of clusters, from the smallest, where all objects are in one cluster, through to the largest set, where each observation is in its own cluster. One must often also define a distance measure between clusters or groups of miRNA and the linkage methods used are single, complete, average, distance between centroids and Ward Linkage. Hierarchical clustering methods produce a tree or dendrogram. The partitions are obtained from cutting the tree at different levels. The tree can be built in two distinct ways bottom-up (agglomerative clustering) or top-down (divisive clustering). Examples of Hierarchical clustering methods are Self-Organizing Tree Algorithm – SOTA [71] and DIvisive ANAlysis – DIANA [72].

Partitioning methods require the specification of the number of clusters. A mechanism for apportioning objects to clusters must be determined, then data is portioned into a prespecified number K of mutually exclusive and exhaustive groups and iteratively reallocated to clusters until some criterion is met, e.g., minimize within-cluster sums-of-squares. Examples of partitioning methods are k -means and its extension to fuzzy k -means, Partitioning Around Medoids – PAM [72], – Self-Organizing Maps – SOM [73] and model-based clustering, e.g., Gaussian mixtures in [74–76] and McLachlan et al. [77, 78].

An important feature of partitioning methods consists in satisfying an optimality criterion (approximately), however they need an initial K and long computation time. Hierarchical methods are computationally fast (for agglomerative clustering) but rigid, since they cannot later correct for earlier erroneous decisions.

Most methods used in practice are agglomerative hierarchical methods. In large part, this is due to the availability of efficient exact algorithms that implement them.

Model based approaches assume that data are 'generated' from a mixture of K distribution. They try to fit a model to the data and try to get the best fit. A classic example is a mixture of Gaussians (mixture of normals). They take advantage of probability theory and well-defined distributions in statistics.

In microarray experiments is also useful to detect the presence of outliers. Outlier detection is an important step since they can greatly affect the between-cluster distances. Simple tests for outliers should be identifying observations that are responsible for a disproportionate amount of the within-cluster sum-of-squares.

Most features in high dimensional datasets will be uninformative, examples are unexpressed genes, housekeeping genes, 'passenger alterations'. Clustering (and classification) has a much higher chance of success if uninformative features are removed. Simple approaches to feature selection are: selecting intrinsically variable genes or requiring a minimum level of expression in a proportion of samples.

Clustering can be also employed for quality control purposes. The clusters that are obtain from clustering samples/microRNA should be compared with different experimental conditions such as batch or production order of the arrays, batch of reagents, microRNA amplification procedure, technician, plate origin of clones, and so on. Any relationships observed should be considered as a potentially serious source of bias.

15.2.5 miRNA as Biomarkers

miRNAs have a very important role in cancer. Their expression is often dysregulated in malignant cells. Some miRNAs that are temporarily over-expressed in early development and shut off in the normal differentiated state may re-express in cancer, causing a persistent stem cell–like dedifferentiated state. Many miRNAs may act like oncogenes by promoting proliferation and/or repressing apoptosis. Other miRNAs play the role of tumor-suppressors. They have a regulatory function in normal tissues but when they are down-regulated in cancer, they abrogate their tumor-suppressor activity.

Over-expression or lack of expression of specific miRNAs appears to correlate with clinically aggressive or metastatic phenotypes [79, 80].

miRNA expression has tissue specificity and has been used for identifying the tissue in which cancers of unknown primary origin arose [81]. Rosenfeld and colleagues constructed a miRNA-based tissue classifier by measuring miRNA expression levels using a microarray platform in 336 primary and metastatic tumors representing 22 different cancer types. They built and tested a classifier for 48 miRNAs that accurately predicted tissue type in 86% of the test set, including 77% of the metastatic samples. Moreover, the classifier predicted tissue type with 100% accuracy for six of the ten tumor types in the metastatic test set. The authors proposed that their classification system could be applied to cancer of unknown primary origin, defined as histologically confirmed metastatic cancer for which no primary site of disease can be identified.

Cancer classifications previously determined by mRNA expression profiling are now being investigated with miRNAs. One study has directly compared mRNA and miRNA microarray expression data and shown that known molecular subtypes of breast cancer can also be identified using miRNAs, and that expression of processing enzymes and proteins involved in miRNA biogenesis are down-regulated in the more aggressive subtypes [82]. Clinical trials are underway that test different therapies in different breast cancer molecular subtypes as defined by mRNA expression.

Mitchel et al. [83] discovered microRNAs in healthy human plasma that can be traced back to specific tissue (miRNA-15b, miRNA-16, and miRNA-24). In addition, they found that serum is more readily available than plasma and the stability of miRNA compared to the plasma is strongly positively correlated.

They found the baseline levels of miRNA expression in healthy individuals and detected the levels of prostate cancer-expressed miRNAs (miRNA-100, miRNA-125b, miRNA-141, miRNA-143, miRNA-205, and miRNA-296) in serum. miRNA-141 level is specifically elevated in prostate cancer in serum and several experiments illustrated that miRNA-141 is expressed by several common human cancers. They established that tumor-derived miRNAs can be detected in plasma or serum and serve as an effective circulating biomarker of common human cancer types.

A lot of benefits will come from using miRNA to diagnose cancer. miRNA is 97.6% accurate for sensitivity as a biomarker for cancer and 96.3% accurate as a biomarker for the classification of cancer [84]. It means less false positive or false negative cases. It will decrease the delay in diagnosis of cancer because a blood test with miRNA assay or electrophoresis is much cheaper and sufficient for diagnosis.

It will avoid invasive, expensive and/or unnecessary tests to find out if a patient has cancer and what type of cancer. All this will get patients less stressed.

Significant progress has been made on the relationship between miRNAs and cancers and the important function of miRNAs in a variety of cancers has been reviewed by several research groups.

In fact as shown by Lu et al., the miRNA expression profile based on the expression of only 200 miRNA genes successfully classified poorly differentiated tumors confirming in the majority of cases the clinical diagnosis whereas mRNA profiling, based on the expression of about 16,000 protein coding genes, failed to do so [85].

Visone et al. found out that miR-181b is a unique biomarker for CLL since its expression can be monitored throughout the disease course of a patient and this change in the leukemic cells correlate with the overexpression of four genes with great significance in CLL and other cancers (i.e. MCL1, TCL1, BCL2 and AID). Collectively, this information together with the analysis of stable prognostic markers (e.g. ZAP-70 and IGHV mutation status) specify disease progression in chronic lymphocytic leukemia and is associated with clinical outcome [86].

Finally, a significant justification for using miRNAs as biomarkers is that miRNAs have an unusually high stability in formalin-fixed tissues, which means that the miRNA can be stored and extracted with minute degradation. Short miRNAs from older tumors preserved as formalin-fixed paraffin-embedded tissue are less susceptible to chemical modification and degradation over time and have proven satisfactory for miRNA analysis.

15.3 miRNA Target Prediction

In order to determine miRNA functions it is fundamental to find their targets. While miRNA target prediction in plants is rather simple, due to the perfect complementarity that plant miRNAs usually exhibit to their targets, the prediction of miRNA binding sites in animals is much more challenging. In fact, perfect complementarity in animals is usually limited to the 5′ end of the miRNA, which is usually referred to as the seed (~6–8 nt long) [87]. The target sites are usually located in the 3′ UTR sequences of mRNAs. In order to significantly reduce the number of false positives, other determinants are needed due to the fact that the short length of the miRNA seeds raises the probability of finding random matches that don't correspond to functional sites [7]. Such determinants or rules should be primarily inferred from experimentally verified targets, thus showing how important it is to have good sources of data as the basic step in the development of prediction tools. A significant amount of miRNA/target interactions data, usually coming from the literature, is publicly available on web databases, such as Tarbase [88] and miRecords [89]. Information on the binding sites of miRNAs in their verified targets is usually provided by this data. Moreover, high-throughput sequencing of RNAs isolated by crosslinking immunoprecipitation (HITS-CLIP) has recently identified functional RISC interaction sites on mRNAs, allowing the creation of libraries of reliable miRNA binding sites [90]. Data Mining analysis of these sequences could help identifying important discriminant features for the prediction of new binding sites.

In predicting functional targets, miRNA/target interaction rules are generally not sufficient due to the high number of false positives that derive from random matches of the short seed region of miRNAs to false targets. Consequently, other kinds of data are needed to improve prediction algorithms. For instance, target conservation is widely used as a valid additional criterion. High sequence conservation is indeed revealed by the alignment of miRNAs in different species, especially in the seed regions, which often corresponds to high conservation of their targets. Therefore, an help in detecting functional sites could come from the identification of conserved regions in the 3′ UTR of a gene, even though this approach is not useful in the case of non-conserved miRNAs [91]. Several prediction methods exploit thermodynamics properties. Free energy (ΔG) can be used to evaluate the stability of the predicted duplexes. Low values of free energy, usually below −20 kcal/mol, characterize indeed all validated miRNA/target pairs [92]. A low energy value, however, is a necessary but not sufficient condition. Not all energetically favourable miRNA/target duplexes, in fact, are functional. Structural accessibility of the target molecule is another thermodynamic feature used by computational methods. miRNA binding sites shouldn't be involved in any intra-molecular base pairing, and any existing secondary structure should be disrupted in order to make the site accessible to the miRNA [93]. This very complex problem mostly relies on secondary structure prediction computation, which is still one of the challenges of computational biology [94].

Nucleotide composition surrounding the binding sites and the position of the sites in the UTR, as well as the presence of multiple sites on the same UTR, are additional features used by prediction tools. In fact, it is proven that a single miRNA can have more binding sites on the same target and that a target can have multiple sites for different miRNAs [95].

15.3.1 Tools for the Prediction of miRNA Targets

Several computational tools for the prediction of miRNA targets are currently available on the web [96].

In this subsection we will review the basic concepts behind the most popular ones: TargetScan, miRanda, Pictar, Diana-microT, RNA22, RNAHybrid, StarMir and PITA.

TargetScan is one of the most popular tools for miRNA targets prediction. It's a sophisticated algorithm based on both conservation and base pairing rules [91, 95] that searches for miRNA seed matches on UTRs, considering different kinds of seeds and making use also of secondary structure prediction in order to calculate the free energy of the predicted duplexes. In addition, it considers the presence of multiple sites for the same miRNA on a target as positive contribution to the score of the prediction. Through sequence alignment, TargetScan also takes into account the conservation on different species for the identification of the most probable targets. All the predictions, computed for different species like human, mouse and rat, are available on the TargetScan website.

miRanda is another web tool that performs predictions and it's based on an alignment algorithm which uses a weighted matrix aimed at promoting the binding of the seed of the miRNA rather than its 3′ end. It also uses the free energy of predicted duplexes and the conservation criteria to select the most probable targets. Its website allows predictions on human, mouse and rat [97, 98].

Another popular tool for the prediction of miRNA targets on vertebrates, nematodes and flies is PicTar [99]. Its algorithm is trained to identify binding sites for a single miRNA and multiple sites regulated by different miRNAs acting cooperatively. It makes use of a pairwise alignment algorithm in order to find sites conserved in many species (7 Drosophila species and 8 vertebrate species), considering also the clustering and co-expression of miRNAs together with ontological information, such as the time and tissue specificity of miRNAs and their potential targets, to enhance its predictions.

Diana-MicroT implements an algorithm that is trained to identify targets with a single binding site for a miRNA [100]. Its sequence alignment algorithm focuses on the search for miRNA/target duplexes characterized by central bulges and paired 5′ and 3′ ends.

A different approach is instead adopted by the web tool RNA22. It performs the analysis of miRNA sequences to find intra- and inter-species patterns of conserved sequence features [101]. The algorithm generates the reverse complement of the most significant patterns and searches for their instances in the UTRs in order to identify the target islands supported by a minimum number of pattern hits. A target island is defined as any hot spot where the reverse complement of mature miRNA patterns aggregate. It then computes the pairing of each target island with each candidate miRNA and evaluates the thermodynamic stability of the duplex obtained.

The miRNA target prediction tool RNAHybrid is conceived as an extension of the RNA secondary structure prediction algorithm by Zuker and Stiegler to two sequences [92]. Hybridization of the miRNA to the target is considered through an energetically optimal criterion, i.e. yielding the Minimum Free Energy (MFE), but absolutely avoiding intra-molecular base pairing and multiloops. The algorithm used adopts dynamic programming, forcing the perfect match of the seed. Bulges and internal loops are restricted to a constant maximum length in either sequence.

The computation of the structural accessibility of the targets is instead the main feature of the tools StarMir and PITA. StarMir is based on the target's secondary structure as predicted by the tool Sfold [102]. The miRNA/target interaction is modelled as a two-step hybridization reaction: the nucleation at an accessible site and the hybrid elongation to disrupt the local secondary structure of the target and form the complete duplex. PITA is based on a slightly different model which computes the difference between the free energy gained from the formation of the miRNA/target duplex and the energetic cost of unpairing the target to make it accessible to the miRNA [103].

In spite of the rather successful predictions of effective miRNA targets performed by the tools mentioned above, the problem still remains a big challenge. The high number of false positives and the use of conservation criteria clearly show our partial knowledge in the targeting mechanisms. Combining Data Mining, Pattern Discovery and Machine Learning techniques together with thermodynamics and the availability of more reliable experimental data, will allow the improvement of predictions and enhance our knowledge and understanding of RNAi.

15.4 Functional Annotation of miRNAs

As discussed in the previous section, our knowledge about the molecular rules that underlie miRNA targeting is still incomplete, hence the huge number of false positives that target prediction tools can produce. Functional analysis of miRNAs may help to identify the most probable targets and to uncover non trivial relationships between miRNAs and other molecular actors, such as genes and transcription factors, allowing the development of regulatory models describing complex processes and the effects of their dysregulation. There are several tools available online which collect and integrate miRNA-related data retrieved from different sources in order to infer miRNA functions.

In this section we are going to describe the most popular tools for functional analysis of miRNAs, that we divided in three categories: tools for miRNA/phenotype associations, tools integrating target prediction with expression data and tools for the generation and the analysis of network models of miRNA function.

15.4.1 miRNA/Phenotype Associations

Several tools provide users with manually curated information about the involvement of miRNAs in diseases and biological processes. Some of them also make use of computational predictions, statistics and data mining features in order to filter the data and infer new knowledge.

miR2Disease and the Human microRNA Disease Database are manually curated databases based on experimental data. They aim at providing a comprehensive resource of miRNA deregulation in various human diseases [104, 105]. These web based tools offer user friendly interfaces to query the information on miRNA/disease relationships. miR2Disease also allows researchers to contribute to the data contents through a submission page.

The authors of the Human microRNA Disease Database performed some analysis on their dataset and found that there is a negative correlation between the tissue-specificity of a microRNA and the number of diseases associated to it. They also found that miRNAs that are close in the genome, like members of the same clusters, are often associated with the same diseases. This suggests that neighboring miRNAs might be regulated by common regulators, and that they might regulate different genes involved in the same pathways. Finally, the analysis revealed that miRNAs which are conserved in other species, tend to be significantly associated with diseases with a higher probability.

miReg is also a manually curated miRNA Regulation Resource that provides users with regulatory relationships among validated upstream regulators like transcription factors or drugs, downstream targets, associated biological processes, experimental conditions or disease states and dysregulation of the miRNA in those conditions [106]. All the collected data is described in the literature and the corresponding references are provided together with other useful links about the studied miRNAs. The website has a user-friendly interface browseable through different options.

A further step in the integration of heterogeneous information about miRNA is miRo', a web environment that provides users with miRNA–phenotype associations in humans [107]. It integrates data from various online sources, such as databases of miRNAs and targets, Gene Ontology terms and diseases into a unified database equipped with a flexible query interface and data mining facilities. miRo' allows both simple and advanced queries and introduces a new layer of associations between genes and phenotypes inferred based on miRNAs annotations.

miRNAs are connected to diseases, GO processes and functions through their validated and predicted targets (miRecords, miRanda, PicTar, TargetScan) [89, 91, 97, 99, 108].

The simple search allows the selection of a single miRNA, process, function, disease or tissue and quickly displays the corresponding information. By selecting a miRNA, for example, the user can obtain the list of diseases, processes and functions in which the miRNA is potentially involved through its targets. Moreover, a list of tissues expressing the miRNA and the corresponding expression values is given. These are obtained from the Mammalian microRNA Atlas [109].

Similarly, by selecting a process, a disease, a function or a tissue, the user obtains a list of miRNAs related to the selected item. In all cases, detailed information about the miRNAs, the targets and the source of predictions are given, together with links to the original data sources.

The advanced search allows users to perform more complex queries through the introduction of specific constraints that data must satisfy. For example, it is possible to search for all the miRNAs which are involved in a group of diseases and processes or for all the diseases related to a group of miRNAs, genes, processes and functions. The results are given in a table with details about the miRNA/target predictions. Furthermore, this advanced query tool allows to identify new potential associations between diseases, processes and functions inferred based on miRNA annotations. For example, a disease d and a process p which are not linked through any common gene might be associated through a miRNA which regulates a gene gd, involved in d, and a gene gp, involved in p.

miRo' is also equipped with a special Data Mining module which allows clustering of miRNAs that are associated to the same set of terms. Chosen a set of up to five miRNAs and an association criteria (i.e. process or disease), the system will find all the subsets of the selected miRNAs which are closely associated to groups of processes or diseases. This feature may help to identify a set of miRNAs acting cooperatively to carry out certain biological functions. Moreover, a specificity score allows to evaluate the relationships between the miRNAs and their annotation terms.

In a similar way, the tool FAME uses computational target predictions in order to automatically infer the processes affected by human miRNAs [110]. The website provides a simple menu for retrieving offline computed data. By choosing a miRNA from the list, the user obtains two tables reporting the most significantly associated Gene Ontology processes and KEGG Pathways, respectively.

For each miRNA-process/miRNA-pathway association, a score, a p-value, a q-value and an enrichment factor are given, together with the list of target genes involved in the process/pathway.

In the paper, the authors used their method to identify 68 miRNA families and 27 genomic clusters regulating 21 gene co-expression clusters in diverse human stem cell lines. They found out that clusters enriched with the targets of a specific miRNA tend to be anti-correlated with the miRNA expression, whereas clusters depleted of miRNA targets are co-expressed with it.

15.4.2 miRNA Target Prediction Consensus and Gene Expression Data Integration

Most of the available tools for miRNA functional analysis make use of heterogeneous information, and their classification into categories, based on their purposes and the kind of data that they use, is not an easy task. However, there is a well distinct class of tools which make use of miRNA and gene expression data, either retrieved from public sources or provided by users. As discussed in Sects. 15.2 and 15.3, miRNA and gene expression profiling is an important source of information in the study of miRNA functions. In this section we introduce miRonTop, MAGIA and Diana-miRExTra, three tools that combine target prediction with expression data.

miRonTop is an online application allowing the detection of miRNAs that significantly affect gene expression at a large scale [111]. It is a java web tool that integrates DNA microarrays or high-throughput sequencing data with target predictions in order to identify the potential implication of miRNAs on a specific biological system.

Users have to provide a table summarizing a large-scale gene expression study in a tab-delimited file, and select the prediction software to be used, among miRbase, miRanda, TargetScan, PicTar or the exact seed (7-mer/8-mer) match.

The program then performs an enrichment analysis of the predicted targets, for each miRNA considered in the expression table, according to the selected prediction tool across the DOWN and the UP gene sets. The significance is evaluated using the hypergeometric distribution.

MAGIA is a web-based tool which allows to retrieve and browse miRNA target predictions for human miRNAs, based on a number of different algorithms (PITA, miRanda and TargetScan), setting cuttoffs on prediction scores, with the possibility of combining them with Boolean operators [112]. The query output is a table including the list of predicted target genes or transcripts with different prediction scores according to the methods chosen by the user. For each prediction several external links are provided.

The tool also includes an analysis framework. Given as input miRNA and gene expression profiles (MATCHED or UNMATCHED expression data) it provides different statistical measures of profiles relatedness and algorithms for expression profiles combination.

For unmatched expression data, MAGIA employs a meta-analysis approach based on a p-value combination, while one of four different measures of relatedness (Spearman and Pearson correlation, mutual information, and a variational Bayesian model) can be adopted for the analysis of matched profiles.

The results are reported in a web page containing different sections. For the top 250 most probable functional miRNA–mRNA interactions according to the association measure selected by the user, the interactive bipartite regulatory network obtained through the analysis is reported along with the corresponding browsable table of relationships.

Finally, Diana-miRExTra is a web-based tool that allows the detection of overrepresented motifs (hexamers) on the 3′ UTRs of deregulated genes, in order to identify miRNAs responsible for such deregulation [113].

The input consists of two lists: a list of changed genes and a list of unchanged genes (background). Moreover, the web server offers the option to use evolutionary information in order to refine results.

Instead of a gene list the user may provide a list of genes with associated fold change values (or any other metric used in high-throughput experiments). Optionally, the user may provide a list of miRNAs of interest to calculate results only for hexamers corresponding to these miRNAs.

The tool compares the distributions of all possible hexamers on the 3′UTR sequences between changed and unchanged genes. A one-sided Wilcoxon Rank Sum test is used in order to identify hexamers that are present significantly more often in the set of changed genes compared to the background of unchanged genes. A p-value for each motif is calculated signifying the probability that the changed and unchanged sets are produced by the same distribution and the differences between them are due to chance alone. DIANA-mirExTra provides a combinatorial hexamer score that takes into account the whole active region of the 8 first nucleotides of the miRNA.

15.4.3 miRNA, Gene Expression and Networks

The third class of miRNA functional analysis tools that we consider provides users with network oriented data. Networks constitute an effective tool for modelling complex biological systems and since miRNAs play a central role in many processes and pathways, it is important to have tools able to integrate miRNA related data into networks. In this subsection we briefly introduce four different tools which combine miRNA related data with other information such as transcription factors or gene expression in order to create interaction networks which model and describe the molecular systems involving miRNA regulation. Most of these tools also offer computational facilities for the visualization and the analysis of such networks.

The first tool that we describe is strictly connected to Diana-miRExTra, introduced in the previous section, and is called Diana-miRPath. It is a web-based computational tool developed to identify molecular pathways potentially altered by the expression of single or multiple microRNAs [114]. The user can select either a single miRNA or multiple miRNAs and specify the tools used for the prediction of targets, among Diana-MicroT, PicTar and TargetScan [115]. The software then performs an enrichment analysis of the predicted miRNA targets comparing them to all known KEGG pathways. The output consists of a list of pathways in which the miRNA is potentially involved through its target genes. For each association an enrichment p-value is given. When working on multiple miRNAs, the algorithm also performs an enrichment analysis of the Union and Intersection target sets. The graphical output of the program provides an overview of the parts of the pathway modulated by microRNAs, facilitating the interpretation and presentation of the analysis results. A direct link to the Diana-miRPath analysis is also provided in Diana-miRExTra for the targets of each miRNA belonging to the set of 'changed' genes.

MIR@NT@N is a tool which predicts regulatory networks and sub-networks including conserved motifs, feedback loops (FBL) and feed-forward loops (FFL) [116]. It integrates Transcription Factors, miRNAs and genes into a unified model and allows the identification and the analysis of molecular interaction networks within a given biological context.

The MIR@NT@N database integrates information from multiple available databases: PAZAR, JASPAR and oPOSSUM for TF regulations, miRBase, MicroCosm and microRNA.org for miRNA target predictions, UniHI for protein-protein interactions and Ensembl for gene annotations [117–124]. The tool is based on a meta-regulation network model that illustrates interactions between the considered three biological entities, transcription factors, microRNAs and protein-coding genes.

The tool allows to perform two types of query. The first type allows to search for novel key actors in a biological context. This query includes three sections. The first one is called Transcription Factor regulation which statistically predicts potential TFs regulating a list of miRNAs, or conversely miRNAs regulated by a list of TFs.

The second section is called miRNA regulation and allows the prediction of significant targets of a list of miRNAs or the miRNAs targeting a list of genes. The third section is called Regulation Network. It allows to reconstitute meta-regulation networks together with the detection of regulatory motifs such as FBL or FFL, by combining both TF and miRNA regulation predictions.

Users can also provide a list of miRNA-gene interactions experimentally inferred from microarray data combining genes and miRNA expression, or a list of published TF-miRNA interactions.

The second type of query provides an overview on any TF, gene or miRNA, including their interactions. It has two types of search called Quick Search and Quick Network.

The first one rapidly retrieves information on any actor, its regulators and/or targets, while Quick Network generates regulation networks from a list of actors presumed to be involved in a particular biological context, and also allows the extraction of sub-networks including regulatory motifs. The output is an exportable interaction graph recapitulating all predicted interactions and which is linked to external resources.

Based on these predictions, the user can generate networks and further analyze them to identify sub-networks, including motifs such as FBL and FFL. In addition, networks can be built from lists of molecular actors in a given biological process to predict novel and unanticipated interactions.

miRConnX is a web tool for the identification of gene network motifs involving transcription factors and miRNAs [125]. Users have to provide a document with a gene expression profile. Optionally, a document with a miRNA expression profile can be provided.

The output consists of the graphic visualization of networks involving miRNAs, transcription factors and miRNA-regulated genes.

Details about the miRNA/gene and the transcription factor/miRNA interactions are provided in tables, reporting the effect (activation/repression), the identified FFL motifs, if any, the strength of the interaction and several links to other resources about the corresponding miRNA and genes, like Gene Ontology, miRo' and miR2Disease.

The tool uses a pre-compiled network, which is derived from transcription factors binding predictions, miRNA target predictions and literature evidences. All the connections in this network correspond to direct, predicted or verified interactions. Another network based on the input expression data is then created by using a statistical association measure. This network connects transcription factors and miRNAs and doesn't discriminate between direct and indirect interactions. The two networks are superimposed via an integration function. The result is a directed network, which is a smaller version of the pre-compiled network, refined by the user provided expression data. Since the expression profiles can be related to a certain disease or phenotype, the resulting network is representative of the condition of interest.

Finally, we describe miRScape, a Cytoscape plugin for annotating networks with miRNAs. Cytoscape is a software environment for the visualization and analysis of biological networks [126–129]. It has a basic set of features for data integration and visualization, while additional features are available as plug-ins. miRScape is the first Cytoscape plug-in allowing the mining of biological networks annotated with miRNAs. The data is retrieved from miRò, thus miRScape represents a bridge connecting miRò and Cytoscape. Given a network, previously loaded into Cytoscape, miRScape allows to identify relationships among genes, processes, functions and diseases at the miRNA level and annotating them as attributes of each network node. These annotated networks may be further analyzed by using mining features available as plug-ins on Cytoscape allowing to find for examples hubs, interesting motifs and so on.

mirScape is equipped with two modules, available on two different panels. The first panel allows users to perform a "Search by Gene" query. Once a set of nodes in the network have been selected, users can choose the kind of data to be retrieved from miRò, which can be processes, functions, and diseases in which the selected genes are involved and the miRNAs regulating them. The result is the annotation of the network with the obtained information. The "Search by miRNA" panel allows the selection of a set of miRNAs and the source of miRNA target information, which can be TargetScan, PicTar, miRanda and miRecords (validated interactions). Moreover, it is possible to choose to annotate the nodes with information about the related diseases, processes and functions.

Once the information has been retrieved from miRò, the new miRNA nodes are added to the network and connected to target gene nodes, if they are present in the network. The annotation function allows to store such acquired data as network attributes. However, the annotation function can be used as a stand-alone tool, storing in the network all the information retrievable from miRò.

15.5 Conclusions

MiRNA and, more in general, ncRNA research is in its golden age. It is clear that miRNAs are involved in a variety of fundamental processes and that their dysregulation can be related to cancer and many other diseases. Evidence shows that they don't act as single actors but cooperate among themselves and together with other molecules, like transcription factors, to regulate gene expression and, indirectly, carry out specific functions.

The number of computational tools for the analysis of miRNA related data is continuously increasing. They range from simple look-up resources to more sophisticated analysis tools. Some of them are based on manually curated information but the vast majority makes also use of computationally predicted data. Although miRNA profiling is a valuable diagnostic and prognostic tool itself, allowing the classification of samples and the identification of biomarkers, the central data in the analysis pipeline is the target gene, through which the miRNA is connected to all the other data. Indeed, miRNAs exert their functions by directly regulating the expression of their target genes and most genes are well annotated with the processes, diseases and pathways in which they are involved. Thus, miRNAs inherit these annotations, but this only represents a first step in their functional analysis. Much effort is needed to uncover the real role of miRNAs in the great number of processes and diseases in which they are potentially involved and this is the ultimate goal of most of the computational tools reviewed in this chapter.

Some of them are focused on specific kinds of data, while others try to provide a complete view of the environment in which miRNAs operate and offer modules for the analysis of the complex relationships in that they intertwine with the other molecular actors.

The increase of precision in the data produced by the use of new technologies for the measurement of gene expression and high-throughput sequencing, involves the need for more sophisticated software tools for the analysis of this data.

As in a bottom-up schema, the collected raw data constitutes a first layer. The upper layers consist of tools for the annotation of this data, often focused on specific aspects. This annotated data constitutes the input for the top layer tools, whose aim is the integration of heterogeneous information in order to produce general models of miRNA functions in the context of complex processes. These tools must be equipped with powerful analysis facilities, helping researchers to formulate concrete functional hypotheses and guiding them to design the correct experiments to perform hypotheses validation. Then, the data produced with these experiments represents a feedback for the refinement of the analysis pipeline.

The final key point is the integration of public data with user data, and this is already partly fulfilled by some of the reviewed tools. In fact, many users typically get original data from their experiments, thus it is important to have tools able to combine this data with the other information stored in databases, in order to produce more reliable models specific to user needs.

References

1. Jovanovic M, Hengartner MO (2006) miRNAs and apoptosis: RNAs to die for. Oncogene 25:6176–6187
2. Wienholds E, Plasterk RHA (2005) MicroRNA function in animal development. FEBS Lett 579:5911–5922
3. Xiao C, Rajewsky K (2009) MicroRNA control in the immune system: basic principles. Cell 136(1):26–36
4. Small EM, Olson EN (2011) Pervasive roles of microRNAs in cardiovascular biology. Nature 469(7330):336–342
5. Lau P, de Strooper B (2010) Dysregulated microRNAs in neurodegenerative disorders. Semin Cell Dev Biol 21(7):768–773
6. Iorio MV, Croce CM (2009) MicroRNAs in cancer: small molecules with a huge impact. J Clin Oncol 27(34):5848–5856
7. Bartel DP (2009) MicroRNAs: target recognition and regulatory functions. Cell 136(2):215–233
8. Didiano D, Hobert O (2006) Perfect seed pairing is not a generally reliable predictor for miRNA-target interactions. Nat Struct Mol Biol 13(9):849–851
9. Didiano D, Hobert O (2008) Molecular architecture of a miRNA-regulated 3′ UTR. RNA 14(7):1297–1317
10. Yin JQ, Zhao RC, Morris KV (2008) Profiling microRNA expression with microarrays. Trends Biotechnol 26:70–76
11. Chen C, Ridzon DA, Broomer AJ, Zhou Z, Lee DH, Nguyen JT, Barbisin M, Xu NL, Mahuvakar VR, Andersen MR et al (2005) Real-time quantification of microRNAs by stem–loop RT-PCR. Nucleic Acids Res 33:e179. doi:10.1093/nar/gni178
12. Shi R, Chiang VL (2005) Facile means for quantifying microRNA expression by real-time PCR. Biotechniques 39:519–525
13. Hafner M, Landgraf P, Ludwig J, Rice A, Ojo T, Lin C, Holoch D, Lim C, Tuschl T (2008) Identification of microRNAs and other small regulatory RNAs using cDNA library sequencing. Methods 44:3–12
14. Krichevsky AM, King KS, Donahue CP, Khrapko K, Kosik KS (2003) A microRNA array reveals extensive regulation of microRNAs during brain development. RNA 9:1274–1281
15. Liu CG, Calin GA, Meloon B, Gamliel N, Sevignani C, Ferracin M, Dumitru CD, Shimizu M, Zupo S, Dono M, Alder H, Bullrich F, Negrini M, Croce CM (2004) An oligonucleotide microchip for genome-wide microRNA profiling in human and mouse tissues. Proc Natl Acad Sci USA 101:9740–9744
16. Zhao JJ, Hua YJ, Sun DG, Meng XX, Xiao HS, Ma X (2006) Genome-wide microRNA profiling in human fetal nervous tissues by oligonucleotide microarray. Childs Nerv Syst 22:1419–1425
17. Schmittgen TD, Jiang J, Liu Q, Yang L (2004) A high-throughput method to monitor the expression of microRNA precursors. Nucleic Acids Res 32:e43
18. Chen C, Ridzon DA, Broomer AJ, Zhou Z, Lee DH, Nguyen JT, Barbisin M, Xu NL, Mahuvakar VR, Andersen MR, Lao KQ, Livak KJ, Guegler KJ (2005) Real-time quantification of microRNAs by stem-loop RT-PCR. Nucleic Acids Res 33:e179
19. Schmittgen TD, Lee EJ, Jiang J, Sarkar A, Yang L, Elton TS, Chen C (2008) Real-time PCR quantification of precursor and mature microRNA. Methods 44:31–38
20. Bissels U, Wild S, Tomiuk S, Holste A, Hafner M, Tuschl T, Bosio A (2009) Absolute quantification of microRNAs by using a universal reference. RNA 15:2375–2384
21. Hall N (2007) Advanced sequencing technologies and their wider impact in microbiology. J Exp Biol 210(Pt 9):1518–1525
22. Church GM (2006) Genomes for all. Sci Am 294(1):46–54
23. Zak DE, Aderem A (2009) A systems view of host defense. Nat Biotechnol 27(11):999–1001
24. Geiss GK et al (2008) Direct multiplexed measurement of gene expression with color-coded probe pairs. Nat Biotechnol 26:317–325
25. Kulkarni MM (2011) Digital multiplexed gene expression analysis using the NanoString nCounter system. Curr Protoc Mol Biol Chapter 25:Unit25B.10

26. Lopez-Romero P, Gonzalez MA, Callejas S, Dopazo A, Irizarry RA (2010) Processing of Agilent microRNA array data. BMC Res Notes 3:18
27. Hua YJ, Tu K, Tang ZY, Li YX, Xiao HS (2008) Comparison of normalization methods with microRNA microarray. Genomics 92:122–128
28. Pradervand S, Weber J, Thomas J, Bueno M, Wirapati P, Lefort K, Dotto GP, Harshman K (2009) Impact of normalization on miRNA microarray expression profiling. RNA 15:493–501
29. Risso D, Massa MS, Chiogna M, Romualdi C (2009) A modified LOESS normalization applied to microRNA arrays: a comparative evaluation. Bioinformatics 25:2685–2691
30. Sarkar D, Parkin R, Wyman S, Bendoraite A, Sather C, Delrow J, Godwin AK, Drescher C, Huber W, Gentleman R, Tewari M (2009) Quality assessment and data analysis for microRNA expression arrays. Nucleic Acids Res 37:e17
31. Peltier HJ, Latham GJ (2008) Normalization of microRNA expression levels in quantitative RTPCR assays: identification of suitable reference RNA targets in normal and cancerous human solid tissues. RNA 14:844–852
32. Chang KH, Mestdagh P, Vandesompele J, Kerin MJ, Miller N (2010) MicroRNA expression profiling to identify and validate reference genes for relative quantification in colorectal cancer. BMC Cancer 10:173
33. Sato F, Tsuchiya S, Terasawa K, Tsujimoto G (2009) Intra-platform repeatability and inter-platform comparability of microRNA microarray technology. PLoS One 4:e5540
34. Benes V, Castoldi M (2010) Expression profiling of microRNA using real-time quantitative PCR, how to use it and what is available. Methods 50:244–249
35. Mestdagh P, Van VP, De WA, Muth D, Westermann F, Speleman F, Vandesompele J (2009) A novel and universal method for microRNA RT-qPCR data normalization. Genome Biol 10:R64
36. Vandesompele J, De PK, Pattyn F, Poppe B, Van RN, De PA, Speleman F (2003) Accurate normalization of real-time quantitative RT-PCR data by geometric averaging of multiple internal control genes. Genome Biol 3:RESEARCH0034
37. Git A, Dvinge H, Salmon-Divon M, Osborne M, Kutter C, Hadfield J, Bertone P, Caldas C (2010) Systematic comparison of microarray profiling, real-time PCR, and next-generation sequencing technologies for measuring differential microRNA expression. RNA 16:991–1006
38. Pfaffl MW, Tichopad A, Prgomet C, Neuvians TP (2004) Determination of stable housekeeping genes, differentially regulated target genes and sample integrity: BestKeeper–excel-based tool using pairwise correlations. Biotechnol Lett 26:509–515
39. Andersen CL, Jensen JL, Orntoft TF (2004) Normalization of real-time quantitative reverse transcription-PCR data: a model-based variance estimation approach to identify genes suited for normalization, applied to bladder and colon cancer data sets. Cancer Res 64:5245–5250
40. Wang B, Wang XF, Howell P, Qian X, Huang K, Riker AI, Ju J, Xi Y (2010) A personalized microRNA microarray normalization method using a logistic regression model. Bioinformatics 26:228–234
41. Baskerville S, Bartel DP (2005) Microarray profiling of microRNAs reveals frequent coexpression with neighboring miRNAs and host genes. RNA 11:241–247
42. Liang RQ, Li W, Li Y, Tan CY, Li JX, Jin YX, Ruan KC (2005) An oligonucleotide microarray for microRNA expression analysis based on labeling RNA with quantum dot and nanogold probe. Nucleic Acids Res 33:e17. doi:10.1093/nar/gni019
43. Wang H, Ach RA, Curry B (2007) Direct and sensitive miRNA profiling from low-input total RNA. RNA 13:151–159
44. Rao Y, Lee Y, Jarjoura D, Ruppert AS, Liu CG, Hsu JC, Hagan JP (2008) A comparison of normalization techniques for microRNA microarray data. Stat Appl Genet Mol Biol 7: Article22
45. Chiogna M, Massa MS, Risso D, Romualdi C (2009) A comparison on effects of normalisations in the detection of differentially expressed genes. BMC Bioinform 10:61
46. Sun Y, Koo S, White N, Peralta E, Esau C, Dean NM, Perera RJ (2004) Development of a microarray to detect human and mouse microRNAs and characterization of expression in human organs. Nucleic Acids Res 32:e188. doi:10.1093/nar/gnh186
47. Castoldi M, Schmidt S, Benes V, Hentze MW, Muckenthaler MU (2008) miChip: an array-based method for microRNA expression profiling using locked nucleic acid capture probes. Nat Protoc 3:321–329
48. Garzon R, Volinia S, Liu CG, Fernandez-Cymering C, Palumbo T, Pichiorri F, Fabbri M, Coombes K, Alder H, Nakamura T, Flomenberg N, Marcucci G, Calin GA, Kornblau SM, Kantarjian H, Bloomfield CD, Andreeff M, Croce CM (2008) MicroRNA signatures associated with cytogenetics and prognosis in acute myeloid leukemia. Blood 111(6): 3183–3189
49. Miska EA, Alvarez-Saavedra E, Townsend M, Yoshii A, Sestan N, Rakic P, Constantine-Paton M, Horvitz HR (2004) Microarray analysis of microRNA expression in the developing mammalian brain. Genome Biol 5:R68. doi:10.1186/gb-2004-5-9-r68
50. Tian Z, Greene AS, Pietrusz JL, Matus IR, Liang M (2008) MicroRNA-target pairs in the rat kidney identified by microRNA microarray, proteomic, and bioinformatic analysis. Genome Res 18:404–411
51. Bolstad BM, Irizarry RA, Astrand M, Speed TP (2003) A comparison of normalization methods for high density oligonucleotide array data based on variance and bias. Bioinformatics 19:185–193
52. Steinhoff C, Vingron M (2006) Normalization and quantification of differential expression in gene expression microarrays. Brief Bioinform 7:166–177

53. Huber W, von Heydebreck A, Sultmann H, Poustka A, Vingron M (2002) Variance stabilization applied to microarray data calibration and to the quantification of differential expression. Bioinformatics 18(Suppl 1):S96–S104
54. Xiong H, Zhang D, Martyniuk CJ, Trudeau VL, Xia X (2008) Using generalized procrustes analysis (GPA) for normalization of cDNA microarray data. BMC Bioinformatics 9:25
55. Bargaje R, Hariharan M, Scaria V, Pillai B (2010) Consensus miRNA expression profiles derived from interplatform normalization of microarray data. RNA 16:16–25
56. Do JH, Choi DK (2006) Normalization of microarray data: single-labeled and dual-labeled arrays. Mol Cells 22:254–261
57. Garzon R, Garofalo M, Martelli MP, Briesewitz R, Wang L, Fernandez-Cymering C, Volinia S, Liu CG, Schnittger S, Haferlach T, Liso A, Diverio D, Mancini M, Meloni G, Foa R, Martelli MF, Marcucci C, Croce CM, Falini B (2008) Distinctive microRNA signature of acute myeloid leukemia bearing cytoplasmic mutated nucleophosmin. Proc Natl Acad Sci USA 105(10):3945–3950
58. Perkins DO, Jeffries CD, Jarskog LF, Thomson JM, Woods K, Newman MA, Parker JS, Jin J, Hammond SM (2007) MicroRNA expression in the prefrontal cortex of individuals with schizophrenia and schizoaffective disorder. Genome Biol 8:R27. doi:10.1186/gb-2007-8-2-r27
59. Chen Y, Dougherty E, Bittner ML (1997) Ratio-based decisions and the quantitative analysis of cDNA microarray images. J Biomed Opt 2:364–374
60. Tusher VG, Tibshirani R, Chu G (2001) Significance analysis of microarrays applied to the ionizing radiation response. Proc Natl Acad Sci USA 98(9):5116–5121
61. Mukherjee S, Roberts SJ, van der Laan M (2005) Data-adaptive test statistics for microarray data. In: The ninth annual international conference on research in computational molecular biology, Cambridge, MA, pp 237–238
62. Martin DE, Demougin P, Hall MN, Bellis M (2004) Rank Difference Analysis of Microarrays (RDAM), a novel approach to statistical analysis of microarray expression profiling data. BMC Bioinform 5(1):148
63. Breitling R, Armengaud P, Amtmann A, Herzyk P (2004) Rank products: a simple, yet powerful, new method to detect differentially regulated genes in replicated microarray experiments. FEBS Lett 573(1–3):83–92
64. Smyth GK (2004) Linear models and empirical bayes methods for assessing differential expression in microarray experiments. Stat Appl Genet Mol Biol 3(1):Article 3
65. Kerr MK, Martin M, Churchill GA (2000) Analysis of variance for gene expression microarray data. J Comput Biol 7:819–837
66. Lee ML, Lu W, Whitmore GA, Beier D (2002) Models for microarray gene expression data. J Biopharm Stat 12:1–19
67. Li C, Wong WH (2001) Model-based analysis of oligonucleotide arrays: model validation, design issues and standard error application. Genome Biol 2:research0049.1–0049.12
68. Irizarry RA, Hobbs B, Collin F, Beazer-Barclay YD, Antonellis KJ, Scherf U, Speed TP (2003) Exploration, normalization, and summaries of high density oligonucleotide array probe level data. Biostatistics 4(2):249–264
69. Tsai CA, Chen JJ (2004) Significance analysis of ROC indices for comparing diagnostic markers: applications to gene microarray data. J Biopharm Stat 14(4):985–1003
70. http://bioconductor.org/packages/release/bioc/html/RankProd.html
71. Dopazo J, Carazo JM (1997) Phylogenetic reconstruction using an unsupervised growing neural network that adopts the topology of a phylogenetic tree. J Mol Evol 44:226–233
72. Kaufman L, Rousseeuw PJ (1990) Finding groups in data: an introduction to cluster analysis. Wiley, New York
73. Kohonen T (1995) Self-organizing maps, vol 30, Springer series in information sciences. Springer, Berlin
74. Fraley C, Raftery AE (1998) How many clusters? Which clustering methods? Answers via model-based cluster analysis. Comput J 41:578–588
75. Fraley C, Raftery AE (1999) MCLUST: software for model-based cluster analysis. J Classif 16:297–306
76. Fraley C, Raftery AE (2002) Model-based clustering, discriminant analysis, and density estimation. J Am Stat Assoc 97:611–631
77. McLachlan GJ, Bean RW, Ben-Tovim JL, Zhu JX (2005) Using mixture models to detect differentially expressed genes. Aust J Exp Agric 45:859–866
78. McLachlan GJ, Bean RW, Ben-Tovim JL (2006) A simple implementation of a normal mixture approach to differential gene expression in multiclass microarrays. Bioinformatics 22:1608–1615
79. Ma L, Teruya-Feldstein J, Weinberg RA (2007) Tumour invasion and metastasis initiated by microRNA-10b in breast cancer. Nature 449:682–688
80. Tavazoie SF, Alarcon C, Oskarsson T, Padua D, Wang Q, Bos PD, Gerald WL, Massague J (2007) Endogenous human microRNAs that suppress breast cancer metastasis. Nature 451:147–152
81. Rosenfeld N, Aharonov R, Meiri E, Rosenwald S, Spector Y, Zepeniuk M, Benjamin H, Shabes N, Tabak S, Levy A, Lebanony D, Goren Y, Silberschein E, Targan N, Ben Ari A, Gilad S, Sion-Vardy N, Tobar A, Feinmesser M, Kharenko O, Nativ O, Nass D, Perelman M, Yosepovich A, Shalmon B, Polak-Charcon S, Fridman E, Avniel A, Bentwich I, Bentwich Z, Cohen D, Chajut A, Barshack I (2008) MicroRNAs accurately identify cancer tissue origin. Nat Biotechnol 26:462–469
82. Blenkiron C, Goldstein LD, Thorne NP, Spiteri I, Chin SF, Dunning MJ, Barbosa-Morais NL, Teschendorff AE, Green AR, Ellis IO et al (2007)

MicroRNA expression profiling of human breast cancer identifies new markers of tumour subtype. Genome Biol 8:R214
83. Mitchell PS, Parkin RK, Kroh EM, Fritz BR, Wyman SK, Pogosova-Agadjanyan EL, Peterson A, Noteboom J, O'Briant KC, Allen A, Lin DW, Urban N, Drescher CW, Knudsen BS, Stirewalt DL, Gentleman R, Vessella RL, Nelson PS, Martin DB, Tewari M (2008) Circulating microRNAs as stable blood-based markers for cancer detection. Proc Natl Acad Sci USA 105:10513–10518
84. Lange J (2010) microRNA profiling on automated biochip platform reveals biomarker signatures from blood samples. Nat Methods 7. doi:10.1038/nmeth.f.281
85. Lu J, Getz G, Miska EA, Alvarez-Saavedra E, Lamb J, Peck D, Sweet-Cordero A, Ebert BL, Mak RH, Ferrando AA, Downing JR, Jacks T, Horvitz HR, Golub TR (2005) MicroRNA expression profiles classify human cancers. Nature 435:834–838
86. Visone R, Veronese A, Rassenti LZ, Balatti V, Pearl DK, Acunzo M, Volinia S, Taccioli C, Kipps TJ, Croce CM (2011) miR-181b is a biomarker of disease progression in chronic lymphocytic leukemia. Blood 118(11):3072–3079
87. Bartel DP (2004) MicroRNAs: genomics, biogenesis, mechanism, and function. Cell 116:281–297
88. Sethupathy P et al (2006) TarBase: a comprehensive database of experimentally supported animal microRNA targets. RNA 12:192–197
89. Xiao F et al (2009) miRecords: an integrated resource for microRNA-target interactions. Nucleic Acids Res 37(Database issue):D105–D110
90. Chi SW et al (2009) Argonaute HITS-CLIP decodes microRNA–mRNA interaction maps. Nature 460(7254):479–486
91. Lewis BP, Burge CB, Bartel DP (2005) Conserved seed pairing, often flanked by adenosines, indicates that thousands of human genes are microRNA targets. Cell 120(1):15–20
92. Rehmsmeier M et al (2004) Fast and effective prediction of microRNA/target duplexes. RNA 10:1507–1517
93. Hofacker IL (2007) How microRNAs choose their targets. Nat Genet 39(10):1191–1192
94. Mathews DH (2006) Revolutions in RNA secondary structure prediction. J Mol Biol 359(3):526–532
95. Grimson A et al (2007) MicroRNA targeting specificity in mammals: determinants beyond seed pairing. Mol Cell 27:91–105
96. Mazière P, Enright AJ (2007) Prediction of microRNA targets. Drug Discov Today 12(11–12):452–458
97. John B et al (2004) Human MicroRNA targets. PLoS Biol 2(11):1862–1879
98. Betel D et al (2007) The microRNA.org resource: targets and expression. Nucleic Acids Res 36(Database issue):D149–D153
99. Krek A et al (2005) Combinatorial microRNA target predictions. Nat Genet 37(5):495–500
100. Kiriakidou M et al (2004) A combined computational-experimental approach predicts human microRNA targets. Genes Dev 18(10):1165–1178
101. Miranda KC et al (2006) A pattern-based method for the identification of MicroRNA binding sites and their corresponding heteroduplexes. Cell 126:1203–1217
102. Long D et al (2007) Potent effect of target structure on microRNA function. Nat Struct Mol Biol 14(4):287–294
103. Kertesz M et al (2007) The role of site accessibilità in microRNA target recognition. Nat Genet 39(10):1278–1284
104. Jiang Q, Wang Y, Juan L et al (2009) miR2Disease: a manually curated database for microRNA deregulation in human disease. Nucleic Acids Res 37(Database issue):D98–D104
105. Lu M, Zhang Q, Deng M et al (2008) An analysis of human microRNA and disease associations. PLoS One 3(10):e3420
106. Barh D, Bhat D, Viero C (2010) miReg: a resource for microRNA regulation. J Integr Bioinform 7(1).
107. Laganà A, Forte S, Giudice A et al (2009) miRò: a miRNA knowledge base. Database (Oxford) 2009:bap008. doi:10.1093/database/bap008
108. Ashburner M, Ball CA, Blake JA et al (2000) Gene ontology: tool for the unification of biology. The Gene Ontology Consortium. Nat Genet 25:25–29
109. Landgraf P, Rusu M, Sheridan R et al (2007) A mammalian microRNA expression atlas based on small RNA library sequencing. Cell 129:1401–1414
110. Ulitsky I, Laurent LC, Shamir R (2010) Towards computational prediction of microRNA function and activity. Nucleic Acids Res 38(15):e160
111. Le Brigand K, Robbe-Sermesant K, Mari B et al (2010) MiRonTop: mining microRNAs targets across large scale gene expression studies. Bioinformatics 26(24):3131–3132
112. Sales G, Coppe A, Bisognin A et al (2010) MAGIA, a web-based tool for miRNA and Genes Integrated Analysis. Nucleic Acids Res 38(Web Server issue):W352–W359
113. Alexiou P, Maragkakis M, Papadopoulos GL et al (2010) The DIANA-mirExTra web server: from gene expression data to microRNA function. PLoS One 5(2):e9171
114. Papadopoulos GL, Alexiou P, Maragkakis M et al (2009) DIANA-mirPath: integrating human and mouse microRNAs in pathways. Bioinformatics 25(15):1991–1993
115. Maragkakis M, Alexiou P, Papadopoulos GL et al (2009) Accurate microRNA target prediction correlates with protein repression levels. BMC Bioinform 10:295
116. Le Béchec A, Portales-Casamar E, Vetter G et al (2011) MIR@NT@N: a framework integrating transcription factors, microRNAs and their targets to identify sub-network motifs in a meta-regulation network model. BMC Bioinform 12:67

117. Portales-Casamar E, Arenillas D, Lim J et al (2009) The PAZAR database of gene regulatory information coupled to the ORCA toolkit for the study of regulatory sequences. Nucleic Acids Res 37(Database issue):D54–D60
118. Portales-Casamar E, Thongjuea S, Kwon AT et al (2010) JASPAR 2010: the greatly expanded open-access database of transcription factor binding profiles. Nucleic Acids Res 38(Database issue):D105–D110
119. Ho Sui SJ, Mortimer JR, Arenillas DJ et al (2005) oPOSSUM: identification of over-represented transcription factor binding sites in co-expressed genes. Nucleic Acids Res 33(10):3154–3164
120. Kozomara A, Griffiths-Jones S (2011) miRBase: integrating microRNA annotation and deep-sequencing data. Nucleic Acids Res 39(Database Issue):D152–D157
121. Griffiths-Jones S, Saini HK, van Dongen S et al (2008) miRBase: tools for microRNA genomics. Nucleic Acids Res 36(Database Issue):D154–D158
122. Griffiths-Jones S, Grocock RJ, van Dongen S et al (2006) miRBase: microRNA sequences, targets and gene nomenclature. Nucleic Acids Res 34(Database Issue):D140–D144
123. Griffiths-Jones S (2004) The microRNA registry. Nucleic Acids Res 32(Database Issue):D109–D111
124. Chaurasia G, Malhotra S, Russ J et al (2009) UniHI 4: new tools for query, analysis and visualization of the human protein-protein interactome. Nucleic Acids Res 37(Database issue):D657–D660
125. Huang GT, Athanassiou C, Benos PV (2011) mirConnX: condition-specific mRNA-microRNA network integrator. Nucleic Acids Res 39(Web Server issue):W416–W423
126. Ferro A, Giugno R, Laganà A et al (2009) miRScape: a Cytoscape plugin to annotate biological networks with microRNAs. NETTAB 2009 conference, Catania, 10–12 June 2009
127. Smoot ME, Ono K, Ruscheinski J et al (2011) Cytoscape 2.8: new features for data integration and network visualization. Bioinformatics 27(3):431–432
128. Cline MS, Smoot M, Cerami E et al (2007) Integration of biological networks and gene expression data using Cytoscape. Nat Protoc 2(10):2366–2382
129. Shannon P, Markiel A, Ozier O et al (2003) Cytoscape: a software environment for integrated models of biomolecular interaction networks. Genome Res 13(11):2498–2504

Working Together: Combinatorial Regulation by microRNAs

16

Yitzhak Friedman, Ohad Balaga, and Michal Linial

Abstract

MicroRNAs (miRNAs) negatively regulate gene expression level of mRNA post-transcriptionally. Deep sequencing and large-scale screening methods have yielded about 1,500 miRNA sequences in human. Each miRNA contains a seed sequence that is required, but not sufficient, for the correct matching with its targets. Recent technological advances make it possible to capture the miRNAs with their cognate mRNAs at the RISC complex. These experiments have revealed thousands of validated mRNA-miRNA pairing events. In the context of human stem cells, 90% of the identified transcripts appear to be paired with at least two different miRNAs.

In this chapter, we present a comprehensive outline for a combinatorial regulation mode by miRNAs. Initially, we summarize the computational and experimental evidence that support a combined effect of multiple miRNAs. Then, we describe miRror2.0, a platform specifically convened to consider the likelihood of miRNAs cooperativity in view of the targets, tissues and cell lines. We show that results from miRror2.0 can be further refined by an iterative procedure, calls Psi-miRror that gauges the robustness of the regulation. We illustrate the combinatorial regulation projected onto graphs of human pathways and show that these pathways are amenable to disruption by a small set of miRNAs. Finally, we propose that miRNA combinatorial regulation is an attractive regulatory strategy not only at the level of single target, but also at the level of pathways and cellular homeostasis. The joint operation of miRNAs is a powerful means to overcome the low specificity inherent in each individual miRNA.

Y. Friedman • M. Linial (✉)
Department of Biological Chemistry, Institute of Life Sciences, The Hebrew University of Jerusalem, Jerusalem 91904, Israel
e-mail: michall@cc.huji.ac.il

O. Balaga
School of Computer Science and Engineering, The Hebrew University of Jerusalem Jerusalem 91904, Israel

U. Schmitz et al. (eds.), *MicroRNA Cancer Regulation: Advanced Concepts, Bioinformatics and Systems Biology Tools*, Advances in Experimental Medicine and Biology 774,
DOI 10.1007/978-94-007-5590-1_16,

Keywords
microRNA • Database • Prediction tools • 3′-UTR • Genomics • Deep sequencing • Regulatory pathway • Bioinformatics

Abbreviations

AGO	Argonaute
DB	database
DIS	disconnecting score
GO	gene ontology
HITS-CLIP	high-throughput sequencing of RNAs isolated by cross-linking immunoprecipitation
PAR-CLIP	photoactivatable-ribonucleoside-enhanced crosslinking and immunoprecipitation
miRNA (miR)	microRNA
ML	machine learning
MS	mass spectrometry
ncRNA	non-coding RNA
RISC	RNA-induced silencing complex
SILAC	stable isotope labeling by amino acids in cell culture
UTR	untranslated region.

16.1 General Overview

MicroRNAs (miRNAs) are short non-coding RNAs (ncRNAs) that negatively regulate gene expression post-transcriptionally [1]. Recent miRNA detection techniques confirmed the presence of hundreds of miRNAs in healthy and diseased tissues [2, 3]. An estimate across animal genomes suggests that almost 1% of the genes in human and *C. elegans* consists of miRNAs. These estimates are derived from a combination of computational and experimental methods [4, 5].

In human and other metazoa, miRNA plays a role as an additional layer of post-transcriptional regulation [6]. Mechanistically, miRNAs exert their function via base-pair complementarity at the RNA-induced silencing complex (RISC) [7]. The binding of miRNA to mRNA leads to gene silencing. Silencing of a gene by miRNA leads to a change in the mRNA stability, enhanced degradation and to some degree also translational arrest [1, 8, 9]. It was originally proposed that the impact of miRNAs in animals is primarily at the translational level [10–12]. However, the current view argues that most of the miRNA effects are attributable to the post-transcriptional enhancement of mRNA degradation (through blocking of cap binding proteins, deadenylation and more).

A coherent picture of miRNA regulation is still highly fragmented, mainly due to gaps in the understanding of miRNA modes of action *in-vivo* [13]. Nevertheless, ample evidence indicates that deregulation of miRNAs leads to pathogenesis (*i.e.*, obesity, cancer, neurodegenerative diseases). In fact, for viral infection and cancer, a coordinated change in the relative expression levels of miRNA-sets were shown to be a strong indicator of the pathological state [7, 14, 15]. Less is known on the role of miRNAs under normal physiology conditions, chronic metabolic stress and ageing [16, 17].

miRNAs are best known for the regulation of stem cell differentiation [18], immunological cell function [19], organogenesis [20], cell identity [21], apoptosis [22] and more. The study of miRNAs in the context of cancer biology shows that a disruption in miRNA biogenesis leads to tumorigenesis [23] and to a drastic change in the relative expression of a large number of mRNAs [24]. Furthermore, several miRNAs directly regulate cell cycle genes and thus induce oncogenic activity [3, 25, 26]. In other instances, the activity of miRNA resembles tumor suppression [27]. Interestingly, many of the miRNAs in human are located at fragile sites [28] in agreement with the prevalence of miRNAs in cancer progression.

16.1.1 Outline

This chapter focuses on the notion of combinatorial miRNA action. We consider this hypothesis by appealing to experimental data as well as computational evidence. We provide evidence that the concept of 'miRNAs working together' is valid for the different levels of cell regulation from an individual target to a set of targets and ultimately, at biological pathways. We present miRror2.0 as a computational, statistical platform that incorporates the concept of cooperativity when analyzing experimental results. We present the notion that within cells, the disruption of a network is best achieved by a coordinated action of a series of miRNAs. We conclude by suggesting some general trends and architectural design principles for cellular regulation by miRNAs.

16.1.1.1 Classification and Nomenclatures

In this section, we focus on the complete set of human miRNAs. There are over 1,500 miRNAs in humans and about 750 in mouse [29]. We introduce some classifications and accepted notations for the inventory of miRNAs:

(i) *Chromosomal organization*: Each miRNA is annotated as 'isolate' or 'cluster.' Specifically, a miRNA that is located in the vicinity of another one (within 10 kb) belongs to a cluster. miRNA that belong to a cluster are encoded in a polycistronic transcript. Over 40% of human miRNAs are organized in such genomic clusters. The remaining miRNAs are considered 'isolates'.

(ii) *Degree of seed sequence overlap*: Two miRNAs that overlap in their seed sequences (6–8 nucleotides at the 5′ region of the mature miRNA) belong to the same family. The term 'single' refers to the appearance of the seed only once. The definition of a family (according to miRBase [29]) is across species. For our discussion, we follow a species centric view of a family. Illustratively, consider a seed sequence of a miRNA that appears once in human, once in mouse but not in the fly. This sequence is viewed as 'single' for human or mouse, but as a 'family' from a cross-taxa perspective.

(iii) *Degree of pre-miR overlap*: The hairpin sequence of the pre-miR (ranges from 70 to 120 nucleotides) includes the mature miRNA sequence. miRNAs that share sequence identity beyond the seed region may be identical in the entire miRNA sequence (22–24 nucleotides), or even at the entire primary transcript (called pri-miR). Such classification is not limited to sequence identity and also considers secondary structure resemblance [29].

(iv) *IsomiR classes*: Along the miRNA biogenesis and maturation process [30], two strands of the stem-loop structure are produced [31]. The accumulated short reads from deep sequencing experiments indicated the presence of IsomiRs. The IsomiRs refer to miRNAs variants that derive from the chromosomal location but are modified, mainly at the 3′ and 5′ tails to produce rich variants of mature miRNAs [32]. In such experiments, often the two complementary strands are identified, albeit at drastically different expression levels. The strands are called 'guide/mature' and 'passenger/star' strands. Traditionally, the non-preferred strand sequences were indicated as miR* (star, antisense) [33].

(v) *Genomic identity*: Most miRNAs are located in intergenic regions, similar to any coding gene. Some miRNAs overlap genes and their sequence is at the same position as other coding or non-coding gene. A substantial number of miRNAs are located in introns of host genes (mirtrons, [34]). The different genomic location is indicative of variation in the maturation and regulation process [35, 36].

(vi) *Evolutionary evidence*: miRNAs can be classified according to their degree of evolutionary conservation. While many miRNAs are human specific, some orthologs are found only in chimpanzees [37]. Other miRNAs span the entire animal tree (from human to hydra) [38]. The presence of miRNAs that are evolutionary conserved is in accord with their role in basic physiology and cell fate functions.

Figure 16.1 illustrates two forms of miRNA classifications (a chromosomal location and a family assignment) and their relation.

The nomenclature of miRNAs is still evolving. The growth in the number of identified miRNAs is attributed to several deep sequencing datasets [33]. Results from this technology have led to the expansion in the number of known miRNAs, mainly by identifying the complementary strand (denoted -5p and -3p to identify the

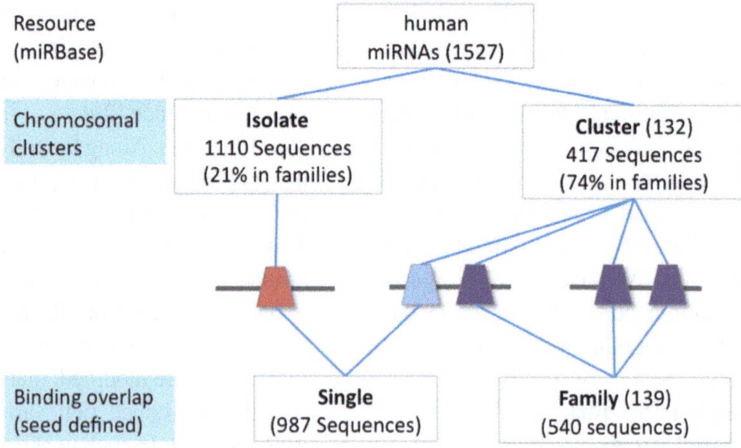

Fig. 16.1 Classifications of the human miRNAs. The human miRNAs are classified according to chromosomal organization and family assignments. Focusing on the 1,527 human miRNAs (miRBase Ver. 18, November 2011) indicates 132 chromosomal clusters (covering 417 miRNAs). The rest are isolated miRNAs (separated by >10 kb). Genes that share the same seed in human are included in the same family. There are 139 families, covering 540 miRNAs. 310 of the miRNAs belong to families that intersect with clusters. The rest (230 miRNAs) are found among the 'isolates'

Table 16.1 Nomenclature of miRNAs

Family name	# ID stem-loop Cross-taxa	# ID stem-loop Human	miR/mir[a] Example	Comment on nomenclature
mir-515	144	42	hsa-miR-527	Unique mature sequence
mir-101	48	2	hsa-mir-101-1	Same mature sequence as hsa-mir-101-2, different genomic location
mir-30	122	6	hsa-mir-30a	High sequence similarity to hsa-mir-30b (*i.e.*, same seed)
mir-19	82	3	hsa-miR-19b-3p	Related seed, indicating the arm of the stem

Examples of the naming of miRNAs and the information these names suggest
[a]miR is used to indicates the mature miRNA sequence. 'mir' indicates the stem-loop precursor

directionality of the sequence as the 5′-arm and the 3′-arm of the stem, respectively) (Table 16.1). Table 16.1 illustrates the guidelines for navigating among the different miRNA names. Unfortunately, different names may be used for the same sequence by different database. For example, microRNA.org [39], PicTar [40] and miRDB [41] refer to the same sequence as hsa-miR-19b-2-5p, hsa-miR-19b and hsa-miR-19b-2*, respectively. Agreement on canonical names is necessary for the task of comparing the performance of individual miRNA-target predictors. Inconsistency in miRNA naming is an unfortunate reality. A reduced consistency between different miRNA-target predictors is partially a result of the inconsistent use of miRNA naming by different miRNA-target DBs.

16.2 Pairing Between a miRNA and Its Target

The most studied recognition signal for miRNA-target pairing is restricted to 6–8 nucleotides (the seed and the immediate vicinity). With 1,500 miRNAs in human and about 10,000 candidate target genes (excluding their alternative splicing variants), the network of interactions is quite complex. Critically, the determinants of binding specificity are poorly understood. Contrary to the known activity of transcription factors, individual miRNAs often attenuate the expression of their direct targets very modestly (*e.g.*, 25% decrease in expression level) [42]. Consequently, a definitive identification of miRNA mappings to their genuine targets is unlikely to succeed comprehensively with current technology [43]. The fraction of false positives across all prediction methods is high, and the number of false negatives is unknown [44, 45]. As a rough estimate, each miRNA is assumed to attenuate tens to hundreds of targets. More importantly, distinguishing between a direct and an indirect miRNA–target interaction remains a crucial challenge in the field.

16.2.1 Computational miRNA-Target Predictions

Currently, miRBase (version 18) is the most exhaustive collection of miRNAs with over 18,000 mature miRNA sequences in over 100 organisms [33]. The 1,500 miRNAs from human and 750 from mouse are estimated to target about half of the genes in human and rodents [46].

There are many databases, algorithms and resources that provide predictions of miRNAs and their direct targets. We will not explicitly discuss the differences among these major resources. We introduce the main resources that were applied while developing the concept of 'combinatorial regulation' and the associated software platform. However, it is important to note that several of these tools are meta-servers that combine results from a variety of individual resources [47].

Currently, there are over a dozen miRNA-target resources [48, 49]. While all resources use the knowledge of seed sequence complementarity, some algorithms add a weight to account for imperfect hybridization, context dependent features (*e.g.*, accessibility of binding sites), species conservation, thermodynamic stability of the miRNA-mRNA duplex and any combination of the above. The stable miRNA-target prediction databases include: (i) *TargetScan* database [50]; (ii) *microCosm* which is based on the miRanda algorithm [51]; (iii) *PicTar* (with two settings, according to the degree of evolutionary conversation) [40]; (iv) *DIANA–MicroT* [52]; (v) *PITA* (with dual settings for stringency and coverage) [53]; (vi) *MirZ* [54]; (vii) *microRNA.org* that allows analysis of multiple miRNAs acting on the same gene-target using the miRanda algorithm [39]; (viii) *miRDB* resource [41]; (ix) *TargetRank* (either conserved or all miRNAs) [55]; (x) miRNAMap2 [56]; (xi) RNA22 *(cbcsrv.watson.ibm.com/rna22.html)*, and (xii) the meta-predictor MAMI (http://mami.med.harvard.edu/). A total of 15 sets of predictions are available for human miRNAs. While this is far from an exhaustive list, the most stable and up-to-date DBs for miRNA-target prediction are listed.

Additional descriptive features associated with miRNAs include the distribution of miRNA binding sites, positioning of the binding sites on the mRNA sequences, transcript length, and energy of the secondary structures of the transcript [57]. These features are often excluded in the prediction algorithms, due to their sparse characterization in existing *in-vivo* studies and the difficulty to generalize these features across animals. Nevertheless, recent miRNA-prediction tools are based on the use of the more thoroughly characterized features of miRNAs in conjunction with algorithms that use statistical models such as a Bayesian models or machine learning (ML). In such schemes, hundreds of examples (negative and positive) are used to suggest the best separation between the true and false predictions. The consistency with validated results is highest among predictors that applied ML technologies [58]. Several of the resources *(MirZ, microRNA.org)* and miRBase [33] provide miRNA expression profiles for a large number of tissues and cell lines.

Rigorous assessment studies showed that the consistency among major miRNA-target prediction tools is rather poor, reflecting a huge fraction of false positives associated with each of them [59]. The ability of the most established predictors (PITA, DIANA-microT4.0, Miranda, Microcosm, TargetScan5.0, TargetScanS, Pictar and MirZ-ElMMo and RNA22) to explain the profile of the down-regulated genes in cells following overexpressing of individual miRNA has been reported. For example, the precision in recovering the results from hsa-miR-1 overexpression in HeLa cells [42] was 23–50% with a sensitivity level of 6–20%. The union or intersection of any five of these predictors had negligible effect on the overall success and, in fact, a slight reduction in the performance was noted.

16.2.2 Experimental Data for miRNAs in Cellular Systems

In recent years, experimental protocols and tailored chemical probes were developed for the study of miRNA regulation within a cellular context. Their development was necessary since the biochemical approaches initially used to identify miRNAs were biased towards the most abundant miRNAs. Other recently developed methods (*e.g.,* those which provide detailed analysis of the factors determining the level for the regulation of a specific miRNA towards a specific target using, *e.g.*, 3′-UTR segment-luciferase reporter systems will not be discussed but are reviewed in [60]).

The identification of miRNAs in a broad spectrum of metazoan calls for applying comparative genomics technologies. The evolutionary signal from sequence conservation and structural consideration was brought into consideration. Hence, the collection of reported miRNAs has expanded and was doubled between 2009 and 2011. We will only mention methods that provide a global view on the cellular regulation by miRNAs [61].

Functional confirmation of the miRNAs and their target genes is mostly based on *in-vitro* studies in which a specific miRNA is introduced to cells. In a mirror view, a candidate miRNA gene is knocked down (*e.g.,* using anti-miR). A few hours after the cell manipulation (usually 12–72 h), global gene expression profiling is performed using transcriptomic DNA microarray. The differential expression levels of genes relative to mock-transfected cells are then recorded. The signal that is sought is a negative correlation between the overexpression of the miRNA of interest and the targeted genes. The results of such experiments are collected in the major gene expression archives including GEO [62] and ArrayExpress [63]. Over 30 large-scale experiments of this type were carried out (some with few controls and a minimal reproducibility in the experimental design). Major concern in such experiments is that a shift in the balance between the authentic binding sites and off-targets cannot be avoided [64].

In a more physiological paradigm, cells are exposed to some predetermined condition (hypoxia, glucose starvation, heat shock, drug) and a change in the expression profile of the miRNAs is monitored. The result from these cases is a list of a few hundred candidate miRNAs. Applying the commercially available miRNA DNA-array platforms, large-scale analyses are presented. In recent years, an experimental design has emerged which uses size fractionated RNA from treated cells as input for deep sequencing, thereby collecting a complete set of 'small ncRNAs', among them the miRNAs [65]. Several large-scale studies of human tissues in health and disease were reported using this approach. For example, a catalogue of the miRNAs was compiled by RNA-Seq from cancerous and normal cervical tissues [66], melanoma [67], human stem cells [68] and more. These experiments detected coordinated change in the expression of groups of miRNAs [5]. We expect to see more experiments that follow such experimental setting.

In human embryonic stem cells (hESC), the regulation of miRNAs was monitored by immunoprecipitation (IP) of the Argonaute (AGO) proteins [69]. Variations of this approach were reported using tagged AGO proteins. In such settings, the short miRNAs were eluted from the complexes and characterized by sequencing [70]. A similar approach was followed with other RISC proteins [71].

The methods described above focus on a miRNA paired with its cognate mRNA target. The potential effect of miRNAs at the proteome level is typically studied using Mass spectrometry (MS) based methods [11, 72]. Briefly, cells are metabolically labeled (*e.g.*, stable isotope labeling by amino acids, SILAC) to allow a direct comparison of protein expression. While the results from such methodologies are non-conclusive, the expression level of hundreds of proteins appears to be affected [72]. It is likely that miRNAs do modify, whether directly or as a secondary effect, the expression levels of many proteins. This is probably achieved through attenuation of ribosomal initiation, elongation or translational rate.

16.2.3 Validation of miRNAs-Targets

Computational predictions and the experimental results are both compiled in the TarBase database [73]. TarBase is a manually compiled repository for achieving the validated miRNA-target pairs. It relies mainly on *in-vitro* miRNA overexpression experiments. Additionally, literature-based records for miRNA-targets are collected in miRecords [74]. An assessment using the TarBase benchmark [73] with ~50 manually validated instances and the results of a quantitative Mass Spectrometry (MS) experiment confirmed the limited agreement between the target prediction algorithms [58]. A gain in prediction success was achieved by combining several prediction resources (using TarBase as a benchmark). However, the consistency in prediction of the different algorithms is rather poor [75, 76].

Many experimentalists tend to use their favorite miRNA-target predicting algorithm and resources (often with insufficient justification). The difference in the prediction results and hence, in the biological interpretation is a result of a substantial fraction of false positives present by all existing methods. To provide a functional relevance to miRNA regulation, several tools analyze the match of miRNA-targets in view of the cell processes, GO (gene ontology), diseases [77] and pathways [78, 79].

16.3 Missing Pieces in Understanding miRNA Regulation

Some aspects of miRNA regulation in cells remain open. For example, we lack an understanding of the kinetics and rate limiting steps throughout the maturation of the pri-miRNA into their active miRNA form [80, 81]. The mechanistic details of miRNA-target recognition remain elusive. For example, the functional relevance of the complementary strand in the pre-miRNA duplex, coined passenger miRNA, is a matter of debate. Initially, it was thought that this strand is simply destined for degradation. However, in some cases, the expression of the passenger miRNAs is associated with tissue specificity and developmental stages. RNA editing of miRNAs provides an additional level of diversity. IsomiRs are miRNAs sequences that are slightly different from those encoded by the DNA. It is postulate that isomiRs are active components of miRNA-based regulation [31]. An even broader questions concerns miRNAs in the context of the mRNAs and related miRNAs. The following aspects in miRNA biology remain for further investigations:

(i) The dynamic aspects of the regulation. Kinetic parameters are missing for degradation, turnover and extrusion of miRNAs [82].
(ii) The AGO occupancy in the cells. The link between induction of mRNA degradation and the AGO occupancy remains to be determined [83]. Specifically, the overlap between the miRNAs that are associated with a specific AGO (*e.g.,* AGO-1 and AGO-2) is minimal [84].
(iii) The quantitative nature of miRNA regulation. Specifically, the balance of miRNAs and mRNAs can be considered from the perspectives of titration [85], accessibility of binding sites and competition [86]. miRNAs in cells are probably sequestered by a "sponge" effect [87]. In such scenarios, a competition on binding sites may dominate the balance of miRNAs, AGO binding sites and eventually the cellular response [88].
(iv) The localization of miRNA regulation. Most of the proteins necessary for miRNA gene silencing are localized to P-bodies. However, the knowledge regarding sub-cellular partition of RISC, mRNA and miRNA within the cells is very limited [89].

16.4 Working Together

Several of these open questions can be approached by a quantitative consideration of mRNAs and miRNAs guided by the concept of competing endogenous RNA (ceRNA) [90]. Accordingly, miRNA-binding sites are assumed to regulate the availability of miRNAs. A prediction from the ceRNA hypothesis claims that the induction of genes having specific miRNA binding sites will indirectly lead to a reduction in the potency of such miRNAs. This hypothesis is supported by experimental data [86]. Based on the ceRNA concept, viruses, pseudogenes and even duplicated genes should be considered as elements that may titrate out the miRNAs, leading to a relief of basal repression.

We will henceforth focus our discussion to the notion of combinatorial activity of miRNAs and the view of miRNAs 'working together' as part of a broader cellular design principle.

16.4.1 Evidence for miRNAs Working Together

The concept of combinatorial regulation by miRNAs was validated experimentally. Manipulating a target gene by adding multiple distinct miRNA binding sites on the same transcript augmented the regulation levels [8, 91, 92]. In a cellular context, a parallel overexpression of 2–3 miRNAs resulted in a synergistic effect on the transcriptional level of some candidate genes [93]. For example, in pancreatic cells, for the known target of miR-375, a combined addition of miR-124 and let-7b led to synergy in target inhibition [40]. Similarly, the expression of miR-16, miR-34a and miR-106b altered the cell cycle while no effect on the cell cycle is monitored by each of these miRNAs, separately [85]. The regulation of the tumor suppressor Fus1 in cancer cells depends on the presence of at least three miRNAs (miR-93, miR-98, miR-197) working together [94]. Importantly, introducing several miRNAs not only affected specific candidate targets but also had a measurable effect on specific pathways. For example, a complete block in cell cycle was achieved by a combination of three miRNAs, while the impact of each of these miRNAs alone was less pronounced [95]. A synergetic effect on cell death and the oncogenic properties of multiple miRNAs acting on the same target was recently established [96]. Despite the growing number of instances reported, the generality of the combinatorial phenomenon is yet to be fully established [97].

Some genes are known to have many (predicted) miRNA binding sites. Many of these genes are cell cycle regulators. Recently, several systematic analyses were performed in order to validate the cooperative action of miRNAs on candidate genes [61]. Cells were manipulated to express a luciferase reporter gene for the 3′-UTR of p21 which is a cyclin-dependent kinase inhibitor 1 (also called p21/WAF1). From over 250 different miRNAs that were tested, about 30 miRNAs showed a direct attenuation of the reporter gene [98]. Similar experiments for the CCND1 gene (G1/S-specific cyclin-D1) revealed similar results with seven miRNAs cooperatively attenuating the expression of CCND1, as established by using a 3′-UTR luciferase reporter construct [99].

16.4.2 Tools for the Detection of Combinatorial Regulation by miRNAs

The goal of most described miRNA-target predicting tools (Sect. 16.2.1) is to predict one-to-one relations, namely to determine a miRNA that matches a transcript (at a single or multiple sites). However, as discussed in Sect. 16.4.1, miRNAs most likely act as an ensemble that directly and specifically alters the expression of multiple gene-targets. Conversely, a collection of genes that are targeted coordinately in some experimental settings can be used as input to uncover the set of miRNAs that is most likely responsible for their measured level of gene expression. Along this principle, the MiRonTop [100], Diana-mirExTra [101] and GeneSet2miRNA [102] were developed.

We expanded this notion in the miRror Suite platform. miRror Suite transforms noisy miRNA predictions into a rational unified analysis. The miRror Suite is centered on the miRror2.0 algorithm [103]. The implementation is based on projecting most existing prediction tools into a unified statistical platform. Thus miRror2.0 can predict a coherent list of miRNAs that best explain the observed, complex signature of hundreds of down-regulated genes from experimental data. While our discussion on miRror2.0 focuses on miRNAs from human, the system supports the analysis of other animals (mouse, rat, fly, worm and zebrafish). There are a number of optional parameters that allow the miRror2.0 tool to restrict the analysis to more specific requests. For example, the prediction can operate on preselected tissues or cell-lines (from about 100 options). We demonstrate the generality of the miRror application and its inherent flexibility.

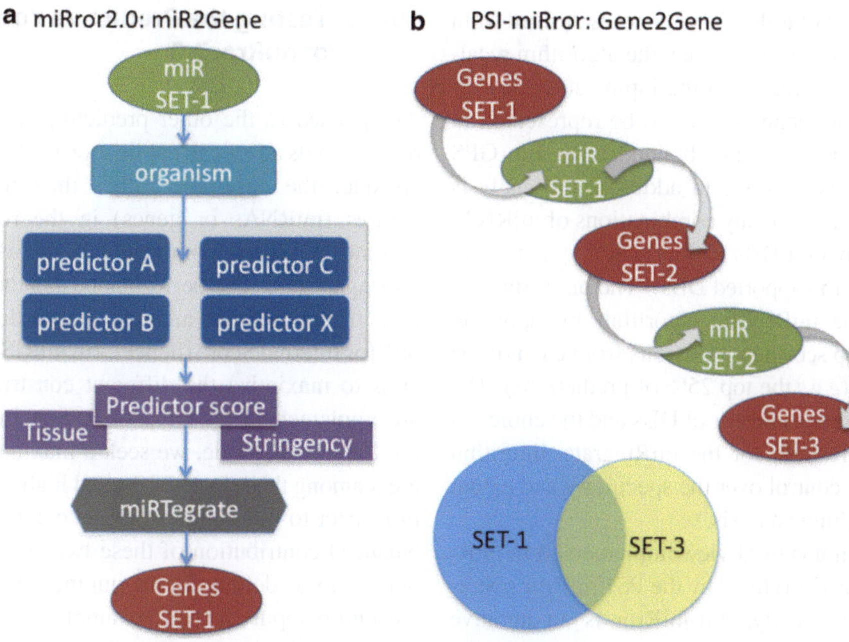

Fig. 16.2 The workflow of the miRror2.0 platform. (**a**) miRror2.0 in the miR2Gene mode. There are two main modes of operations: the miR2Gene and the Gene2miR. The input for these modes is a set of miRNAs or genes of any size. Following a selection of an organism (human, mouse, rat, worm, fly and zebrafish) and the operational mode (miR2Gene, Gene2miR), some optional choices are available: (i) the tissue of interest or the preferred cell-lines. The information is processed from the atlas of gene expressions [63]. (ii) Selection of all genes or only highly expressed subset (above a predetermined value, typically it reduces the list of genes by ~30%); (iii) the top scoring miRNA binding sites according to each DB. For each DB scoring method, a fraction that accounts for the top (*i.e.,* 10, 25, 50% and all predictions) can be selected for the analysis. (iv) Select any combination of the DBs (15 in the case of human). (v) To initiate the miRror2.0 search, several free parameters that determine the stringency of the procedure are selected. These parameters include the choice of P-value threshold, the minimal number of supporting DBs and the minimal number of input 'hits'. By changing these parameters, a relaxed or a strict search protocol is activated. (**b**) PSI-miRror in a Gene2Gene operational mode with two iterations (from Set-1 to Set-3). The *Venn diagram* shows the overlap of the input gene list (Set-1) and the output gene list (Set-3). The Venn allows focusing on genes that were removed from the input set or those that were added to the output set

In practice, the miRror platform is used to connect a gene list to the minimal set of preferred miRNAs or a miRNA collection to a set of genes (Fig. 16.2a). We refer to these analyses as Gene2miR and miR2Gene, respectively. The core of the statistical basis underlying miRror is the miRtegrate algorithm (Fig. 16.2a). In a nutshell, for the Gene2miR mode, miRtegrate calculates the probability of matches between the experimental gene set and all miRNAs. This is done by comparing the gene set to the complete gene list that is reported by each of the miRNA-target prediction DBs. The probability of the miRNA's interaction with the input gene set as opposed to the rest of the genes in each DB is calculated. Calculating a P-value for the set of input (miRNA or Genes) is performed according to the hypergeometric distribution [103]. The reported result is any set with ranked probabilities and scores that meet the statistical threshold (*e.g.*, P-value =0.01, corrected for multiple tests).

Figure 16.2a illustrates the key principles of the miRror2.0 platform. In reality, the number of genes or miRNAs that result from any large-scale transcriptomic experiments is in the 10s–100s for miRNAs and the 100s–1,000s for genes. The platform is based on a large number of parameters allowing control over the statistical threshold of miRtegrate (which effectively translates to operation stringency). For example, in a case where the set of miRNAs or genes is derived from a specific tissues or cell line, the algorithm recalculates the likelihood of the input set in view of the candidate genes known to be represented in the selected tissues as obtained from Bio-GPS (http://biogps.gnf.org). In addition, the analysis can be restricted to any combinations of miRNA-target prediction DBs (for human, any combination of the 15 supported DBs). The platform may activate the miRror2.0 algorithm by applying only the top scoring predictions from each of the DBs used (*e.g.,* the top 25% of predictions). The flexibility in the number of DBs and the choice of P-value threshold for the miRtegrate algorithm allows full control over the specificity and extent of the resulting analysis.

The combinatorial view implemented in miRror2.0 is further refined by the PSI-miRror operation. Schematically, PSI-miRror is an iterative protocol that aims to refine the input sets (Genes or miRNAs) by increasing the coherence of the input miRNA set to the set of genes and iteratively refining the list of the genes by re-applying the miRror2.0 cycle.

Figure 16.2b illustrates the operation of PSI-miRror in Gene2Gene mode. PSI-miRror can be activated in four modes: Gene2Gene and miR2miR but also miR2Gene and Gene2miR. Figure 16.2b shows that an input Gene Set (Set-1) results in a list of miRNAs that is then fed to an additional iteration that results in an intermediate gene set (Set-2). By analogy to PSI-BLAST [104], the procedure halts when no additional refinement is achieved. In most instances, genes are added or removed from the original list. The application of PSI-miRror is attractive for testing hypotheses. Genes that were added along the iterations of PSI-miRror and are reported in the final set (Set-3, Fig. 16.2b) are candidates for further investigation and experimental validation. The intuition is that genes or miRNAs that are not coherent with the experimental results (*e.g.*, due to the indirect effect of miRNAs) will be removed by the PSI-miRror operation, while coherent genes or miRNAs that were missed, will be added. Often, the intersection of the initial set and the final set is the most coherent set that can be further analyzed (Fig. 16.2b, Venn diagram).

16.4.3 Testing the Predictive Power of miRror2.0

As opposed to the other predicting tools, miRror2.0 and its advanced application of PSI-miRror consider the ensemble rather than individual entities (miRNAs or genes) in the regulatory scheme. A crucial component is the associated scoring system. The performance and the predictions from miRror2.0 are ranked according to the miRror Internal Score (henceforth miRIS). miRIS aims to maximize the different constraints that are implemented by miRror2.0. Specifically, in the Gene2miR mode, we seek a maximal agreement among the selected DBs and high sensitivity in respect to the input. miRIS is composed of a balanced contribution of these two components. Sensitivity is defined as the number of hits from the entire input list. For example, consider an overexpression experiment of a specific miRNA in cells and assume that 400 genes were down-regulated (as measured by a DNA microarray). miRror application at the Gene2miR mode is applied after setting the desired level of stringency (determined by a P-value threshold) and selecting the number of predictors for the analysis. For this illustration, we assume that 12 predicting DBs are selected. miRIS is associated with any of the predicted miRNAs. For example, for a miRNA on the list, only 6 DBs support the prediction and only 200 out of the 400 genes in the input set are marked as relevant genes (which we refer to as 'hits'), the calculated miRIS for this miRNA is therefore 0.5. A miRIS of 0.75 is calculated once all the 400 genes are reported (*i.e.*, maximal sensitivity) or if only 200 hits are reported, but with a full agreement of all 12 DBs. As shown by this example, miRIS combines DB consistency and sensitivity into a single score.

The validity of the concept of miRNAs 'working together' by miRror platform was tested. To this end, we took advantage of the growing number of experiments in which (i) miRNAs were introduced into cell cultures; (ii) the entire transcriptome is compared to control (often cells introduced with a scrambled sequence or a mock transfection). We analyzed data from such experiments from a variety of cell lines. The analyses

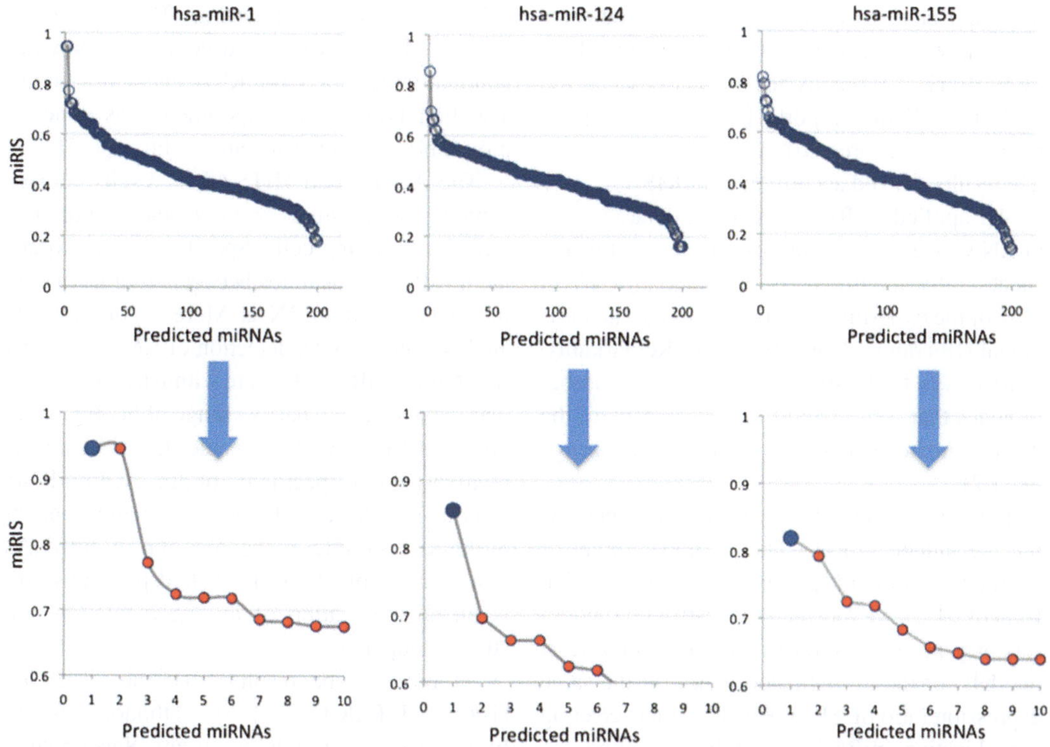

Fig. 16.3 Ranked analysis on miRror2.0. Data were collected from the GEO [62] and extracted from the SOFT files. The Affymetrix platform datasets include: GSE6207 (HepG2 cells, hsa-miR-124-24hr) and GSE22002 (HeLa cells, hsa-miR-1, hsa-miR-155). The results are shown in view of the miRror2.0 results, ranked according to miRIS for the top 200 predictions (*top*) and the top 10 predictions (*bottom*). The correct miRNA from the over-expression experiments are marked *blue*

combined data from DNA microarray platforms (Affymetrix, Agilent) and quantitative mass spectrometry experiments. Specifically, we focused on experiments that report on miRNAs that were overexpressed in cells. Under such controlled conditions, we tested whether miRror can successfully identify the actual overexpressed miRNA using solely the gene expression profiles.

A growing number of such experiments are reported for which the whole transcriptome is compared to that of controlled cells. We assess the ability to recover the evidence of the transfected miRNA from the global transcriptomic profiles of the down-regulated genes (at a threshold >1.2-fold). Importantly, for some experiments hundreds of genes were used as input without reduction in the performance. For example, while only 270 genes were down regulated in an experiment of hsa-miR-145 transfection (DLD-1 cells, GSE18625), for hsa-miR-335 (LM2-Lung cells, GSE9586) this number was almost 10,000. Success was determined according to miRIS. Namely, reporting the position of the correct (the overexpressed) miRNA in the ranked list of all miRNA predictions.

Figure 16.3 shows the results from miRror for overexpression experiments of hsa-miR-124 (GSE6207), hsa-miR-155 and hsa-miR-1 (GSE2002). Ten percent of the down-regulated genes in these experiments (1,700–3,300 genes) were used as input for miRror2.0. Zooming on the top ten predictions (Fig. 16.3, bottom) shows that the actual miRNA is recovered as the top prediction (from 200 best predicted miRNAs). Moreover, the ten top miRNA predictions show a sharp drop in score. In some cases an additional miRNA reaches very significant miRIS (*e.g.*, has-miR-1). Interestingly, the extent of down-regulated genes (*i.e.*, fold change) is not a significant indicator for a successful recovery of the relevant miRNA by miRror2.0. We consider a success when the miRror prediction reports the correct (experimentally over-expressed) miRNA among the top five results, ranked according to miRIS.

We applied miRror to about 30 large-scale miRNA over-expression experiments. miRror successfully identified the relevant miRNA in 70% of the experiments. The success of the individual DBs ranges from 20 to 60%. Remarkably, miRror was fairly stable regarding the number of genes that were loaded, from 1 to 50% of the down-regulated genes (at a moderate fold change of ≥1.2).

In order to assess the high success in recovering the hidden miRNA from a noisy signal of hundreds of unfiltered genes (at a subtle repression level of >1.2), we repeated the tests with random sets of genes (genes must be reported in at least one DB) or the up-regulated genes (at the same expression ratio of ≥1.2). We show that selecting the objective miRNA failed by repeating the miRror protocol on randomized sets (multiple randomization of identical group size).

We attribute the source of stability in miRNA identification to (i) the predetermined statistical threshold that is applied for a dozen of miRNA-target predictors; (ii) the obligatory demand for a minimal consistency ≥2 DBs and (iii) a requirement of a minimal agreement on number of hits from the input gene list for each proposed miRNA.

16.4.4 Measurements of Direct Binding by miRNAs

The method of HITS-CLIP [45] was developed as a way to directly monitor protein–RNA interactions in living cells [105]. Briefly, the method is based on trapping by cross-linking RNA–protein complexes of interacting molecules that are within a minimal molecular distance. The protocol allows the collection of trapped molecules which are then subjected to trimming of the RNA hanging tails. The result of this protocol is a collection of minimally sized fragments that are suitable for deep sequencing. The use of the HITS-CLIP on AGO-based complexes provides genome wide miRNA–mRNA interaction maps. The first HITS-CLIP experiment was done on mouse brain under stringent conditions [45].

The AGO based HITS-CLIP results address some of the questions on the *modus operandi* of miRNAs in living cells. Specifically, the experiment (ideally) separates between a direct and an indirect effect of miRNAs. More importantly, the analysis only allows detection of RNA segments (mRNA or miRNA) that are within a short molecular distance and a narrow range of 50–60 nucleotides of the mRNA molecule. Considering the relatively high specificity of the AGO-mRNA (relying on the correlation among independent biological samples), the number of miRNAs that were trapped and identified per transcript is an approximation of *in-vivo* regulation and AGO-occupancy.

At present, the results from the CLIP-Seq [106] and PAR-CLIP [76] methodologies are limited to only few cellular settings. Nevertheless, some trends for the combinatorial activity of the miRNAs can already be demonstrated. Actually using HITS-CLIP [107], only 10% of the genes were regulated by a single miRNA. Results from the recent CLIP based methods (HITS-CLIP, PAR-CLIP and CLIP-Seq) [58] show that each of the genes that were identified to be subjected to miRNA regulation is in fact targeted by multiple miRNAs. These experiments allow, for the first time, the construction of a miRNA-mRNA interaction map, which supports the notion of combinatorial, cooperative action by miRNAs on targeted transcripts [45].

A collection of large-scale CLIP experiments is compiled in the StarBase database [108]. StarBase includes thousands of experimentally confirmed miRNA-target interactions and complementary data from other AGO proteins. In addition, it provides a genome browser for the reads that were collected during the deep sequencing phase. Figure 16.4 illustrates the complexity in CLIP-experimental interpretation. The deep sequencing reads are illustrated as 'piles' (Fig. 16.4a). Using a consensus of prediction algorithms and in some cases, direct sequencing

a

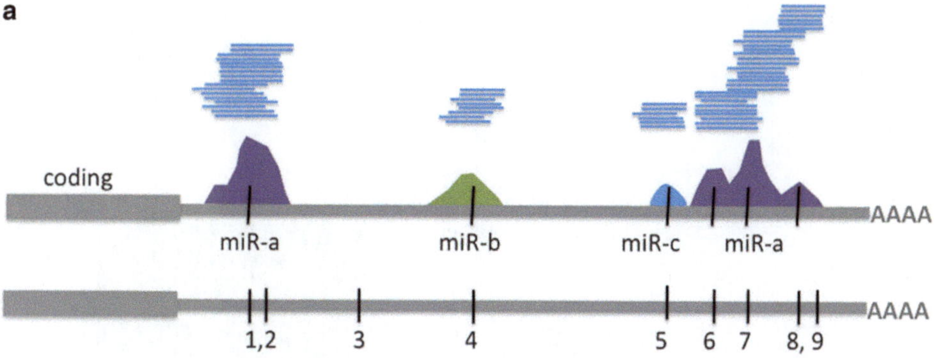

b

Binding	1	2	3	4	5	6	7	8	9
Predictor A									
Predictor B									
Predictor C									
Predictor D									
HITS-CLIP	miR-a	miR-d	miR-e	miR-b	miR-c	miR-a	miR-a	miR-a	miR-f
Intensity	++++	-	-	++	+	++	+++	+	-
Comment	OL	OL						OL	OL

Fig. 16.4 Interpretation of CLIP-based experiments. Data analyzed from StarBase [108]. (**a**) The deep sequencing reads are illustrated and the 'piles' along the 3′-UTR of the gene. (**b**) Each of the binding sites (1–9) is associated with predicting DB (marked **A–D**). Note that some binding sites are more potent than other (*marked by + symbol*). The consistency between the predicting DBs is only partial (see binding site 3,4). There are multiple binding sites (at different extent) for the same miRNA (miR-a, *purple*). The predicting DBs indicate miRNAs that overlap on the sequence of their binding sites (overlap, OL). The OL sites are often excluded due to accessibility argumentation

of miRNAs, the match of the miRNA and the gene is reported. In Fig. 16.4a three different miRNAs regulate the presented gene (indicated as miR-a, miR-b and miR-c). However, the miRNA-target prediction algorithms (Fig. 16.4b, marked as Predictors A-D), predict nine miRNA binding sites (1–9). Recall that many of the binding sites were not validated experimentally (miR-e and miR-f). Additional information that became evident from the CLIP-based experiments concerns the intensity of the reads for each miRNA (indicated schematically by the + sign, Fig. 16.4b). The simplistic illustration (Fig. 16.4a) emphasizes the challenge in formulating a miRNA combinatorial view: (i) Some binding sites are more potent than others. (ii) The consistency between the different predictors is limited. (iii) Multiple binding sites for the same miRNA differ in the intensity of the reads (miR-a, purple). (iv) The miRNA-target prediction algorithms often support overlapping binding sites. In reality, the overlap with other binding sites may be excluded due to accessibility argumentation. (v) There is no direct evidence for cooperative binding on one molecule, instead, the scheme (Fig. 16.4a) is most likely a reflection of miRNAs bound to the population of mRNAs.

16.4.5 Looking Through the miRror – Predictions Versus Experimental Reality

The coherence between the miRNA-targets that are based on gene expression (Fig. 16.3) and those obtained from the CLIP-based experiments is surprisingly low [106]. The gene expression data and the CLIP data are collected from live

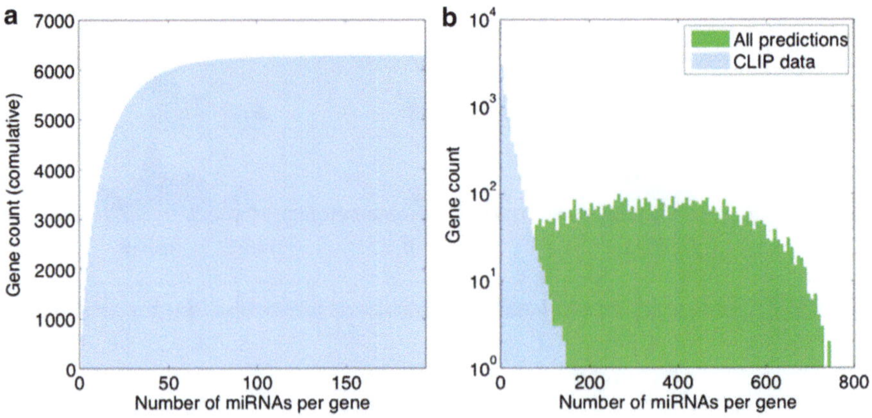

Fig. 16.5 Statistical analysis from CLIP-based experiments. (**a**) A cumulative view of the number of miRNAs regulating each gene, from CLIP data. The steep climb demonstrates how prevalent regulation by multiple miRNAs appears to be. (**b**) A comparison of CLIP data and predicted miRNA-targets. The combination of all 15 human predicting DBs is shown. The miRNA-targets that are covered by the CLIP data are shown separately in *light blue*. The distributions from the CLIP data and the union of the DB predictions are very different. It emphasizes the gap between the computational view on the targeting potential of genes by miRNAs and the observed gene targeting from the most up-to-date CLIP experiments

cells. Still, over-expression of specific miRNA may be prone to non-physiologically high miRNA concentrations. On the other hand, a bias in the CLIP data may reflect the inability to capture transient interactions of AGO and mRNAs. Moreover, the identity of the apparently trapped miRNAs in the CLIP assays is largely based on the set of computational predicting tools that suffer from high false positive rates. It is suggested that the readouts of mRNA suppression (measured by gene expression profiles) and the initial pairing (measured by CLIP technologies) are complementary but non-overlapping assessments of the regulation by miRNAs. An analysis of the features that govern miRNA–target match is critical to improving the prediction power of the methods (both computational and experimental).

Figure 16.5 shows the statistical analysis of the data collected from StarBase [108]. The cumulative representation (combining several experiments) with a gene centric view on all the data shows that 50% of the 6,200 targeted genes are associated with up to eight miRNAs (Fig. 16.5a) and 90% of the genes are targeted by up to 35 miRNAs. A few genes are even targeted by more than 100 miRNAs. Interestingly, when all the potential sites that are predicted by all 12 human miRNA-target DBs (the union of all), the analysis shows that the experimental CLIP data captures only a relative small fraction of the predicted set (Fig. 16.5b).

Figure 16.6 summarizes the potential of miRror2.0 to be used as an assessment tool for combinatorial regulation. Several tests are carried out to examine the coherence of the experimental data and miRror results. The results of the scheme illustrated in Fig. 16.6a are discussed (Fig. 16.3). The scheme in Fig. 16.6b is applied to the CLIP data from StarBase [108]. Specifically, the ensemble of miRNAs that is associated with each targeted gene was collected for each gene from the thousands of genes that are targeted by at least two miRNAs (Fig. 16.6b). For 98% of the genes miRror successfully identified the relevant gene. For 81%, the correct prediction was among the top 10% of predictions (typically from a ranked list of >1,000 predictions). This finding supports the predictive power of miRror2.0 for genes that are targeted by a high number of miRNAs. Furthermore, when considering genes that were targeted by >20 miRNAs, the performance of miRror reached 98% and a complementary view was noted for genes that were targeted by <10 miRNAs (the performance dropped slightly to 90% success).

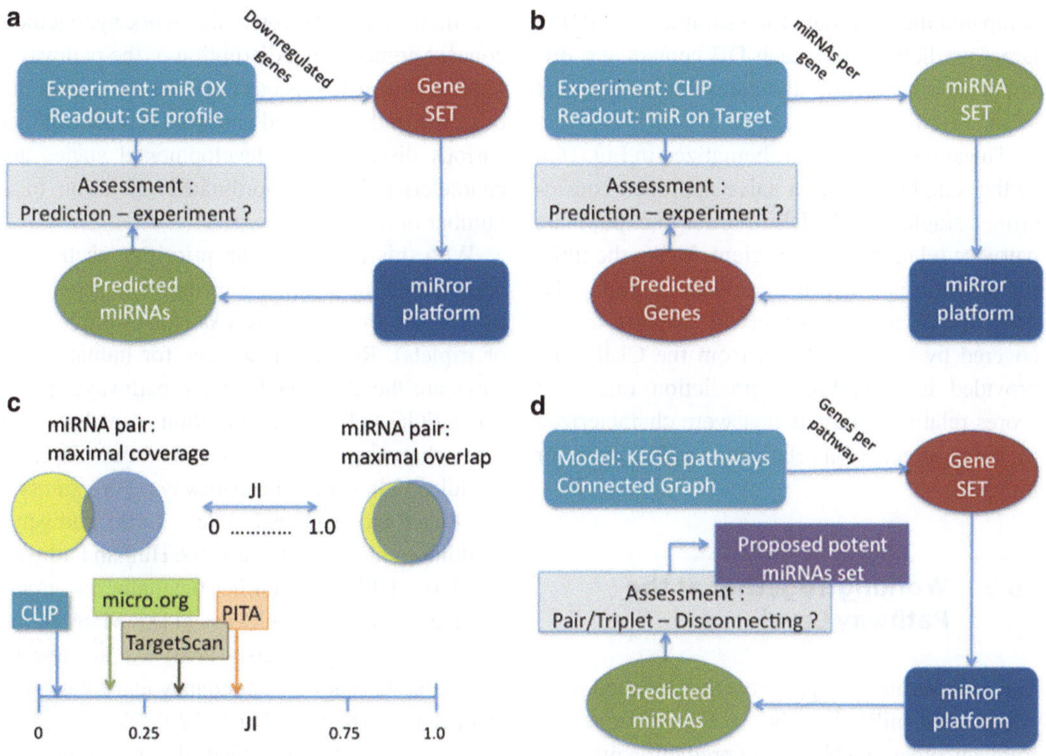

Fig. 16.6 Assessment of the combinatorial nature of miRNAs. (**a**) Using Gene2miR mode to assess the recovery of over-expression of miRNAs from the repressed genes from large-scale transcriptomic profiling platforms. (**b**) Using miR2Gene mode to assess the recovery of a gene from the collection of the available CLIP-based experimental data. (**c**) Assessment using the Jaccard Index (JI) of pairs of miRNAs. Data were from CLIP-based experiments and from the prediction according to selected predicting DBs. (**d**) Using miRror2.0 at a Gene2miR mode for KEGG human pathway graphs and testing the ability of small sets of miRNA to disrupt the connectivity of the pathway. For details see text

The tools and methodologies developed for the validation of miRror2.0 and the large CLIP dataset allowed us to test the minimal mode of combinatorial regulation – namely the concept of 'miRNA pairs'. Figure 16.6c shows the protocol applied in formulating the concept of miRNA-pairs. Pairs are the simplest form of multiple miRNA co-regulation and thus are a natural starting point for a computational assessment. We wish to see how well each of the major prediction DBs matches the biological experimental data. The number of genes that are reported from the multiple CLIP data (compiled by StarBase [108]) is >6,200. A vast majority of them are regulated by multiple miRNAs (Fig. 16.5). There are two general modes that comply with regulation by pairs (Fig. 16.6c): (i) A pair of miRNAs that expands the set of targets, thus allowing a better coverage of the relevant genes. (ii) Each miRNA in the pair tightly overlaps the targets of the other miRNA. In this case, the pair of miRNAs acts in 'backup' mode, with a high degree of redundancy. The two extreme scenarios are formulated using the Jaccard Index (JI). Intuitively, JI is a simple measure for comparing the similarity (intersection) and diversity (union) of the sample sets. A low JI value is indicative of the expansion mode while high JI indicates the backup mode.

From this naïve view on all pairs of miRNAs that were reported in CLIP data, several observations are worth noting: (i) The data from the CLIP experiments are dominated by a very low JI. Although only less than 20% of the pairs have no shared targets, the JI is extremely low. When compared the same data for a sample of miRNA-target predicting DBs, each DB centers at a different JI (the arrows indicates the average JI value, Fig. 16.6c).

The analyses that are schematized in Fig. 16.6 further emphasize that a naive approach considering a single miRNA DB in order to extrapolate pairwise relations is insufficient. Using the miRror platform is a step toward such an extrapolation. Specifically, we observed that the targets covered by many miRNAs from the CLIP data provided us with higher prediction rates and scores relative to targets that were characterized by being targets of a relatively small number of miRNAs.

16.5 Working Together at the Pathway Level

The interpretation of gene sets that resulted from coordinated miRNAs (Fig. 16.6 miR2Gene), or from any other miRNA-target prediction protocol should be analyzed within a cellular context. A number of tools were developed that cover aspects of protein and functional interaction (STRING [109]), regulatory pathways (Reactome [110]) and functional annotations (PANDORA [111] and DAVID [112]).

The pathway representation best describes the biological processes in cells. The human regulatory pathways are compiled by the KEGG resource [113]. KEGG pathways are a collection of manually drawn pathway maps. These maps represent knowledge on the molecular interaction and reaction networks for domains including human diseases, organismal systems, cellular processes, and environmental information processing. The collection covers ~100 pathway maps for human.

The notion of 'miRNAs working together' is tested in view of metabolic and regulatory pathways. Specifically, regulation of miRNAs was suggested at the level of pathway or biological process [114]. We assess the possibility that cooperative action by a small, selected group of miRNAs can alter the expression of genes that belong to the same pathway, without sacrificing specificity. An extension of the 'working together' concept argues that a disruption of the pathway's topology by miRNAs has the potential to alter the outcome of the targeted pathway. It is known that various diseases and developmental stages are characterized by a coordinated alteration of a number of miRNAs.

With this idea, one can prioritize each pathway according to its susceptibility to regulation by a small group of miRNAs (for example pairs or triplets). Reliable resources for human pathways are the PID (NCI human pathways) [115] and KEGG [116]. The definition of pathways is somewhat vague as some resources describe modules rather than full pathways. For example, the Reactome database covers 1,100 pathways (cellular modules) [117] and the Human Pathway Database (HPD), that unifies the major resources, includes over 1,000 pathways [118]. Many of the pathways were previously analyzed in view of their modular nature, redundancy and robustness against perturbations [119, 120]. While recent studies have acknowledged the usefulness of miRNAs on regulation pathways [121, 122], adding the connectivity of genes in the pathway is a key determinant that was largely ignored. The intuition for the pathway disruption approach is that a quantitative change in a set of miRNAs is expected to alter the pathway outcome [90, 123].

A test case (Fig. 16.6d) using the 100 regulatory, disease oriented and metabolic pathways from KEGG that cover about 4,500 human genes revealed an intriguing principle of miRNA regulation. Analysis of the pathway via the concept of individual miRNA prediction databases results in hundreds of potential miRNAs. Therefore, matching a small number of miRNAs to a pathway is virtually impossible with the current prediction DBs. However, the potential of a small group of coordinated miRNAs to alter the integrity of human pathway can be challenged using the miRror2.0 combinatorial tool. The motivation of our approach is in assessing the potential of a set of miRNAs to disrupt a graph that represents a cellular pathway.

For the scheme in Fig. 16.6d, we start from a pathway and end up with a selected list of miRNA sets (*i.e.*, pairs and triplets that work together) which preferentially disrupt the integrity of the pathway. The following steps are taken: (i) All high quality human pathways are converted to undirected graphs in which the nodes are the genes (or complexes) and the edges are the regulatory interactions. (ii) Each pathway is converted to a gene set that is subjected to miRror2.0 to determine the ranked list of possible miRNAs. Refinement by PSI-miRror further limits the list of possible miRNAs to the most relevant set. (iii) Designing a disconnecting score (DIS) that captures the degree of network disruption (for example, the partition into connected components, edge elimination and the like). (iv) Applying an exhaustive search for all pairs and triplets from all candidate miRNAs combinations. Finally, providing the sets that maximally impact graph connectivity (resulting in a high DIS score). As the miRNA combinatorial space is vast, for a pathway that reported by miRror2.0 to have 40 miRNAs, about 10,000 possible miR-Triplets need to be ranked (by DIS score) in order to identify the best sets. For a pathway with 50 potential miRNA candidates, the search space for miRNA triplets is 19,600. Of course, the topology of the pathway graph is a key determinant in this scheme. Note that the number of genes in the human KEGG pathways ranges from 10 to 250 and the number of initial candidates for miRNA disruption (according to miRror output) range from 2 to 60.

Several conclusions are derived from this approach. Most notably, 85% of all KEGG based pathways are amenable to disruption by a small miRNA set of pairs or triplets (the same results apply to PID pathways [115]). Analyzing all pathways revealed that typically, only 4–5 miRNAs are associated with the most potent set of miRNA pairs and triplets for each pathway. Several biological interpretations from the miRNA cooperative pathway disruption scheme can be drawn. Genes such as MAPK1, EGFR, AKT3, SRC that are prevalent in tens of regulatory pathways, are almost always included in the set of disrupted genes. Most likely, these proteins serve as connectors in the pathway graphs. Thus, targeting these signaling genes will lead to a substantial disruption of many pathways. Most surprisingly, the selected miRNA sets (pairs and triplets) with a maximal capacity to disrupt pathways show a minimal overlap. Therefore, it seems that a critical factor in selecting the most influential miRNA combinations is the graph topology rather than the identity of the individual nodes in the pathway graph.

16.6 Concluding Remarks

The concept of miRNAs working together is not new. In this chapter, we present experimental evidence while emphasizing tradeoffs in adopting a combinatorial mode of regulation for living cells under changing conditions.

We present a tool that incorporates the concept of 'working together' and describe some tests in view of the current experimental knowledge. The miRror Suite is a platform that empowers experimental biologists in gaining insights from a broad range of experimental protocols. It is based on a many-to-one and many-to-many approach. Namely, a group of miRNAs as an input leads to a minimized set of genes that best explain the observed gene expression profile. Similarly, this applies for a set of genes as input. The many-to-many optimization is performed by the PSI-miRror approach that provides a refined set of molecules by iterative application of the PSI-miRror algorithm. The miRror Suite provides an integrative, statistically based platform and exposes miRIS: a combined scoring system for a successful prediction of miRNA combinatorial regulations. miRror performance is discussed in [124].

Multiple layers of regulations in the cell are coordinated in governing cellular phenotypes. Most notable are: epigenetic chromatin marks, the transcription machinery of gene expression, the translation process, the degradation of transcripts and proteins, the metabolic balance and more. miRNAs constitute an additional layer of regulation that was carefully studied in stem cells, viral infection, cancer progression and other pathologies. It is likely that regulation of miRNAs is a key strategy of the cell as it strives to maintain robust homeostasis. Under this assumption, a modest modulation executed in a combinatorial mode can be manifested by a substantial change in cell physiology and phenotype. We illustrate the combinatorial regulation concept at the level of the individual target (Sect. 16.4.4), at the level of a set of genes whose expression were moderately changed (Sect. 16.4.3), and finally at the level of human pathway integrity (Sect. 16.5). While many questions remain to be solved, we expect the ensemble-oriented tools will prove essential to the biological interpretation of miRNA data.

Acknowledgements We thank Guy Naamati whose contribution to the miRror platform development was seminal. We thank Nati Linial for insightful ideas. We thank Solange Karsenty for maintenance of the miRror website. We thank Manor Askenazi for a critical reading and editing of the manuscript. A student fellowship (O.B) is awarded by the SCCB, the Sudarsky Center for Computational Biology. We apologize to those that we could not cite. In numerous instances we replace the primary citations by review articles due to space constraints. This study is partially supported by the ISF 592/07, the BSF 2007219 and the EU Framework VII of Prospects.

References

1. Bartel DP (2004) MicroRNAs: genomics, biogenesis, mechanism, and function. Cell 116:281–297
2. Landgraf P, Rusu M, Sheridan R et al (2007) A mammalian microRNA expression atlas based on small RNA library sequencing. Cell 129:1401–1414
3. Iorio MV, Ferracin M, Liu CG et al (2005) MicroRNA gene expression deregulation in human breast cancer. Cancer Res 65:7065–7070
4. Grad Y, Aach J, Hayes GD et al (2003) Computational and experimental identification of *C. elegans* microRNAs. Mol Cell 11:1253–1263
5. Friedlander MR, Chen W, Adamidi C et al (2008) Discovering microRNAs from deep sequencing data using miRDeep. Nat Biotechnol 26:407–415
6. Bushati N, Cohen SM (2007) microRNA functions. Annu Rev Cell Dev Biol 23:175–205
7. Bartel DP (2009) MicroRNAs: target recognition and regulatory functions. Cell 136:215–233
8. Brodersen P, Voinnet O (2009) Revisiting the principles of microRNA target recognition and mode of action. Nat Rev Mol Cell Biol 10:141–148
9. Gregory RI, Chendrimada TP, Cooch N et al (2005) Human RISC couples microRNA biogenesis and posttranscriptional gene silencing. Cell 123:631–640
10. Pillai RS, Bhattacharyya SN, Filipowicz W (2007) Repression of protein synthesis by miRNAs: how many mechanisms? Trends Cell Biol 17:118–126
11. Baek D, Villen J, Shin C et al (2008) The impact of microRNAs on protein output. Nature 455:64–71
12. Bandyopadhyay S, Mitra R (2009) TargetMiner: MicroRNA target prediction with systematic identification of tissue specific negative examples. Bioinformatics 25:2625–2631
13. Filipowicz W, Bhattacharyya SN, Sonenberg N (2008) Mechanisms of post-transcriptional regulation by microRNAs: are the answers in sight? Nat Rev Genet 9:102–114
14. Cullen BR (2009) Viral and cellular messenger RNA targets of viral microRNAs. Nature 457:421–425
15. Alvarez-Garcia I, Miska EA (2005) MicroRNA functions in animal development and human disease. Development 132:4653–4662
16. Liang R, Bates DJ, Wang E (2009) Epigenetic control of MicroRNA expression and aging. Curr Genomics 10:184–193
17. Schroen B, Heymans S (2012) Small but smart-microRNAs in the centre of inflammatory processes during cardiovascular diseases, the metabolic syndrome, and ageing. Cardiovasc Res 93:605–613
18. Zhang B, Pan X, Anderson TA (2006) MicroRNA: a new player in stem cells. J Cell Physiol 209:266–269
19. Xiao C, Rajewsky K (2009) MicroRNA control in the immune system: basic principles. Cell 136:26–36
20. Zhao Y, Samal E, Srivastava D (2005) Serum response factor regulates a muscle-specific microRNA that targets Hand2 during cardiogenesis. Nature 436:214–220
21. Liao R, Sun J, Zhang L et al (2008) MicroRNAs play a role in the development of human hematopoietic stem cells. J Cell Biochem 104:805–817
22. Cheng AM, Byrom MW, Shelton J et al (2005) Antisense inhibition of human miRNAs and indications for an involvement of miRNA in cell growth and apoptosis. Nucleic Acids Res 33:1290–1297
23. Kumar MS, Lu J, Mercer KL et al (2007) Impaired microRNA processing enhances cellular transformation and tumorigenesis. Nat Genet 39:673–677
24. Volinia S, Calin GA, Liu CG et al (2006) A microRNA expression signature of human solid tumors defines cancer gene targets. Proc Natl Acad Sci U S A 103:2257–2261
25. He L, Thomson JM, Hemann MT et al (2005) A microRNA polycistron as a potential human oncogene. Nature 435:828–833
26. Calin GA, Ferracin M, Cimmino A et al (2005) A MicroRNA signature associated with prognosis and progression in chronic lymphocytic leukemia. N Engl J Med 353:1793–1801
27. Lee YS, Dutta A (2007) The tumor suppressor microRNA let-7 represses the HMGA2 oncogene. Genes Dev 21:1025–1030
28. Calin GA, Sevignani C, Dumitru CD et al (2004) Human microRNA genes are frequently located at fragile sites and genomic regions involved in cancers. Proc Natl Acad Sci U S A 101:2999–3004

29. Griffiths-Jones S, Grocock RJ, van Dongen S et al (2006) miRBase: microRNA sequences, targets and gene nomenclature. Nucleic Acids Res 34:D140–D144
30. Cullen BR (2004) Transcription and processing of human microRNA precursors. Mol Cell 16:861–865
31. Fernandez-Valverde SL, Taft RJ, Mattick JS (2010) Dynamic isomiR regulation in Drosophila development. RNA 16:1881–1888
32. Berezikov E, Robine N, Samsonova A et al (2010) Deep annotation of Drosophila melanogaster microRNAs yields insights into their processing, modification, and emergence. Genome Res 21:203–215
33. Kozomara A, Griffiths-Jones S (2011) miRBase: integrating microRNA annotation and deep-sequencing data. Nucleic Acids Res 39:D152–D157
34. Rodriguez A, Griffiths-Jones S, Ashurst JL et al (2004) Identification of mammalian microRNA host genes and transcription units. Genome Res 14:1902–1910
35. Barik S (2008) An intronic microRNA silences genes that are functionally antagonistic to its host gene. Nucleic Acids Res 36:5232–5241
36. Ruby JG, Jan CH, Bartel DP (2007) Intronic microRNA precursors that bypass Drosha processing. Nature 448:83–86
37. Berezikov E, Thuemmler F, van Laake LW et al (2006) Diversity of microRNAs in human and chimpanzee brain. Nat Genet 38:1375–1377
38. Liu N, Okamura K, Tyler DM et al (2008) The evolution and functional diversification of animal microRNA genes. Cell Res 18:985–996
39. Betel D, Wilson M, Gabow A et al (2008) The microRNA.org resource: targets and expression. Nucleic Acids Res 36:D149–D153
40. Krek A, Grun D, Poy MN et al (2005) Combinatorial microRNA target predictions. Nat Genet 37:495–500
41. Wang X (2008) miRDB: a microRNA target prediction and functional annotation database with a wiki interface. RNA 14:1012–1017
42. Lim LP, Lau NC, Garrett-Engele P et al (2005) Microarray analysis shows that some microRNAs downregulate large numbers of target mRNAs. Nature 433:769–773
43. Doench JG, Sharp PA (2004) Specificity of microRNA target selection in translational repression. Genes Dev 18:504–511
44. Enright AJ, John B, Gaul U et al (2003) MicroRNA targets in Drosophila. Genome Biol 5:R1
45. Chi SW, Zang JB, Mele A et al (2009) Argonaute HITS-CLIP decodes microRNA-mRNA interaction maps. Nature 460:479–486
46. Bentwich I (2005) Prediction and validation of microRNAs and their targets. FEBS Lett 579:5904–5910
47. Sethupathy P, Megraw M, Hatzigeorgiou AG (2006) A guide through present computational approaches for the identification of mammalian microRNA targets. Nat Methods 3:881–886
48. Rajewsky N (2006) microRNA target predictions in animals. Nat Genet 38(Suppl):S8–S13
49. Sethupathy P, Corda B, Hatzigeorgiou AG (2006) TarBase: a comprehensive database of experimentally supported animal microRNA targets. RNA 12:192–197
50. Lewis BP, I-h S, Jones-Rhoades MW et al (2003) Prediction of mammalian microRNA targets. Cell 115:787–798
51. John B, Enright AJ, Aravin A et al (2004) Human MicroRNA targets. PLoS Biol 2:e363
52. Maragkakis M, Reczko M, Simossis VA et al (2009) DIANA-microT web server: elucidating microRNA functions through target prediction. Nucleic Acids Res 37:W273–W276
53. Kertesz M, Iovino N, Unnerstall U et al (2007) The role of site accessibility in microRNA target recognition. Nat Genet 39:1278–1284
54. Hausser J, Berninger P, Rodak C et al (2009) MirZ: an integrated microRNA expression atlas and target prediction resource. Nucleic Acids Res 37:W266–W272
55. Nielsen CB, Shomron N, Sandberg R et al (2007) Determinants of targeting by endogenous and exogenous microRNAs and siRNAs. RNA 13:1894–1910
56. Hsu S-D, Chu C-H, Tsou A-P et al (2008) miRNAMap 2.0: genomic maps of microRNAs in metazoan genomes. Nucleic Acids Res 36:D165–D169
57. Long D, Lee R, Williams P et al (2007) Potent effect of target structure on microRNA function. Nat Struct Mol Biol 14:287–294
58. Alexiou P, Maragkakis M, Papadopoulos GL et al (2009) Lost in translation: an assessment and perspective for computational microRNA target identification. Bioinformatics 25:3049–3055
59. Martin RC, Liu PP, Goloviznina NA et al (2010) microRNA, seeds, and Darwin?: diverse function of miRNA in seed biology and plant responses to stress. J Exp Bot 61:2229–2234
60. Orom UA, Lund AH (2009) Experimental identification of microRNA targets. Gene 451:1–5
61. Thomson DW, Bracken CP, Goodall GJ (2011) Experimental strategies for microRNA target identification. Nucleic Acids Res 39:6845–6853
62. Barrett T, Edgar R (2006) Mining microarray data at NCBI's Gene Expression Omnibus (GEO)*. Methods Mol Biol 338:175–190
63. Parkinson H, Sarkans U, Kolesnikov N et al (2011) ArrayExpress update–an archive of microarray and high-throughput sequencing-based functional genomics experiments. Nucleic Acids Res 39:D1002–D1004
64. van Dongen S, Abreu-Goodger C, Enright AJ (2008) Detecting microRNA binding and siRNA off-target effects from expression data. Nat Methods 5:1023–1025
65. Creighton CJ, Reid JG, Gunaratne PH (2009) Expression profiling of microRNAs by deep sequencing. Brief Bioinform 10:490–497
66. Witten D, Tibshirani R, Gu SG et al (2010) Ultra-high throughput sequencing-based small RNA discovery and discrete statistical biomarker analysis in a collection of cervical tumours and matched controls. BMC Biol 8:58

67. Stark MS, Tyagi S, Nancarrow DJ et al (2010) Characterization of the melanoma miRNAome by deep sequencing. PLoS One 5:e9685
68. Bar M, Wyman SK, Fritz BR et al (2008) MicroRNA discovery and profiling in human embryonic stem cells by deep sequencing of small RNA libraries. Stem Cells 26:2496–2505
69. Goff LA, Davila J, Swerdel MR et al (2009) Ago2 immunoprecipitation identifies predicted microRNAs in human embryonic stem cells and neural precursors. PLoS One 4:e7192
70. Easow G, Teleman AA, Cohen SM (2007) Isolation of microRNA targets by miRNP immunopurification. RNA 13:1198–1204
71. Zhang L, Ding L, Cheung TH et al (2007) Systematic identification of *C. elegans* miRISC proteins, miRNAs, and mRNA targets by their interactions with GW182 proteins AIN-1 and AIN-2. Mol Cell 28:598–613
72. Selbach M, Schwanhausser B, Thierfelder N et al (2008) Widespread changes in protein synthesis induced by microRNAs. Nature 455:58–63
73. Papadopoulos GL, Reczko M, Simossis VA et al (2009) The database of experimentally supported targets: a functional update of TarBase. Nucleic Acids Res 37:D155–D158
74. Hua Y-J, Tang Z-Y, Tu K et al (2009) Identification and target prediction of miRNAs specifically expressed in rat neural tissue. BMC Genomics 10:214
75. Mendes ND, Freitas AT, Sagot MF (2009) Current tools for the identification of miRNA genes and their targets. Nucleic Acids Res 37:2419–2433
76. Hafner M, Landthaler M, Burger L et al (2010) Transcriptome-wide identification of RNA-binding protein and microRNA target sites by PAR-CLIP. Cell 141:129–141
77. Jiang Q, Wang Y, Hao Y et al (2009) miR2Disease: a manually curated database for microRNA deregulation in human disease. Nucleic Acids Res 37:D98–D104
78. Backes C, Meese E, Lenhof HP et al (2010) A dictionary on microRNAs and their putative target pathways. Nucleic Acids Res 38:4476–4486
79. Papadopoulos GL, Alexiou P, Maragkakis M et al (2009) DIANA-mirPath: integrating human and mouse microRNAs in pathways. Bioinformatics 25:1991–1993
80. Saj A, Lai EC (2011) Control of microRNA biogenesis and transcription by cell signaling pathways. Curr Opin Genet Dev 21:504–510
81. Winter J, Jung S, Keller S et al (2009) Many roads to maturity: microRNA biogenesis pathways and their regulation. Nat Cell Biol 11:228–234
82. Chatterjee S, Grosshans H (2009) Active turnover modulates mature microRNA activity in Caenorhabditis elegans. Nature 461:546–549
83. Khan AA, Betel D, Miller ML et al (2009) Transfection of small RNAs globally perturbs gene regulation by endogenous microRNAs. Nat Biotechnol 27:549–555
84. Beitzinger M, Peters L, Zhu JY et al (2007) Identification of human microRNA targets from isolated argonaute protein complexes. RNA Biol 4:76–84
85. Linsley PS, Schelter J, Burchard J et al (2007) Transcripts targeted by the microRNA-16 family cooperatively regulate cell cycle progression. Mol Cell Biol 27:2240–2252
86. Arvey A, Larsson E, Sander C et al (2010) Target mRNA abundance dilutes microRNA and siRNA activity. Mol Syst Biol 6:363
87. Ebert MS, Neilson JR, Sharp PA (2007) MicroRNA sponges: competitive inhibitors of small RNAs in mammalian cells. Nat Methods 4:721–726
88. Seitz H (2009) Redefining microRNA targets. Curr Biol 19:870–873
89. Liu J, Valencia-Sanchez MA, Hannon GJ et al (2005) MicroRNA-dependent localization of targeted mRNAs to mammalian P-bodies. Nat Cell Biol 7:719–723
90. Salmena L, Poliseno L, Tay Y et al (2011) A ceRNA hypothesis: the Rosetta Stone of a hidden RNA language? Cell 146:353–358
91. Hon LS, Zhang Z (2007) The roles of binding site arrangement and combinatorial targeting in microRNA repression of gene expression. Genome Biol 8:R166
92. Brennecke J, Stark A, Russell RB et al (2005) Principles of microRNA-target recognition. PLoS Biol 3:e85
93. Tu K, Yu H, Hua YJ et al (2009) Combinatorial network of primary and secondary microRNA-driven regulatory mechanisms. Nucleic Acids Res 37:5969–5980
94. Du L, Schageman JJ, Subauste MC et al (2009) miR-93, miR-98, and miR-197 regulate expression of tumor suppressor gene FUS1. Mol Cancer Res 7:1234–1243
95. Ivanovska I, Cleary MA (2008) Combinatorial microRNAs working together to make a difference. Cell Cycle 7:3137–3142
96. Mu P, Han YC, Betel D et al (2009) Genetic dissection of the miR-17 92 cluster of microRNAs in Myc-induced B-cell lymphomas. Genes Dev 23:2806–2811
97. Zhou YM, Ferguson J, Chang JT et al (2007) Inter- and intra-combinatorial regulation by transcription factors and microRNAs. BMC Genomics 8:396
98. Wu S, Huang S, Ding J et al (2010) Multiple microRNAs modulate p21Cip1/Waf1 expression by directly targeting its 3′ untranslated region. Oncogene 29:2302–2308
99. Jiang Q, Feng MG, Mo YY (2009) Systematic validation of predicted microRNAs for cyclin D1. BMC Cancer 9:194
100. Le Brigand K, Robbe-Sermesant K, Mari B et al (2010) MiRonTop: mining microRNAs targets across large scale gene expression studies. Bioinformatics 26:3131–3132
101. Alexiou P, Maragkakis M, Papadopoulos GL et al (2010) The DIANA-mirExTra web server: from gene expression data to microRNA function. PLoS One 5:e9171

102. Antonov AV, Dietmann S, Wong P et al (2009) GeneSet2miRNA: finding the signature of cooperative miRNA activities in the gene lists. Nucleic Acids Res 37:W323–W328
103. Friedman Y, Naamati G, Linial M (2010) MiRror: a combinatorial analysis web tool for ensembles of microRNAs and their targets. Bioinformatics 26:1920–1921
104. Altschul SF, Madden TL, Schaffer AA et al (1997) Gapped BLAST and PSI-BLAST: a new generation of protein database search programs. Nucleic Acids Res 25:3389–3402
105. Darnell RB (2011) HITS-CLIP: panoramic views of protein-RNA regulation in living cells. Wiley Interdiscip Rev RNA 1:266–286
106. Wen J, Parker BJ, Jacobsen A et al (2011) MicroRNA transfection and AGO-bound CLIP-seq data sets reveal distinct determinants of miRNA action. RNA 17:820–834
107. Alves L, Niemeier S, Hauenschild A et al (2009) Comprehensive prediction of novel microRNA targets in Arabidopsis thaliana. Nucleic Acids Res 37:4010–4021
108. Yang JH, Li JH, Shao P et al (2011) starBase: a database for exploring microRNA-mRNA interaction maps from Argonaute CLIP-Seq and Degradome-Seq data. Nucleic Acids Res 39: D202–D209
109. Jensen LJ, Kuhn M, Stark M et al (2009) STRING 8–a global view on proteins and their functional interactions in 630 organisms. Nucleic Acids Res 37:D412–D416
110. Vastrik I, D'Eustachio P, Schmidt E et al (2007) Reactome: a knowledge base of biologic pathways and processes. Genome Biol 8:R39
111. Rappoport N, Fromer M, Schweiger R et al (2010) PANDORA: analysis of protein and peptide sets through the hierarchical integration of annotations. Nucleic Acids Res 38:W84–W89
112. da Huang W, Sherman BT, Tan Q et al (2007) The DAVID Gene Functional Classification Tool: a novel biological module-centric algorithm to functionally analyze large gene lists. Genome Biol 8:R183
113. Kanehisa M (2002) The KEGG database. Novartis Found Symp 247:91–101, discussion 101–103, 119–128, 244–152
114. Lall S, Grun D, Krek A et al (2006) A genome-wide map of conserved microRNA targets in *C. elegans*. Curr Biol 16:460–471
115. Schaefer CF, Anthony K, Krupa S et al (2009) PID: the Pathway Interaction Database. Nucleic Acids Res 37:D674–D679
116. Ogata H, Goto S, Fujibuchi W et al (1998) Computation with the KEGG pathway database. Biosystems 47:119–128
117. D'Eustachio P (2010) Reactome knowledgebase of human biological pathways and processes. Methods Mol Biol 694:49–61
118. Chowbina SR, Wu X, Zhang F et al (2009) HPD: an online integrated human pathway database enabling systems biology studies. BMC Bioinform 10(Suppl 11):S5
119. Ideker T, Sharan R (2008) Protein networks in disease. Genome Res 18:644–652
120. Stelling J, Sauer U, Szallasi Z et al (2004) Robustness of cellular functions. Cell 118:675–685
121. Cui Q, Yu Z, Purisima EO et al (2006) Principles of microRNA regulation of a human cellular signaling network. Mol Syst Biol 2:46
122. Gusev Y (2008) Computational methods for analysis of cellular functions and pathways collectively targeted by differentially expressed microRNA. Methods 44:61–72
123. Peter ME (2010) Targeting of mRNAs by multiple miRNAs: the next step. Oncogene 29:2161–2164
124. Balaga O, Friedman Y, Linial M (2012) Toward a combinatorial nature of microRNA regulation in human cells Nucl. Acids Res 40:9404–9416

ERRATUM

MicroRNAs in Human Cancer

Thalia A. Farazi, Jessica I. Hoell, Pavel Morozov,
and Thomas Tuschl

U. Schmitz et al. (eds.), *MicroRNA Cancer Regulation: Advanced Concepts, Bioinformatics and Systems Biology Tools*, Advances in Experimental Medicine and Biology 774,
DOI 10.1007/978-94-007-5590-1_1, © Springer Science+Business Media Dordrecht 2013

DOI 10.1007/978-94-007-5590-1_17

The chapter 'MicroRNAs in Human Cancer' was based on the review article 'miRNAs in human cancer' (Farazi TA, Spitzer JI, Morozov P, Tuschl T. J Pathol. 2011 Jan;223(2):102-15. doi:10.1002/path.2806. Epub 2010 Nov 18) originally published in the Journal of Pathology in January 2011, with the exception that authors updated citations and included several minor revisions. Unfortunately, the citation of the original source was inadvertently left out and authors apologize for the oversight.

The online version of the original chapter can be found at
http://dx.doi.org/10.1007/978-94-007-5590-1_1

Index

U. Schmitz et al. (eds.), *MicroRNA Cancer Regulation: Advanced Concepts, Bioinformatics and Systems Biology Tools*, Advances in Experimental Medicine and Biology 774,
DOI 10.1007/978-94-007-5590-1,

FSC
www.fsc.org
MIX
Papier aus verantwortungsvollen Quellen
Paper from responsible sources
FSC® C105338

If you have any concerns about our products,
you can contact us on
ProductSafety@springernature.com

In case Publisher is established outside the EU,
the EU authorized representative is:
Springer Nature Customer Service Center GmbH
Europaplatz 3, 69115 Heidelberg, Germany

Printed by Libri Plureos GmbH
in Hamburg, Germany